Ann,

This book is taki[illegible] life of its own with [illegible] writing and all. Similar in a way to our groups' interactions - fun, creative, supportive. Thank God we are all females and can do this so well. You are too cool! Keep inspiring all of us to hang in and beyond.

Lovin' stuff
Toni

Ann,

I have been so slow to write that many things have happened since our last meeting together. I feel a little frightened for our future as Dean Dunn has announced his resignation but am hopeful that President Castor will set the tone for continued advancement of women. I am serving on both the WF Physicians Group Board & Executive Committee & have been asked to serve on the search committee for a Cardiothoracic surgeon. I am simultaneously excited & overwhelmed. Your mentorship is both appreciated & missed!

- Sally Houston

Dear Ann,

I hope that you enjoy reading through this book, but if not, I hope your husband does – at least he'll understand all of it! I can't begin to thank you for everything you've done for the McClintock Society, and for me personally. It has been my distinct pleasure to have gotten to know you during your time here at USF. You have been a mentor, a role model and a friend to all us McClintockorians (I have to give credit to Joan Christie, who coined this term!). We will all miss you tremendously – although we know that you'll continue to expand and enlighten junior women faculty at Indiana College of Medicine. As I said before, our loss is most certainly their gain. Congratulations on your move up the ladder – if anyone can "break the glass ceiling" it's you!

Love,

Jeanne

12/94

The Dynamic Genome

Something to remember us by. We will have no trouble remembering you. You have been an inspiration to us - those lucky ones chosen to attend the Professional Development Seminar and become McClintock founding members! What better way to show us how its done but to go out and get a great job befitting your many talents. We'll be hungry for news of you, so keep in touch - and know that you have planted seeds at USF that will continue to grow.

Much luck,
Verena

Dear Ann

Best of luck in Indiana - thank you for your support, kindness and encouraging words. I am very grateful for your help and hope to keep in touch!

Elizabeth Warner

1.4.95

Dear Ann,

Knowing you has been inspiring, enormously helpful & really fun. You are one of the most consummate natural strategists & effective executives I've ever known. The other is Dr. Spellacy.

I'm grateful you chose to mentor us. You can be assured that we will try to execute the obligations that accompany our rather privileged positions — to ourselves, to our professional juniors & to our institution. You thought you were eating dinner & dishing data* but you were really modelling behaviors, traits & understandings we needed... you've given us such a Jumpstart — could we go to dinner somewhere once a year? or every 5 years? Jeannie's right — you will be a Dean. I'm curious about what would be next. I'll always be grateful for your professional support & for key pieces of advice, Like the suggestion that my son sleep in his clothes so he'd make it to school on time.

Love,

Anna

P.S. * I know you knew what you were doing all along.

BARBARA McCLINTOCK'S IDEAS IN THE CENTURY OF GENETICS

EDITED BY

NINA FEDOROFF
Carnegie Institution of Washington

DAVID BOTSTEIN
Stanford University School of Medicine

Cold Spring Harbor Laboratory Press 1992

It is with pleasure that we acknowledge those who participated in the adventure of assembling this volume. We thank the many colleagues who took time to talk with us as we struggled to identify—and find—the individuals whose lives and thoughts have most clearly been influenced by McClintock. We thank the authors for the fun we had talking with them before and as they wrote, for their promptness, and for the delight of reading what they had written. Special thanks go to Ron Phillips for assistance with the contribution of Charles Burnham.

We are deeply grateful to the staff of Cold Spring Harbor Laboratory Press for making this book a reality. We thank publisher John Inglis for endless patience, courtesy, and enthusiasm, managing editor Nancy Ford for overseeing the process (and trying her best to please everyone's esthetic sensibilities), and editor Dorothy Brown and editorial assistant Joan Ebert for cheerfully chasing authors and polishing what they wrote.

Printed in the United States of America
Book and cover design by Emily Harste

Photo Credits
Front cover: Photograph of maize ear by Connie Jewel (Carnegie Institution of Washington); *Back cover:* Photograph of Queen Anne's Lace by Barbara McClintock; *Back cover flap:* Photograph courtesy of Ron Phillips (University of Minnesota, St. Paul); p. 6 (*center, bottom*), p. 72 (*top*), p. 212 (*center, bottom*), p. 360 (*bottom*): CSHL Archives; p. 72 (*center, bottom*): Marjorie M. Bhavnani; p. 6 (*top*): Marcus Rhoades; p. 204 (*top*): Mary-Lou Pardue; p. 360 (*top*): Swedish Institute, Stockholm

Quote Credits
p. 5: *Annu. Rev. Genet.* (vol. 18, pp. 1–29, 1984); p. 71 (*bottom*): *Cold Spring Harbor Symp. Quant. Biol.* (vol. 41, pp. 13–47, 1952); p. 71 (*top*): Unpublished report completed January 1949; p. 203: *Stadler Genet. Symp.* (vol. 10, pp. 25–47, 1978); p. 359: *Brookhaven Symp. Biol.* (vol. 8, pp. 58–74, 1956)

Reprint Credits
The following articles have been reproduced with permission: pp. 7–12 from *Proc. Natl. Acad. Sci.* (vol. 17, 1931); pp. 45–69 from *Annu. Rev. Genet.* (vol. 18, © 1984 by Annual Reviews, Inc); pp. 73–107 from *Cold Spring Harbor Symp. Quant. Biol.* (vol. 16, 1952); pp. 185–201 from *Brookhaven Symp. Biol.* (vol. 8, 1956); pp. 205–212 from *Genes, Cells, and Organisms* (vol. 17, 1987); pp. 335–357 from *Stadler Genet. Symp.* (vol. 10, 1978); pp. 361–380 from the Nobel Foundation (1984)

Library of Congress Cataloging-in-Publication Data

The Dynamic genome: Barbara McClintock's ideas in the century of genetics / edited by Nina Fedoroff, David Botstein.
p. cm.
Includes bibliographical references and index.
ISBN 0-87969-422-X
1. Genetics. 2. McClintock, Barbara, 1902– . 3. Women geneticists—United States—Biography. I. Fedoroff, Nina. II. Botstein, David.
QH430.D96 1992
575.1'092—dc20 92-10074
[B] CIP

All Cold Spring Harbor Laboratory Press publications may be ordered directly from Cold Spring Harbor Laboratory Press, 10 Skyline Drive, Plainview, New York 11803. Phone: 1-800-843-4388 (Continental U.S. and Canada). All other locations: (516) 349-1930. FAX: (516) 349-1946.

Contents

Introduction

Barbara McClintock was born within a few years of the rediscovery of Mendel's Laws. Her life, her discoveries, and her insights span the history of genetics in this century of genetics. She carried out her graduate studies at Cornell University as a member of the maize genetics group that flourished under R.A. Emerson in the 1920s. She became a dominant figure in that talented group, making a remarkable series of technical and conceptual advances in maize cytogenetics that brought it to a place of eminence. Marcus Rhoades, Barbara's recently deceased long-time friend and colleague, set these early contributions into historical perspective in a review written in 1984 and reproduced here. Throughout her long career, McClintock maintained a fully integrated cytogenetic approach, deriving fundamental insights from the interplay of crossing, counting, and observing. She is uniquely gifted in her ability to discern relationships between the behavior of chromosomes, observed under the microscope, and the properties of the whole organism. McClintock's work is both powerful and elegant. Her insights into the structure and behavior of chromosomes, often gained by applying techniques of her own devising, won the admiration of her peers and earned her early recognition. She was elected to membership in the National Academy of Sciences in 1944 and to the presidency of the Genetics Society of America in 1945. The influence of her early work is greater than that of any of her peers, with the possible exception of Alfred Sturtevant. Had she done no more, McClintock would have become a major figure in the history of genetics.

But Barbara McClintock did do much more, of course. After joining the Carnegie Institution of Washington's genetics group at Cold Spring Harbor in 1942, McClintock commenced the studies on the consequences of dicentric chromosome formation and breakage that led her to the discovery of transposable elements. The results of her genetic and cytological experiments revealed the existence of genetic elements that were capable both of moving from place to place in the genome and of exerting control over expression of other genes. In the parlance of the historian of science, these discoveries were ahead of their time and Barbara found herself in an anomalous and unique position. She was universally respected and admired as one of the great geneticists of her era, yet the reaction to her latest, perhaps most profound, discoveries and insights was often uncomprehending or indifferent and not infrequently dismissive or even hostile. As Barbara recounts in the Introduction to her collected papers, she eventually concluded "...that no amount of published evidence would be effective." McClintock continued her research, wrote up the results with data, diagrams, and discussion, and put the reports in her filing cabinet. She regularly summarized her results in the yearbooks of the Carnegie Institution of Washington. Fortunately, she correctly assessed the importance of her work and took care to preserve her material. In due course, transposable elements were discovered in other organisms and made molecular. The generality of mobile genetic elements and the concept of a dynamic genome were eventually understood and widely accepted. In 1983, she received an unshared Nobel prize, still the highest honor and most potent recognition our society can bestow upon a scientist.

As Barbara's 90th birthday approached, we invited some of her many friends and colleagues to write essays for the occasion. We left the subject and form to each author and so the gifts we have collected are of all kinds. We chose the order in which they appear, and it reflects, in a rough way, the chronicle of McClintock's career. Each essay's place marks the time at which Barbara's life or her ideas intersected those of the author. We have subdivided the whole into parts that mark first the progress of McClintock's work and later the generalization of her discoveries and the penetration of her ideas into the mainstream of genetics.

Barbara McClintock has become something of a legend in her own time, an almost mythical figure. Perhaps because she is at once so sharp an intellect and so private a person, the mythopoetic process began early in her career. With the belated award of the Nobel prize for her discovery of transposition almost four decades earlier, Barbara became a public figure. The myth inevitably acquired new dimensions, and it has continued to evolve over the ensuing years, aided by a steady stream of popular pieces, ranging from interviews to books. Myths are made by people, of course, and represent their way of seeing, understanding, perhaps coming to terms with the complex, the extraordinary. But they are *not* the literal, fragmented, contradictory reality which they represent.

There is a certain kinship between a myth and what we call a *model* in science. Models are abbreviated abstractions by which scientists seek to convey to others their own ways of organizing and extracting meaning from bits of data. Models often include diagrams or drawings that depict the author's abstract notions in a kind of cartoon form. These cartoons, so helpful and yet so misleading in their simplicity, increasingly supplant the intellectual abstractions. So it is with a delicious sense of irony that we heard many of the scientists with whom we discussed this project express an urgent desire to "set the record straight," by which they meant exploding one aspect or another of the McClintock myth. The irony is that both Barbara's contemporaries and those who came later often criticized her sharply for what they saw as her unwillingness to communicate. By this, they mostly meant she had not made cartoons to reify the genetic abstractions from which she inferred the existence of controlling elements that transpose from place to place in the genome.

Indeed, although Barbara often diagrammed chromosomes, she never drew cartoons, and she readily admits frustration with her failure to communicate. Perhaps the responsibility is hers, perhaps not. Among the essays that follow are the voices of some who had the patience and took the time to read what she wrote. It does indeed take some patience to read McClintock's prose and a great deal of concentration too: she didn't waste a word. And it helps to start at the beginning: she didn't repeat herself much. Yet by all reports, the rewards are substantial. They take the reader well beyond (mere) substance and into the realm of the esthetic. Words like "beautiful" and "elegant" have a persistent way of creeping into discourse on McClintock's science. McClintock's 73 papers are few by contemporary standards, but they are all well worth the reading. Those that touch on transposition have been collected in one volume titled *The Discovery and Characterization of Transposable Elements: The Collected Papers of Barbara McClintock*, published by Garland Press (1987).

What about the myth of great insight disbelieved, or worse yet, ignored? That is the one nobody seems to like. In the Introduction to her collected papers, Barbara comments on the deafening silence that greeted her astonishing announcement, complete with convincing evidence, that genes could move. Mel Green quotes Sturtevant admitting bafflement, but *not* disbelief—such was her extraordinary stature as a scientist. In what follows, you will catch glimpses of how she earned her reputation, even though this volume comes too late to capture the voices of many who knew the young McClintock. More to the point, the maize genetic literature bears witness to McClintock's immediate impact. Transposition became an active and productive area of maize genetics in the decades immediately following her discovery.

The rest of the genetics community eventually caught on as well, of course. Transposable elements surfaced first in bacteria and later in virtually every organism geneticists study. What emerges powerfully in this kaleidoscope of contributions by her intellectual heirs, especially those who themselves stumbled on, or deliberately sought, transposition in a host of other organisms, is the difference it made that she had already been there, walked the path before. Some read McClintock first and some read her afterward. Some sought her counsel, some did not. But no matter, it *was* well known, however imperfectly understood and skeptically regarded, that McClintock had shown that genes transpose. So, too, did her earlier work on chromosomes prove remarkably prescient. Perhaps equally important, she shook a few shoulders and more than a few minds, engendering the courage to break free and see the familiar with new eyes, rearrange the pieces in novel ways, remaining faithful only to what is really there, not the dogma of the day.

And the real person behind the myth? You will glimpse that, too, in fragments, small bits of raw data: tough tennis player, intimidating intellect, generous colleague, tireless listener. These small insights are all we will get. Barbara is indeed a private person.

McClintock's ideas, like her discoveries, truly span the century of genetics. Some were timely and quickly recognized. Others were well ahead of their time, but eventually caught on. Some still pull us into the future. What of her theories of gene regulation in development? Take note. It matters little that the details were wrong at such an early date. McClintock was way out ahead with the notion that there are genetic elements separate from genes that regulate their expression in development. Not bad. And what about her ideas on evolution? Those of us who have heard her talk about evolution know that she already knew, a long time ago, that genes are all interconnected. Molecular developmental biology is just starting to draw the lines connecting the dots that are the genes, and we can already guess that soon the interconnections will comprise a dense three-dimensional network. In the face of such intricate interconnectedness, Barbara always asks, how is it that the whole organism changes over time, evolves? Should one gene mutate, by what means and mechanisms are adjustments made in every part of the complex whole?

Still ahead of us at 90—we celebrate you, Barbara!

Nina Fedoroff
David Botstein

Baltimore, Maryland
February, 1992

CYTOGENETICS

If Emerson was the towering figure in classical maize genetics, McClintock played an even more seminal role in the development of cytogenetics. She was the dominant figure in the talented Cornell group of cytogeneticists and she, more than anyone else, was responsible for a long series of remarkable findings.

M.M. Rhoades 1984

Top from left: Charles Burnham, Marcus Rhoades, Rollins Emerson, and Barbara McClintock. *Kneeling:* George Beadle (Cornell University 1929)

Barbara McClintock and Harriet Creighton (Cornell University 1931)

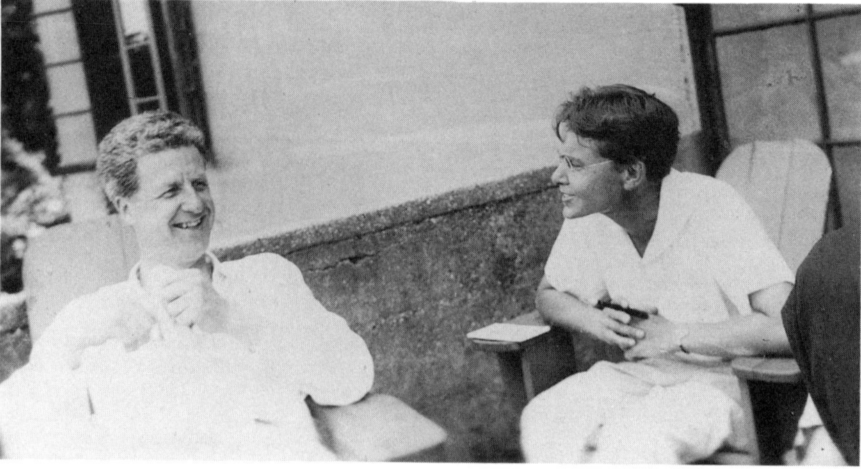

L.C. Dunn and Barbara McClintock (CSHL 1942)

A CORRELATION OF CYTOLOGICAL AND GENETICAL CROSSING-OVER IN ZEA MAYS

BY HARRIET B. CREIGHTON AND BARBARA McCLINTOCK

BOTANY DEPARTMENT, CORNELL UNIVERSITY

Communicated July 7, 1931

A requirement for the genetical study of crossing-over is the heterozygous condition of two allelomorphic factors in the same linkage group. The analysis of the behavior of homologous or partially homologous chromosomes, which are morphologically distinguishable at two points, should show evidence of cytological crossing-over. It is the aim of the present paper to show that cytological crossing-over occurs and that it is accompanied by genetical crossing-over.

In a certain strain of maize the second-smallest chromosome (chromosome 9) possesses a conspicuous knob at the end of the short arm. Its distribution through successive generations is similar to that of a gene. If a plant possessing knobs at the ends of both of its 2nd-smallest chromosomes is crossed to a plant with no knobs, cytological observations show that in the resulting F_1 individuals only one member of the homologous pair possesses a knob. When such an individual is back-crossed to one having no knob on either chromosome, half of the offspring are heterozygous for the knob and half possess no knob at all. The knob, therefore, is a constant feature of the chromosome possessing it. When present on one chromosome and not on its homologue, the knob renders the chromosome pair visibly heteromorphic.

In a previous report[1] it was shown that in a certain strain of maize an interchange had taken place between chromosome 8 and 9. The interchanged pieces were unequal in size; the long arm of chromosome 9 was increased in relative length, whereas the long arm of chromosome 8 was correspondingly shortened. When a gamete possessing these two interchanged chromosomes meets a gamete containing a normal chromosome set, meiosis in the resulting individual is characterized by a side-by-side synapsis of homologous parts (see diagram, figure 1 of preceding paper). Therefore, it should be possible to have crossing-over between the knob and the interchange point.

In the previous report it was also shown that in such an individual the only functioning gametes are those which possess either the two normal chromosomes (N, n) or the two interchanged chromosome (I, i), i.e., the full genom in one or the other arrangement. The functional gametes therefore possess either the shorter, normal, knobbed chromosome (n) or the longer, interchanged, knobbed chromosome (I). Hence, when such a plant is crossed to a plant possessing the normal chromosome complement, the presence of the normal chromosome in functioning gametes of the former will be indicated by the appearance of ten bivalents in the prophase of meiosis of the resulting individuals. The presence of the interchanged

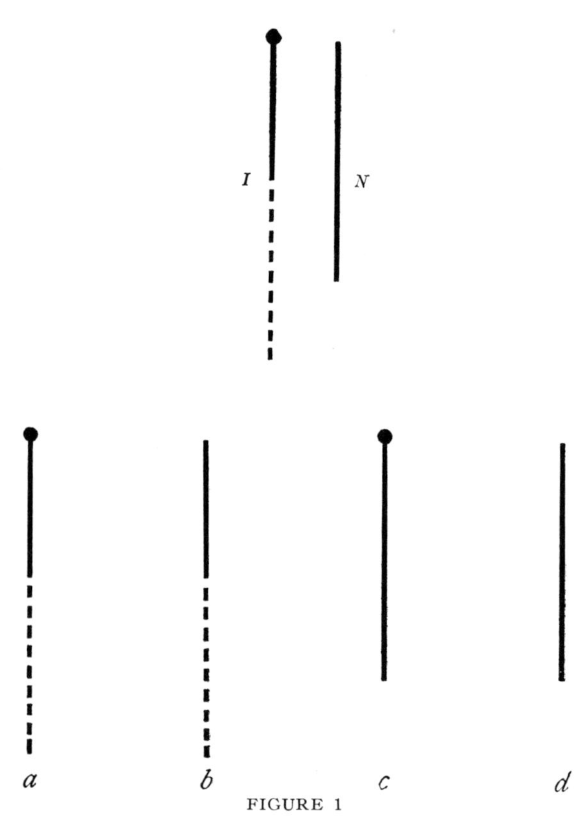

FIGURE 1

Above—Diagram of the chromosomes in which crossing-over was studied. (Labeled as in figure 1, preceding paper.)

Below—Diagram of chromosome types found in gametes of a plant with the constitution shown above.

a—Knobbed, interchanged chromosome.
b—Knobless, interchanged chromosome.
c—Knobbed, normal chromosome.
d—Knobless, normal chromosome.
a and *d* are non-crossover types.
b and *c* are crossover types.

chromosome in other gametes will be indicated in other F_1 individuals by the appearance of eight bivalents plus a ring of four chromosomes in the late prophase of meiosis.

If a gamete possessing a normal chromosome number 9 with no knob, meets a gamete possessing an interchanged chromosome with a knob, it is clear that these two chromosomes which synapse along their homologous parts during prophase of meiosis in the resulting individual are visibly different at each of their two ends. If no crossing-over occurs, the gametes formed by such an individual will contain either the knobbed, interchanged chromosome (*a*, Fig. 1) or the normal chromosome without a knob (*d*, Fig. 1). Gametes containing either a knobbed, normal chromosome (*c*, Fig. 1) or a knobless, interchanged chromosome (*b*, Fig. 1) will be formed as a result of crossing-over. If such an individual is crossed to a plant possessing two normal knobless chromosomes, the resulting individuals will be of four kinds. The non-crossover gametes would give rise to individuals which show either (1) ten bivalents at prophase of meiosis and no knob on chromosome 9, indicating that a gamete with a chromosome of type *d* has functioned or (2) a ring of four chromosomes with a single conspicuous knob, indicating that a gamete of type *a* has functioned. The crossover types will be recognizable as individuals which possess either (1) ten bivalents and a single knob associated with bivalent chromosome 9 or (2) a ring of four chromosomes with no knob, indicating that crossover gametes of types *c* and *b*, respectively, have functioned. The results of such a cross are given in culture 337, table 1. Similarly, if such a plant is crossed to a normal plant possessing knobs at the ends of both number 9 chromosomes and if crossing-over occurs, the resulting individuals should be of four kinds. The non-crossover types would be represented by (1) plants homozygous for the knob and possessing the interchanged chromosome and (2) plants heterozygous for the knob and possessing two normal chromosomes. The functioning of gametes which had been produced as the result of crossing-over between the knob and the interchange would give rise to (1) individuals heterozygous for the knob and possessing the

TABLE 1

KNOB-INTERCHANGED / KNOBLESS-NORMAL × KNOBLESS-NORMAL, CULTURE 337 AND KNOBBED-NORMAL CULTURES A125 AND 340

CULTURE	PLANTS POSSESSING 2 NORMAL CHROMOSOMES: NON-CROSSOVERS	PLANTS POSSESSING 2 NORMAL CHROMOSOMES: CROSSOVERS	PLANTS POSSESSING AN INTERCHANGED CHROMOSOMES: NON-CROSSOVERS	PLANTS POSSESSING AN INTERCHANGED CHROMOSOMES: CROSSOVERS
337	8	3	6	2
A125	39	31	36	23
340	5	3	5	3
Totals	52	37	47	28

TABLE 2

KNOB-*C-wx* / KNOBLESS-*c-Wx* × KNOBLESS-*c-wx*

C-wx		*c-Wx*		*C-Wx*		*c-wx*	
Knob	Knobless	Knob	Knobless	Knob	Knobless	Knob	Knobless
12	5	5	34	4	0	0	3

interchanged chromosome and (2) those homozygous for the knob and possessing two normal chromosomes. The results of such crosses are given in cultures A125 and 340, table 1. Although the data are few, they are consistent. The amount of crossing-over between the knob and the interchange, as measured from these data, is approximately 39%.

In the preceding paper it was shown that the knobbed chromosome carries the genes for colored aleurone (*C*), shrunken endosperm (*sh*) and waxy endosperm (*wx*). Furthermore, it was shown that the order of these genes, beginning at the interchange point is *wx-sh-c*. It is possible, also, that these genes all lie in the short arm of the knobbed chromosome. Therefore, a linkage between the knob and these genes is to be expected.

One chromosome number 9 in a plant possessing the normal complement had a knob and carried the genes *C* and *wx*. Its homologue was knobless and carried the genes *c* and *Wx*. The non-crossover gametes should contain a knobbed-*C-wx* or a knobless-*c-Wx* chromosome. Crossing-over in region 1 (between the knob and *C*) would give rise to knobless *C-wx* and knobbed-*c-Wx* chromosomes. Crossing-over in region 2 (between *C* and *wx*) would give rise to knobbed-*C-Wx* and knobless-*c-wx* chromosomes. The results of crossing such a plant to a knobless-*c-wx* type are given in table 2. It would be expected on the basis of interference that the knob and *C* would remain together when a crossover occurred between *C* and *wx;* hence, the individuals arising from colored starchy (*C-Wx*) kernels should possess a knob, whereas those coming from colorless, waxy (*c-wx*) kernels should be knobless. Although the data are few they are convincing. It is obvious that there is a fairly close association between the knob and *C*.

To obtain a correlation between cytological and genetic crossing-over it is necessary to have a plant heteromorphic for the knob, the genes *c* and *wx* and the interchange. Plant 338 (17) possessed in one chromosome the knob, the genes *C* and *wx* and the interchanged piece of chromosome 8. The other chromosome was normal, knobless and contained the genes *c* and *Wx*. This plant was crossed to an individual possessing two normal, knobless chromosomes with the genes *c-Wx* and *c-wx*, respectively. This cross is diagrammed as follows:

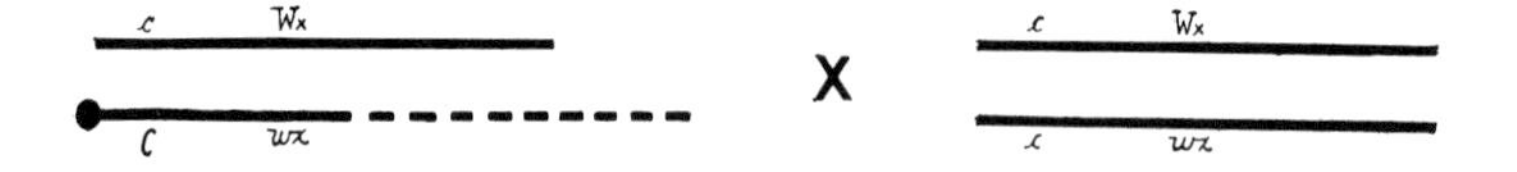

The results of the cross are given in table 3. In this case all the colored kernels gave rise to individuals possessing a knob, whereas all the colorless kernels gave rise to individuals showing no knob.

The amount of crossing-over between the knob and the interchange point is approximately 39% (Table 1), between *c* and the interchange approximately 33%, between *wx* and the interchange, 13% (preceding paper). With this information in mind it is possible to analyze the data given in table 3. The data are necessarily few since the ear contained but few kernels. The three individuals in class I are clearly non-crossover types. In class II the individuals have resulted from a crossover in region 2,

TABLE 3

KNOB-*C*-*wx*-INTERCHANGED / KNOBLESS-*c*-*Wx*-NORMAL × KNOBLESS-*c*-*Wx*-NORMAL / KNOBLESS-*c*-*wx*-NORMAL

PLANT NUMBER	KNOBBED OR KNOBLESS	INTERCHANGED OR NORMAL	
Class I, *C*-*wx* kernels			
1	Knob	Interchanged	
2	Knob	Interchanged	
3	Knob	Interchanged	
Class II, *c*-*wx* kernels			
1	Knobless	Interchanged	
2	Knobless	Interchanged	
Class III, *C*-*Wx* kernels			*Pollen*
1	Knob	Normal	*WxWx*
2	Knob	Normal	....
3	....	Normal	*WxWx*
5	Knob	Normal	....
6	Knob	......	....
7	Knob	Normal	....
8	Knob	Normal	....
Class IV, *c*-*Wx* kernels			
1	Knobless	Normal	*Wxwx*
2	Knobless	Normal	*Wxwx*
3	Knobless	Interchanged	*Wxwx*
4	Knobless	Normal	*Wxwx*
5	Knobless	Interchanged	*WxWx*
6	Knobless	Normal	*WxWx*
7	Knobless	Interchanged	*Wxwx*
8	Knobless	Interchanged	*WxWx*
9	Knobless	Normal	*WxWx*
10	Knobless	Normal	*WxWx*
11	Knobless	Normal	*Wxwx*
12	Knobless	Normal	*Wxwx*
13	Knobless	Normal	*WxWx*
14	Knobless	Normal	*WxWx*
15	Knobless	Normal	*Wx*—

i.e., between *c* and *wx*. In this case a crossover in region 2 has not been accompanied by a crossover in region 1 (between the knob and *C*) or region 3 (between *wx* and the interchange). All the individuals in class III had normal chromosomes. Unfortunately, pollen was obtained from only 1 of the 6 individuals examined for the presence of the knob. This one individual was clearly of the type expected to come from a gamete produced through crossing-over in region 2. Class IV is more difficult to analyze.

Plants 6, 9, 10, 13, and 14 are normal and *WxWx;* they therefore represent non-crossover types. An equal number of non-crossover types are expected among the normal *Wxwx* class. Plants 1, 2, 4, 11 and 12 may be of this type. It is possible but improbable that they have arisen through the union of a *c-Wx* gamete with a gamete resulting from a double crossover in region 2 and 3. Plants 5 and 8 are single crossovers in region 3, whereas plants 3 and 7 probably represent single crossovers in region 2 or 3.

The foregoing evidence points to the fact that cytological crossing-over occurs and is accompanied by the expected types of genetic crossing-over.

Conclusions.—Pairing chromosomes, heteromorphic in two regions, have been shown to exchange parts at the same time they exchange genes assigned to these regions.

The authors wish to express appreciation to Dr. L. W. Sharp for aid in the revision of the manuscripts of this and the preceding paper. They are indebted to Dr. C. R. Burnham for furnishing unpublished data and for some of the material studied.

[1] McClintock, B., *Proc. Nat. Acad. Sci.*, **16**, 791–796 (1930).

Recollections of Barbara McClintock's Cornell Years

HARRIET B. CREIGHTON
Wellesley College, Wellesley, Massachusetts 02181

The celebration of a 90th birthday is still quite an event, even now, when the life expectancy for women has increased year after year. It does not surprise me that Barbara McClintock has made it to 90 and that surely she will add on quite a few more years. She cannot quit now. There are too many intriguing problems that she will keep on pursuing.

Lately, a number of people have asked me to tell them about Barbara's early career. They seem to think that because I did my graduate work at Cornell University with her long ago I can provide some clues about what made her a successful scientist. Some of these people are trying to interest young women in enjoying and studying the sciences by writing the stories of successful women scientists. Others are women teachers in high schools or in colleges who hope to gain some insights into ways to get more of their students to become scientists. I am sympathetic to their concerns. Recognizing that at the time I was working with Barbara I paid little attention to what was going on, now that I look back, I have some observations that may be useful.

Shaping a Scientist

The first and most important factor in shaping the development of a scientist, obviously, must be intellectual ability. There is probably more of it around than we recognize and undoubtedly much among young women. But ability alone is not enough. There must be the determination to pursue wherever curiosity leads with full attention and perseverance. In some scientific work, physical strength and endurance are required. Barbara had all of these characteristics. Let me give two examples, the first of which is in some sense trivial. On her 28th birthday, we played tennis and I was badly beaten. This was not unusual. Her birthday preceded my 21st birthday by a few days. I still felt young and remember thinking that she was pretty old to be playing so vigorously, particularly at the end of a long day's work in the cornfield and in the laboratory. Barb won by going after every shot of mine and by returning the ball where I could hardly ever reach it. Sometimes I did not even try. That brought a scathing crack from Barb about my laziness. There was no point in trying to get her to see that my plan was to husband my energy and to use it when I had a fair chance of scoring. Built into her was the conviction that always you gave full physical and mental effort to whatever pursuit you cared about. And she cared about winning at tennis.

A second example of McClintock's drive dates from a time in June at Cornell when the corn seedlings in the Plant Breeding Garden were about a foot high. Each of us had our space where we grew our pedigreed stocks. A once-in-a-century torrential rain hit Ithaca, lasting for hours, and by nighttime, the water rushing down through the gorges was washing out tons of rock. I had to go out in this weather at about 3 in the morning because my mother, who lived down near the mouth of one of the raging creeks, telephoned me to come and get her before she was flooded out. Having moved her to higher ground and being wide awake, I decided to drive around to see what the damage was on the campus. I drove to the main road above the Plant Breeding Garden and down in the garden I saw the beam of auto headlights and a person moving around. So I drove down, and there was Barb working in 2–3 inches of water that was running between the rows of her corn plants, washing them out. With her hands, she was piling mud on their roots and replanting some if she knew exactly from which spot they had washed away. It was of crucial importance to save every plant because each had been numbered at the time she had taken root tip samples to determine the chromosome number. All of us also had genetic stocks that were distinctive and the basis of our research and we stood to lose too. But only Barbara imagined what the flooding could do down in the garden hollow and only she got there long before dawn to save her plants. If she lost them, she would have blamed only herself, not nature or fate, for not having made the maximum effort to save her research.

Cultivating a Scientist

A person can have imagination, determination, physical strength, and mental ability to pursue a course and yet not be able to achieve the goal. Many men and women with ability and drive have faltered and failed. To make progress toward their goals, there must be a confluence of favorable factors. Below I describe four factors that seem to have helped Barbara's career get started.

The first factor was the nature of the Departments of Botany and of Plant Breeding at Cornell in the 1920s and early 1930s. Barbara earned her bachelor's degree in botany from the College of Agriculture in 1923. Her Ph.D. degree in botany and genetics was granted by the University in 1927. For several years, she was a research assistant and then an instructor in the Botany Department until she left Cornell in 1932. Over all of these years, there was very little change in the faculty or in the course offerings of these departments. One change of some significance was the 1932 move of the Botany Department and the Plant Breeding Department from separate, inadequate, old buildings into the large new Plant Science Building. I think that what I saw and felt from the autumn of 1929 when I came to Cornell until I left in 1934 was representative of the whole preceding decade when Barbara and others were developing corn cytogenetics. On that assumption, let me describe the situation.

Supporting a Scientist

The Department of Botany at Cornell was in the College of Agriculture—the one "classical" science there. Chemistry, Physics, Zoology, and Animal Physiology were in the College of Liberal Arts of the privately supported part of Cornell University. Departments of applied plant and animal sciences such as

Plant Breeding were of course in the Agricultural College. The botany professors were all about the same age, young, and enthusiastic. Their research interests and course offerings included anatomy, morphology, taxonomy, physiology, cytology, phycology, and paleobotany. As an undergraduate, McClintock was exposed to classical plant science and also to current thinking about plant structure, function, and evolution including field botany.

All of the botany professors enjoyed teaching and were conscientious about the laboratory offerings as well as their lectures. All of them were publishing their research. They were active in their professional societies and they encouraged graduate students to attend the societies' annual meetings and to present papers. Three members of the department wrote books for the then prestigious McGraw-Hill Company's, *Publications in the Agricultural and Botanical Sciences*. Lester W. Sharp, with whom Barbara studied both as an undergraduate and as a graduate student, published an *Introduction to Cytology* in 1926. He followed it in 1934 with an expanded and revised edition. It was while he was occupied in producing these books that Barbara was engaged to help in the laboratory teaching of his cytology course. She also assisted and participated in a course in microtechnique given jointly by Sharp and the plant anatomist Arthur Eames primarily for graduate students from all of the plant science departments. Barbara had gotten her start in microscopy, microtechnique, and photomicrography from Sharp, who freely shared with her all he knew. One thing that anyone working with McClintock learned was how to set up a microscope and its illumination. If at any time she came to look at a slide on your microscope and did not immediately make some adjustment, you felt triumphant relief.

All of the botany professors had graduate students in whose progress they took a continuing interest. In the 1920s, graduate students either paid their own way or luckily obtained an assistantship. In the Botany Department, there were, at most, ten teaching assistantships. Most years, there were only one or two women holding these jobs. There were probably more male applicants and probably fewer females were recommended by their undergraduate teachers. The Cornell professors were not against women. Sharp was by no means the only one encouraging them. The head of the department, Karl Wiegand, a taxonomist, with his wife, a trained botanist, had taught at Wellesley College for several years. They had no doubt that young women could be able and serious students, some of them well qualified to pursue advanced degrees. For a long period of time, Wellesley graduates were recommended for assistantships and were accepted. One of them, Rita Ballard, married the plant anatomy professor, Arthur Eames. After getting her Master's degree, she worked with him on his research and in the writing of his plant anatomy book in the McGraw-Hill series. She made all of the illustrations. I mention Mrs. Eames because she worked in the laboratory almost every day. Obviously, she was not just a pair of hands, but a colleague of her husband. And there were two other wives whose advanced training in botany and whose conspicuous presence in the laboratories made it evident that women could carry on scientific work. I think that it is not surprising that Barbara found this Botany Department a supportive academic home in her undergraduate and graduate years.

The second favorable factor in scientific progress was in the Department of

Plant Breeding. Here was where the only genetics courses in the University were taught. Not too many undergraduates took the introductory course, but the many graduate students in the biological departments of the Agricultural College took it and the advanced course in general genetics. Women were a rarity in these courses, and so when Barbara took them and lapped them up, she was noticed. The head of the department, Rollins A. Emerson, was at the time Dean of the Graduate School. He took on few graduate students (George Beadle and Marcus Rhoades in the late 1920s), but encouraged those interested in genetics (as distinct from plant breeding) to work with A.C. Fraser, another member of the department. Both Emerson and Fraser assumed that women were perfectly capable of carrying out the intellectual and the physical labor that is a necessary part of corn genetic studies. Emerson frequently called in his teenage daughter to help him when the pollinating season was at its peak. I am sure that he knew from his midwest farm background that women could work in the corn genetics fields as well as men. It was obvious that Barbara was academically outstandingly capable. She was accepted by the faculty and by the plant breeding graduate students. All were men, many of whom came from foreign countries where women were not usually found in genetics. She was given space in the Plant Breeding garden and any pedigreed corn stocks she needed.

The third favorable factor that was supportive of McClintock's undergraduate and graduate work was the unplanned result of crowding in the old botany building, Stone Hall, in the Agricultural College prior to the opening of the new Plant Science Building. The room assigned to graduate students had been a classroom. It was unpartitioned. Bookcases acquired from somewhere made possible the fashioning of semiprivate cubicles. Barbara had space there as Dr. Sharp's assistant and one of Dr. Eames' students was there. The rest of the space was assigned to students from any department who needed to use Botany Department microscopes. Thus, Marcus Rhoades had a microscope table and a slide preparation table and so did George Beadle, but their principal location was in another building where Plant Breeding Department was located. When Charles Burnham came from Wisconsin on one of the rare National Research Council postdoctoral fellowships, he was also squeezed in. I shared Beadle's cubicle, which was so small that when we were there at the same time one person had to get up and get the chair out of the way if the other person wanted to get out. How could this jammed-up inadequate arrangement be of benefit to anyone? Because, when any one of us working on corn found anything puzzling or exciting on a microscope slide, every one could come immediately to look. Interpretations could be launched and argued and thinking could certainly be stimulated. The excitement of seeing what someone else had found carried each of us over the times when we were not turning up anything new. All of us profited from the quick, easy, and open exchange of observations and interpretations.

To a great extent, we lost much of this interaction when we all moved into the big new Plant Science building. Although both Botany and Plant Breeding Departments were located there, long distances separated our individual research spaces. What one person found on a microscope slide had to be pretty exciting to make the person carry the slide across the room to someone else's microscope. The old crowding had a great deal to be said for it and we had not realized it. Barbara and all of us had benefited.

The fourth part of the confluence of factors favorable for the development of McClintock's early career is very different from the first three. It was her good fortune to have found a comfortable place to live with a congenial and supportive person. Dr. Esther Parker, M.D., opened her home to Barb. She was an insightful person who provided encouragement when it was needed. She unobtrusively watched after Barb's health and tried to keep her from exhausting herself. Dr. Parker's mother provided a homey stability by seeing to their meals, laundry, and house cleaning. A cottage on Cayuga Lake provided cool nights for sleep in the hot humid Ithaca summers. Compared with living in a rented room or an apartment with all of the annoying and time-consuming chores of living, this arrangement was enjoyable and allowed for concentration on research in the very busy summer season.

Stumbling Blocks

Having described several of the factors that seem to have provided a favorable climate for McClintock's early scientific development does not mean that the furtherance of her career was inevitable. There was little the Botany and Plant Breeding professors who had been so supportive could do now to help her get a position commensurate with her ability, achievements, and promise. The late 1920s and early 1930s were not the best of times to get a job. At Cornell, there was no place for her because there was no possibility of enlarging departments and no imminent retirements. Emerson managed to help George Beadle and Marcus Rhoades to get jobs in the Plant Breeding Departments. They could continue their corn genetics research along with some corn breeding and teaching. Such jobs were not open to women because of an unchallenged tradition that if a woman had to go out to talk to farmers about corn problems they would not listen.

There were no openings in any undergraduate colleges or universities for a geneticist. The subject was usually taught to juniors and seniors and the faculty member had to teach several other courses. Most institutions had no facilities for growing experimental plants. Barbara did not want to teach undergraduates and she would very probably have been uncomfortable and frustrated in that part of the academic world. For the next few years, she lived and worked on fellowships like the Guggenheim or in temporary positions like the one that Lewis Stadler was able to devise at the University of Missouri. Stadler's pioneering use of X-rays to produce mutations in corn provided an abundance of chromosomal abnormalities that he was anxious to have Barbara try to elucidate.

Dedication Is Rewarded

At the end of a decade through which only a dedicated person would, or could, have persisted and carried on research, an opening came at the Carnegie Institution of Washington's Department of Genetics at Cold Spring Harbor, New York. The Datura group, Alfred F. Blakeslee and Sophia Satina, had retired and moved to Smith College. Milislav Demerec, a Cornell-trained corn geneticist was the new head. Here were laboratories, microscopes, fields, housing, and, most important, intellectual encouragement. Here was recognition of McClintock's worth as a scientist and continued support for her to follow her investigations wherever they led. Here again, there was a confluence of favorable factors.

Barbara McClintock: Reminiscences*

CHARLES BURNHAM
c/o Dr. Ron Phillips, University of Minnesota
St. Paul, Minnesota 55108

I first worked with Barbara McClintock in an area called the "hole" at Cornell University. After receiving my Ph.D. in genetics from the University of Wisconsin in June 1929, I began a National Research Council Fellowship for residence at Cornell University. I was to spend the summers of 1929 and 1930 at Cornell and the winter of 1929–1930 at Bussey Institute at Harvard. The research proposal was to study crosses between Dr. Royal A. Brink's *semisterile-1* and two semisterile plants that were among the progeny of a red aleurone *waxy* (*pr wx*) line selected as a standard normal line (semisteriles have 50% aborted pollen and 50% seed set). While studying the *waxy* gene, Dr. Brink had found one line with semisterile plants, and this line was called *semisterile-1*. When the tassel samples of the plants in the normal (*pr wx*) line were examined later, three of them were found to be semisterile. Two of them, *semisterile-2* and *semisterile-3*, had been crossed with plants in the *semisterile-1* line. These crosses were made in 1927. When grown in 1928 at the University of Wisconsin, the progeny in both crosses had plants with 75% sterility, and these were self-pollinated and also crossed with normal plants. We had followed the sterility but had done no cytological studies. The next step was to do the cytology. Dr. Brink gave me the material to use for the proposed research for the National Research Council Fellowship. This seed was sent to Cornell University for planting in 1929, and I drove to Ithaca in June of that year.

At Cornell University

I believe it was on a Sunday morning that someone directed me to the area known as the "hole," several acres of level land surrounded by banks or low hills, an area not subject to late frosts in the spring or early frosts in the fall. I met Barbara McClintock there and together we located my plants growing from the seed I had sent. She was then doing cytological studies of a triploid maize plant found by L.F. Randolph and had already published in the journal *Science* a short note that included an idiogram of the ten chromosomes of corn. This idiogram was based on her study at the first division of the microspores to form the tube and generative nuclei of trisomic plants with ($2n + 1$) chromosomes. Microspores on $2n + 1$ plants are of two kinds: One has one of each of the 10 chromosomes, and the other has 11 chromosomes, i.e., one of the 10 chromosomes is represented twice. Each chromosome has a centromere that divides the chromosome into two arms. At one stage of the division,

*These reminiscences supplement material in Helen Fox Keller's book, *A Feeling For The Organism*. I did not know about her book until it appeared.

McClintock could measure the total length and arm lengths. Different chromosomes have different total lengths and arm lengths, and an idiogram is based on these differences. She used genetic marker stocks for crosses with the different trisomics. If the marker is carried by the extra chromosome, the genetic ratio is modified. Thus, Barbara combined her cytological and genetic studies. She was growing this material in the field in 1929 and 1930.

McClintock's laboratory was on the second floor of Stone Hall, and her work space on the south side of the room was parallel to several partitions. A small table was set up at the end of the first partition for my work space, and so I was working opposite Barbara McClintock. The room had a wood floor, and to reduce the shaking when people walked across the floor, sponge rubber had to be placed under the microscope and under each table leg. When Barbara found critical figures under the microscope, she showed them to me. Using the acetocarmine stain, I could not see the chromosomes at first. However, I finally discovered that by reducing the light under the microscope, I could see them. I learned the technique from her. She understood the microscope and how to adjust it to get the best resolution. She had a keen sense of the normal contents of cells so that if anything unusual were present she saw it as well—not just what she happened to be studying. Barbara also improved the acetocarmine smear technique. She discovered that application of heat, just short of boiling, to the slide after applying the coverslip greatly improved the contrast in staining between the chromosomes and cytoplasm.

The results of pollen classification were recorded on the blackboard daily. In the progeny of the *pr wx* stock crossed with 75% sterile plants from *semisterile-1* crossed with *semisterile-2*, there was a ratio of 1 normal:2 semi-sterile:1 75% sterile; the semisterile class was really one *semisterile-1* plus one *semisterile-2*. There were a few intermediates. Pollen was also classified for the *waxy* gene. McClintock discovered that indirect lighting showed the difference clearly when the $KI+I_2$ solution was used: *waxy* pollen is brick red and normals are blue. It was soon evident that there was linkage between the *waxy* gene and sterility. From her trisomic study, McClintock had identified the chromosome carrying *waxy* as chromosome 9; hence, this was one of the chromosomes involved in *semisterile-2*. The plants with 75% sterility had two rings with four chromosomes each. The plants with intermediate degrees of sterility later proved to have an extra chromosome, one of those in a ring. Contrary to Keller's book, plants with intermediate sterility did not "disturb" me. They were something to be explained. They were trisomic. McClintock had found that by pulling away the soil from the base of the plant, saturating the ground with water, and covering it with a wet burlap sack, roots came to the surface after a few days and root tips were obtained. The smear technique had not been developed for root tips, but I believe Randolph had developed the procedure for embedding them in paraffin, microtome sectioning, and staining. Some of the cells had division figures in which the number of chromosomes could be determined.

In the progeny from the *pr wx* stock crossed with 75% sterile plants from the cross of *semisterile-1* with *semisterile-3*, most of the plants were semisterile. A few had low sterility. While thinking about this one night, I realized that if one of the chromosomes in *semisterile-1* was also involved in *semisterile-3*, this might explain the results. Six chromosomes would be in the ring. The next morning I went to the laboratory early and by the time the others had arrived,

I had cytological preparations that showed the ring of six. In the progeny from the 75% sterile plant involving *semisterile-3*, plants with intermediate sterility had narrower and shorter leaves, typical of McClintock's trisomic for chromosome 7. McClintock had found in her material one line that formed quadrivalents in crosses with normal. She had done only the cytology. When she examined the pollen, she found that those plants were semisterile. I believe that Rhoades later published a paper on this semisterile.

Rollins A. Emerson, head of the Department of Plant Breeding at Cornell, had the complete collection of genetic marker stocks, and he gave me several semisterile lines. Several of his graduate students were working on various traits he had accumulated in his studies. H.W. Li was working with dwarfs and J.D. Hofmeyr was working with barren stalk lines (earless plants). Harold S. Perry, later at Duke University, was also working at Cornell, as was Marcus Rhoades in 1930. George Beadle was studying male steriles cytologically in one of the areas in McClintock's laboratory. Several proved to have cytological abnormalities. He often discussed the cytology of one male sterile, termed *A*-synaptic. I knew the *A* factor for anthocyanin color but could not see what it had to do with synapsis. Finally, one day while working at the microscope, it dawned on me that *A*-synaptic was lack of pairing, i.e., asynapsis. As the season progressed, McClintock checked her trisomic crosses for aleurone color by peeling down the husks and counting the colored and colorless kernels before the ears were mature. The $2n + 1$ for chromosome 9 showed a trisomic ratio. As more information became available, we often needed to make crosses with some of Emerson's stocks. Everyone shared in the use of these stocks and we also shared our stocks. Those were exciting times.

Barbara began to study the chromosomes at the earlier stages of meiosis. The chromosomes at the pachytene stage were long and paired. In *semisterile-2* plants, there was a 4-armed cross-shaped configuration. In each arm, the homologs were paired but changed partners at the center of the cross. That point indicates the point at which a segment of the long arm of chromosome 8 had exchanged with a segment of the long arm of 9. This laid the foundation for subsequent studies of chromosomal changes in maize.

McClintock became an expert at identifying the different chromosomes at the pachytene stage. She worked closely with the cytologists at CIMMYT (Centro International de Mejoramiento de Maiz y Trigo) in studies of the different races of maize found in Mexico and Central America. During a later visit to Cornell, after she discovered transposable elements, she showed me an album full of color pictures of the variations in flower color patterns of the touch-me-not that grows in different moist sites, which she interpreted as another example.

At the California Institute of Technology

At the end of the summer of 1930, I went to the California Institute of Technology to work with E.G. Anderson, who was working with semisteriles in corn that had been induced by X-rays. Gertrude Geertsen was doing the cytology; Sterling Emerson and Alfred H. Sturtevant were studying *Oenothera* species; and Walter Lammerts was working with crosses between *Nicotiana* species, tetraploid and diploid lines derived from backcrosses that segregated for new traits. When in my second year at Caltech I was asked to teach a course in cytological technique, McClintock suggested that I have the students make slides using the acetocarmine smear technique on corn sporocytes. I remember that all the students were successful in getting the chromosome count. I was studying the cytology of a line of maize with low sterility. It had a chain of four chromosomes attached to the nucleolus. McClintock came to E.G. Anderson's Caltech laboratory at Temple City near Pasadena. She looked at my slides and the next day, true to form, came back with the complete explanation. I then confirmed her analysis by finding the pachytene figures which showed that the terminal one or two chromomeres of the satellite of chromosome 6 had exchanged with a long segment of the long arm of chromosome 1. It is now designated *T1-6b*.

At the University of Missouri

The next year, I went to the University of Missouri at Columbia. Lewis J. Stadler was studying *Sorghum versicolor*, which has five chromosome pairs. I never was able to obtain a cell in which all of the chromosomes could be followed at pachytene. Barbara McClintock was also there at that time doing the cytology on corn plants that Stadler had produced by X-ray treatment. Included were chromosomal interchanges, inversions, and deletions from crosses using X-rayed normal pollen on recessive mutants. Mutants among the progeny resulted from deletions in the normal chromosomes that were at the site of the mutation. Pachytene analysis identified the deletion and the site of the gene for that trait.

Barbara's workplace was at a different location from mine but we continued to exchange information. I was attempting to establish a multiple interchange stock which, when crossed with normal, would produce plants that would have all 20 chromosomes in one ring. I remember Stadler saying "some little trick will do it." Later, one of my students at the University of Minnesota, Lawrence L. Inman, discovered the trick and demonstrated in his thesis how additional interchanges could be added without increasing the sterility beyond 75%. Subsequent work established two stocks, one of which, when crossed with normal, would produce plants with a ring of 10, and the other stock would produce a ring of 12. One interchange chromosome in this latter stock was also involved in the other interchange stock. When crossed with each other, the plants have a ring with 20 chromosomes. The chromosomes involved in the multiple interchange stocks are 1-5-6-7-8 and 3-2-4-9-10-8.

The year the AAAS science meetings were held in New York (1949), I attended with one of my students, Mahmoud A. Ibrahim. Barbara was there, but we did not have a chance to talk. A half hour or so before train time, she came to the station waiting room and discussed with us her findings about the *Ac* and *Ds* controlling elements. When Mahmoud and I settled in our seats on the train, I wrote down what we could remember about the discussion. I wrote to her later asking for answers to some questions—which she answered.

Epilogue

Barb, I learned my cytological technique from you, and much more. For example, Emma L. Fisk had published a paper on the chromosomes of corn, but they were only globs. When I started work in your laboratory, I believe you showed me that the shapes of chromosome pairs at metaphase or early anaphase-1 at meiosis depended on the position of the centromere: One with a near median position is much different from one with the centromere near one end. Also, that at diakinesis, chromosome 6 was attached to the nucleolus; there were five chromosomes longer than 6, numbers 1 to 5 and four shorter ones, 7, 8, 9, and 10. The ring of four in *semisterile-1* left only three longer than 6, and thus two of the long pairs were in the ring. In *semisterile-2*, there were only two short ones outside; the ring had two short ones.

I remember how excited you were when you began studying *semisterile-2* at the pachytene stage. Not only did you see the cross-shaped configuration which indicated the breakpoints in 8 and 9, but there was a darkly stained large knob on the end of the short arm of 9, from the *pr wx* stock. You had a special explanation for it, but a day or so later you realized it was a part of the chromosome pattern of 9 in that stock. This opened the field for studying maize cytology.

That summer of 1929, it seemed as though new information was being added almost every day. What you were getting from the trisomics, Beadle's, and others' linkage results, and what I was getting from the progeny from crosses with the two 75% sterile lines, one from *S.S.-1* x *S.S.-2* and the other from *S.S.-1* x *S.S.-3*. It was obvious that a set of semisteriles in which the chromosomes had been identified could be used for crosses with any unknown semisterile to identify the chromosomes involved. The diakinesis stage could be used. E.G. Anderson was also studying semisteriles. The set I finally used included *semisterile-1* (*T1-2a*), *2-4d*, *3-7c*, *5-7c*, *semisterile-2* (*8-9a*), and *8-10b*. If 6 were involved, the ring would be attached to the nucleolus. Anderson used a slightly different set for his studies.

You were the enthusiastic leader, thrilled to see the cytology of maize being combined with genetics. There was a sense of teamwork. We were all participating. About that time, was it 1929?, Dr. Brink was studying the linkage of a new dominant trait, *Ragged* (*Rg*). Lo and behold it showed linkage with genes in what had been considered to be two linkage groups. Now there was one group missing. I had linkage tests in progress between *semisterile-2* and *japonica stripe* (*j*). That marker proved to be in 8, which had the missing group.

I have continued working with chromosome interchange and inversion stocks, mainly as tools to study complex traits, or to study chromosome behavior. Several of my students used them for their thesis studies. The barley breeder at the University of Minnesota began X-ray treatments on barley, looking for favorable mutants. The heads of plants from treated seeds were planted in single rows. Four plants in each row were examined for pollen abortion. Sporocytes from each plant with aborted pollen were examined. Many had a ring of 4, one a ring of 6, and one had long chromosomes at metaphase-1. Plants with a ring of 4 had only 25% pollen abortion, a result of directed orientation higher than 50%. A set of chromosomal interchanges was established for chromosome identification as in maize. One student did a thesis on chromosomal interchanges in the tomato. An identification set was established in follow-up studies. Linkage studies were conducted in Datura, using interchange stocks established by the Blakeslee group. Lines with a ring of four ranged from no pollen abortion or very low to 50%. Orientation of the rings are mostly zig-zag, i.e., directed, producing only microspores with normal or the interchange chromosomes. Crossing over between the centromere and the interchange breakpoint accounts for the sterility. I have continued working after retirement. In 1990, I made the final test crosses to identify stocks which, when crossed on normal, would produce plants with two interdependent rings of four. They are expected to produce two viable chromosome combinations, rather than four.

In closing, Barb, some of your scientific fervor rubbed off on me, and in turn on my students. Four of my graduate students have received special honors. Three of them have been elected to the National Academy of Sciences: Dr. Oscar Miller at the University of Virginia, Dr. Jack Axtell at Purdue University, and recently, Dr. Ronald Phillips, my successor at the University of Minnesota. The fourth is Dr. Ken Kasha at the University of Guelph, Canada, who was elected to the corresponding organization, the Royal Society of Canada.

Barbara McClintock: Recollections of a Graduate Student

HELEN V. CROUSE
Talahassee, Florida 32312

I discovered who Barbara McClintock was during my senior year (1934–1935) at Goucher College. One day late in the term, Professor Ralph Cleland brought into cytology class a recently published paper written by a young woman working at Cornell University that he considered to be of unusual importance. It was McClintock's paper on nucleolus formation in maize. Cleland outlined the essentials of McClintock's cytogenetic strategy and suggested that we examine the paper for ourselves. What a revelation! I had never seen such beautiful chromosomes. Nor did I realize that decisive answers to fundamental cytological questions could be obtained by rearranging chromosome segments. I was delighted and thought that some day I should like to work with this remarkable scientist.

The opportunity to meet Barbara McClintock did not arise until 3 years later. In the meantime, I had accepted a fellowship to study zoology at Smith College, where I completed a Master's Degree in 1936. Because of the dearth of biology courses taken at Goucher, I had to carry a very heavy program at Smith. My thesis project, of my own choosing, was far too broad in scope; I sought to determine the effects of heat shock on variously aged germ cells in *Drosophila*. I believed, naively, that I could shed light on some of the issues concerning Dauer modifications raised by Richard B. Goldschmidt and Viktor Jollos! Instead, I exhausted myself and felt unable to accept a fellowship for graduate study the following year at Cornell University.

I therefore decided to take a job as research assistant in Charles W. Metz's laboratory at Johns Hopkins. Metz had lectured to our cytology class at Goucher about his work on *Sciara* (fungus flies). The chromosome behavior he described was so unorthodox as to be almost unbelievable. Thus, it was with considerable interest that I undertook my new position.

Metz held a dual appointment; he was Professor of Zoology and also a member of the Carnegie group, then housed at the School of Hygiene. He spent most of his time at the Homewood campus, his office and the *Sciara* laboratory being in Gilman Hall. This location provided me with an opportunity to attend a variety of lectures and seminars, including class lectures for the graduate students in botany and zoology. Both departments at that time boasted a number of heavyweights, including Herbert Spencer Jennings, S.O. Mast, and Duncan Johnson. Jennings' lectures on the genetics of the protozoa were outstanding. In addition to the regular staff, a stream of visitors passed through. Early in 1938, the corn geneticist Marcus M. Rhoades appeared. Rhoades, who was then employed at the USDA in Washington, discussed his work on the newly discovered Dotted (*Dt*) gene. I knew that Rhoades had gotten his Ph.D. with R.A. Emerson at Cornell so I asked him about Barbara McClintock. He said she still grew her corn every summer in Ithaca but that she also had an academic appointment at the University of Missouri. Smiling, he told me that visitors were anathema to her! As luck would have it, within a few weeks L.F. Randolph, Professor of Botany at Cornell, came by to see Metz. I seized the opportunity and asked if I might visit his laboratory during the summer. He said yes and that I should simply send a postcard giving him the date of my arrival.

Cornell University: I Meet Barbara McClintock

I followed Randolph's instructions and very early one June morning set out for Ithaca in my old Ford coupe which burned a quart of oil every hundred miles. I reached Ithaca shortly before noon only to discover that Randolph was in the hospital and was scheduled for thyroid surgery that very day. I looked in vain for Barbara McClintock's name. Fortunately, I spotted "A. Lebedeff " on the Directory. Lebedeff had sent a manuscript to Metz earlier that year and Metz had asked me to read it. Lebedeff was very cordial, but when I inquired about McClintock, he said her laboratory was up under the roof and that she did not want to see anyone. I asked if he would show me the way. He hesitated but finally acquiesced. When we arrived at the door, I stepped up and boldly knocked. The door opened and there stood a wisp of a woman looking straight at me from under an opaque green eye shade. At that moment I realized I was alone in the doorway: Lebedeff had vanished! Before I could utter a word, McClintock said she was expecting me. Mrs. Randolph had apprised her of the situation and had asked her to look out for me. She said she had in fact gone to a great deal of trouble rounding up a microscope and lamp, etc.—all this during her busy season. She also said she had obtained a rented room for me nearby and that we could drop my bag off on the way to lunch at Dr. Parker's.

Esther Parker was a much revered local physician with whom McClintock lived when she was in Ithaca. Dr. Parker was a dynamic, fun-loving person who made me feel welcome as soon as we entered through the kitchen door. For lunch, she fed us corn flakes and country cream with lots of fresh-picked strawberries: a delightful prelude to three memorable days!

My mornings were usually spent alone in the laboratory making corn slides and examining them under the microscope. The weather was clear throughout my visit, so McClintock spent the mornings pollinating in the field. During the afternoon, we either discussed my slides or examined plants together. There were many diagrams of ring chromosomes drawn on flat stones picked up here and there as we walked the corn rows. McClintock's great opus on ring chromosome behavior published in *Genetics* (August, 1938) had not yet appeared. In the evenings, we relaxed with Dr. Parker at her cottage on the lake, where three beloved Irish setters were in residence. The importance of these animals to Dr. Parker was made eminently clear to me one evening when we ate dinner in a downtown restaurant. At least four groups of diners stopped by our table to chat and to give Dr. Parker the bones and trimmings from their steaks and chops! We had quite a sizeable bundle as we made our way joyously to the local cinema to see "Alexander's Ragtime Band."

By the end of my 3-day visit, I told McClintock that I wanted to do graduate work with her. She was very discouraging and said women in science were not considered for university jobs. Her own position at the University of Missouri was due entirely to the good offices of Lewis J. Stadler. The Missouri Department of Botany really did not want her, and she felt she did not belong. I refused to be discouraged and persisted in believing that one day I would indeed study with Barbara McClintock.

En route to visit with friends in Garrison, New York, I discovered I had walked off with the laboratory key. I mailed it back with apologies and mentioned that I had left behind my gray Waterman's fountain pen. Later, I discovered the pen in my car. On my arrival at the Marine Biological Laboratory, Woods Hole, Massachusetts, I was overwhelmed to find a parcel from McClintock that contained a new Waterman's pen. I went directly to Mr. Panis, the silversmith in Falmouth, and asked him to make a pair of copper bookends featuring not his usual shore birds but corn plants in full tassel! The resulting creation was magnificent.

Back to Johns Hopkins University: But We Keep in Touch

Following my return to Johns Hopkins at the end of June, I had a letter from McClintock inviting me to attend with her the Genetics Society meeting to be held in Woods Hole the last week of August. It was a great gathering. Not only did I meet L.J. Stadler and R.A. Emerson, but also Sewall Wright among others. I talked to Stadler about the possibility of doing graduate work at the University of Missouri. My preparation had been in zoology; Stadler was in field crops; and McClintook was in botany. It looked as though I should have to enroll in zoology. McClintock wanted me to enroll in the fall of 1938. I hesitated because I had not yet discussed with Metz the possibility of my departure from Johns Hopkins. When I did take up the matter, he said he could not replace me on such short notice, so I decided to stay on for another year. McClintock visited me in Baltimore that fall on her way to Missouri and we discussed the *Sciara* project I hoped to undertake. I was surprised to learn that she had once been very interested in *Sciara*, at a time when Metz among others was trying to locate a position for her. Later that autumn I went to Missouri for an interview with Mary Jane Guthrie, Professor and Chairman of the Zoology Department. In a four-way discussion with Guthrie, Stadler, and McClintock, it was agreed that I should apply for a fellowship in zoology and do my thesis under the supervision of McClintock and Stadler. With this agreement, I returned to Hopkins and immediately began irradiating *Sciara* in order to recover the rearrangements I would need for my project. By midwinter I learned that I had been awarded a fellowship in zoology, and by April I knew that X-autosome translocations in *Sciara* were viable and transmissible through the male and female germ line.

The University of Missouri: Access to McClintock

On arrival at the University of Missouri in September 1939 (the University's Centennial), I found an atmosphere very different from that of Johns Hopkins. Instead of red brick Georgian buildings set off from the city by beautifully landscaped lawns, parks, and gardens, our Botany-Zoology building (Lefevre

Hall) was at the outer edge of the "White Campus" (white stone buildings) and faced directly on to Hitt Street. Opposite were modest residential homes; behind stretched a flat plain and a scattering of buildings belonging to the School of Agriculture. A rustic note resounded each time classes changed; the undergraduates, as they crossed campus, would exuberantly practice hog-calling!

In the fall of 1939, the new Genetics Building was still under construction. The Botany and Zoology Departments were housed in Lefevre Hall, Zoology in the east end, Botany in the west. Just east of Zoology, there was a small separate building called "Wildlife" that was surrounded by a garden; it had been built originally for Professor Bennett and students engaged in the study of wildlife management. McClintock's office, a delightful light-filled area, occupied the upper level of this small building. Her office could be reached from the garden by its own set of steps and was thus very private and quite removed from the din of student traffic. Here, she was ensconced in solitary splendor and provided with the best equipment available. She had furnished her office with two good-looking, very comfortable arm chairs. Present also was her splendid phonograph. I had learned in Ithaca and on our trip to Woods Hole how much McClintock enjoyed music and how well she could whistle and play the accordion. Her office was literally her home, both in "Wildlife" and later in her three-room suite on the top floor of the new Genetics Building. My niche, assigned to me by my official adviser Miss Guthrie, was a corner of the Zoology Animal Room located adjacent to the east exit of Lefevre Hall. Thus, my access to McClintock was very convenient. McClintock's other two students, Spencer Brown and Ed Weaver, were teaching assistants in the Botany Department.

The town of Columbia, Missouri, was small and isolated. It offered little in the way of diversion except for football games in the fall. Nearby Stephens College had a good Drama Department and they often ran foreign films. McClintock's one diversion that I was aware of was having lunch downtown daily at Jack's Latch Cafeteria, followed by a visit to the local weather station located on the second floor of the Post Office building. She was interested in weather forecasting and tried to convince the two men who worked at the station of the superiority of forecasting by air mass analysis. I accompanied her to the station on several occasions and thought the men looked forward to her daily visits. They were very cordial and shared with her their latest charts and data. Occasionally, she joined a group of us for supper at the Virginia Café, where you could get an excellent western steak for fifty cents. Spencer Brown was always present, so there was never a dull moment. And Ray Dawson, a member of the Botany Faculty, who was full of spirit, often joined us.

Since Miss Guthrie was my official adviser, she helped arrange my course of study. Because I was expected to complete the Prelims by year's end, I attended as many lectures on zoology as possible. I enrolled in Daniel Mazia's Cellular Physiology. Mazia, a protégé of L.V. Heilbrunn, was well versed not only in the contemporary literature of physiology and developmental biology, but also in the classic papers. He loved ideas and the history of ideas. His lectures were delivered in a unique, nearly spell-binding manner, and his laboratory experiments were most interesting and well designed. I also attended the lectures in embryology by Donald H. Barron. Barron, a midwesterner by birth, was a colorful Anglophile who had come recently from Sir Joseph Barcroft's laboratory. His British accent, manner, and dress invited a certain amount of teasing. One evening at supper, Spencer Brown quipped, "A new novel by Donald H. Barron: 'I'd Rather Be British Than Bright.'" Barron's lectures were well attended, and the large audience included both graduates and undergraduates. One of Miss Guthrie's cytology classes surprised me when she spent 1 hour on the nucleolus without even mentioning McClintock's brilliant paper on the organizer region!

In addition to Mazia's cell physiology, I enrolled in three chemistry courses, including Addison T. Gulick's physiological chemistry, qualitative analysis with Gerald E. Breckenridge, and advanced inorganic chemistry with Lloyd Thomas. The work with Thomas was very demanding, but it was the most rewarding course I had ever taken. Thomas's audience must have numbered 250—all young men. At the final exam, I saw at least 50 of them give up and depart within the first 10 minutes, and after the first hour, I doubt that there were as many as 100 remaining.

The Zoology Department did not have a seminar or colloquium. In other words, no effort was made to discuss the work that was being done by either the faculty or student members of the department. Neither were there any presentations by guest lecturers. In fact, I saw no visiting scientists in Zoology the entire time I was in Missouri. This was in marked contrast to Stadler's department, where geneticists and cytogeneticists passed through fairly frequently. When the new Genetics Building was completed, there were two visitors in residence: Dr. Carl P. Swanson from Harvard and Dr. and Mrs. Carl Lindegren from Illinois. Moreover, from the beginning of the fall term (1939), Stadler arranged a seminar for the graduate students in genetics. I remember that McClintock's sessions were excellent and exceedingly comprehensive. She dealt with the overall problem of heterochromatin and with position effects in *Drosophila*. These seminar meetings not only exposed us to new ideas, but also enabled the students of zoology to get to know those in botany and field crops, etc.

In addition to the zoology and chemistry courses, I and some six or seven other zoology students were enrolled by Miss Guthrie in Cryptogamic Botany. Our instructor was a young Ph.D. from Harvard whose name I have forgotten. His daily performance was unbelievable: He would *read* verbatim from Smith's "Cryptogamic Botany," including the words "comma," "semi-colon," etc. When the zoology contingent took up the matter with Miss Guthrie, she was furious. I am sure that Tucker, Chairman of Botany, heard from her that very day, and I suspect that the young man from Harvard was soon sent packing.

I do not recall just when we moved into the new Genetics Building, but for me it was a godsend. While I was housed in the zoology animal room, I had no way to control the temperature of the *Sciara* cultures, and there was always the danger of mite infestation from the various mammal residents, including bats. In the new building, I was assigned a beautiful ground-level room with two large windows facing northward, through which I could look across a broad expanse of land to Lefevre Hall. Opposite my room was the autoclave center. Between my room and the adjacent laboratory, there was a controlled-temperature room independently accessible to both laboratories. What delightful accommodations for *Sciara* and for me!

Ed McCleary, an older student who worked with Stadler on the chemistry of anthocyanin production in the corn leaf, shared the temperature-controlled room, and Spencer Brown had the laboratory on the other side. Brown and I liked each other from the day we met and we remained friends for life. He was a brilliant and unusually sophisticated boy who had come from the University of Minnesota in 1938 with the express purpose of working with McClintock. I do not know what had happened during Brown's first year, but, by the time I arrived in 1939, he was not working with McClintock on a corn problem. Instead, he seemed to be trying his hand at various projects with Daniel Mazia. I understood from the grapevine that McClintock had convinced him he could not make a decent corn slide. Finally, by the end of the fall term of 1940, the Botany Department decided it would be best if Brown transferred to Berkeley. He completed his Ph.D. there with Ledyard Stebbins and, some years later, was made a professor in the same department. McClintock's other botany student, Ed Weaver, was trying to induce gene mutations on chromosome V in the vicinity of the brown midrib (*bm1*) locus that would yield the same phenotypes McClintock had obtained through homozygous deficiencies produced during the course of her ring chromosome study. Whether Weaver ever finished this project I do not know.

I believe that Brown, Weaver, and I all had the same difficulty in our association with McClintock: Her quick intelligence, superior powers of analysis, and prodigious technical skills were overwhelming. The three of us were veritably frozen in awe and admiration. Personally, I shall never forget the day when I called her to look at a new translocation. After the briefest glance in the microscope, she found, in addition to the translocation, a second rearrangement that I had overlooked!

At the end of December, 1939, the genetics group was dealt a dreadful blow: Stadler was offered and accepted a visiting professorship at the California Institute of Technology. Rumor was rife as to whether this might be a prelude to a permanent offer. If so, would Stadler choose to remain in Pasadena and, in that event, what would happen to the genetics group at Missouri? This question festered for months. It was obvious to everyone that no one at the University of Missouri could replace Stadler. His was a towering intellect, and his personal qualities were an inspiration to every student who had ever known him. Consternation persisted throughout the second semester; the situation was resolved only when W.C. Curtis, Dean of Arts and Science, threatened to fire the entire genetics group if Stadler did not come back. Given these conditions, he chose to return.

Toward the end of April, 1940, I received a letter from Metz telling me he had been made Chairman of the Department of Zoology at the University of Pennsylvania and that he would be pleased to have me continue graduate work with him if for any reason I should wish to leave Missouri. He said furthermore that I would be welcome to spend the summer as a member of the Penn group at the MBL in Woods Hole. I gave his proposition serious thought. Because it was still uncertain what Stadler would do and because I had been awarded by Goucher College the Dean Van Meter Alumna Fellowship for 1940–1941, I decided to go to Woods Hole. When I apprised Miss Guthrie of my decision she was quite annoyed and said it would have a negative effect on the chance of women being awarded future fellowships by the University of Missouri. Early in the summer, I received at Woods Hole a letter from Stadler saying how disappointed he was that I had departed Missouri, so I decided to return in September.

During my second (final) year at the University of Missouri, I took only two courses. One course for graduate students given first term by Miss Guthrie dealt with the reproductive cycle of female mammals. Miss Guthrie and her students had done a great deal of work over the years on reproductive cycles in bats native to the Ozarks. The other course, offered second term by McClintock for graduate students, was a cytogenetics course with projects involving both laboratory and greenhouse.

These two courses stood in stark contrast to each other. There was no laboratory work in Miss Guthrie's course. She lectured once a week for three solid hours from 7 to 10 p.m. with no break. Each week, we received two or three pages of single-spaced references and were expected to know the contents of these references and give proof of such knowledge in a 10-minute quiz at the beginning of the next class meeting. The anxiety that this course produced was so great that it disrupted the ovulatory cycle of some of the female students.

McClintock's course was great fun. At the outset, she met with each of us individually in the genetics greenhouse and assigned to us a group of corn plants that were our first "unknown." We had to determine the chromosome constitution of each plant. This meant that we had to sample the tassel at the appropriate stage of development, make sporocyte preparations, read the slides under our own microscope, and show McClintock the results in progress. I remember very well the day she gave me my plants. As I recall, there were eight or ten of them; they were all morphologically distinct from each other and from the normal. At once, I said I thought they were trisomics. She denied it, but the look of surprise on her face led me to believe I was close to the truth. My hunch was based on some reading I had done in A. Müntzing's book on polyploidy and plant evolution. He had described in detail the effects of aneuploidy on plant morphology, physiology, and reproduction. Actually, my plants were the progeny of a triploid and some of them had one, two, or as many as five extra chromosomes. To diagnose the chromosome makeup of any one plant meant diagramming many pachytene nuclei. Recognizing chromosome VI was easy because of its attachment to the nucleolus, but then numbers V and VII, then X, etc. could soon be spotted. I thoroughly enjoyed the game. When it was over, I knew all ten chromosomes very well and I could readily distinguish B types from the A types.

In addition to studying corn microsporogenesis, we learned how to make root tip squashes of a variety of plants, including normal and colchicine-treated material. We also followed the progression of endosperm and embryo development in corn kernels 6 and 8 days after pollination.

The second part of our cytogenetics course was a group meeting each Saturday morning from 9 to 12, where we examined permanent slides from McClintock's own collection; they showed various modes of chromosome behavior that her research had elucidated. McClintock spent Friday evenings setting up these slides under ten brand new research scopes that had been purchased just for our class. What a demonstration: ring chromosomes undergoing division; dicentric and acentric formation in paracentric inversion material; the chromatid breakage–fusion–bridge cycle, etc. Each slide led us back to one of McClintock's great papers.

McClintock Leaves Missouri

Most of the research for my thesis was complete by Christmas, 1940; but all of the data had to be summarized, salivary chromosome maps drawn, photomicrographs made of each translocation, and the writing of the entire text accomplished. I began to get leverage on some of these tasks during the spring of 1941 and received a boost when I discovered that McClintock had given me an "A" in cytogenetics. I was settling in for the hard pull when McClintock came by and said she was leaving the University of Missouri for good; her car was all packed and she was going back east. I was so upset by this startling announcement I hardly comprehended the situation. Suddenly she was gone, never to return.

During the ensuing 5 months I worked doggedly on the thesis. Stadler was helpful in every way possible. When I learned that the American Association of University Women had awarded me the Elizabeth Avery Colton Fellowship for 1941–1942, he arranged a postdoctoral appointment in Theodosius Dobzhansky's laboratory at Columbia University. Getting the thesis approved by Miss Guthrie was tedious and exhausting to say the least. Finally, after more than a month wasted in trivial changes in punctuation, Stadler told me to tell her she could take it or leave it and that I was going to New York. He assured me I had his full support in this confrontation. He also advised me jokingly to remove my lipstick before the encounter and to appear as bedraggled as possible. I cannot remember exactly what she said, but she temporized again, saying of course the thesis would have to be approved by McClintock. The whereabouts of McClintock at that time was unknown, but I believe I mailed the tome to her at Cornell University. She was not in Ithaca but at Cold Spring Harbor. Finally, she had the thesis in hand, read it immediately, and wired her approval. As soon as Miss Guthrie was convinced of the support of my mentors, she scheduled my oral examination for the next morning at 8. The oral went well and in a few days' time I was on my way to New York. By then, it was nearly November, 1941.

I had to return to Missouri the following spring for the official awarding of my degree. When Miss Guthrie offered me an instructorship in her department I declined it, although I had no other job in sight. Thus ended my bittersweet experience at the University of Missouri.

Cold Spring Harbor: We Meet Periodically

After I departed graduate school, McClintock continued to loom large in my life. No one else was so dedicated to discovering how chromosomes work. And no one else was so successful. Certain of her papers I regularly used as texts in my genetics class. She was always supportive of my teaching, and on at least three occasions, at Goucher College and later at Columbia University, she came to my class and lectured on her work. Over the years, we exchanged reprints and occasional letters, and when I was living in New York City, I went periodically to visit her at Cold Spring Harbor. I witnessed the monumental effort involved in substantiating the behavior of transposing genetic elements. No one was more delighted than I when finally in 1983, the long overdue accolade of honors was showered upon her.

Neurospora Chromosomes

DAVID D. PERKINS
Stanford University, Stanford, California 94305-5020

"You may think they are small when I show you pictures of them. But when you look at them, they get bigger and bigger and bigger."

Many who are familiar with the main body of Barbara McClintock's work may not be aware of her important contribution to fungal cytogenetics. Although sparsely documented, her work was crucial in demonstrating that fungal chromosomes and meiosis can be worked with effectively, that they conform to a pattern that is typically eukaryotic, and that meiotic cytology can provide critical information to complement genetic analysis in fungi. Her brief excursion into *Neurospora* cytology demonstrated a virtuosity that continues to evoke admiration.

George Beadle invited McClintock to come from Cold Spring Harbor to Stanford in 1944 with the object of looking at *Neurospora* chromosomes. The *Neurospora* group at Stanford was then chiefly preoccupied with biosynthetic pathways, gene-enzyme relationships, and practical applications such as bioassay. Mutant genes were being mapped, and there was curiosity about the number of chromosomes. Earlier in his career, Beadle had been concerned with mutants affecting meiosis in maize, and with crossing over in *Drosophila* inversions and attached-X chromosomes. It was therefore natural for him to be interested in cytogenetic aspects of *Neurospora* and to turn for help to McClintock, whom he had known at Cornell University when they were students together under R.A. Emerson.

Three years prior to the invitation, Beadle and Edward Tatum had published their epochal paper reporting the first biochemical mutants. This first brought *Neurospora* to the attention of a broad audience. The organism had already become known to geneticists in the 1930s, however, from the work of B.O. Dodge and Carl Lindegren. By 1939, crossing over and segregation in the *Neurospora* ascus were being diagrammed in texts such as A.H. Sturtevant and Beadle's *An Introduction to Genetics*. The fact that all four products of meiosis could be recovered was of special interest to a generation of geneticists for whom meiotic crossing over was both a basic tool and a central problem for study. The ability to map centromeres on the basis of segregation patterns in the linear ascus added to the attractiveness of *Neurospora*. Haploidy was still a novelty.

The first introduction to *Neurospora* for both Beadle and McClintock was probably in 1929, when Dodge gave a seminar at Cornell. Dodge was puzzled by the fact that genes sometimes showed first-division segregation patterns in individual asci, and sometimes second. Students at the Cornell seminar were familiar with the work of C.B. Bridges and with E.G. Anderson's recent analysis of crossing over in *Drosophila* attached-X chromosomes. According to

Beadle (1966), they explained to Dodge that the chromosomal basis of second-division segregation must be crossing over at the four-chromatid stage in the region between gene and centromere.

McClintock's introduction to *Neurospora* cytology came some years later, quite by accident, while she was at the University of Missouri. A student who was interested in fungi had somehow obtained *Neurospora*, and he brought her a slide that contained asci with excellent meiotic figures. Additional acetocarmine squash preparations were made and examined. This brief experience convinced her that it was indeed possible to work with *Neurospora* chromosomes (McClintock 1961). Before going to Stanford in 1944, she had also looked at chromosomes in a strain of *Neurospora tetrasperma* provided by Dodge (McClintock 1945).

Fungal Chromosomes before McClintock

Prior to McClintock's work, knowledge of fungal chromosomes was rudimentary and confused. In part, this was due to their small size. The DNA content of the *Neurospora* genome, for example, is only 0.5% that of maize, and the two smallest chromosomes each contain less DNA than the chromosome of *Escherichia coli* (Orbach et al. 1988). The confusion arose in part because the material was routinely embedded and sectioned, which made chromosome counts unreliable. Already, before the turn of the century, mycologists such as R.A. Harper had examined nuclear divisions in the asci of various fungi with the object of establishing when it was that fertilization and reduction occurred. The fact that asci in most species were eight-spored was puzzling and led to controversy. Harper thought he saw two successive nuclear fusions followed by two reduction divisions in the ascus. Harper's interpretation was supported by Helen Gwynne-Vaughan, who termed the process *brachymeiosis*. Other mycologists disagreed, holding that there was only one nuclear fusion followed by an orthodox meiosis, with a single reduction division that was followed by a postmeiotic mitosis. The controversy went on for years (see Olive 1950).

Following the discovery of the sexual phase of *Neurospora* by Shear and Dodge in 1927, stained sections were again used to examine the asci. A study of *Neurospora tetrasperma* by Dodge (1927) was remarkably accurate with respect to nuclei, spindles, spindle-pole bodies, and ascospore formation, but it failed to provide information on chromosomes. McClintock (1945) cites two previous studies that attempted unsuccessfully to determine the chromosome number.

McClintock's Work with Neurospora

Unlike those who had preceded her, McClintock had the advantage of a broad knowledge of both genetics and cytology, and long experience working with meiotic chromosomes. Methodology for *Neurospora* was adapted from that which she had perfected for maize. She used squashes rather than sections, so that the chromosome complement remained intact. Following trials, orcein was selected as the preferred dye. (A legend has McClintock giving instructions, "Bring the dye to a boil in acetic acid, then stop boiling just before it changes color.")

Neurospora proved to have been a fortunate choice for meiotic cytology, just as it had been for genetics. The *Neurospora* ascus develops into a single unpartitioned giant cell (~175 x 20 μm) within which meiosis and an additional mitosis take place in a common cytoplasm. Only after the three divisions have been completed are ascospore walls laid down around the eight resulting nuclei. Nuclei and chromosomes are much larger in the ascus than at any other stage of the life cycle, especially during the prolonged first meiotic prophase. The largest *Neurospora* bivalent can attain a length of 20 or 25 μm at late pachytene. (In contrast, mitotic nuclei in the vegetative mycelium are about 2 μm in diameter.) Many fungi would have proved far less tractable for meiotic cytology. The asci of *Aspergillus* and *Saccharomyces*, for example, are minute by comparison with those of *Neurospora* (for comparisons, see Fig. 11 in Emerson 1966).

During the 10 weeks of her 1944 visit, McClintock laid a sound foundation for fungal meiotic cytology. She described her observations in a 1945 paper and summarized them as follows:

> Neurospora offers adequate and in some respects unique opportunities for cytogenetic research. The chromosomes were followed from the nuclear division preceding zygote formation through the division in the ascospore. Chromosome morphology was considered with reference to the absolute and relative sizes of the seven chromosomes in various division cycles, the centromere positions, the nucleolus chromosomes, the pachytene chromomere morphology and the presence of heterochromatin. Chromosome behavior was followed with reference to the atypical timing of chromosome synapsis, the elongation of the chromosomes during a prolonged "pachytene," chiasma formation and the general behavior of the chromosomes in the two meiotic mitoses and the two subsequent equational mitoses. Several reciprocal translocations were investigated and their usefulness for special studies indicated.
>
> No distinctively unique features of chromosome organization were recognized. The presence of translocations between non-homologous chromosomes following irradiation treatment and the behavior of these translocated chromosomes in the meiotic stages of heterozygous asci are likewise indicative of the orthodox organization of the Neurospora nuclei and chromosomes.

McClintock returned to *Neurospora* for brief periods in 1946 and 1953 during visits to the California Institute of Technology, where Beadle had moved from Stanford and where she supervised the thesis work of Jesse Singleton. The results of her work during those visits were published only as brief reports in the Carnegie Year Book (McClintock 1947, 1955). In 1946, while working with Singleton, she obtained excellent photographs. She showed these as lantern slides on at least three occasions: in talks in 1947 at a Genetics Society Symposium, in 1948 at the New York Microscopical Society, and in 1961 at the first *Neurospora* Information Conference. The quotation at the head of this essay is taken from a transcript of the 1961 meeting.

It remained for Singleton to provide a full account of chromosome morphology and behavior in his 1948 thesis and a 1953 paper. Thirty five of McClintock's photographs were used to complement his own photographic documentation.

As an aid to identifying the individual chromosomes, McClintock made sketches showing chromomere patterns and their variations. Some of these unpublished sketches are reproduced in Figure 1. For many purposes, drawings of this type are more informative than photographs, the usefulness

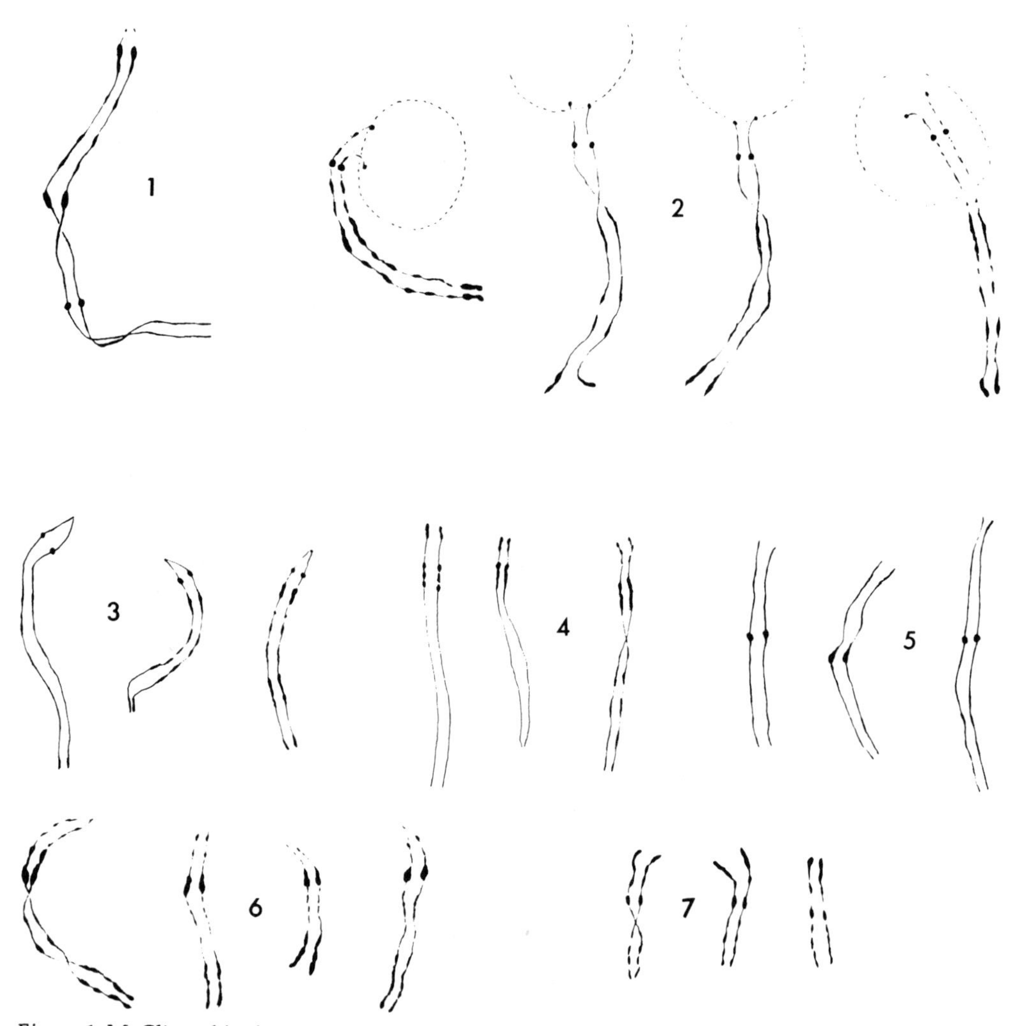

Figure 1 McClintock's sketches of the seven *Neurospora crassa* chromosomes at late pachytene.

of which is severely limited because focal depth is minimal at the high magnification required for working with such small chromosomes. The limitation is especially apparent at pachytene, where the most critical observations are made on unbroken nuclei that are not completely flattened. Only a fraction of the complement can be seen at any one focal level in such a nucleus, even though it may be possible to identify and trace all of the chromosomes by focusing up and down. Multiple photographs taken at successive levels are required (see, e.g., Barry 1967; Barry and Perkins 1969).

The small size of the *Neurospora* nucleus and chromosomes can be appreciated by comparing the pachytene complements of *Neurospora* and maize in Figure 2. Following convention, McClintock assigned numbers to the *Neurospora* chromosomes according to decreasing length. The lengths she determined for the pachytene complement are given in Table 1, together with other measurements and correlates based on the subsequent work of others.

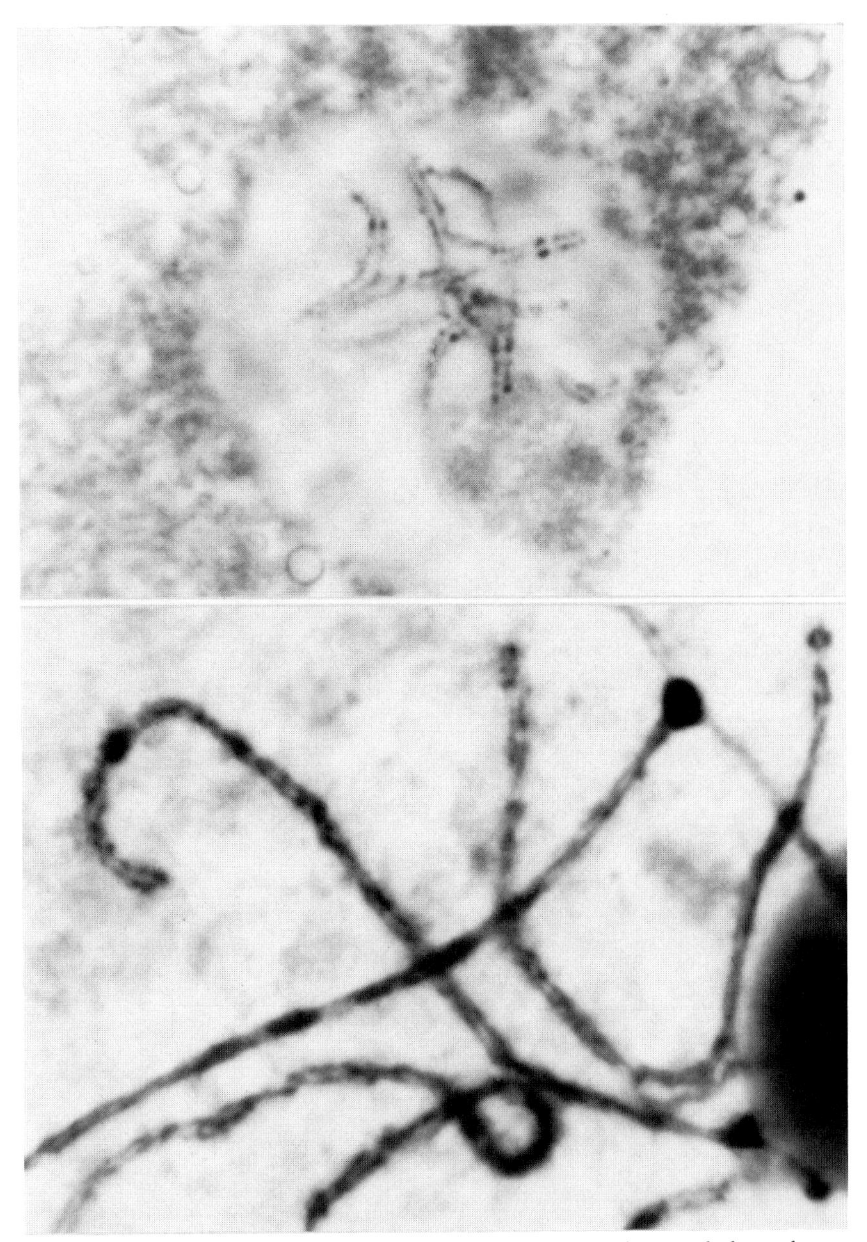

Figure 2 Pachytene chromosomes of *Neurospora* above and maize below, shown at the same magnification. The small crescent-shaped bivalent at 11 o'clock in the *Neurospora* nucleus is chromosome 6, which has a 1C DNA content less than that of *E. coli*. Magnification, 3100. (Photographed by E.G. Barry and J.R. Singleton.) (Reprinted, with permission, from Perkins 1979.)

Although *Neurospora* meiosis conformed in all essential respects to that of plants and animals, significant minor differences were noted. Synapsis occurred between homologs that were not yet fully extended, and elongation followed pairing rather than preceding it. Homologously paired pachytene chromosomes were aligned in parallel, but rather than intimately touching each other, they were separated by a conspicuous intervening space along

Table 1 *Neurospora crassa* Chromosome Lengths, Linkage Groups, DNA Migration, Numbers of Mapped Genes, and Numbers of Rearrangement Breakpoints

Chromosome number[a]	Pachytene length (μm) aceto-orcein[b]	Pachytene length (μm) synaptonemal complex[c]	Linkage group[d]	Rank by DNA migration[e]	Mapped loci[f]	Rearrangement breakpoints[g]
1	15.0 (25%)	13.3 (23%)	I	1	184 (27%)	195 (29%)
2[h]	11.1 (18%)	9.5 (16%)	V	2	118 (17%)	90 (13%)
3	8.7 (14%)	8.7 (15%)	III	4	78 (11%)	81 (12%)
4	7.3 (12%)	7.9 (14%)	IV	3	110 (16%)	84 (12%)
5	7.3 (12%)	7.0 (12%)	VI	(6 or 7)	55 (8%)	89 (13%)
6	6.2 (10%)	6.4 (11%)	II	5	81 (12%)	86 (13%)
7	5.6 (9%)	5.6 (10%)	VII	(6 or 7)	73 (11%)	58 (8%)
Total	61.2	58.4			694	683

[a]Based on decreasing length at late pachytene (McClintock 1945; Singleton 1948, 1953). Ranking the two shortest chromosomes has been difficult. With the exception of nos. 3, 4, and 5, chromosomes can readily be distinguished qualitatively on the basis of chromomere patterns at pachytene (Singleton 1948; Figure 3 of Perkins and Barry 1977).

[b]Based on Figure 1 of McClintock (1945). The values obtained by Singleton (1953) were somewhat larger, with means ranging from 19 μm for chromosome 1 to 7 μm for chromosome 7.

[c]Reconstructed from thin sections of pachytene nuclei (Table 2 of Gillies 1979).

[d]Chromosome assignments of genetic linkage groups are based on correlations between intergroup marker linkages and pachytene chromosome pairing in translocation heterozygotes or on linkage of genetic markers to the heteromorphic cytological marker *satelliteless* (see Perkins and Barry 1977).

[e]Determined by pulsed-field gel electrophoresis of intact chromosomal DNA (Orbach et al. 1988). Correspondence of genetic linkage groups to individual DNA bands is based on molecular hybridization when probed with cloned genes that represent each linkage group. Bands representing the two smallest chromosomes have not been resolved.

[f]Perkins (1990 and unpubl.).

[g]Perkins and Barry (1977 and unpubl.).

[h]Contains the nucleolus organizer region with about 200 copies of genes specifying ribosomal RNA. This rDNA makes up most of the ~8% repetitive DNA in the genome (Krumlauf and Marzluf 1980). It is not clear whether it contributes proportionately to measured pachytene length.

their length, with no or little twisting. As a result, the bivalents had the appearance of railroad tracks. A rationale for these observations was provided only later, following the discovery of the synaptonemal complex (SC) in 1956. The width of the SC central element is essentially the same in organisms with vastly different genome sizes (see Westergaard and Wettstein 1972). When *Neurospora* is compared with maize or lily, the distance between paired homologs appears to be much greater at the higher magnification required for observing the tiny chromosomes of *Neurospora*, even though the distance is the same for all three organisms. Hence, the appearance of railroad tracks. The role of the SC in pairing and elongation has been most fully illustrated for a close relative of *Neurospora*, *Sordaria*, using a series of nuclei reconstructed from thin sections (Zickler 1977).

In addition to describing the chromosome complement and the behavior of chromosomes during meiosis and ascus development, McClintock's 1945 paper initiated the study of chromosome rearrangements in fungi. The genetic behavior of several Stanford University mutants had suggested that they contained chromosomal abnormalities. Several of these putative rearrangement strains were crossed to normal wild type, and developing asci were examined cytologically. McClintock's drawing of a pachytene nucleus from one of the crosses shows chromosome pairing typical of a heterozygous reciprocal translocation (Fig. 3A). This translocation, now designated *T(I;II)4637 al-1*, has seen many uses since that time. It was employed first to assign genetic linkage groups to identified chromosomes (St. Lawrence 1953; Barry 1967), and it was combined with other translocations to construct an efficient linkage-tester strain (Perkins et al. 1969). The tester was in turn used to demonstrate correspondence between recombination nodules and crossing-over events (Gillies 1979). Because one of its breakpoints is in the *albino-1* locus, the translocation was chosen for use as recipient in transformation experiments that led to molecular cloning of the *al-1* gene (Schmidhauser et al. 1990). Most recently, strains of the translocation have been used to obtain intact synaptonemal complex spreads (B.C. Lu, pers. comm.). Different views of the translocation are shown in Figure 3.

McClintock's study of rearrangements was not limited to the examination of pachytene pairing configurations. She also considered the different possible modes of disjunction from translocation quadrivalents and inferred their consequences for ascospores in the linear asci. The spores are expected to contain duplications and deficiencies with predictable patterns and frequencies, depending on the location of breakpoints relative to centromeres. Analysis of rearrangements is aided in *Neurospora* because mature viable ascospores are black, whereas ascospores with sizeable deficiencies are inviable and remain unpigmented. Patterns of unpigmented ascospore pairs in individual asci thus provide a visual clue that can be used as an aid in diagnosing the type of rearrangement. McClintock was the first to apply this type of analysis to tetrads of meiotic products in fungi, using as an example intact linear asci from a cross of translocation x normal sequence. This approach was elaborated by Singleton (1948) and by St. Lawrence (1953). Years later, during a 1967 visit to California, she encouraged my efforts to extend the method to unordered asci, which can easily be obtained in large numbers as octets of ascospores that have been shot from the perithecia. The method proved successful beyond expectation (Perkins 1974).

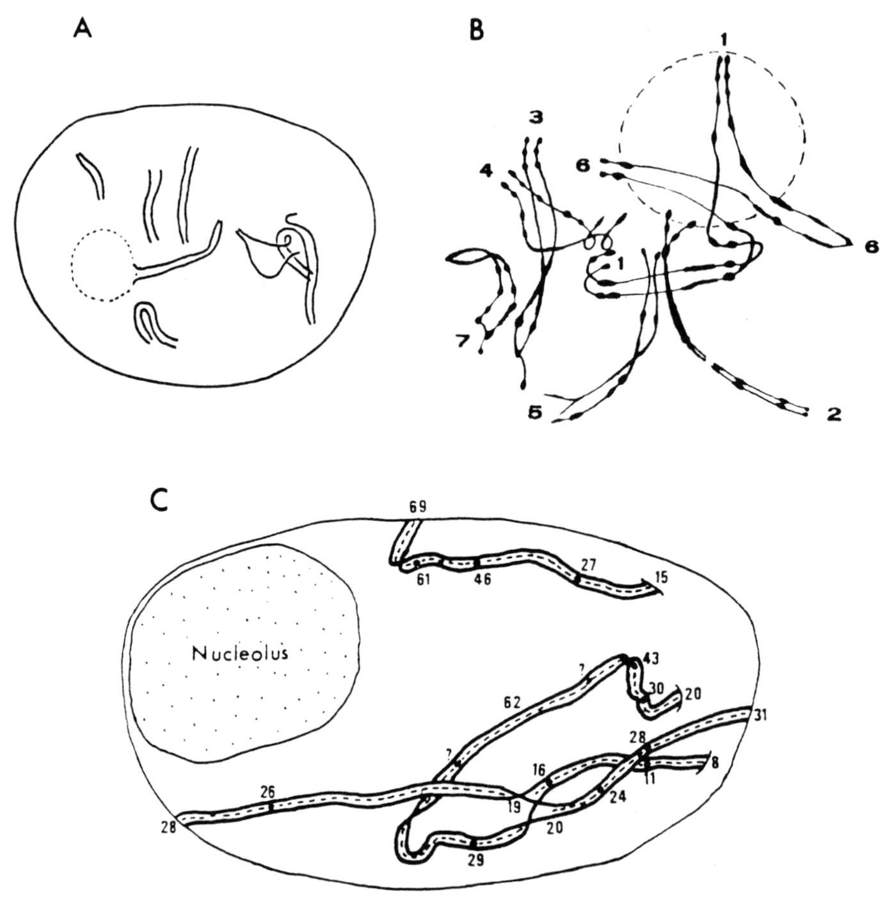

Figure 3 Diverse views of *T(I;II)4637 al-1* in heterozygous asci at pachytene. (*A*) McClintock's sketch (1945); aceto-orcein. (*B*) Camera lucida drawing by Barry (1967); aceto-orcein. (This is accompanied by photographs in the original publication.) Numbers identify the chromosomes and chromosome arms. (C) Synaptonemal complex reconstruction from thin sections (Gillies 1979). Only the quadrivalent and the chromosome 7 univalent are shown in the drawing. Numbers indicate the sections.

Neurospora Chromosomes after McClintock

Although McClintock's venture into *Neurospora* cytology was only a minor incident in her own career, it was of major importance for fungal cytogenetics, providing impetus and introducing effective methods. The squash technique was soon applied to other fungi, where it dispelled any lingering doubts about brachymeiosis (for review, see Cutter 1951; Olive 1953, 1965). Work with *Neurospora* was continued by disciples. McClintock supervised the thesis research not only of Singleton (1948), but also of Patricia St. Lawrence (1953). In turn, both of them instructed Edward Barry (1960) in McClintock's methods. In 1961, Singleton came to Stanford for a 2-month visit during which he and Barry examined a variety of *Neurospora* rearrangements while seeking to improve techniques. Their experience reinforced McClintock's observation that the quality of cytological preparations depended in an important way on the genetic background of the strains that were used. In 1962, Singleton submitted an ambitious proposal to the National Science Foundation for studying *Neurospora* chromosome rearrangements. The grant was awarded, but he died of cancer before it could be activated. St. Lawrence's interests were diverted away from cytology to other projects in *Neurospora* genetics. Thus, Barry emerged as the primary person committed to extending the studies that McClintock had begun (see Barry 1966, 1967, 1969; Barry and Perkins 1969).

Different techniques and approaches were also introduced by other workers, complementing the aceto-orcein procedure. An iron-hematoxylin squash technique originally developed by B.C. Lu for *Coprinus* was adapted to *Neurospora* (see Raju and Newmeyer 1977; Raju 1978). This proved to be extremely effective for examining chromosomes, nuclei, spindles, and organelles during ascus development, and the technique has been used extensively to examine wild type, developmental mutants, and aneuploids (for reviews, see Raju 1980, 1992). However, iron-hematoxylin is inferior to aceto-orcein for revealing the fine chromomere detail of pachytene bivalents. The DNA-specific fluorochrome acriflavine promises to be useful in this respect (Raju 1986a). Another view of meiotic chromosomes has been provided by synaptonemal complex reconstructions (Gillies 1972, 1979; Bojko 1989, 1990) and spreads (B.C. Lu, pers. comm.). Pulsed-field gel electrophoresis has made it possible to separate intact chromosomal DNAs and to do molecular karyotyping (Orbach et al. 1988).

Convergent Studies in Maize and Neurospora

This is not an appropriate occasion for reviewing advances that have been made using these techniques. I shall conclude simply by pointing out four examples of cytogenetic findings in *Neurospora* that parallel and extend previous results in maize.

Brink and Cooper (1932) described a maize translocation in which a terminal segment of one chromosome had become detached from its normal position and attached by its broken end to an essentially terminal position in a nonhomologous chromosome. Terminal rearrangements of this type are readily recognized in a haploid organisms such as *Neurospora*, where patterns of aborted ascospores are diagnostic and where structurally heterozygous crosses produce a class of progeny consisting of viable partial disomics. Thirty genetically terminal translocations have been identified, mapped, and put to various uses in *Neurospora* (Perkins and Barry 1977 and unpubl.).

Rhoades (1968) described a translocation in which an interstitial segment from one maize chromosome had been inserted in another chromosome. Insertional translocations of this type are readily recognized in *Neurospora*, where more than 30 have been characterized. As with terminal translocations, one class of viable progeny from a structurally heterozygous cross contains a defined chromosome segment in duplicate. Duplications obtained as segregants from terminal and insertional rearrangements have been valuable not only for mapping, but also for studying mating type, vegetative incompatibility, regulation, dominance, dosage effects, chromosome instability, and repeat-induced point mutation (Perkins 1986 and unpubl.; for review, see Perkins and Barry 1977).

Beadle (1931) described a mutant in maize called *polymitotic*, in which postmeiotic divisions continue to occur in the haploid microspore in the absence of chromosome replication. As a result, the ten chromosomes are distributed at random into daughter nuclei. Similar behavior is shown by the recessive mutagen-sensitive mutant *mus-8* in *Neurospora,* where ascospore walls are never formed and nuclear divisions continue in the ascus until most nuclei contain few chromosomes, or only one (Raju 1986b).

In a classic paper, McClintock (1934) described a maize translocation that transferred part of the nucleolus organizer region (NOR) to a nonhomologous chromosome. Each NOR segment was shown to be capable of forming a nucleolus, and relative activities of the two segments were described. A counterpart of McClintock's translocation was identified in *Neurospora* and has been subjected to genetic, cytological, and molecular analyses (Perkins et al. 1984, 1986; Butler and Metzenberg 1989). Control of activity of the displaced NOR segment was shown to involve methylation of rDNA repeats, segments in both positions were shown to undergo frequent changes in size during the sexual phase, and restoration of quasinormal sequence was shown to occur as a result of meiotic crossing over between segments in the normal and the displaced locations.

It is satisfying to see from such examples how organisms apparently as different as maize and *Neurospora* are basically so similar, and how knowledge of one can contribute to understanding of the other. Nowhere is this better illustrated than in the studies of *Neurospora* chromosomes that were initiated by Barbara McClintock.

Acknowledgments

Work on *Neurospora* cytogenetics in my laboratory has been supported since 1956 by research grant AI-01462 from the National Institutes of Health. I am indebted to Edward Barry for the drawings used in Figure 1.

References

Barry, E.G. 1960. "A complex chromosome rearrangement in *Neurospora crassa.*" Ph.D. thesis, Stanford University, Stanford, California. (*Diss. Abstr. 21:* 3233 [1961].)

———. 1966. Cytological techniques for meiotic chromosomes in *Neurospora. Neurospora Newsl. 10:* 12.

———. 1967. Chromosome aberrations in Neurospora and the correlation of chromosomes and linkage groups. *Genetics 55:* 21.

———. 1969. The diffuse diplotene stage of meiotic prophase in *Neurospora. Chromosoma 26:* 119.

Barry, E.G. and D.D. Perkins. 1969. Position of linkage group V markers in chromosome 2 of *Neurospora crassa. J. Hered. 60:* 120.

Beadle, G.W. 1931. A gene in maize for supernumerary cell divisions following meiosis. *Cornell Univ. Agric. Exp. Stn. Mem. 135:* 1.

———. 1966. Biochemical genetics: Some recollections. In *Phage and the origins of molecular biology* (ed. J. Cairns et al.), p. 23. Cold Spring Harbor Laboratory, Cold Spring Harbor, New York.

Bojko, M. 1989. Two kinds of "recombination nodules" in *Neurospora crassa. Genome 32:* 309.

———. 1990. Synaptic adjustment of inversion loops in *Neurospora crassa. Genetics 124:* 593.

Brink, R.A. and D.C. Cooper. 1932. A structural change in the chromosomes of maize leading to chain formation. *Am. Nat. 66:* 310.

Butler, D.K. and R.L. Metzenberg. 1989. Premeiotic change of nucleolus organizer size

in *Neurospora. Genetics 122:* 783.
Cutter, V.M., Jr. 1951. The cytology of the fungi. *Annu. Rev. Microbiol. 5:* 17.
Dodge, B.O. 1927. Nuclear phenomena associated with heterothallism and homothallism in the ascomycete *Neurospora. J. Agric. Res. 35:* 289.
Emerson, S. 1966. Mechanisms of inheritance. 1. Mendelian. In *The fungi: An advanced treatise* (ed. G.C. Ainsworth and A.S. Sussman), vol. 2, p. 513. Academic Press, New York.
Gillies, C.B. 1972. Reconstruction of the *Neurospora crassa* pachytene karyotype from serial sections of synaptonemal complexes. *Chromosoma 36:* 119.
———. 1979. The relationship between synaptonemal complexes, recombination nodules and crossing over in *Neurospora crassa* bivalents and translocation quadrivalents. *Genetics 91:* 1.
Krumlauf, R. and G.A. Marzluf. 1980. Genome organization and characterization of the repetitive and inverted repeat DNA sequences in *Neurospora crassa. J. Biol. Chem. 255:* 1138.
McClintock, B. 1934. The relation of a particular chromosomal element to the development of the nucleoli in *Zea mays. Z. Zellforsch. Mikrosk. Anat. 21:* 294.
———. 1945. *Neurospora.* I. Preliminary observations of the chromosomes of *Neurospora crassa. Am. J. Bot. 32:* 671.
———. 1947. Cytogenetic studies of maize and *Neurospora. Carnegie Inst. Washington Year Book 46:* 146.
———. 1955. Mutations in maize and chromosomal aberrations in *Neurospora. Carnegie Inst. Washington Year Book 53:* 254.
———. 1961. (Unpublished transcript of a talk given at the first *Neurospora* Information Conference, La Jolla, California.)
Olive, L.S. 1950. A cytological study of ascus development in *Patella melaloma* (Alb. & Schw.) Seaver. *Am. J. Bot. 37:* 757.
———. 1953. The structure and behavior of fungus nuclei. *Bot. Rev. 19:* 439.
———. 1965. Nuclear behavior during meiosis. In *The fungi: An advanced treatise* (ed. G.C. Ainsworth and A.S. Sussman), vol. 1, p. 143. Academic Press, New York.
Orbach, M.J., D. Vollrath, R.W. Davis, and C. Yanofsky. 1988. An electrophoretic karyotype of *Neurospora crassa. Mol. Cell. Biol. 8:* 1469.
Perkins, D.D. 1974. The manifestation of chromosome rearrangements in unordered asci of *Neurospora. Genetics 77:* 459.
———. 1979. *Neurospora* as an object for cytogenetic research. *Stadler Genet. Symp. 11:* 145.
———. 1986. Determining the order of genes, centromeres, and rearrangement breakpoints in *Neurospora* by tests of duplication coverage. *J. Genet. 65:* 121.
———. 1990. *Neurospora crassa* genetic maps, June 1989. *Genet. Maps 5:* 3.9. (Also in *Fungal Genet. Newsl. 36:* 105 [1989].)
Perkins, D.D. and E.G. Barry. 1977. The cytogenetics of *Neurospora. Adv. Genet. 19:* 133.
Perkins, D.D., N.B. Raju, and E.G. Barry. 1984. A chromosome rearrangement in *Neurospora* that produces segmental aneuploid progeny containing only part of the nucleolus organizer. *Chromosoma 89:* 8.
Perkins, D.D., D. Newmeyer, C.W. Taylor, and D.C. Bennett. 1969. New markers and map sequences in *Neurospora crassa,* with a description of mapping by duplication coverage, and of multiple translocation stocks for testing linkage. *Genetica 40:* 247.
Perkins, D.D., R.L. Metzenberg, N.B. Raju, E.U. Selker, and E.G. Barry. 1986. Reversal of a *Neurospora* translocation by crossing over involving displaced rDNA, and methylation of the rDNA segments that result from recombination. *Genetics 114:* 791.
Raju, N.B. 1978. Meiotic nuclear behavior and ascospore formation in five homothallic species of *Neurospora. Can. J. Bot. 56:* 754.
———. 1980. Meiosis and ascospore genesis in *Neurospora. Eur. J. Cell Biol. 23:* 208.
———. 1986a. A simple fluorescent staining method for meiotic chromosomes of *Neurospora. Mycologia 78:* 901.
———. 1986b. Postmeiotic mitoses without chromosome replication in a mutagen-sensitive *Neurospora* mutant. *Exp. Mycol. 10:* 243.
———. 1992. Genetic control of the sexual cycle in *Neurospora. Mycol. Res. 96:* (in press).

Raju, N.B. and D. Newmeyer, 1977. Giant ascospores and abnormal croziers in a mutant of *Neurospora crassa. Exp. Mycol. 1:* 152.

Rhoades, M.M. 1968. Studies on the cytological basis of crossing over. In *Replication and recombination in genetic material* (ed. W.J. Peacock and R.D. Brock), p. 229. Australian Academy of Science, Canberra.

St. Lawrence, P. 1953. "The association of particular linkage groups with their respective chromosomes in *Neurospora crassa.*" Ph.D. thesis, Columbia University, New York. (*Diss. Abstr. 14:* 7 [1954].)

Schmidhauser, T.J., F.R. Lauter, V.E.A. Russo, and C. Yanofsky. 1990. Cloning, sequence, and photoregulation of *al-1*, a carotenoid biosynthetic gene of *Neurospora crassa. Mol. Cell. Biol. 10:* 5064.

Singleton, J.R. 1948. "Cytogenetic studies on *Neurospora crassa.*" Ph.D. thesis, California Institute of Technology, Pasadena.

———. 1953. Chromosome morphology and the chromosome cycle in the ascus of *Neurospora crassa. Am. J. Bot. 40:* 124.

Westergaard, M. and D. von Wettstein. 1972. The synaptinemal complex. *Annu. Rev. Genet. 6:* 71.

Zickler, D. 1977. Development of the synaptonemal complex and the "recombination nodules" during meiotic prophase in the seven bivalents of the fungus *Sordaria macrospora* Auersw. *Chromosoma 61:* 289.

*The Early Years of Maize Genetics**

MARCUS M. RHOADES
Indiana University, Bloomington, Indiana 47405

This chronicle of the early history of maize genetics unavoidably reflects a personal bias regarding the significance of the work of various students of the maize plant. It is history seen through my eyes. Unfortunately, personal recollections and impressions cannot be documented but, if skillfully and judiciously woven into the fabric of a narrative, may transform it from a prosaic account into a more interesting one. The problem remains that different individuals will have varying interpretations of the cause and course of events and of the role played by leading protagonists. Objectivity and an open mind are essential prerequisites in describing the growth and development of a scientific field. The story given here has been read by several knowledgeable colleagues. Nothing I have written has aroused strong dissent, but this does not, of course, mean that they agree with all that I have said. Needless to say, I alone am responsible for this narrative. I have tried to be both just and objective in judgments and assessments. I hope I have met with some measure of success in giving credit where credit is due and that any errors of omission and commission will be forgiven.

I accepted the invitation to write this account of the beginnings of maize genetics when I realized with something of a shock that I, now in my eighties, am one of the few geneticists living whose investigations with maize have extended from the 1920s to the present. It was my good fortune to have known well the two individuals whose contributions overshadowed those of all others, namely R.A. Emerson and Barbara McClintock. It is generally acknowledged that Emerson was the spiritual father of maize genetics, and some consider this to be his greatest achievement. It was he who in the formative years was largely, if not solely, responsible for the remarkable esprit de corps that prevailed among maize workers throughout the world. Emerson was trusted, admired, and respected. Completely unselfish, truly an honorable man, he was able to elicit an unrivaled degree of cooperative endeavor from fellow workers. In addition to his leadership, Emerson was a superb investigator whose genetic experiments were exemplars of design and execution. He was the dominant figure among maize geneticists until the advent of cytogenetics and Barbara McClintock.

McClintock and I labored in the same vineyard. No one is more aware than I of how important were her investigations; these shall be discussed in the following pages. In 1950 (86) I wrote: "In particular I should like to acknowledge my great indebtedness to the pioneering work of Dr. Barbara McClintock. The identification of the pachytene chromosomes and their association with specific linkage groups came about entirely from her brilliant

*This chapter is reproduced, with slight modification, with permission from the *Annual Review of Genetics* (vol. 18, copyright 1984 by Annual Reviews Inc.)

and illuminating studies. Maize cytogenetics surely would not occupy its present high estate were it not for her remarkable contributions." Today, a third of a century later, my judgment of her scientific achievements remains unchanged. If Emerson was the towering figure in classical maize genetics, McClintock played an even more seminal role in the development of cytogenetics. She was the dominant figure in the talented Cornell group of cytogeneticists and she, more than anyone else, was responsible for a long series of remarkable findings.

I was both an observer and a participant in that exciting and productive period when cytogenetics had its day of glory. I learned a great deal of historical lore from Professor Emerson as he reminisced about maize genetics and maize geneticists. Not only was I his student, but for the better part of three years I lived in the Emerson home with free run of house and kitchen—literally a member of the family. Many was the time I listened enthralled as he related past experiences. In short, I knew well these two remarkable individuals and in the following pages I hope to enliven what otherwise would be a dull and factual recounting of past achievements by introducing some personal recollections. Fallible in some respects may be my memory, but I vividly recall many incidents of my student years.

The acquisition of the now impressive body of knowledge about maize heredity was greatly aided by the admirable cooperation of maize geneticists; they unselfishly shared breeding stocks, disclosed unpublished information, freely discussed experiments and research strategies, and, most commendable of all, permitted hard-won linkage data to appear in review or summary publications where the only acknowledgement of their contributions was a line in a table or a footnote in the text. In the first third of the twentieth century, the task of placing unlinked genes and ascertaining map positions was more laborious and time-consuming than now. Today, the use of the B-A translocations first found by Herschel Roman (89, 90), and the waxy-marked reciprocal translocations developed by E.G. Anderson (1, 2) and A.E. Longley (60), makes chromosomal assignment less arduous, even though the precise determination of map position still requires orthodox 3-point tests (see also 74). This gathering of routine information is essential for progress but the individual rewards are meager. The unparalleled generosity of the maize geneticists in providing unpublished crossover data needed to flesh out the linkage maps is evidenced by the following statistic: of the 859 recombination values listed in the first comprehensive linkage summary, Memoir 180 of the Cornell University Agricultural Experiment Station (37) published in 1935, nearly 70%, or more than two-thirds, were previously unpublished. The exemplary behavior of the maize workers is a welcome contrast to the secretive, competitive atmosphere said to exist today in much of modern molecular biology. I know of no group of scientists more cooperative, more altruistic. I am proud to be a member of this select society and to have shared their friendship. It is to them all that this paper is dedicated.

Genesis

Every student who has had a course in biology is taught that the science of genetics had its origin in Gregor Mendel's hybridization experiments with *Pisum* where the well-known laws governing the transmission of heritable characteristics were first discovered. Surprisingly few geneticists are aware, however, that Mendel also studied inheritance in maize hybrids as he sought to determine if the rules found for *Pisum* were of general validity. Since the birth of genetics and the beginning of maize genetics are interrelated, let us examine the historical record in some detail.

Mendel's experiments in hybridizing pea varieties differing in alternative traits were begun in 1856, when he was 34 years old, and completed in 1863. At this time he had deduced from extensive F_2 and backcross data that there were specific rules governing the disjunction and recombination of characters from one generation to the next. He was now prepared to announce his conclusions. On the 8th of February and on the 8th of March, 1865, he did so in two lectures before the Brünn Society for the Study of Natural Sciences.[1] Later that year, he accepted the invitation to publish his lectures in the 1865 Proceedings of the Society (69), which appeared in 1866.

It is from Mendel's April 18, 1867, letter (70) to the famous botanist Carl von Nägeli that we learn that the published version of his paper was unchanged from the one he read before the members of the Brünn Society. His manuscript was neither reviewed nor subjected to editorial revision. Apparently, it was the editorial policy of this provincial journal to publish verbatim the draft provided by seminar speakers. This was a fortunate circumstance for Mendel. The minds of his contemporaries were not receptive to his unorthodox ideas and it is highly questionable if Mendel's maiden publication would have been published if acceptance had hinged upon approval by reviewers. The results of his historic experiments appear in an unrefereed publication and what Correns (26), de Vries (96, 97), and Tschermak (95) found awaiting them when his paper came to light in 1900 was pure Mendel.

Mendel was especially anxious that his paper come to the attention of von Nägeli of Munich, whose wide botanical interests included extensive observations on *Hieracium* hybrids. To insure that this would occur, he sent Nägeli in December of 1866 a reprint of his paper accompanied by a letter. Nägeli answered and thus began a sporadic but historic correspondence that ended in 1873. Mendel's letters were found among Nägeli's effects and after his death came to the attention of Carl Correns, a former student of Nägeli. Correns recognized the great significance of these letters and had them published in the original German. It is to the English translations by Piternick & Piternick (70) that I refer. In his December 31, 1866, letter, Mendel said he recognized that the conclusions drawn from the *Pisum* data had to be confirmed by comparable experiments with other plants before it could be

[1] The record is not clear regarding the reception accorded Mendel's lectures to members of the Brünn Society for the Study of Natural Sciences. Hugo Iltis (43) writes: "The minutes of the meeting inform us that there were neither questions nor discussion. The audience dispersed and ceased to think about the matter." A somewhat different reaction by his listeners comes from the following quotation from Robert Olby (73): "In a recent publication Dr. J. Sajner has drawn attention to the reports of Mendel's 'Versuche' lectures which appeared in the daily newspaper Brünner Neuigkeiten on the 9th February and 10th March, 1865. In addition to mentioning discovery of constant numerical relationships, the report refers to the lively participation of the audience which it regarded as evidence of the success of the lecture." If the reputed lively discussion indicated that Mendel's audience understood and grasped the significance of his results, they stood alone. Perhaps the only reasonable conclusion that can be drawn from these conflicting accounts is that reports of what actually transpired at scientific meetings were no more accurate in 1865 than they are today.

concluded that the *Pisum* rules had general applicability. None of the prior investigators working with segregating populations from hybrid parents had used his analytical methodology; consequently, the published record neither confirmed nor refuted the general validity of the laws operating in *Pisum*. Mendel put it simply when he wrote concerning the general relevance of his findings: "A decision can be reached only when new experiments are performed in which the degree of kinship between the hybrid forms and their parental species are precisely determined rather than simply estimated from general impressions." To this end, in 1865, he had begun hybridization experiments with *Hieracium*, *Cirsium*, and *Geum* and the following year added several new species. In all, he worked with 26 different genera, among them *Zea*. Subsequent to 1865, Mendel, following the advice of Nägeli, spent most of his time and energy on *Hieracium* hybrids. It is questionable if Nägeli grasped the significance of the conclusions drawn from the *Pisum* studies or that he was interested in the results of Mendel's experiments with plants other than *Hieracium*. The selection of *Hieracium* was the worst possible choice. It was a disaster. In antithesis to *Pisum*, uniform, nonsegregating progenies were found even in highly heterozygous material. Unknown to Mendel and to his contemporaries, apomixis was widespread throughout the genus. Mendel, unable to account for the refractory *Hieracium* data, was forced to conclude that the laws derived from *Pisum* were of limited application.

That Mendel worked with maize is known only from his correspondence with Nägeli. There is no mention of his experiments with maize in his 1865 publication. Mendel wrote in his April 18, 1867, letter as follows: "Hybrids of *Zea Mays major* (with dark red seeds) + *Z. Mays minor* (with yellow seeds) and of *Zea Mays major* (with dark red seeds) + *Z. Cuzko* (with white seeds) will develop during the summer. Whether *Zea Cuzko* is a true species or not I do not dare to state. I obtained it with this designation from a seed dealer. At any rate, it is a very aberrant form." Approximately three years later, on July 3, 1870, he wrote again to Nägeli saying, "Of the experiments of previous years, those dealing with *Matthiola annua* and *glabra*, *Zea* and *Mirabilis* were concluded last year. Their hybrids behave exactly like those of *Pisum*." No data were given, just the simple statement that segregation in maize followed the principles discovered in *Pisum*.

In the above maize crosses the strain with dark red seed presumably, almost certainly, had the dominant *P* gene for red coloration of the pericarp tissue and the other two strains carried the recessive *p* allele, which produces a colorless pericarp. It is unlikely that the pigmentation of the "dark red seeds" was in the aleurone and not in the pericarp. The development of aleurone color depends upon the complementary interaction of the dominant alleles of four different loci and Mendel could have encountered F_2 ratios of kernels with colored to colorless aleurones such as 3:1, 9:7, 27:37, depending upon how many of the aleurone genes were heterozygous. And, too, the phenomenon of xenia, whereby owing to double fertilization the effects of dominant genes contributed by the pollen parent are expressed in the endosperm of kernels of parental ears, would have introduced a complicating factor that Mendel was ill-prepared to comprehend. Double fertilization was unknown until it was independently described by Nawaschin (72) and by Guignard (40) in 1899. It was a phenomenon unheard of by Mendel and it is difficult to believe that he would confidently write to Nägeli saying that *Zea* behaved exactly like *Pisum* if indeed he had been studying the complex inheritance of aleurone pigmentation.

However, it matters not what traits Mendel investigated; what is important is that he was the first to demonstrate experimentally that the principles discovered in *Pisum* also held for *Zea*. Insofar as we know, Mendel carried out no further studies with maize but he undeniably was the first maize geneticist.

Mendel in a few short years did what may be unique in the annals of science. Unaided, largely self-taught with no rigorous scientific training, a novice in research, in a period of seven years he designed, executed, and analyzed the data from a series of experiments on pea hybrids that led to the discovery of basic laws underlying the transmission of hereditary characters and single-handedly founded a new field of science. His 1865 paper was the first of two publications on his hybridization studies, but what a masterpiece it is. It laid the foundation upon which the magnificent body of knowledge comprising the field of genetics has been erected. This now historic paper attracted no attention from the wider scientific community, although copies were distributed to more than 120 societies, universities, academies, and research centers throughout the world (43). It was not praised, it was not criticized; it was, the crowning indignity, ignored. The lack of recognition, the confusing *Hieracium* data, and the ever-increasing demands on his time by his newly assumed administrative duties as Abbot were together responsible for Mendel abandoning all hybridization work after 1870. So complete was the divorce from his hybridization studies that no mention was made of them in the announcement of his death in 1884. Mendel was before his time, his accomplishments in research unhonored to the end; recognition of his work was posthumous. His time finally came, as he purportedly predicted it would in a comment made late in his life to his colleague Niessl (43, p. 282).

Rediscovery

Not only did maize play a part in Mendel's experiments, but it, as well as *Pisum*, was also used by both Hugo de Vries and Carl Correns in their studies, which led to the rediscovery of Mendelian principles of inheritance (26, 96, 97). Prior to 1900, de Vries and Correns investigated the phenomenon of xenia or the immediate effect of the pollen parent on the phenotype of the maize endosperm. Crosses were made between races of maize differing in contrasting endosperm characters such as sugary versus starchy and colored versus colorless. Both men were puzzled by the segregation of contrasting traits on the parental ears rather than in the next generation as anticipated. It was not until the independent cytological observations of double fertilization by Nawaschin and by Guignard in 1899 that a rational and satisfactory mechanism for xenia was provided. Now that the basis for xenia was clear, it was obvious to both investigators that maize showed the same pattern of inheritance as that found in *Pisum* by Mendel. Yellow versus white endosperm and starchy versus sugary endosperm both proved to be due to a single factor. Many biologists were aware of the great variability present in diverse races of maize and a number of intervarietal hybrids were made by plant breeders and botanists, but none other than these two perceived that heredity in maize followed Mendelian laws. Correns and de Vries were among the pioneer maize geneticists. Not only had they demonstrated Mendelian segregation in maize but Correns had unknowingly found what could be considered the first example of linkage, albeit cryptic, in the progeny of a controlled mating. Although his earlier crosses between starchy and sugary strains had given 3:1 ratios in the F_2, he reported in 1902 (27) an apparent exception to Mendelian segregation, again using the starchy-sugary pair of contrasting characters. Self pollinations of starchy-sugary hybrids gave ears with only 16% sugary kernels and not the 25% he had found earlier. This F_2 ratio is the same as that found by Emerson a few years later (see further). It is now known through Emerson's subsequent thorough analysis that he and probably Correns had unwittingly crossed a sugary with a starchy race whose chromosome 4 had a gametophyte factor linked to the *Su* locus. If only *Ga* pollen grains function when *Ga Su/ga su* plants are self pollinated, and if *Ga* and *ga* megaspores are equally viable, 16% of sugary kernels in the F_2 generation is expected with 32% recombination between *Ga* and *Su*. *Ga* is a cryptic gene, detected only by distorted ratios of linked loci. It must be appreciated that Correns's findings were made before the establishment of the chromosome theory of heredity, before it was known that all genes are not independently assorted and that some are coupled or linked. Understandably, Correns did not realize that he had evidence of linkage and, at the turn of the century, could not have been expected to do so. Undeniably, Correns and de Vries played a vital role in the early days of maize genetics but they had little or no impact in the later years when the impressive corpus of knowledge about maize heredity was obtained. If maize had not been used by Mendel, and by Correns and de Vries in their rediscovery of Mendel's Laws, it is probable that this would have had little effect on the later development of maize genetics. Too many factors were in its favor for it to be ignored. The wealth of variability, the separation on the stalk of the male and female inflorescences, which greatly facilitated hybridization, and the large number of investigators working with maize because of its economic importance all served to insure the rapid advance of maize genetics. It was widely perceived to be a favorable organism for experimental purposes and, immediately following the announcement of the rediscovery of Mendel's Laws, maize became the object of considerable attention.

Laying the Foundations of Maize Genetics

Rollins Adams Emerson was a young assistant professor of horticulture at the University of Nebraska when news of the rediscovery of Mendel's Laws burst upon the scientific world in 1900. Today we cannot recapture the intellectual excitement created but it must have been intense, at least among biologists. Young Emerson was caught and excited by the promise of this new field of research and was soon engaged in studying heredity in plants of horticultural interest. He at this time did little if any work with maize; most of his efforts were spent on inheritance of the common garden bean. He became a student of maize in an unpredictable way. Part of his academic responsibilities was to teach an undergraduate course for horticultural students that consisted of lectures and laboratory exercises. Desiring to acquaint the students with material illustrating the recently discovered Mendelian Laws of Heredity, he, following Correns's earlier example, made a cross of a starchy by a sugary variety, selfed the F_1 plants, and distributed the selfed ears to the students along with the request that they score and count the starchy and sugary kernels. When the students handed in their results, Emerson was distressed to find that their ratios of starchy:sugary were far from the 3:1 ratio found in Correns's first crosses. According to Professor Emerson's account of this incident, as related to me years later, he said to himself "Gad! They can't even count." Emerson never or rarely, if ever, cursed and "Gad" was a strong expletive used only to express extreme unhappiness or annoyance—as close to cursing as he ever came. However, upon checking the students' work, he found that they had both scored and counted correctly; there was indeed a highly significant deficiency of sugary kernels. Emerson was both embarrassed and perplexed by this failure to illustrate the operation of Mendelian segregation. Something was amiss. He had to find out what it was. And so began the genetic studies with maize to which he devoted himself to the end of his days. Although he did not publish his analysis of the cause of the aberrant sugary ratios until many years later, Emerson (35) found that in making the class hybrid he had inadvertently used as the starchy parent a popcorn variety carrying a gametophyte locus that controlled pollen tube functioning. The gametophyte factor and the sugary gene were located on chromosome 4 and it was the loose linkage of the two that was responsible for the observed ratios found by the students.

It is not known whether or not Emerson was aware of Correns's 1902 publication on crosses having an aberrant F_2 sugary ratio when his interest was piqued by his Nebraska undergraduates finding a similar distortion of the sugary ratios in their material. I judge that he was not, because it seems wholly out of character for him to devote his time and interest to a problem earlier reported by Correns, since he could logically assume that the latter was engaged in further investigating the matter. Later, of course, Emerson became cognizant of Correns's data, but I judge this was some time after he had become a full-time student of maize.

While the rediscovery of Mendel's Laws was enthusiastically greeted by some biologists, others were skeptical and vocally critical of this fledgling field of science. Perhaps, it was argued, the inheritance of superficial qualitative characters such as pigmentation was in a Mendelian fashion but was there not abundant evidence for all to see of the "blending" inheritance of really important quantitative traits like size, shape, and yield? The Mendelists replied that it mattered not if one studied a trait governed by one or a thousand genes individually with small effects, all genes obeyed the same laws of heredity; they did not march to different drummers. Although Emerson devoted an increasing amount of time to the study of simply inherited variations, he early began a collaborative study with E.M. East on the important and unresolved problem of inheritance of quantitative characters because it was the apparent nonMendelian transmission of these complex traits that led many biologists to question the significance of Mendelian heredity. In 1913, he and East published a masterful analysis of quantitative inheritance in maize (36). They collected a vast amount of data in the parental, the F_1, F_2, and F_3 generations on ear length, diameter, and row number, weight and breadth of kernels, plant height, tillering, numbers of nodes per stalk, and internode length. From these raw data they calculated the mean and coefficient of variability for each character in every generation and concluded: "In general, thus, it may be said that the results secured in the experiments with maize were what might well be expected if quantitative differences were due to numerous factors inherited in a strictly Mendelian fashion." Following the establishment and acceptance of the chromosome theory of heredity, the problem of quantitative inheritance became somewhat academic, but it was a burning issue in the early years. The data of Emerson and East convinced many that the genes controlling quantitative characters are inherited in precisely the same fashion as genes for simple qualitative traits.

Another paper of importance in advancing the status of maize genetics was the 1911 publication by East & Hayes (31). Their extensive studies were on a wide range of characteristics, including endosperm color and texture where xenia occurs, on floury-flinty endosperm where a dosage of two floury or two flinty alleles in this triploid tissue is dominant to one of the other, on pod corn or Tunicate, on pericarp color, on several plant abnormalities, and on the complementary interaction of the *C* and *R* genes in aleurone color formation. Some of the traits were monogenically determined, while others had a more complicated genetic base. They concluded that the inheritance of all characters studied could be readily interpreted by Mendelian principles. These early studies with maize helped to establish the general validity of Mendelian heredity and also indicated that maize was a favorable organism for genetic research.

But there were others who were unwilling to accept Mendelism in all or in part. Typical of those who accepted Bridges's (9) convincing demonstration of the validity of the chromosome theory of heredity but held reservations about some of the genetic dogma was R.A. Harper, a well-known professor of botany at Columbia University. Although a mycologist, Harper carried out a series of experiments with starchy-sugary hybrids in maize. He found (41), as had others before him, that some families gave ears with good 3:1 starchy-sugary ratios while other crosses gave sugary kernels that were not translucent but had a pseudo-starchy appearance. To Harper, this change in expression indicated that the sugary gene had become contaminated by its intimate association with the starchy gene at the time of meiotic pairing and argued against the concept of gametic purity inherent in Mendelian theory. Harper wrote as follows: "On *a priori* grounds, it would seem unlikely that such complex and labile compounds as we may suppose constitute the germ plasms should enter into such close physical relations of fusion without a greater or less amount of mixing and interaction which would more or less permanently alter their character." In short, Harper accepted the chromosome theory of heredity but questioned whether the factors affecting different forms of the same trait were recovered unmodified in the gametes from hybrids after being in intimate association during meiosis. A defensible position, but one for which Harper had no good evidence.[2]

Harper's 1920 position was similar in some respects to that held by Morgan in the first decade of the century before he found the white-eyed mutant

[2] We now know from the work of C.C. Lindegren on *Saccharomyces* (53) that in heterozygous asci one allele can infrequently be converted to the other (gene conversion) and from the work of R.A. Brink (11) with maize that exposure of a sensitive R^r allele to a mutagenic R^{st} allele results in a heritable modification of the activity of the sensitive R^r allele (paramutation) but that these exceptional cases do not invalidate the essential correctness of Mendel's rule that heterozygotes produce germ cells in which the two parental alleles are found unchanged and in the same frequency. Gene contamination can occur, but it is of more theoretical than practical significance.

Drosophila whose inheritance was sex-linked. Convinced from his own experiments of the validity of Mendelian heredity, he quickly became the most influential of all geneticists as head of the famous school of *Drosophila* at Columbia University. It should be remarked that Harper and Morgan were colleagues at Columbia, working in the same building although in different departments. In retrospect, it would appear that Harper missed a golden opportunity. He had no convincing evidence of gene contamination from his starchy-sugary crosses. Indeed, D.F. Jones (45) had shown that pseudo-starchy kernels, similar to those found by Harper, were due to modifying genes and that the sugary gene extracted from such kernels emerged unchanged. Harper in his time was a well-known botanist but today is largely a forgotten figure. Had he accepted the mounting evidence of the rich promise of this new field of science and devoted himself to genetic experiments with plants at the same time and in the same building that the Morgan school was having such spectacular success in their genetic investigations with *Drosophila*, he, too, might have achieved immortality as one of the influential pioneers in the history of genetics. But he chose a different road.

No publication did more for the status of maize genetics than the Cornell Memoir published by Emerson in 1921 (34) on the genetic relations of plant color. Emerson, in an extensive series of carefully designed and executed experiments, showed that the bewildering complexity of the inheritance of plant pigmentation resulted from the interaction of the alleles present at the *A*, *B*, *Pl*, and *R* loci. Modifying genes sometimes made classification difficult but, once the genetic basis was perceived, Emerson was able to make order out of a chaotic mass of observations. It was an intellectual feat and established Emerson as a sophisticated and skillful investigator. It placed maize genetics on a firm footing. Emerson's plant color memoir is a paradigm of lucidity, critical analysis, and rigorous testing of hypotheses. The careful reading of this paper is still a must for every maize geneticist.

Maize geneticists in the first two decades of this century were chiefly concerned with the discovery, description, map location, and accumulation of mutant traits. These affected all parts of the plant. Especially valuable were the aleurone and endosperm mutants because the phenomenon of xenia greatly facilitated genetic investigations, particularly linkage determinations and map positioning. The density of mutant loci on those chromosomes with a good aleurone gene, such as C on chromosome 9, is not fortuitous, but stems in part from the fact that unplaced genes were invariably tested for linkage with established aleurone and endosperm mutants. Other studies actively pursued included complementary and duplicate factor interaction, allelism tests of mutants having a similar phenotype (albinos, glossy seedlings, virescents, dwarfs, etc), and multiple alleles. These investigations laid the foundation of basic information upon which later studies were dependent. Several score of new mutants were described in a series of papers appearing in the *Journal of Heredity* under the rubric "Heritable Characters in Maize."

Among the more interesting of the early studies was that of Emerson (32), who showed that the red-white sectors on ears with variegated pericarp were caused by somatic mutation of the unstable P^{vv} gene. Nearly 40 years later, Brink & Nilan (15) found that this instability was an example of McClintock's *Ac-Ds* system of controlling elements. The P^{vv} locus consists of the genetic element Modulator (*Mp* = *Ac*), lying adjacent to the P^{rr} allele and inhibiting its functioning. Removal of *Mp* from its association with P^{rr} results in restoration of its activities producing a sector of red pericarp color.

First Examples of Linkage

The seven pairs of alternative characters studied by Mendel in *Pisum* were independently inherited. This led him to conclude that all traits, no matter how many, would segregate randomly. The discovery by Bateson & Punnett (3) of the correlated inheritance in *Lathyrus* of two sets of alternative traits made it clear that this conclusion had to be modified. The term *gametic coupling* was chosen by Bateson & Punnett to describe their example of correlated inheritance, which they hypothesized came from a differential rate of replication of specific cell types formed after random segregation had taken place. Formally, the replication hypothesis was a possible explanation, but there was no supporting cytological evidence and it never gained wide acceptance. It is ironical that Bateson recognized at once the significance and potential of Mendel's 1865 paper when he first encountered it. Bateson more than anyone else grasped the rich promise inherent in Mendel's discovery of the principles underlying the transmission of characters from parent to offspring and it was he who eloquently argued Mendel's case before the scientific world. Bateson clearly had an open and receptive mind and yet, after having done much to advance Mendelism in the early 1900s, he was among the last to accept the chromosome theory of heredity. Even after the evidence favoring the chromosome theory became overwhelming, he refused for many years to believe that correlated inheritance (his gametic coupling) was the consequence of two loci residing in the same chromosome.

Other reports of linked or correlated inheritance followed on the heels of the Bateson-Punnett finding. One of them was the paper by Collins & Kempton (23) on the coherence or positive correlation of waxy endosperm and aleurone color in the F_2 generation of maize plants differing in these traits. The 3:1 F_2 ratios for starchy:waxy endosperm and for colored:colorless aleurone indicated monogenic control for each pair of traits. One cross was between a strain dominant for one trait and recessive for the other with plants recessive for the first trait and dominant for the second (repulsion phase). F_2 progenies were also obtained from crosses where one parent had both dominant traits and the other was waxy, colorless (coupling phase). They found that the characters associated in the parents had a strong tendency to appear together in later generations. The following year, Collins (21) reported further data on the linked inheritance of the same sets of characters. As in the 1911 paper, the waxy (*wx*) and the color (C) genes did not segregate independently in the F_2 to give a 9:3:3:1 ratio. The parental combinations of traits were positively correlated in the F_2 generation. Since Collins accounted for his data by gametic coupling, as did Bateson & Punnett for *Lathyrus*, it may be assumed that he, too, believed in the differential multiplication of specific cell types arising after segregation. It should be remarked that in 1911–12 the chromosome theory of heredity had not been firmly established. Its validity was an open question in the minds of many geneticists and there was no compelling reason to believe that the correlated inheritance of waxy endosperm and aleurone color was the result of their determinants being situated in the same chromosome. The *C*-*Wx* data reported by Collins & Kempton (23) and Collins (21) represent the first authentic linkage in maize, but these investigators never received due credit for it. Their data are not cited by Emerson et al (37) or by others insofar as I am aware. The 22% of recombination between the *C* and *Wx* loci calculated from Collins's data is close to the standard value for this region of chromosome 9.

Later, Lindstrom (54) published data showing linkage between the *R* locus for aleurone color and the golden gene for reduced chlorophyll. A gene for luteus seedling was closely linked to *R* and less so to golden. This is the first example of a linkage group in maize with more than two genes. Emerson (33) reported linkage between *Y* and *Pl*. Many more linkages were published in the 1920s and different linkage groups established in which the linear order and map positions of the mutant loci were slowly being determined. In Lindstrom's 1928 review of the state of maize genetics (55) he lists ten linkage groups, not all of which proved to be independent in subsequent tests. Nothing was known at this time about which one of the haploid set of ten chromosomes bore a specific group of linked genes. This determination awaited the advent of cytogenetics.

Centers of Maize Research and the Training of Maize Geneticists

Nearly all of the first generation of maize geneticists were products of either the Emerson school at Cornell or the East school at Harvard. Both schools were world-recognized centers of excellence in plant genetics. When Emerson left Nebraska in 1914 to head the Department of Plant Breeding at Cornell, he was accompanied by two graduate students, Ernest G. Anderson and Eugene W. Lindstrom, both of whom were to play a prominent role in the early development of maize genetics. Following in their footsteps were two more Nebraskans, George F. Sprague and George W. Beadle, who arrived at the Cornell campus in the mid-1920s and became Emerson's students. Both Sprague and Beadle were exceptional individuals and Emerson remarked that he couldn't decide which of the two was the better. Suffice it to say that fame came to both. It is noteworthy that Emerson and four of his ablest and most productive students were from Nebraska. Cornell's rise as a genetics center stemmed in large part from its Nebraska connection. This influx of talented midwesterners led to major advances in the status of maize genetics in the teens and early 20s of this century. Indeed, after obtaining their doctorates, the Nebraskans continued to influence the quality of genetics at Cornell by encouraging their promising undergraduates to obtain their PhDs under Emerson's direction.

It is ironical that, having trained so many able students who gained fame elsewhere, neither Cornell nor Harvard saw fit to keep the services of one of their promising graduates and thus maintain a hard-won tradition. Such able men as H.K. Hayes, D.F. Jones, R.A. Brink, and P.C. Mangelsdorf were East's students[3] but only Mangelsdorf returned to the Harvard faculty and he was

[3] Technically, R.A. Emerson was East's student since he obtained his ScD from Harvard with East as chairman of his committee. However, Emerson was older than East by several years and was a well-established geneticist when he went to Harvard for the sole purpose of getting his doctorate, which was awarded after only one year in residence. East never thought of Emerson as a student but as a distinguished colleague.

not brought back until 1940, years after East had retired, and Harvard by then had irretrievably lost its eminent position in maize genetics. At Cornell the story was similar. The roster of Emerson's students includes such distinguished geneticists as E.G. Anderson, E.W. Lindstrom, M. Demerec, G.F. Sprague, G.W. Beadle, and other able investigators but all went elsewhere. None was retained. Inevitably, when Emerson retired, Cornell no longer remained the mecca of maize genetics. It is true that Cornell never completely abandoned maize genetics, but much of the former glory was never regained. Emerson hoped that one of his finishing students would be given a faculty appointment and kept at Cornell, but this never happened. I recall his telling me that more than once he had gone to the dean urging just such a step be taken only to have the dean reply "Let him go, we can always bring him back." Such attempts were later but unsuccessfully made; the departed students had new loyalties and new obligations that could not be ignored.

Concomitant with the decline of Cornell and Harvard as centers of maize genetics was the rise of new schools headed by their former students. Regarding the Harvard alumni, Hayes went to Minnesota where, with the help of able colleagues like F.R. Immer, Leroy Powers, I.J. Johnson, and C.R. Burnham, he developed a department of agronomy and plant genetics famed for the distinction achieved by its graduates in pure and applied genetics. Mangelsdorf spent six years at the Connecticut Experiment Station, followed by 13 years at Texas A. and M., before joining the Harvard faculty, and Brink took a position at Wisconsin where over the years he trained many excellent students, beginning with Charles R. Burnham, who played a leading role in the study of reciprocal translocations. With Emerson's students the story is much the same. Lindstrom went first to Wisconsin and then to Iowa State as founder of the Department of Genetics; Beadle was at Stanford before going to the California Institute of Technology as head of biology and finally became president of the University of Chicago. Demerec went to the Cold Spring Harbor Laboratory of the Carnegie Institution of Washington, where he was director for many years. Anderson became a member of the faculty of Cal Tech in 1928 and remained there until his retirement in 1961. I could go on, but the list is long and the point has been made that both Cornell and Harvard trained an exceedingly able group of first-generation maize geneticists. These men produced a second generation and they a third and so on. For example, I am a first-generation maize geneticist, having gotten my degree with Emerson. Drew Schwartz, my colleague at Indiana, is in the second generation since he was my student at Columbia. He in turn trained Michael Freeling, now in California at Berkeley, who is in the third generation and Freeling's students, some of whom have their PhDs, are in the fourth generation. And so the torch is passed from one student generation to another.

One very prominent member of the first generation of maize geneticists who did not work for the PhD degree under either Emerson or East was Lewis J. Stadler. After obtaining an MS degree from the University of Missouri, he came to Cornell and to the Department of Plant Breeding. He left after a few months, having made a poor impression on the plant breeding faculty. Emerson was among those who thought young Stadler was not promising graduate material. He later became a great admirer of Stadler and in speaking of the man and his work would frequently exclaim "Gad! I sure missed that one"; and indeed he did, because Stadler became one of the more influential and productive maize geneticists. Refusing to give up graduate work despite the rebuff at Cornell, he returned to Missouri and obtained his PhD degree in 1922. The chairman of his committee was W.H. Eyster, a former student of Emerson. Eyster obtained his PhD degree in 1920 and went to Missouri as an assistant professor of botany, where he remained until 1924. Eyster was on Stadler's committee and served, apparently for some obscure administrative reason, as chairman when Stadler defended his thesis, but in no sense was Stadler trained by Eyster and in no way should he be considered as Stadler's mentor. Stadler's thesis research dealt with field plot techniques, but his interest turned to the problem of gene structure and mutation. He was convinced that an understanding of gene structure required the induction of mutations, which could then be analyzed. To this end, he decided to investigate the mutagenic effects of X-irradiation on barley and maize kernels. In 1928 and 1930, he published two articles announcing positive results (92, 93). H.J. Muller was also investigating the biological effects of X-rays using *Drosophila* as his experimental organism. *Drosophila* with its short generation time enabled Muller to obtain positive results before Stadler had completed his experiments with plants. Muller published first (71) and is justly known as the pioneer in the induction of gene mutation, but Stadler's work received wide recognition. The two men were good friends and mutual admirers. There was a significant difference in the methodology of the two. Stadler was using X-rays to induce mutations in order to study gene structure and the nature of mutational changes. Instead of studying random mutation occurring at various loci, he concentrated on a few loci with particularly favorable attributes. Muller was more interested in the spectrum of induced mutational events. Both found that X-rays were mutagenic but they differed in one respect. Stadler found no convincing evidence that the induced mutations were other than deficiencies while Muller thought he had some cases of true gene mutation. This difference was never satisfactorily resolved.

Stadler, who spent his entire scientific career at Missouri, was not an employee of the university but of the US Department of Agriculture. Through a cooperative agreement with the university, he had professorial rank and all of the privileges pertaining thereto. Among these was the training of graduate students. A number of very able students came to work with him. Their research in addition to his own brought wide recognition to the university. Few genetic laboratories have produced students the equal of Herschel Roman, John Laughnan, Seymour Fogel, Margaret Emmerling, and M.G. Neuffer. Maize genetics has added lustre to the name of the University of Missouri and, unlike some institutions that watched their strength wither away with seeming indifference, Missouri has successfully maintained its position as a leading center.

In this account, I have placed Emerson and East in the parental generation of maize geneticists because they founded the two schools from which came the next generation. Some, at least, of the fame they rightfully possess derives from the achievements made by students of the schools they headed. But others of their generation played important roles. Guy M. Collins belongs in the same generation with Emerson and East but he did not have a university post. Collins worked for the US Department of Agriculture in Washington, D.C., where he had no opportunity of having graduate students. Nevertheless, he and his associates, James H. Kempton and Albert E. Longley, were for many years engaged in research with maize. The group headed by Collins did not participate in the main thrust of the USDA research on corn and they were somewhat isolated from other maize workers. This may have been due to the strained relations between them and F.D. Richey, who was in charge of corn investigations for the Bureau of Plant Industry of the US Department of Agriculture. Collins's group was administratively separated from that headed by Richey and the impression I gained during my five years in the department was that the integrated approach to corn research sought by Richey was unattainable. Nevertheless, some important contributions came from this laboratory. The first recognized linkage, between *C* and *Wx*, was reported by Collins & Kempton (23). Collins also described the dominant *Tu* gene (22), which allegedly played a key role in the evolution of maize. Among the mutants studied by Kempton are adherent tassel, branched silkless, brachytic, dead leaf margin, and the duplicate genes white sheath 1 and 2 (cited in 37). They had joint papers on the lineate stripe mutant (24) and on variability of linkage values between two endosperm genes (25).

Longley was a cytologist, an indefatigable worker who spent long hours peering down the tube of his microscope. Some idea of his industry is gained from the fact that the knurls on the knob of the fine adjustment focus of the Zeiss he used for many years were completely worn away. At first he worked on the cytology of a variety of plants, but later he concentrated on maize and its relatives. He published a series of papers (58, 59) on the chromosome morphology of pachytene chromosomes in various races of maize and teosinte. Knob positions and frequencies, arm ratios, and number of B chromosomes were determined. Longley was an excellent observer; he could follow the course of individual chromosomes through a tangled mass with unsurpassed skill. I marveled that it was possible. It is my conviction that Longley's influence would have been greater if, at the beginning of his career, he had been associated with cytogeneticists. Working in the relative isolation of Collins's laboratory, he was handicapped. But he was a gifted cytologist who did significant work.

Although this chronicle is concerned with maize genetics and not plant breeding, mention must be made of the role of geneticists in the development of hybrid corn. The basic concept of producing hybrid seed corn came from the minds of two geneticists who were studying inbreeding depression. Both George H. Shull and Edward M. East found a steady decline in vigor in the early generations of inbreeding that, however, ceased as the lines became homozygous. Intercrosses of different inbred lines gave F_1 progenies that in some crosses had a yield significantly exceeding that of the parent varieties. Shull (91) was impressed by this heterotic vigor and suggested a method of hybrid seed production that is essentially the same as that so widely used today. Because the homozygous inbred lines were much reduced in vigor, the high cost of producing hybrid seed for the farmer to plant made this method commercially impractical and H.K. Hayes & East (42) proposed that the production of varietal hybrids was a better alternative. Shull's method was not widely used until Donald F. Jones (44, 46) recommended that high-yielding double cross hybrids could be produced by crossing different vigorous single crosses. This modification solved the problem of the high cost of hybrid seed and Shull's method was widely accepted. Throughout America little but hybrid corn is grown. It was a spectacular success.

Today, the planting by the farmer of double cross hybrid seed to produce his commercial crop is on the wane because single cross hybrids from elite inbred lines of sufficient vigor and grain productivity have been developed. The somewhat higher cost of single cross seed is more than compensated for by the greater yield of single cross hybrids. It has been said that the development of hybrid corn did more than anything else to convince the average farmer of the importance of experiment station research. Prior to the 1930s, many growers were skeptical, even scornful, of the value of much agricultural research, but this attitude quickly and dramatically changed with the advent of hybrid corn. They became enthusiastic supporters of experiment station research at both the state and federal level. Hybrid corn cannot claim all of the credit for this change in the farmer's attitude, but it played a very important role. The farmers recognized a good thing when they saw it and hybrid corn was a sensational success. To paraphrase a popular TV ad, "When the experiment station speaks, the farmer listens."

Arranged in alphabetical order are the maize investigators who made important contributions to basic genetic knowledge in the first three decades of this century: E.G. Anderson, G.W. Beadle, T. Bregger, H.E. Brewbaker, R.A. Brink, A.M. Brunson, A.A. Bryan, C.R. Burnham, W.A. Carver, G.N. Collins, M. Demerec, E.M. East, R.A. Emerson, W.H. Eyster, A.C. Fraser, W.B. Gernert, M. Hadjinov, H.K. Hayes, J.D.H. Hofmeyr, W.A. Huelson, C.B. Hutchinson, H.W. Li, M.T. Jenkins, D.F. Jones, J.H. Kempton, P. Kvakan, E.W. Lindstrom, A.E. Longley, E.B. Mains, P.C. Mangelsdorf, Barbara McClintock, W.J. Mumm, H.S. Perry, I.F. Phipps, M.M. Rhoades, A.D. Suttle, S. Singh, W.R. Singleton, G.F. Sprague, L.J. Stadler, G.N. Stroman, A. Tavčar, J.B. Wentz, and C.M. Woodworth. Publications by the above individuals are listed in J. Weijer's maize bibliography (98).

Cooperative Efforts

The Maize Genetics Cooperation, as it later was known, began informally in a smoke-filled room. At an annual meeting of the Genetics Society of America, Professor Emerson rose at the end of one of the sessions to announce that a "cornfab" would be held that evening in his hotel room and that interested maize geneticists were welcome. I well remember that meeting. It was held in Emerson's New York City hotel room in late December of 1928. The dozen or so maize workers present sat on the few chairs available, the bed, and the floor while Emerson led the discussions on the current state of the maize linkage maps.

At this time, the ten linkage groups corresponding to the haploid number of ten chromosomes had not been definitively determined, let alone assigned to a cytologically recognizable chromosome. Furthermore, the linear order of many genes known to be in the same linkage group was ambiguous. The uncertainty as to whether a locus fell to the right or left of a reference gene was indicated on the linkage map by connecting the two alternative sites by arced lines, thus giving rise to the descriptive term *rainbow maps*. Emerson and Beadle had constructed linkage maps largely from unpublished data generously provided by other investigators. These rainbow maps had been distributed to cooperating maize workers and much of the discussion in the crowded New York hotel room that night centered on these maps. Cornell was serving as a clearing house for research reports but the task of obtaining further data to complete the ordering in each linkage map was far too formidable for one institution. It had to be a collaborative effort. In the course of the discussion, certain individuals volunteered to accept responsibility for a specific linkage group and by the end of the evening all linkage groups had been assigned.

The formal organization of the Maize Genetics Cooperation occurred in August 1932. The maize geneticists attending the 6th International Genetics Congress held at Cornell met and agreed to establish a cooperative enterprise to further the advance of maize genetics. Among the aims of this organization was the collection and dissemination to interested workers of unpublished data and information and the maintenance and distribution of tester stocks. I was asked to serve as custodian and shortly after the Congress ended I issued a request for research items to be included in the first issue of what later became known as the *Maize Genetics Cooperation News Letter*. The *News Letter* was devised to serve as the house organ of the Cooperation. The first issue was only a few pages long, consisting of research notes, comments, and unpublished data. Initially the *News Letters* appeared sporadically but soon were published annually. As the number of investigators using maize as a research organism grew with the years, so did the size of the *News Letter*. These reports of unpublished research have had a profound and stimulatory effect on maize genetics. It is generally recognized that the successful cooperative effort of the maize geneticists was a personal triumph for Professor Emerson. No one questioned his integrity, his unselfishness. So successful was this unique and unparalleled synergistic effort that it attracted the attention of other groups of investigators, who started comparable news letters. Imitation is the sincerest form of flattery and the *Maize News Letter* was the progenitor of them all. In the foreword of the first issue of the *Drosophila Information Service*, the editor, Milislav Demerec, who obtained his PhD degree in maize genetics as Emerson's student, wrote that the success of the maize letter had prompted him to start a similar publication for *Drosophila*.

These cooperative efforts culminated in 1935 with the publication of a linkage summary appearing as Memoir 180 of the Cornell University Agricultural Experiment Station. The preparation of the linkage summary was begun by Emerson and his colleague, A.C. Fraser, but progress was desultory. It was not until George Beadle arrived at Cornell to study with Emerson that the pace quickened. Beadle threw himself into the project with such enthusiasm and effectiveness that his contribution outstripped Fraser's and Emerson decided that, in justice, Beadle should be second author. The paper by Emerson, Beadle, and Fraser is more than a detailed compilation of published and unpublished linkage data. It is also a concise summary of the state of maize genetics in 1935. Included are: *(a)* brief descriptions and the genetic basis of all known mutant phenotypes, linked or unlinked, *(b)* inter- and intra-allelic interactions, *(c)* the inheritance of plant, aleurone, and pericarp colors, and *(d)* gametophytic characters; indeed, all of the diverse, often bizarre phenotypes affecting every part of the maize plant are listed. Today, nearly 50 years after it first appeared, the dog-eared, dilapidated copy in my laboratory, held together by scotch tape generously applied, attests to the frequent use of this publication. It is a highly prized reprint. Even now, it is unequaled as a source reference.

The Rise of Cytogenetics

By the mid-1920s, an impressive body of knowledge had been obtained about maize genetics but the early cytological investigations were confined to determining the true chromosome number in somatic metaphases of root tips (38, 39, 48–52, 56, 57, 75, 76, 82). Although there was a consensus that 20 was the diploid number, the problem of accounting for the supernumerary chromosomes present in certain races of maize had not been solved. The early cytological studies had done little to advance the status of maize genetics. A more sustained and rigorous cytological approach was needed. This came about in the following way.

Frederick D. Richey, in charge of corn investigations for the US Department of Agriculture, was an agronomist by training but a firm believer in a synergistic relationship between genetics and plant breeding. The primary function of most of the USDA corn breeders was the development of high yielding hybrids but a few, notably Lewis J. Stadler, were permitted to spend most of their time on more theoretical problems. All were encouraged to conduct some genetic investigations and some, particularly George F. Sprague and Merle T. Jenkins, were able to do a surprising amount although burdened with large breeding programs. Richey, an advocate of the concept that a sound and enlightened breeding program was based on a fuller understanding of a cultivar's genetic potentialities, was anxious that the USDA play a significant role in the advancement of maize genetics. The decision was reached that the services of a highly trained cytologist would complement the work of the geneticists and do more to advance maize genetics than would any other single appointment. Although on the payroll of the federal government, the new appointee would be stationed at the Agriculture Experiment Station of a land grant college. Cornell was the chosen institution and Lowell F. Randolph was selected as the cytologist.

Randolph obtained his PhD degree in 1921 under the direction of Lester W. Sharp at Cornell. His thesis dealt with the cytology of different types of maize chloroplast. He was a competent, well-trained cytologist when he joined the USDA in 1922. The detailed morphological-cytological study of the development of the maize kernel, which he published in 1936 (79), today remains the authoritative work. Another investigation was on the nature and inheritance of the supernumerary, accessory chromosomes (76) found in certain races. He demonstrated that they carried no known genes, that they were all alike, and that they were genetically inert. An unusual phenomenon was their nondisjunction during the development of the male gametophyte. These inert supernumeraries were called B chromosomes, a term which is now widely used to designate all supernumerary chromosomes wherever encountered. Among Randolph's accomplishments are the induction of polyploidy by heat treatment (77); the culturing of immature excised embryos (80), which permitted the production of difficult-to-obtain hybrids; and his cytogenetic studies on tetraploids (78). His research was characterized by thoroughness and objectivity. An unrealized attainment was the cytological identification of the different maize chromosomes. The small size and lack of detail in metaphase chromosomes of sectioned root tips make them unfavorable cytological material. However, their recognition had long been a goal toward which Randolph devoted considerable effort. So bleak was the prospect of individualizing them that, according to Professor Emerson, Edward M. East forsook maize for *Nicotiana* in the mistaken belief that it was better for cytological studies. East's defection happened, of course, before the advent of Belling's acetocarmine smear technique.

Randolph required technical help so funds were provided for a research assistant. Barbara McClintock, a young graduate student in botany, entered his laboratory in 1925 in this capacity. This event had momentous consequences, but it was an ill-fated arrangement that was soon dissolved. The only tangible evidence of their collaborative research was a jointly authored 1926 paper (81) describing the first triploid in maize. Why they ceased to work together is unclear, at least to me, but when I came to Cornell in 1928 they had gone their separate ways. However, from what I heard later, it was not a comfortable working situation. Their personalities were too dissimilar for tension to be avoided. McClintock was quick, imaginative, and perceptive. Almost instantly she grasped the significance of a new observation or recently discovered fact. Randolph, though able, was more methodical and less gifted. McClintock almost certainly became the dominant member of the team and Randolph, the nominal leader, found this irritating, even intolerable. Dissension was unavoidable and McClintock departed. Apparently their personal relationship was strained, but bitterness had not yet developed. Their brief association was momentous because it led to the birth of maize cytogenetics.

McClintock continued to work on the triploid material employing the recently invented acetocarmine smear technique of Belling. Her results constituted her PhD thesis, which was published in 1929 (61). Among the progeny of the triploid were 2n + 1 individuals (primary trisomics) with one additional A chromosome. She determined which of the primary trisomics carried a specific group of linked genes. This could be done even though the chromosomes were not cytologically distinguishable. McClintock soon found that all ten members of the haploid set could be recognized individually in acetocarmine smears of late prophase or metaphase stages at the first microspore mitosis (62). She identified them on the basis of total length, arm ratios, and position of heterochromatic regions. The longest was designated as chromosome 1 and the shortest as chromosome 10.

It was now relatively simple for McClintock to determine in the n + 1 microspores of a primary trisomic, known to give trisomic ratios for a specific group of linked genes, which one of the ten different chromosomes was in duplicate. In this manner she was able to assign seven of the linkage groups to a cytologically identifiable chromosome. One of the remaining three linkage groups was assigned to chromosome 1 by Brink & Cooper (13) and Burnham (17) using two different reciprocal translocations, both involving chromosome 1. A second linkage group was placed in chromosome 8 by Burnham using a trisomic test (cited in 37) and also by McClintock from a cytological deficiency (cited in 37). Chromosome 4 was shown to carry the remaining linkage group by discovering linkage of the *su* locus with a *T4-8* translocation [Anderson, cited in (37)] and a 1:1 starchy:sugary ratio in a testcross of a chromosome 8 primary trisomic [McClintock, cited in (37)]. The linkage groups are now numbered according to the relative length of the chromosome in which they reside. The rapidity with which maize cytogenetics was developing is evident from the fact that in 1928 none of the chromosomes had been cytologically identified and none of the linkage groups was known to be situated in a specific chromosome, but by 1931 all ten linkage groups had been assigned to identifiable chromosomes. This amazing accomplishment, so quickly achieved, was primarily due to McClintock.

The preparation of an idiogram of the maize chromosomes had been for some time a primary concern to Randolph. Studying polar views of flattened metaphase plates in sectioned root tips, he had attempted to individualize the maize complement. Success was limited and progress slow. It was McClintock who capitalized on the use of Belling's new acetocarmine smear technique. In the course of her triploid studies, she had discovered that the metaphase or late prophase chromosomes in the first microspore mitosis were far better for cytological discrimination than were root tip chromosomes in paraffin sections. In a few weeks' time she had prepared an idiogram of the maize chromosomes, which she published in *Science* (62). There had been no communication between Randolph and McClintock for some months. Neither knew of the current work in the other's laboratory. Her *Science* article left him embittered since she, unknown to him, had reached his long-sought goal.

A regrettable aftermath of the strained relationships between Randolph and McClintock was that Emerson came to look upon her with disapproval. I was not privy to the precise complaint Randolph voiced to Emerson about McClintock's behavior but I knew, as did McClintock, that Emerson sympathized with Randolph and viewed McClintock as something of a trouble-maker. They were not on good terms. McClintock felt the cold chill of disapproval and stopped discussing her work with Emerson, so he was unaware of her current findings. This was the situation in 1929, near the end of my first year of graduate study at Cornell. I had come to know McClintock well and was convinced of her unusual talents. The initial split between Randolph and McClintock had happened before my time. I could not judge who was at fault and blamed neither one nor the other, but I felt that Emerson should not be uninformed about her research and I saw to it that he wasn't. Her work was so remarkable that Emerson soon became one of her strongest supporters.

One of the early studies in maize cytogenetics was on the relationship between semi-sterility and interchanges involving heterologous chromosomes. Brink (10) and Brink & Burnham (12) found that semi-sterility, the abortion of 50% of the ovules and pollen, had an unusual pattern of inheritance. They suspected some type of chromosomal translocation was involved but had no cytological evidence in support of this surmise. Burnham, who obtained his PhD degree under Brink at Wisconsin in 1929, came to Cornell as a postdoctoral fellow to work with Emerson. His semi-sterile material was planted at Cornell in late spring 1929. Burnham could not have picked a more propitious time to arrive on the Cornell campus. McClintock had recently found that the pachytene chromosomes in microsporocytes were far superior to those of microspores. When Burnham's plants reached meiosis he, under McClintock's tutelage, quickly found that the meiocytes of plants with 50% abortion of ovules and pollen had a ring of four chromosomes at diakinesis, i.e., they were heterozygous for a translocation. This finding created great excitement among the Cornell cytogeneticists and, for Burnham, it marked the beginning of a productive life-long study of chromosomal interchanges.

The years at Cornell from 1928 to 1935 were ones of intense cytogenetical activity. Progress was rapid, the air electric. Somewhat surprising in retrospect is the small number of investigators involved. There are only five individuals in the now-famous photograph of maize workers taken at the Cornell experimental field in the summer of 1929. The quintet included Professor Emerson, McClintock, Beadle, Burnham, and Rhoades. Emerson was not a cytogeneticist but the others were. The only cytogeneticist at Cornell not in the photograph is L.F. Randolph. There were a number of graduate students in genetics at Cornell, including H.S. Perry, H.W. Li, and J.D.J. Hofmeyr, but their problems involved little or no cytogenetics. The number of cytogeneticists at Cornell at any one time was never large. Harriet Creighton joined the group in the fall of 1929 and Virginia Rhoades in 1931, but their coming did not swell the ranks since Beadle and Burnham were leaving. Although few in number, their morale was high. There was no shortage of energy and enthusiasm. They were exciting years, flawed only by the Great Depression. No positions were available to any of this group of able young investigators and it was some time before all gained suitable employment, even though two of them, Beadle and McClintock, were destined to become Nobel Laureates.

In depth of knowledge of its genetic constitution and in its cytological resolution, maize remained unsurpassed as an experimental organism until the significance of the banded nature of the giant salivary chromosomes of *Drosophila* was recognized and the technical advantages afforded by the polytene chromosomes for cytological investigations were exploited. The salivary chromosomes had a resolving power for cytological detail that permitted the detection of minute rearrangements and deficiencies that would have escaped observation in maize. However, for a few short years maize cytogenetics reigned supreme, and the young maize cytogeneticists trained at Cornell capitalized on their golden opportunity. Genetic problems could now be attacked from both the genetical and cytological fronts. That progress was explosive in the next few years is evident from the following list of accomplishments between 1929 and 1935. The list includes the nature of the discovery and the investigator. The compilation certainly is not complete, since other works could have been selected, but I believe the best have been included. It should be stressed that this list far from represents the totality of cytogenetic research with maize. Much has been learned in subsequent years and today it remains an active field, although molecular genetics has replaced it at the cutting edge. Highly significant though some of the more recent studies are, they are not the concern of this account. These later investigations are discussed in the 1977 reviews by Wayne Carlson, "The Cytogenetics of Corn" (19), by Edward Coe and M.G. Neuffer in "The Genetics of Corn" (20), and in Burnham's book on cytogenetics (18).

The status of maize cytogenetics in the mid-1930s was summarized in the 1935 paper by Rhoades & McClintock (87) and the cytogenetical discoveries listed below are largely taken from that publication. In a few cases the publication date is after 1935, but the work had been done prior to that time.

1. The individualization of the ten maize chromosomes at pachynema and at metaphase of the first microspore mitosis: McClintock (62);
2. The association of different linkage groups with a particular, cytologically identifiable member of the chromosome complement: McClintock & Hill (68), Brink & Cooper (13), and Burnham (17);

3. The association of semi-sterility with heterozygous translocations: Burnham (16);
4. The cytological proof of genetic crossing over:[4] Creighton & McClintock (29), Brink & Cooper (14);
5. Cytological and genetic proof of chromatid crossing over: McClintock (63), Creighton & McClintock (30), Rhoades (84);
6. Cytological determination of the physical location within the chromosomes of genes, using reciprocal translocations, inversions, and deficiencies: McClintock (63, 65), Creighton (28), Rhoades (85), V.H. Rhoades (88), Stadler (94);
7. The genetic control of chromosome behavior: Beadle (4, 5, 7, 8);
8. Evidence that chiasmata are points of genetic crossing over: Beadle (6);
9. Nonhomologous pairing and its genetic consequences: McClintock (65);
10. Instability of ring-shaped chromosomes leading to variegation: McClintock (64);
11. Divisibility of centromeres: McClintock (64);
12. Breakage-fusion-bridge cycle: McClintock (67);
13. Relationship of a particular chromosomal element to the development of the nucleolus: McClintock (66);
14. Cytogenetics of tetraploid maize: Randolph (78);
15. Correlation of heterochromatin with genetic inertness: Randolph (unpublished data);
16. Mutagenic effects of X-irradiation: Stadler (92, 93);
17. Discovery of cytoplasmic male sterility: Rhoades (83).

It is appropriate to end this account of the early history of maize genetics with the year 1935. This was the year in which appeared the famous Cornell Memoir on linkage values (37) that provided an excellent summary of the progress made by students of the maize plant following the rediscovery of Mendel's Laws. Published the same year was a review (87) of the remarkable advances made from 1928 to 1935 in correlating cytological observations with genetic data, giving rise to the new, clearly marked discipline of cytogenetics. It also witnessed the end of the golden age of cytogenetics at Cornell as the principal protagonists departed to continue productive careers elsewhere. Drawing the curtain at this time unfortunately allows no consideration of

[4] In Evelyn Keller's biography of McClintock (47), there is an account of the alleged role T.H. Morgan played in ensuring priority of publication by Creighton & McClintock over Curt Stern on the correlation of cytological and genetical crossing over. I find this to be a flaw in an otherwise excellent book. I cannot accept the statement that Morgan urged Creighton & McClintock to publish their data at once because he knew that Stern was well along in his work on the same problem with *Drosophila*. Morgan is quoted as later saying, "I thought it was about time that corn got a chance to beat *Drosophila*." Not only does Morgan's purported remark suggest that he betrayed the confidence of a former student and colleague, something totally out of character, but it does not reflect the relative status of maize and *Drosophila* cytogenetics in 1931. Maize investigators could study in detail the long, individualized pachytene chromosomes found in carmine smears of male meiocytes, while *Drosophila* workers had to resort to inferior cytological preparations, usually of sectioned somatic cells. In 1931, maize had taken center stage and the fortunes of *Drosophila* were apparently on the wane. The situation was dramatically reversed following the development of carmine smears of the giant, banded salivary chromosomes, but as of 1931 certain studies had been carried out with maize that had no counterpart with *Drosophila*. Fortunately. Stern published his data in 1931, shortly after Creighton & McClintock's paper appeared, so no great damage was done; no one was really scooped and both parties shared the widespread acclaim these papers were accorded. What really transpired may never come to light, but I believe Morgan's name was unjustly maligned.

many important investigations made subsequent to 1935. Among the casualties are the studies leading to recognition of transposable genetic elements and a host of others. These significant works must be left for a later account by another chronicler. If I have succeeded in capturing some of the excitement and temper of those early years, I rest content.

Acknowledgements

Most of this account of the early years of maize genetics was written in January and February of 1983, when I was a visiting professor in the Department of Botany of the University of Florida. I wish to express my deep appreciation of the many courtesies extended to me by William Louis Stern, chairman of the department. I would be remiss if I did not acknowledge with gratitude the patient and efficient assistance of Ellen Dempsey in readying the manuscript for publication.

Literature Cited

1. Anderson, E.G. 1943. Utilization of translocations with endosperm markers in the study of economic traits. *Maize Genet. Coop. Ned Letter* 17: 4–5
2. Anderson, E.G. 1956. The application of chromosomal techniques to maize improvement. *Brookhaven Symp. Biol.: Genet. Plant Breed* 9: 23–26
3. Bateson, W., Saunders, E.R., Punnett, R.C. 1906. *Experimental studies in the physiology of heredity.* Report III, Evolut. Comm. Royal Soc. II of London, p. 9. London: Royal Soc.
4. Beadle, G.W. 1930. Genetical and cytological studies of Mendelian asynapsis in *Zea mays. Cornell Univ. Agric. Exp. Stn. Mem.* 129: 1–23
5. Beadle, G.W. 1931. A gene in maize for supernumerary cell divisions following meiosis. *Cornell Univ. Agric. Exp. Stn. Mem.* 135: 1–12
6. Beadle, G.W. 1932. The relation of crossing over to chromosome association in *Zea-Euchlaena* hybrids. *Genetics* 17: 481–501
7. Beadle, G.W. 1932. A gene for sticky chromosomes in *Zea mays. Z. Indukt. Abstamm. Vererbungsl.* 63: 195–217
8. Beadle, G.W. 1932. A gene in *Zea mays* for the failure of cytokinesis during meiosis. *Cytologia* 3: 142–55
9. Bridges, C.B. 1916. Non-disjunction as proof of the chromosome theory of heredity. *Genetics* 1: 1–52, 107–63
10. Brink, R.A. 1927. The occurrence of semisterility in maize. *J. Hered.* 18: 266–70
11. Brink, R.A. 1973. Paramutation. *Ann. Rev. Genet.* 7: 129–52
12. Brink, R.A., Burnham, C.R. 1929. Inheritance of semisterility in maize. *Am. Nat.* 63: 301–16
13. Brink, R.A., Cooper, D.C. 1932. A strain of maize homozygous for segmental interchanges involving both ends of the P-Br chromosome. *Proc. Natl. Acad. Sci. USA* 18: 441–47
14. Brink, R.A., Cooper, D.C. 1935. A proof that crossing over involves an exchange of segments between homologous chromosomes. *Genetics* 20: 22–35
15. Brink, R.A., Nilan, R.A. 1952. The relation between light variegated and medium variegated pericarp in maize. *Genetics* 37: 519–44
16. Burnham, C.R. 1930. Genetical and cytological studies of semisterility and related phenomena in maize. *Proc. Natl. Acad. Sci. USA* 16: 269–77
17. Burnham, C.R. 1932. An interchange in maize giving low sterility and chain configurations. *Proc. Natl. Acad. Sci. USA* 18: 434–40
18. Burnham, C.R. 1962. *Discussions in Cytogenetics.* Minneapolis: Burgess. 375 pp.
19. Carlson, W. 1977. The cytogenetics of corn. In *Corn and Corn Improvement,* ed. G. F. Sprague, pp. 225–303. Madison, WI: Agronomy 18, Am. Soc. Agron.
20. Coe, E.H., Neuffer, M.G. 1977. The genetics of corn. See Ref. 19, pp. 111–223

21. Collins, G.N. 1912. Gametic coupling as a cause of correlations. *Am. Nat.* 46: 559–90
22. Collins, G.N. 1917. Hybrids of *Zea ramosa* and *Zea tunicata*. *J. Agric. Res.* 9: 383–97
23. Collins, G.N., Kempton, J.H. 1911. Inheritance of waxy endosperm in hybrids of Chinese maize. *C. R. 4e Congr. Intl. Genet., Paris,* pp. 547–57
24. Collins, G.N., Kempton, J.H. 1920. Heritable characters of maize. I. Lineate stripe. *J. Hered.* 11: 3–6
25. Collins, G.N., Kempton, J.H. 1927. Variability in the linkage of two seed characters in maize. *US Dep. Agric. Res. Bull.* 1468: 1–64
26. Correns, C. 1900. G. Mendel's Regel über das Verhalten der Nachkommenschaft der Rassenbastarde. *Ber. Deutsch. Bot. Gesells.* 18: 158–68
27. Correns, C. 1902. Scheinbare Ausnehmen von der Mendels'schen Spaltungsregel für Bastarde. *Ber. Deutsch. Bot. Gesells.* 20: 159
28. Creighton, H.B. 1934. Three cases of deficiency in chromosome 9 of *Zea mays*. *Proc. Natl. Acad. Sci. USA* 20: 111–15
29. Creighton, H.B., McClintock, B. 1931. A correlation of cytological and genetical crossing over in *Zea mays*. *Proc. Natl. Acad. Sci. USA* 17: 492–97
30. Creighton, H.B., McClintock, B. 1932. Cytological evidence for 4-strand crossing over in *Zea mays*. *Proc. 6th Intl. Congr. Genet.* 2: 392
31. East, E.M., Hayes, H.K. 1911. Inheritance in maize. *Conn. Agric. Exp. Stn. Bull.* 167: 1–142
32. Emerson, R.A. 1914. Inheritance of a recurring somatic variation in variegated ears of maize. *Am. Nat.* 48: 87–115
33. Emerson, R.A. 1918. A fifth pair of factors, A a, for aleurone color in maize and its relation to the C c and R r pairs. *Cornell Univ. Agric. Exp. Stn. Mem.* 16: 225–89
34. Emerson, R.A. 1921. The genetic relations of plant colors in maize. *Cornell Univ. Agric. Exp. Stn. Mem.* 39: 1–156
35. Emerson, R.A. 1934. Relation of the differential fertilization genes, Ga ga, to certain other genes of the Su-Tu linkage group of maize. *Genetics* 19: 137–56
36. Emerson, R.A., East, E.M. 1913. The inheritance of quantitative characters in maize. *Bull. Agric. Exp. Stn. Nebr.* 2: 1–120
37. Emerson, R.A., Beadle, G.W., Fraser, A.C. 1935. A summary of linkage studies in maize. *Cornell Univ. Agric. Exp. Stn. Mem.* 180: 1–83
38. Fisk, E.L. 1925. The chromosomes of *Zea mays*. *Proc. Natl. Acad. Sci. USA* 11: 352–56
39. Fisk, E.L. 1927. The chromosomes of *Zea mays*. *Am. J. Bot.* 14: 54–75
40. Guignard, L. 1899. Sur les anthérozoides et la double copulation sexuelle chez les vegétaux angiospermes. *C. R. Acad. Sci. Paris* 128: 864–71
41. Harper, R.A. 1920. Inheritance of sugar and starch characters in corn. *Bull. Torrey Bot. Club* 47: 137–86
42. Hayes, H.K., East, E.M. 1911. Improvement in corn. *Conn. Agric. Exp. Stn. Bull.* 168: 1–31
43. Iltis, H. 1932. *Life of Mendel*, transl. E. Paul, C. Paul. London: Allen & Unwin. 336 pp.
44. Jones, D.F. 1918. The effects of inbreeding and cross-breeding upon development. *Conn. Agric. Exp. Stn. Bull.* 207: 5–100
45. Jones, D.F. 1919. Selection of pseudostarchy endosperm in maize. *Genetics* 4: 364–93
46. Jones, D.F. 1919. Inbreeding in corn improvement. *Breed. Gaz.* 75: 1111–12
47. Keller, E.F. 1983. *A Feeling for the Organism. The Life and Work of Barbara McClintock.* San Francisco: Freeman. 235 pp.
48. Kiesselbach, T.A., Peterson, N.F. 1925. The chromosome number of maize. *Genetics* 10: 80–85
49. Kuwada, Y. 1911. Meiosis in the pollen mother cells of *Zea mays*. *Bot. Mag.* 25: 163–81
50. Kuwada, Y. 1915. Ueber die Chromosomenzahl von *Zea mays*. *Bot. Mag.* 29: 83–89
51. Kuwada, Y. 1919. Die Chromosomenzahl von *Zea mays*. Ein Beitrag zur Hypothese der Individualität der Chromosomen und zur Frage über die Herkunft von *Zea mays*. *J. Coll. Sci. Imper. Univ. Tokyo* 39: 1–48
52. Kuwada, Y. 1925. On the number of chromosomes in maize. *Bot. Mag.* 39: 227–34

53. Lindegren, C.C. 1953. Gene conversion in *Saccharomyces*. *J. Genet.* 51: 625–37
54. Lindstrom, E.W. 1917. Linkage in maize: Aleurone and chlorophyll factors. *Am. Nat.* 51: 225–37
55. Lindstrom, E.W. 1930. The genetics of maize. *Bull. Torrey Bot. Club* 57: 221–31
56. Longley, A.E. 1924. Chromosomes in maize and maize relatives. *J. Agric. Res.* 28: 673–81
57. Longley, A.E. 1927. Supernumerary chromosomes in *Zea mays*. *J. Agric. Res.* 35: 769–84
58. Longley, A.E. 1937. Morphological characters of teosinte chromosomes. *J. Agric. Res.* 54: 835–62
59. Longley, A.E. 1938. Chromosomes of maize from North American Indians. *J. Agric. Res.* 56: 177–96
60. Longley, A.E. 1961. Breakage points for four corn translocation series and other corn chromosome aberrations maintained at the California Institute of Technology. *US Dep. Agric. Agric. Exp. Stn. Res. Bull.* 34: 1–16
61. McClintock, B. 1929. A cytological and genetical study of triploid maize. *Genetics* 14: 180–222
62. McClintock, B. 1929. Chromosome morphology in *Zea mays*. *Science* 69: 629
63. McClintock, B. 1931. Cytological observations of deficiencies involving known genes, translocations and an inversion in *Zea mays*. *Miss. Agric. Exp. Stn. Res. Bull.* 163: 1–30
64. McClintock, B. 1932. A correlation of ring-shaped chromosomes with variegation in *Zea mays*. *Proc. Natl. Acad. Sci. USA* 18: 677–81
65. McClintock, B. 1933. The association of non-homologous parts of chromosomes in the mid-prophase of meiosis in *Zea mays*. *Z. Zellforsch. Mikrosk. Anat.* 19: 191–237
66. McClintock, B. 1934. The relation of a particular chromosomal element to the development of the nucleoli in *Zea mays*. *Z. Zellforsch. Mikrosk. Anat.* 21: 294–328
67. McClintock, B. 1938. The fusion of broken ends of sister half-chromatids following chromatid breakage at meiotic anaphases. *Miss. Agric. Exp. Stn. Res. Bull.* 290: 1–48
68. McClintock, B., Hill, H.E. 1931. The cytological identification of the chromosome associated with the R-G linkage group in *Zea mays*. *Genetics* 16: 175–90
69. Mendel, G.J. 1865. Versuche über Pflanzen-Hybriden. *Verh. Naturforsch. Ver. Brünn* 4: 3–47
70. Mendel, G.J. 1866–1873. Mendel's letters to Carl Nägeli. In *Abh. Math. Phys. Kl. Königlich Saechs. Ges. Wiss.* 29: 189–265. Trans. L. K. Piternick, G. Piternick, 1905, in *Suppl. Genet.* 35(5): 1–29
71. Muller, H.J. 1927. Artificial transmutation of the gene. *Science* 66: 84–87
72. Nawaschin, S. 1899. Neue Beobachtungen über Befruchtung bei *Fritillaria* und *Lilium*. *Bot. Centralbl.* 77: 62
73. Olby, R.C. 1966. *Origins of Mendelism*. New York: Schocken. 204 pp.
74. Patterson, E.B. 1982. The mapping of genes by the use of chromosomal aberrations and multiple marker stocks. In *Maize for Biological Research*, ed. W. F. Sheridan, pp. 85–88. Special Publ. Plant Mol. Biol. Assoc. Grand Forks, ND: Mol. Biol. Assoc.
75. Randolph, L.F. 1928. Chromosome numbers in *Zea mays L. Cornell Univ. Agric. Exp. Stn. Memoir* 117: 1–44
76. Randolph, L.F. 1928. Types of supernumerary chromosomes in maize. *Anat. Rec.* 41: 102
77. Randolph, L.F. 1932. Some effects of high temperature on polyploidy and other variations in maize. *Proc. Natl. Acad. Sci. USA* 18: 222–29
78. Randolph, L.F. 1935. Cytogenetics of tetraploid maize. *J. Agric. Res.* 50: 591–605
79. Randolph, L.F. 1936. Developmental morphology of the caryopsis in maize. *J. Agric. Res.* 53: 881–916
80. Randolph, L.F. 1945. Embryo culture of iris seed. *Bull. Am. Iris Soc.* 96: 33–45
81. Randolph, L.F., McClintock, B. 1926. Polyploidy in *Zea mays L. Am. Nat.* 60: 99–102
82. Reeves, R.G. 1925. Chromosome studies of *Zea mays*. *Proc. Iowa Acad. Sci.* 22: 171–79
83. Rhoades, M.M. 1931. The cytoplasmic inheritance of male sterility in *Zea mays*. *J. Genet.* 27: 71–93
84. Rhoades, M.M. 1933. An experimental and theoretical study of chromatid crossing

over. *Genetics* 18: 535–55

85. Rhoades, M.M. 1936. A cytogenetical study of a chromosome fragment in maize. *Genetics* 21: 491–502
86. Rhoades, M.M. 1950. Meiosis in maize. *J. Hered.* 41: 58–67
87. Rhoades, M.M., McClintock, B. 1935. The cytogenetics of maize. *Bot. Rev.* 1: 292–325
88. Rhoades, V.H. 1935. The location of a gene for disease resistance in maize. *Proc. Natl. Acad. Sci. USA* 21: 243–46
89. Roman, H. 1947. Mitotic nondisjunction in the case of interchanges involving the B-type chromosome in maize. *Genetics* 32: 391–409
90. Roman, H., Ullstrup, A.J. 1951. The use of A-B translocations to locate genes in maize. *Agron. J.* 43: 450–54
91. Shull, G.H. 1909. A pure line method of corn breeding. *Am. Breed. Assoc. Rep.* 5: 51–59
92. Stadler, L.J. 1928. Genetic effects of x-rays in maize. *Proc. Natl. Acad. Sci. USA* 14: 69–75
93. Stadler, L.J. 1930. Some genetic effects of x-rays in plants. *J. Hered.* 21: 3–19
94. Stadler, L.J. 1933. On the genetic nature of induced mutations in plants. II. A haplo-viable deficiency in maize. *Miss. Agric. Exp. Stn. Res. Bull.* 204: 1–29
95. Tschermak, E. 1900. Ueber Künstliche Kreuzung bei *Pisum*. *Ber. Deutsch. Bot. Gesells.* 18: 232–39
96. de Vries, H. 1900. Das Spaltungsgesetz der Bastarde. *Ber. Deutsch. Bot. Ges.* 18: 83–90
97. de Vries, H. 1900. Sur la loi de disjonction des hybrides. *C. R. Acad. Sci. Paris* 130: 845–47
98. Weijer, J. 1952. *A Catalogue of Genetic Maize Types Together with a Maize Bibliography*, pp. 189–425. The Hague: Martinus Nijhoff

TRANSPOSITION

At the time, I did not know that *Ds* could change its location. Realization of this did not enter my consciousness until late this spring (1948), following the harvest of the winter 1947–48 greenhouse crop.

B. McClintock 1949

The knowledge gained from the study of mutable loci focuses attention on the components in the nucleus that function to control the action of the genes in the course of development. ...The primary thesis states that instability arises from alterations that do not directly alter the genes themselves, but affect the functioning of the genic components at or near the locus of alteration.

B. McClintock 1952

Barbara McClintock and Jacques Monod
(CSHL 1947)

Working at the microscope
(CSHL 1947)

Analyzing corn kernel patterns
(CSHL 1963)

CHROMOSOME ORGANIZATION AND GENIC EXPRESSION

BARBARA McCLINTOCK

Department of Genetics, Carnegie Institution of Washington, Cold Spring Harbor, N. Y.

During the past six years, a study of the behavior of a number of newly arisen mutable loci in maize has been undertaken. This study has provided a unique opportunity to examine the mutation process at a number of different loci in the chromosomes. For some of these loci, several independent inceptions of instability have occurred during the progress of this study. The types of mutation that appear, and the types of instability expression, need not be the same at any one locus. In fact, comparisons of the behavior of these different mutable conditions at a particular locus have shown striking diversity, not only with regard to the changes in phenotypic expression that result from mutations at the locus, but also with regard to the manner in which mutability is controlled. Knowledge of the genetic constitutions, with respect to mutable loci already present in the plants in which new mutable loci have arisen, and the subsequent behavior of the newly arisen mutable loci, have provided evidence that allows an interpretation of their mode of origin and also their mode of operation. As a consequence of this study, some rather unorthodox conclusions have been drawn regarding the mechanisms responsible for mutations arising at these loci. The same mechanisms may well be responsible for the origins of many of the observed mutations in plants and animals.

Instability of various loci—whether referred to by the terms mutable loci, mutable genes, or variegation, position effect, etc.—has been known for many years, and many such cases have received considerable study. The conditions associated with the more obvious position-effect phenomena in *Drosophila* are well known. Those associated with instability of phenotypic expressions in other organisms have been less well understood. It is because of the distinctive advantages that the maize plant offers for such a study that it has been possible to obtain precise evidence concerning some of the events associated with the origin and behavior of mutable loci. The first of these advantages relates to the ease of observing the chromosomes, and thus determining the nature of some of the changes that occur in them. The presence of a triploid endosperm in the kernel provides a second advantage. This endosperm, with its outer aleurone layer that can develop pigments, and the underlying tissues that may develop starches of several types, or sugars, or carotenoid pigments, permits the detection of differences in phenotypic expression of various types. Some of these may be quantitatively measured. Thirdly, there are a number of different loci known in which heritable alterations have given rise to changes in the expression of these several endosperm components. The mutations at some of these loci affect characters of both the endosperm and the plant tissues. This applies particularly to those mutations that affect the development of the anthocyanin pigments. In the studies to be described, the presence in the short arm of chromosome 9 of four marked loci that affect endosperm characters has been of particular importance for analyzing the events occurring at mutable loci. The necessity of having such markers will become evident in the discussion. For this study, the accumulated knowledge of the behavior of newly broken ends of chromosomes in maize has been of particular importance. Its significance for interpreting the origin of mutable loci will be indicated in the sections that follow.

THE CHROMATID AND CHROMOSOME TYPES OF BREAKAGE-FUSION-BRIDGE CYCLE

The diagrams of Figure 1 illustrate the mode of origin of newly broken ends of chromosomes at a meiotic mitosis and the subsequent behavior of these ends in successive mitotic cycles. A chromosome with a newly broken end entering a telophase nucleus in the gametophytic or endosperm tissues will give rise in the next anaphase to a chromatid bridge configuration (McClintock, 1941). The bridge is produced because fusion occurs between sister chromatids at the position of previous anaphase breakage. This sequence of anaphase breaks and sister-chromatid fusions will continue in successive mitoses. It has therefore been designated the chromatid type of breakage-fusion-bridge cycle. This cycle is illustrated in A of Figure 1. In the sporophytic tissues, however, this cycle usually does not occur. The broken end entering a telophase

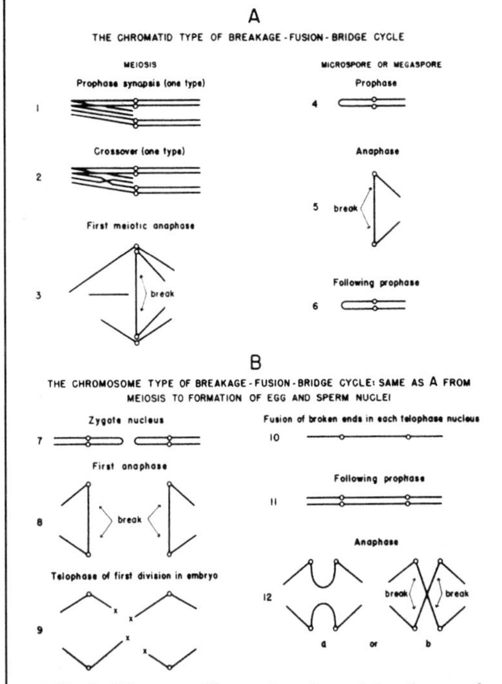

FIG. 1. Diagrams illustrating the origin of a newly broken end of a chromosome at the meiotic anaphase and its subsequent behavior. A. The chromatid type of breakage-fusion-bridge cycle. 1. One type of synaptic configuration at the first meiotic prophase between homologous arms of a pair of chromosomes, one member of which carries a duplication of this arm in the inverted order. 2. The production of a dicentric chromatid as the consequence of a crossover. It is composed of two complete chromatids of this chromosome. 3. Anaphase *I*. Bridge configuration produced by separation of centromeres of the dicentric chromatid. A break in the bridge occurs at some position between the two centromeres. 4. Fusion of sister chromatids at the position of the previous anaphase break is exhibited in prophase of the microspore or megaspore nucleus. 5. Separation of sister centromeres at anaphase in the microspore or megaspore produces a bridge configuration. This bridge is broken at some position between the two centromeres. 6. Fusion of sister chromatids occurs at the position of the preceding anaphase break. Separation of sister centromeres at the next anaphase again produces a bridge which is broken at some position between the two centromeres. This cycle continues in successive mitoses during the development of the gametophyte and the endosperm.

B. The chromosome type of breakage-fusion-bridge cycle. It may be initiated in the sporophyte if each gamete contributes a chromosome which has been broken in the anaphase of the division preceding gamete formation. The zygote nucleus will then contain two such chromosomes. In the prophase of the first division of the zygote, 7, each of these is composed of two sister chromatids fused at the position of the previous anaphase break. In the first anaphase of the zygotic division, 8, these two chromosomes give rise to bridge configurations as the centromeres of the sister chromatids pass to opposite poles. Breaks occur in each bridge at some position between the centromeres. In the telophase nuclei, two chromosomes, each with a newly broken end, are present as diagrammed in 9. The crosses mark the broken ends of each chromosome. Fusion of broken ends of *chromosomes* occurs in each telophase nucleus, 10, establishing a dicentric chromosome. In the next prophase, 11, each sister chromatid is dicentric. At the subsequent anaphase, several types of configurations may result from separation of the sister centromeres, two of which are shown in 12. Separations as shown in *b* of 12 give rise to anaphase bridge configurations. Breaks occur in each bridge at some position between the centromeres. The subsequent behavior of the broken ends, from telophase to telophase, is the same as that given in diagrams 9 to 12.

nucleus heals, and its subsequent behavior resembles that of a normal, nonbroken end of a chromosome. (Note: The chromatid type of breakage-fusion-bridge cycle can continue throughout the development of the sporophytic tissues under certain conditions. These conditions are usually not present in the genetic stocks of maize.) If, however, a chromosome with a newly broken end is introduced into the zygote by each gamete nucleus, the broken ends of the two chromosomes are capable of fusion (McClintock, 1942). This establishes a dicentric chromosome. A different type of breakage-fusion-bridge cycle is thereby initiated. In the telophase nuclei, the fusions now occur between the broken ends of chromosomes rather than between the broken ends of sister chromatids, as described above. This sequence of events has been called the chromosome type of breakage-fusion-bridge cycle, and is illustrated in B of Figure 1. A study of the consequences of these cycles has revealed that they may initiate breakage events in chromosomes of the complement other than those undergoing the cycle. This complication has been of significance, for it appears that these unanticipated alterations of the chromosomes may be responsible primarily for the origin of mutable loci and of other types of heritable change.

Unexpected Chromosomal Aberrations Induced by the Breakage-Fusion-Bridge Cycles

In the course of an experiment designed to induce small internal deficiencies within the short arm of chromosome 9, a number of plants were obtained that had undergone the chromosome type of breakage-fusion-bridge cycle in their early developmental period. The short arm of each chromosome 9 was involved in this cycle. It is known that the cycle will often cease suddenly in certain cells and that these cells are then capable of developing sexually functional branches of the plant. In order to determine the nature of the chromosome changes produced by this cycle, the sporocytes of many of these plants were examined at the pachytene of meiosis. The expected types of altered constitution of the short arm of chromosomes 9 were found. In addition, other quite unexpected types of chromosome aberration appeared in a number of the plants. These alterations had been produced in the early developmental periods when the breakage-fusion-bridge cycles were occurring. With a few exceptions, the chromosome parts in which alterations had been initiated were the knobs and the centromeres, or the nucleolus organizer of chromosome 6. In the majority of cases, either the knob or the centromere of one of the chromosomes 9 that had been undergoing the breakage-fusion-bridge cycle was involved in the structural rearrangement. Nonrandomness was apparent with regard to the other chromosome involved in the aberration. For example, four cases were found in which the centromere of chromosome 9 had fused with the centromere of another chromosome—chromosome 2 in three of the four cases. Chromosome 8 was also very frequently involved in these structural changes.

The breakage-fusion-bridge cycle was obviously responsible for the induction of these alterations in the knobs, centromeres and the nucleolus organizer. That alterations in such elements were occurring without obvious direct participation of the knob or the centromere of the chromosome 9 undergoing the breakage-fusion-bridge cycle has also been indicated. This was made evident by the presence in one plant of an inversion involving the nucleolus organizer and the centromere region in chromosome 6, by an inversion in chromosome 5 in another plant involving the centromere and the knob regions, and by an inversion in chromosome 7 in a third plant involving the centromere region and the knob region in the long arm of this chromosome. In addition, some of the plants examined showed the presence of a ring chromosome that was not composed of segments of chromosome 9, so far as could be determined. It now must be emphasized that it was in the self-pollinated progeny of plants that had undergone the chromosome type of breakage-fusion-bridge cycle in their early developmental period that the initial burst of newly arisen mutable loci appeared. It might be suspected that this burst was a reflection of the mechanism that had produced the alterations mentioned above. If so, the origin of mutable loci would be associated with change in these particular elements of the chromosome complement. It was some time, however, before sufficient evidence had accumulated to allow deductions to be drawn regarding this presumptive relationship. A description of the origin and behavior of some of the representative types of mutable loci should be given before this topic is again considered.

Recognition of the Relation of Mutation to the Mitotic Cycle

Interest in these mutable loci, appearing unexpectedly and in large numbers in the self-pollinated progeny of plants that had undergone the chromosome type of breakage-fusion-bridge cycle in their early developmental periods, was aroused when it was realized that in each case some factor was present which controlled the time or the frequency of mutations. This factor could be altered as a consequence of some event associated with the mitotic process. This was made evident by the appearance of sectors of tissue, derived from sister cells, that exhibited obvious differences in time of mutations, mutation frequency, or both. In many cases, it was also apparent that the mutations themselves arose as a consequence of some event associated with the mitotic cycle. This basic behavior pattern was exhibited by all the various newly arisen mutable loci. It directed attention to the mitotic mechanism as the responsible agent. It was concluded, therefore, that further investigation of these mutable loci might produce some evidence leading to an appreciation of the nature of the responsible mitotic events.

During six years of study of a number of newly arisen mutable loci, some well-established facts have accumulated concerning the processes associated with the origin of mutable loci and their subsequent behavior. Observation of consistent behavior in many mutable loci, where the cytological events associated with a change in phenotype could be determined, and comparison of the behavior of these loci with others in which cytological determinations could not readily be made, have provided an assemblage of interrelated facts upon which the conclusions to be stated later are based.

The Origin of *Ds* and its Behavior

The first evidence of the type of chromosomal event that is associated with the expression of mutability came with the discovery of a locus in the short arm of chromosome 9 at which chromosome breaks were occurring. This was observed in the self-pollinated progeny of one of the plants that had undergone the chromosome type of breakage-fusion-bridge cycle in early development. When first seen, the "mutability" was expressed by the time and frequency of the breaks that occurred at this locus in some cells during the development of a tissue. Also, some change could occur in somatic cells that affected the time and frequency; and this latter event likewise was associated with the mitotic process. The behavior pattern resembled in considerable detail the patterns exhibited by the mutable loci. In this case, however, a mechanism associated with chromosome fusion and subsequent breakage was responsible for the behavior observed. The mutations from recessive to dominant exhibited by the mutable loci would not alone have suggested a chromosome-breakage mechanism as being responsible. Because of this similarity of the patterns of behavior, it was suspected that the basic mechanism responsible for mutations at mutable loci could be one associated with some form of structural alteration at the locus showing the mutation phenomenon. This conclusion was consistent with the very first observations of the behavior of mutable loci. These observations had indicated that the events at mutable loci leading to mutations and also other events controlling their time and frequency of occurrence were associated with alterations that were in some manner produced during the course of a mitotic cycle.

Intensive study of this locus in chromosome 9 at which structural alterations occur at regulated rates and at regulated times in development has been rewarding. A "break" in the chromosome at this locus was the event first recognized. The factor responsible was therefore given the symbol *Ds*, for "Dissociation." The nature of the breakage event was later determined. It arises from dicentric and acentric chromatid formations. The acentric fragment is composed of the two sister chromatids, from the *Ds* locus to the end of the short arm. The complementary dicentric component includes the sister segments from the locus to the centromere plus the long arms of the two sister chromatids. This is the type of recognizable event found to occur most frequently at *Ds*. Other recognizable aberrations, however, may sometimes arise. One of them is the formation of an internal deficiency in the short arm of chromosome 9. Such deficiencies include the regions adjacent to *Ds*, and vary in extent from minute to quite large. Translocations between this chromosome and another chromosome of the complement may arise, with one of the points of breakage at the *Ds* locus. Duplications, or inversions, of segments within chromosome 9 may also be produced, one of the breakage points being at *Ds*.

It was realized early in this study of *Ds* that changes could occur at the locus leading to marked alterations in frequency of the detectable breakage events. The original isolate was showing high frequencies of formation of dicentric chromatids and the associated acentric fragments. Changes arose at the locus, however, as a consequence of some event occurring in a somatic cell. These changes resulted in the appearance, in subsequent cell and plant generations, of lowered frequencies of these events. Such changes in the behavior pattern of *Ds* were called "changes in state"; and the *Ds* with the altered state behaved in inheritance as an allele of the original isolate of *Ds*. A subsequent change could occur, which again was recognized by an altered frequency of detectable breakage events, and which behaved in inheritance as an allele of the initial state, of the derived state, or of other unrelated derived states. By selecting altered states of *Ds*, a series of alleles of the original *Ds* has been isolated. The changes in state of *Ds*, and those occurring at other mutable loci, are of considerable significance in understanding the nature of the events responsible for the patterns of behavior of all mutable loci. A discussion of this significance will be postponed until the behavior of some other mutable loci have been considered. The meaning of the term will then be readily apparent.

Transposition of *Ds*

An important aspect of this study, with regard to the origin of mutable loci and nature of their mutation process, is related to transposition of *Ds* from one location in the chromosome complement to another. The discovery of such transpositions occurred in the course of studies aimed at determining the exact location of *Ds* in chromosome 9. These tests involved linkage relationships. A sequence of six marked loci along the chromosome arm were used, and the linkage studies clearly established the location of *Ds* as shown in Figure 2. This genetically determined location fitted the position of breaks in the chromosome observed in some of the sporocytes of plants having *Ds* in either one or both chromosomes 9. Such chromosome breaks are illustrated in the photographs of microsporocytes at pachytene given in Figures 4 to 8. This was the location of *Ds* when it was first discovered, and has been called the standard location.

In the course of studies of the inheritance behavior of *Ds*, an occasional kernel appeared which showed that *Ds*-type activity—that is, chromosome breakage—was occurring at a new position in the short arm of chromosome 9. Attempts were made to germinate such kernels when they were found. If a plant arose from one, a study was then commenced to determine the new location of the *Ds*-type activity. Over 20 cases of the sudden appearance of *Ds*-type activity in new locations in the short arm of chromosome 9, and several cases of its sudden appearance in other chromosomes of the complement, have been investigated. Within the short arm of chromosome 9, such activity has appeared at various positions. All the isolates studied have shown sharply defined locations of the *Ds*-type activity. In these cases, the cytological determination of breakage position and the genetic determination of location were in agreement. New positions of *Ds*-type activity have appeared between all of the marked loci shown in Figure 2. For example, in four independently arisen cases, the new position of *Ds* has been located between *I* and *Sh*. In two of these, it is to the right of *I*, at or close to the same position in each case—approximately one-fifth the crossover distance between *I* and *Sh*. In the other two it is to the left of *Sh*, with a very low percentage of crossing over between *Ds* and *Sh* in each case.

The mode of detecting new locations of *Ds*-type activity has been selective, in that those arising in the short arm of chromosome 9 are immediately revealed on many of the ears coming from test crosses. *Ds*-type activity has suddenly appeared, however, in other chromosomes of the complement. Only when appropriate genetic markers are present can it be detected readily; and in most tests, such markers have not been present.

Several questions must now be asked. How do new positions of *Ds* activity arise? And what conditions are responsible for their occurrence? The methods used in seeking answers to these questions may be described. In some cases, it could be established that the appearance of *Ds* activity at a new location was associated with its disappearance at the known former location. It has been emphasized that the mechanism under-

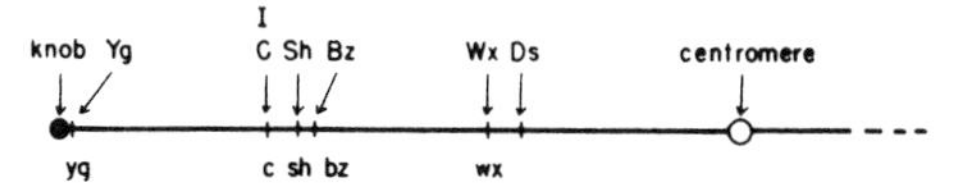

FIG. 2. Diagram showing the approximate locations of the genetic markers in the short arm of chromosome 9 that have been used in this study. In symbolization, dominance is indicated by a capital letter or capitalization of the first letter. Recessiveness is indicated by lower-case letters. The symbols refer to the following plant or endosperm characters: *Yg*, normal chlorophyll; *yg*, yellow-green chlorophyll color in early period of development of the plant. *Sh*, normal endosperm; *sh*, shrunken endosperm. *I*, *C*, and *c* form an allelic series associated with pigment development in the aleurone layer of the endosperm. *I*, inhibitor of aleurone color formation, dominant to *C*. *C*, aleurone color, dominant to *c*, colorless aleurone. The *Bz* factor is associated with development of aleurone and plant color. When homozygous, the recessive, *bz*, (bronze), gives rise to an altered anthocyanin color in the aleurone and plant tissues, from a dark red or purple to a bronze shade. When *Wx* is present, the starch in the pollen and endosperm stains blue with iodine solutions, due to the presence of amylose starch; when only the recessive *wx* (waxy) is present, no amylose starch is formed and with iodine solutions, the starch stains a reddish-brown color. The position of *Ds*, indicated in the diagram, is the standard location (see text).

lying *Ds* events is one that can give rise to translocations, deficiencies, inversions, ring-chromosomes, etc., as well as the more frequently occurring dicentric chromatid formations with reciprocal formation of acentric fragments. It has also been stated that in each such case one breakage point is at the known location of *Ds*. The appearance of *Ds* at new locations is probably associated with such a break-inducing mechanism. This was indicated by extensive analysis of the constitutions of two independent duplications of segments of the short arm of chromosome 9 when a new location of *Ds* activity was also present in this arm. In both cases, only one of the many tested gametes of one of the parent plants carried the particular chromosome aberration with the new location of *Ds*. It was detected in two single aberrant kernels on separate ears coming from similar types of crosses made in two different years. The female parent carried two morphologically normal chromosomes 9, each with the markers *C*, *sh*, *bz*, and *wx*. No *Ds* (or *Ac*, see below) was present in these plants. The male parent (one *Ac* present) carried two morphologically normal chromosomes 9. The markers *I*, *Sh*, *Bz*, *Wx*, and *Ds* (at its standard location) were present in one chromosome 9. The homologous chromosome carried *C*, *sh*, *bz*, *wx*, but no

3

5

3a

a b

5a

a c b

4

4a

c b a

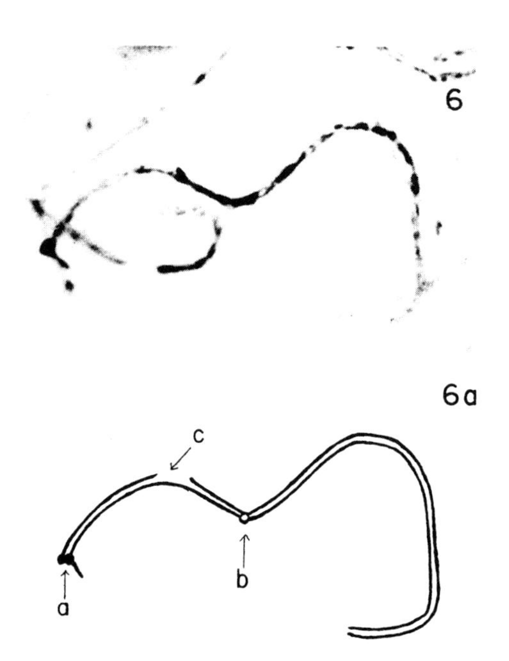

FIG. 3. Photograph of a normal bivalent chromosome 9 at pachytene of meiosis. In the accompanying diagram, 3a, the knob terminating the short arm is indicated by the arrow, *a*. The centromere is indicated by the arrow *b*. Mag. approximately 1800x. Fusion of homologous centromeres appears to occur at pachytene. Consequently, in the diagrams accompanying Figures 3 to 8, this region is indicated as single rather than double.

FIGS. 4 to 7 and accompanying diagrams, 4*a* to 7*a*. Illustrations of the position of breaks at the *Ds* locus as seen at pachytene of meiosis in plants having *Ds* at its standard location in one chromosome 9 and no *Ds* in the homologue. The two homologues are distinguishable. At the end of the short arm of the chromosome 9 having no *Ds*, a segment of deep-staining chromatin extends beyond the knob. The short arm of the chromosome 9 carrying *Ds* terminates in a knob. Magnifications approximately 1800x. In Figure 4, a break at *Ds* occurred in a premeiotic mitosis. The acentric fragment, from *Ds* to the end of the arm, was lost to the nucleus. Consequently, this segment is missing in the bivalent. The homologous segment in the chromosome 9 having no *Ds* is therefore univalent. In making the preparation, this segment was considerably stretched. In the accompanying diagram, arrow *a* points to the knob and the small deep staining segment extending beyond the knob. Arrow *b* points to the centromere region, not clearly shown in the photograph. Arrow *c* points to the position of the break in the chromosome 9 that carried *Ds*. Figures 5 and 6 show the appearance of the bivalent chromosome 9 when a break in the *Ds* carrying chromosome occurred at the meiotic prophase and when the free segment, from *Ds* to the end of the arm, paired with its homologous segment in the chromosome 9 having no *Ds*. In the accompanying diagrams, arrow *a* points to the knobs, arrow *b* points to the centromeres and arrow *c* to the position of the *Ds* break in one of the homologues. Figure 7 is similar to Figures 5 and 6 except that the free fragment, from the position of *Ds* to the end of the arm, did not pair with its homologous segment in the chromosome 9 having no *Ds*. In the accompanying diagram, arrow a^1 points to the knob and the deep-staining chromatin extending beyond the knob in the chromosome having no *Ds*. Arrow a^2 points to the knob of the unpaired acentric fragment. Arrow *b* points to the position of the centromeres, not observable in the photograph. Arrow c^1 points to the broken end of the centric segment, and arrow c^2 points to the broken end of the acentric segment. (For Fig. 7, see next page.)

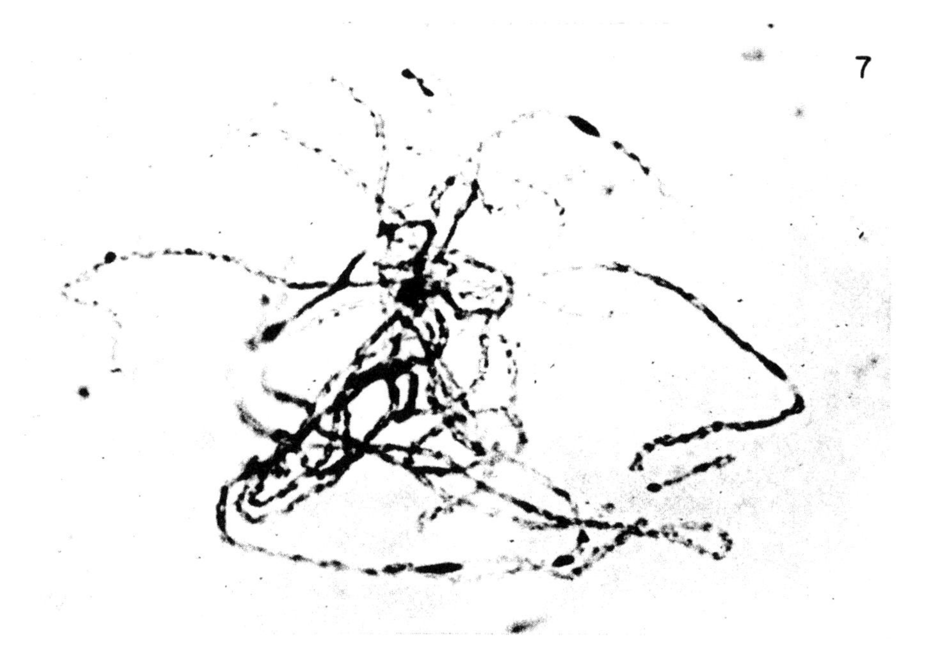
7

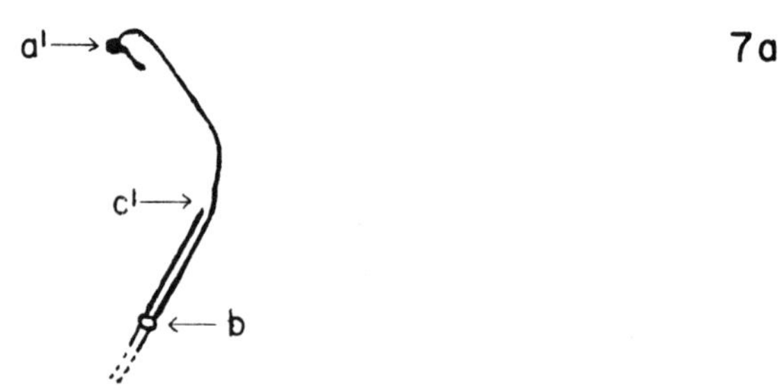
a1
c1
b
7a

a2
c2

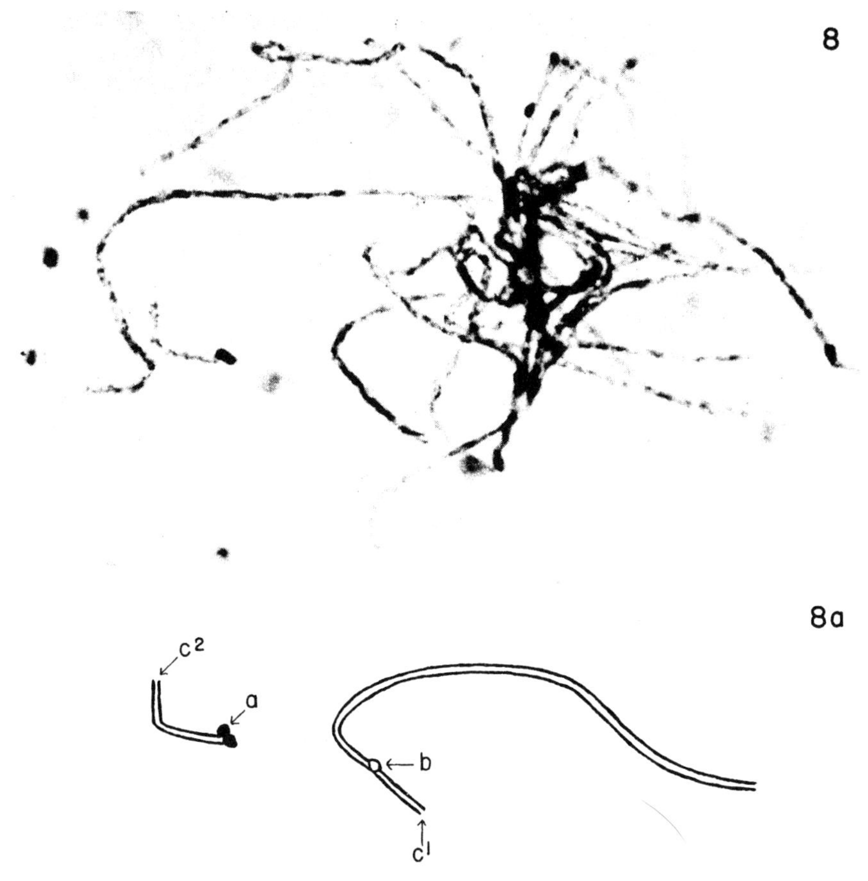

FIG. 8. The chromosome 9 bivalent at pachytene in a plant having *Ds* at its standard location in each chromosome 9. A *Ds* break occurred at the meiotic prophase in each chromosome. Consequently, the bivalent is composed of two free segments: one short acentric segment, from *Ds* to the end of the arm, and a long centric segment, from the position of *Ds* to the end of the long arm. In the accompanying diagram, arrow *a* points to the large knob terminating the short arm of each member of the acentric bivalent segment, and arrow *b* points to the centromeres of the centric bivalent segment. Arrow c^1 points to the broken ends of the centric bivalent segment and arrow c^2 points to the broken ends of the acentric bivalent segment.

Ds. The constitutions of the aberrant chromosomes 9, which were present in an individual pollen grain in each case, are shown in Figure 9. Each has a duplication of a segment of the short arm of chromosome 9. A study of these constitutions reveals that, in each case, the new position of *Ds* activity coincided with the position of one of the breaks that produced the duplication. Also, the position of the second break coincided with the previously known location of *Ds* in the morphologically normal *Ds*-carrying chromosome 9 of the male parent plant. It seems clear from the analysis of both these cases that the breaks were *Ds* initiated, and also that both breaks involved sister chromatids at the same location. Two other cases of newly arisen duplications associated with new positions of *Ds* activity have received study. These have proved to be similar in their modes of origin to the two cases diagrammed. Although these cases suggest that the new positions of *Ds* activity may arise from contacts with the *Ds* that is present in the chromosome complement, they do not constitute evidence that *Ds* is composed of a material substance and that the new positions arise from insertion of this material into new locations.

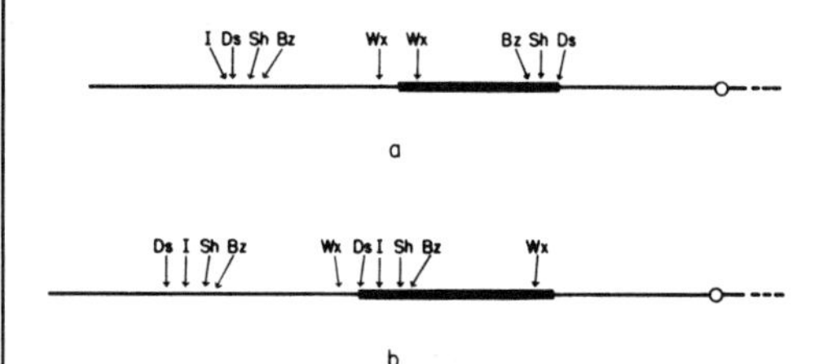

FIG. 9. Diagrams illustrating the constitution of two chromosomes 9, each of which carries a duplicated segment (heavy line). A transposition of *Ds* accompanied the formation of the duplication in each case. For details, see text.

Ac Behavior and Inheritance: General Considerations

Before continuing discussion of the sudden appearance of *Ds* at new locations, it is necessary to consider another heritable factor called Activator (*Ac*). *Ac* is a dominant factor that must be present if any breakage events are to occur at *Ds*. If *Ac* is absent from the nucleus, no detectable events whatsoever occur at *Ds*, wherever it may be located; and no new positions of *Ds* activity appear. Because of this, and because new positions of *Ds* arise only in a few cells, during the period of development of a tissue when breakage events are occurring at *Ds*, it may be concluded that these new positions are one of the consequences of the mechanism that produces chromosome breakage events at *Ds*.

The location of *Ds* in chromosome 9 was first suggested by the altered phenotype of sectors of tissue derived from cells in which a break had occurred at *Ds*. Such sectors will appear if the factorial constitutions of the homologues differ. The chromosome arm carrying *Ds* must also carry dominant markers, and the homologue must carry the recessive alleles. If one chromosome 9, delivered by the male parent, carries *I*, *Sh*, *Bz*, *Wx*, and *Ds* at its standard location, and if the homologue delivered by the female parent carries the recessive alleles *C*, *sh*, *bz*, and *wx*, and has no *Ds*, the events at *Ds* that form a dicentric chromatid and an acentric fragment will lead to elimination of the acentric fragment during a mitotic division. This fragment will form a pycnotic body in the cytoplasm, which subsequently disappears. All the dominant markers carried in this acentric fragment will be removed from the nuclei after such an event has occurred. The sector of tissue derived from these cells will exhibit the collective phenotype of the recessive alleles carried by the homologous chromosome 9 having no *Ds*.

If pollen of plants having one *Ac* factor and carrying *I*, *Sh*, *Bz*, *Wx*, and *Ds* (standard location) is placed upon silks of plants carrying *C*, *sh*, *bz*, and *wx* in their chromosomes 9, but having no *Ds* or *Ac*, two types of kernels will appear, those that received an *Ac* factor and those without *Ac*. The latter kernels will be colorless and non-shrunken, and will show the *Wx* phenotype. No variegation for the characters exhibited by the recessive alleles will appear, as shown in Figure 10. If *Ac* is present in the endosperm, however, the described *Ds* events will occur in some cells during the development of the kernel. This will result in elimination of the segment in the short arm of chromosome 9 carrying the dominant factors. All the cells arising from one in which such an event has occurred will exhibit the *C*, *sh*, *bz*, and *wx* phenotypes. Consequently, these kernels will be variegated. The photographs of kernels in Figures 11 to 13 will illustrate the nature of this variegation.

FIGS. 10 to 15. Photographs of kernels illustrating the effects produced by breaks at *Ds*, and the relation of the presence or absence of *Ac* and the doses of *Ac* on the time of occurrence of *Ds* breaks. The kernels arose from the cross of a plant (♀) carrying *C* and *bz* and having no *Ds* in each chromosome 9, by plants (♂) having *I*, *Bz*, and *Ds* at its standard location in chromosome 9. In Figure 10, no *Ac* is present; the kernel is completely colorless due to the inhibition of aleurone color when *I* is present. In Figures 11 to 13, one *Ac* is present. Breaks at *Ds* occur early in development with consequent loss of *I* and *Bz* to the nuclei. The cells without *I* and *Bz* give rise to sectors showing the *C bz* phenotype. It should be noted that *Bz* substance diffuses through several cell layers, from the *I Bz* areas into the *C bz* areas, producing a *C Bz* phenotypic expression in these genotypically *C bz* cells. Thus, all large areas of the *C bz* genotype are rimmed with the dark aleurone color of the *C Bz* phenotype. Small *C bz* areas may be mostly *C Bz* in phenotype because of this diffusion, and very small *C bz* areas are totally *C Bz* in phenotype. These relationships are clearly expressed in the photographs. In Figures 11 to 13, both large and small *C bz* sectors are present as well as areas in which no *C bz* sectors appear. This irregularity arises from alterations that occur to *Ac* during the development of the kernels. These give rise to sectors with no *Ac* or with altered doses of *Ac* (see text). Figure 14 shows the pattern that may be produced by *Ds* breaks when two *Ac* factors are introduced into the primary endosperm nucleus. There are many small sectors of the *C bz* genotype, all produced by relatively late occurring *Ds* breaks. This results in a heavily speckled variegation pattern. Figure 15 shows the pattern that may appear when three *Ac* factors are present. Only relatively few *C bz* specks, produced by late occurring *Ds* breaks, are present in this kernel.

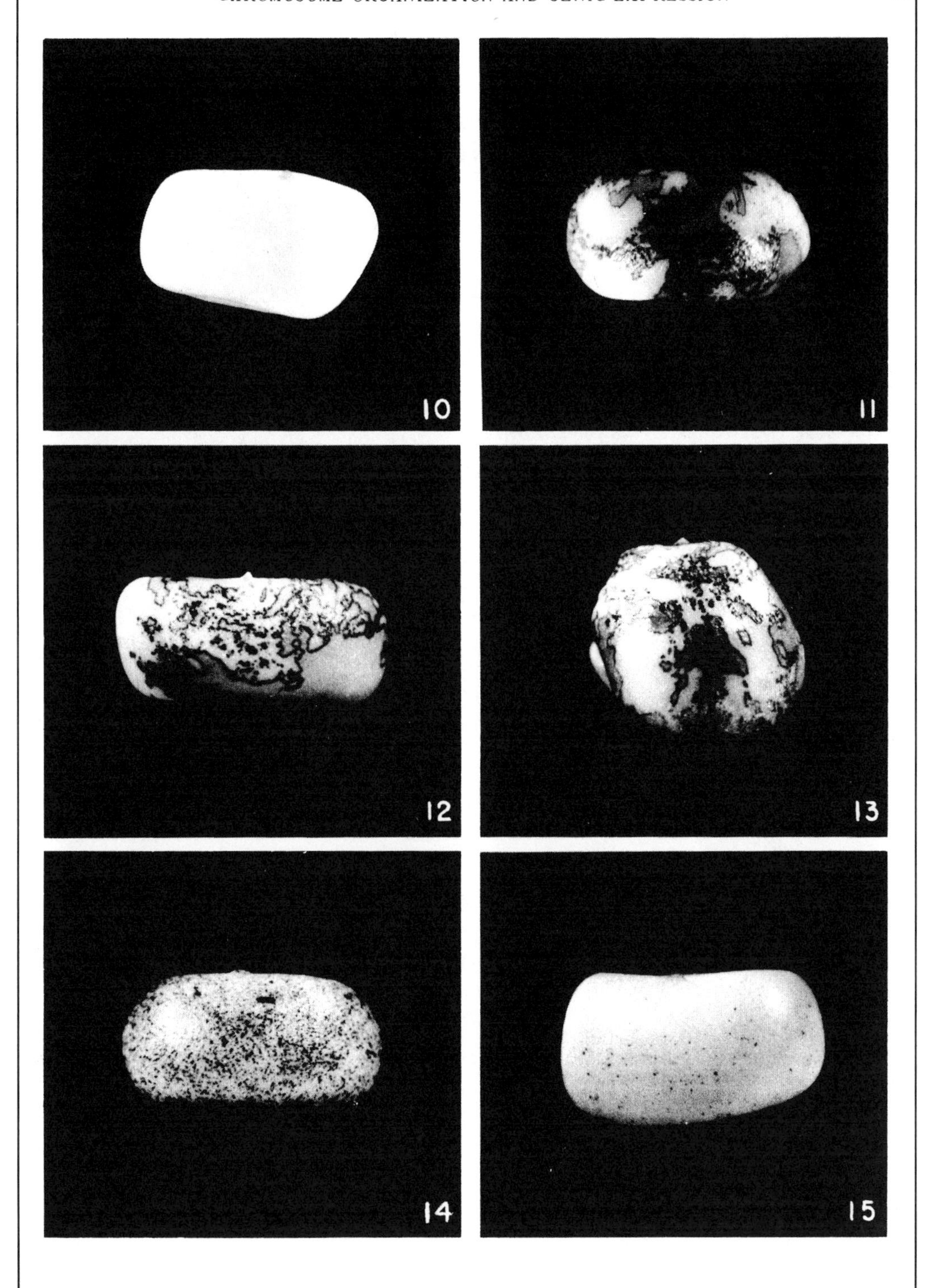
10
11
12
13
14
15

In the early studies, it soon became apparent that the time of occurrence of *Ds* breakage events in the development of a tissue depends upon the dose of *Ac* present. The higher the dose of *Ac*, the later in development will such events at *Ds* occur. Because the endosperm is triploid—the female parent contributing two haploid nuclei derived from the female gametophyte, and the male contributing one haploid nucleus—single to triple doses of *Ac* may readily be obtained after appropriate crosses. The time of response of *Ds* to double and triple doses of *Ac* is shown in Figures 14 and 15.

Initial studies of the inheritance behavior of *Ac* showed that it follows the mendelian laws known to apply to single genetic units. An illustration of this inheritance behavior is given in Figure 16. The ear in this figure was derived from a cross in which the female parent carried the recessive factor *c* in each chromosome 9, but had no *Ds* or *Ac* factors, and the male parent carried in each chromosome 9 the factors *C* and *Ds* (standard location). It also carried a single *Ac* factor. The expected 1-to-1 ratio of colored, nonvariegated kernels (no *Ac* present) to kernels showing sectors of the *c* phenotype (*Ac* present) was apparent on this ear. Efforts to determine the location of *Ac* in the chromosome complement were commenced, but were soon abandoned when it was realized that *Ac* need not remain at any one location in the complement. It can appear at new locations and in different chromosomes. Because of this highly unexpected behavior of a genetic factor, extensive studies were made to determine the mode of inheritance of *Ac* and to learn how these new positions arise. Much has been learned from them about *Ac* behavior and inheritance. The modes of investigating alterations of *Ac* will be considered later. Some of the facts concerning its behavior, however, may be stated here in summary form, by describing the results obtained from several types of experiments.

If plants having one *Ac* factor are self-pollinated, the expected mendelian ratios may appear in the F_2 populations. These are: one with two *Ac*, to two with one *Ac*, to one with no *Ac*. The ratios obtained in one such test were 61 : 145 : 68, which is close to the statistical expectancy. When, however, these F_2 plants having two *Ac* factors are crossed by plants having no *Ac*, the expectation would be that all the progeny would have one *Ac* factor. Usually, most of the plants do have one *Ac* factor; but sometimes there are plants displaying other conditions with respect

FIG. 16. Photograph of an ear produced when pollen from plants having *C* and *Ds* (standard location) in each chromosome 9 and carrying one *Ac* factor is placed on silks of plants carrying *c* in each chromosome 9 and having no *Ds* or *Ac* factors. Approximately half of the male gametes have no *Ac* factor. The kernels arising from the functioning of such gametes are fully colored; no variegation appears. The other half of the male gametes carry *Ac*. The kernels arising from the functioning of these gametes are variegated. They show a number of sectors with the *c* phenotype. Note the approximate 1 to 1 ratio of fully colored kernels to variegated kernels.

to *Ac*. The following unexpected types have appeared: plants having no *Ac* factor; others having two nonlinked *Ac* factors; others having two *Ac* factors that appear to be linked; still others having an *Ac* factor that acts as a single unit in inheritance but gives the same dosage action as would two doses of the *Ac* factor in the parent plant that contributed *Ac*. The dosage action of *Ac* may be altered in other ways, moreover, so that a single *Ac* factor, as determined genetically, may exert an action either less than that of the *Ac* factor contributed by the *Ac*-carrying parent plant or falling between one and two doses of this factor.

In the early studies of *Ac* inheritance, the *Ac* factor was found not to be linked with the genetic markers in the short arm of chromosome 9. In one series of tests of *Ac* inheritance, where a number of *Ac*-carrying plants were derived from a cross between a plant having no *Ac* and a plant having one *Ac*, a single aberrant plant was present in the F_1. Tests of this plant showed that it possessed a single *Ac* factor, which was obviously linked to the markers in the chromosome 9 delivered by the *Ac*-carrying parent plant. In the sister plants, and in the parent plant that contributed *Ac*, no such linkage was evident. Studies were then conducted to examine the linkage relationships of this *Ac* with the marked loci. The very same types of tests for linkage of *Ac* with the genetic markers in chromosome 9 were also conducted with the sister plants. A summary of one set of these comparative tests of *Ac* inheritance is presented in Table 1. In part I of this table, no linkage of *Ac* with *I* or *Wx* is evident; in part II, however, linkage is obvious. The data place *Ac* to the right of *Wx*. Approximately 20 per cent crossing-over appears to have occurred between *Wx* and *Ac*. Actually, this figure is only approximate, for a few kernels in the *Ac-wx* class do not carry *Ac* in chromosome 9. They do not belong in the crossover class. The *Ac* in these kernels had been transposed from chromosome 9 to another chromosome. Also, a few kernels in the *Wx*-no *Ac* class have not lost *Ac* because of crossing over but rather because it was removed from its location in chromosome 9.

The described case of sudden appearance of *Ac* in chromosome 9 is not the only one that has been found and similarly studied. Seven independent cases have so far been identified. In two of these, *Ac* appeared in the short arm of this chromosome: in the first case, a short distance to the left of *Wx*; and in the second case, a short distance to the left of *I*. Studies were then undertaken to investigate not only the linkage behavior of *Ac* but also the types of events that alter *Ac* in these two sharply delimited locations. It was determined that in the majority of the *Ac*-carrying gametes produced by plants having an *Ac* at either of these two stated positions, no change in location or action of *Ac* occurred. Exchanges of *Ac* from one homologue to another took place as a consequence of crossing over, with consistent frequencies in each case. In a few of these gametes, however, the above-described types of aberrant *Ac* conditions were present, as follows: (1) *Ac* was no longer present in chromosome 9, but was carried instead by another chromosome. (2) Two *Ac* factors were present, one at the given location in chromosome 9, and one carried by another chromosome of the complement. (3) The *Ac* factor was unchanged in its location but showed an altered action; in a single dose it could be equivalent to the double dose before the alteration occurred, or could show an increased but not doubled dosage action, or a decreased dosage action.

The behavior of *Ac* was also studied in plants in which the *Ac* factor was present at an allelic position in each chromosome 9. Again, it could be determined that in a few gametes produced by such plants the above-described aberrant conditions with respect to *Ac* position and action were present. In addition, a few gametes were formed that had no *Ac* factor at all.

From what has been said about both *Ds* and *Ac*, it is apparent that with respect to inheritance

TABLE 1

Comparisons of *Ac* inheritance: (I) when *Ac* was not linked to markers in chromosome 9, and (II) when *Ac* was linked to markers in chromosome 9, in crosses of

♀ *C sh bz wx/C sh bz wx*, no *Ds*, no *Ac* by (I) ♂ $\frac{I\ Sh\ Bz\ Wx\ Ds}{C\ Sh\ Bz\ wx\ Ds}$ 1 *Ac*, nonlinked and (II) ♂ $\frac{I\ Sh\ Bz\ Wx\ Ds\ Ac}{C\ Sh\ Bz\ wx\ Ds\ ac}$

Chromosome-9 Constitution of ♂ Gamete	I: Variegated* *Ac* present	I: Nonvariegated* No *Ac*	II: Variegated* *Ac* present	II: Nonvariegated* No *Ac*
I Wx	268	255	928	246
C wx	248	242	164	893
I wx	88	91	62	387
C Wx	84	83	344	100

*Presence of *Ac* detected by sectors of the *C sh bz wx* phenotype in kernels on ears obtained from cross I or cross II. In the absence of *Ac*, no *Ds* events occur; the kernels are therefore nonvariegated.

behavior they are much alike. The same questions concerning mode of appearance in new locations in the chromosome complement apply to *Ac* as to *Ds*. With this relationship in mind, we may now return to further considerations of *Ds*. It should be emphasized again, however, that the described events occurring at *Ds*, wherever it may be located in the chromosome complement, depend upon the presence of an *Ac* factor in the nucleus, regardless of where this latter factor is located in the chromosome complement. New positions of *Ds* activity arise only when *Ac* is also present in the nucleus, and, again, regardless of where *Ac* may be located. In addition, any one altered state of *Ac*—for example, an altered dosage action—affects *Ds* wherever it may be located, and in exactly the same manner for every *Ds*, regardless of its state. In other words, it is the state and the dose of *Ac* that control just when and where *Ds* events will occur, and it is the state of a particular *Ds* that controls the relative frequency of any one type of event that occurs at *Ds*.

The Origin and Behavior of c^{m-1}

In the discussion of the appearance of *Ds* at new locations, the question was raised whether or not this involves the transposition of a material substance from one location to another. The question applies equally to *Ac*. If no material substance is transposed, a serious problem is presented regarding the basic action of any known genetic unit or factor that has been assigned to a particular locus in a particular chromosome. *Ac* clearly produces an obvious, measurable, phenotypic response, wherever it may be located. It shows dosage action, mendelian inheritance, and linkage behavior of the expected type, in any location—with the exception, already mentioned, of a few transpositions, changes in state, and losses of *Ac*. It might be considered that *Ds* and *Ac* represent forms of altered chromosome organization producing somewhat similar effects in each case, much like the Minutes in *Drosophila*. The evidence now to be presented, however, makes that assumption unlikely. This evidence considers the origin and the behavior of many different mutable loci. To begin this part of the discussion, we may consider the origin of mutable c^{m-1}, the first-detected mutable *c* locus that arose in a chromosome carrying a normal-behaving *C* locus (*C*, aleurone color; *c*, recessive allele, colorless aleurone).

The presence of an alteration at the known locus of *C*, which produced c^{m-1}, was detected probably within a few nuclear generations after it occurred. It was present only in one kernel among approximately 4,000 examined that had come from a cross of a single plant, used as a male, to 12 genetically similar female plants. All the other kernels on these ears gave the expected types of phenotypic expression. The male parent carried in both chromosomes 9 the genetic markers *Yg*, *C*, *Sh*, *wx*, and *Ds* (standard location). It also had one *Ac* factor, not linked to these markers. The female parents carried the stable recessive *yg*, *c*, *sh*, and *wx*. No *Ds* was present in their chromosomes 9, and also no *Ac* factor was present in these plants. The types of kernels to be expected from such a cross, and their relative frequencies, are the same as those shown in Figure 16. Approximately half the kernels should show the *C Sh wx* phenotype, with no variegation for the recessive characters since no *Ac* factor is present. The other half should carry *Ac* and thus be variegated. Sectors showing the *c sh* phenotype should be present. In these crosses, the expected classes of kernels appeared with the exception of one kernel. Instead of showing colorless areas in a colored background (resulting from losses of *C* following breakage events at *Ds*), this exceptional kernel showed a colorless background in which colored areas were present. The plant derived from this kernel was tested in various ways in order to determine the reason for this unexpected type of variegation. The early tests indicated, and subsequent tests proved, that mutations were occurring from the recessive *c*, to the dominant *C*, and that the mutable condition had arisen in one of the chromosomes 9 contributed by the *Ac*-carrying male parent plant.

Ds-type activity was also present in the chromosome carrying the new mutable *c* locus. The location of this activity was no longer to the right of *wx*, as would be expected since this was its location in both chromosomes 9 of the male parent plant. The new location was inseparable from that of the mutable *c*. All the recognizable breakage events associated with *Ds*-type activity now happened at this new location. In addition, mutations to *C* occurred. It was soon discovered that the mutations to *C* would appear only if *Ac* were also present in the nucleus, and that the time of occurrence of these mutations was controlled by the state and dose of *Ac* in precisely the same way that the state and dose of *Ac* controls *Ds* breakage events wherever *Ds* may be located. If *Ac* were absent, neither mutations to *C* nor *Ds*-type breakage events would occur. Thus when *Ac* is absent, the behavior of c^{m-1}

is equivalent to the previously known recessive *c*. If, however, *Ac* is again introduced into the nucleus by appropriate crosses, the potential mutability of this recessive is realized, for then mutations to *C* occur. The previously known recessive *c*, used for many years in genetic studies, is unaffected by the presence of *Ac*; and it remains stable, nonmutable, when present in nuclei in which $c^{m\text{-}1}$ is also present and undergoing mutations.

In considering the mode of origin of $c^{m\text{-}1}$ and its behavior, the following points may be reviewed: (1) the appearance of a new recessive that is mutable; (2) its derivation from a normal dominant *C*, which is nonmutable; (3) its appearance in a single gamete of a plant carrying *Ds* and *Ac;* (4) *Ac* control of the new mutable condition in exactly the same manner as *Ds* is *Ac*-controlled; (5) *Ds*-type chromosome breakage events also occurring at this mutable locus; and (6) disappearance of *Ds* from its former location in the same chromosome that carries the new mutable locus. This series of coincidences is striking enough in itself to command consideration of the possibility that this mutable recessive originated by transposition of *Ds* to the locus of the normal *C* factor. It is immediately apparent that, if this is true, the transposition of *Ds* from its former location to the new location created a condition that affects the formation of pigment in the aleurone layer; for no pigment is formed until some event occurs at this locus, and only when *Ac* is present. Previous tests have shown that the same *c* phenotypic expression can also arise if the tissues of the endosperm are homozygous deficient for the segment of chromatin carrying *C* (McClintock, unpublished). This might suggest that the presence of *Ds* has inhibited the normal action of the chromatin materials at the *C* locus. A final and most significant argument for the origin of $c^{m\text{-}1}$ by a transposition of *Ds* to the *C* locus is derived from the fact that a mutation to *C* is associated with the loss of any further recognizable *Ds* events at the immediate location of *C*. It is apparent, therefore, that the mutation-producing event is associated with one involving *Ds* at this locus. All the evidence is consistent with the assumption that $c^{m\text{-}1}$ arose by transposition of *Ds* to the locus of *C*, thereby inhibiting its action, and that removal of *Ds* is associated with removal of this inhibitory effect. The restored activity at the *C* locus is permanent, and its subsequent behavior resembles that present before *Ds* activity appeared at the locus. *Ac* no longer has any effect on its action and behavior, just as it had no effect before *Ds* appeared at the *C* locus to give rise to $c^{m\text{-}1}$.

Additional *Ds*-Initiated Mutable Loci

Simple coincidence rather than a relationship with *Ds* might still be claimed for the origin and behavior of $c^{m\text{-}1}$ if it were the only case of such origin and behavior. Two other cases, similar to $c^{m\text{-}1}$ and involving another marked locus, have appeared independently in the *Ds*- and *Ac*-carrying plants. Both involve the locus in chromosome 9 associated with the bronze phenotype (see Fig. 2). These two independent cases have been designated $bz^{m\text{-}1}$ and $bz^{m\text{-}2}$. The description of the types of events occurring at $c^{m\text{-}1}$ may be applied also to $bz^{m\text{-}1}$ and $bz^{m\text{-}2}$. *Ds*-type breakage events occur at the mutable locus, as well as mutations from *bz* to an apparently full *Bz* expression. Both the *Ds*-type breakage events and the mutations to *Bz* are *Ac*-controlled; for if *Ac* is absent, neither will occur. In these cases also, the time when mutations to *Bz* or chromosome breaks occur is under the control of the state and the dose of *Ac*. Again, as with *C*, previous investigations had shown that a homozygous deficiency of the segment of chromosome including the *Bz* locus will reproduce the known recessive phenotype, *bz* (McClintock, unpublished).

That the presence of *Ds* close to a marked locus in a chromosome may result in frequent changes in the phenotypic expression of the marker has been indicated. Two independent cases of transposition of *Ds* from its standard location to a position near and to the left of *Sh* have been studied (see Fig. 2). In these two similar cases, less than one-half of one per cent crossing over occurs between *Ds* and *Sh*. In both cases, however, many gametes are produced that carry a "spontaneous" mutation from *Sh* to *sh*. These mutations occur only in those chromosomes carrying *Ds* immediately to the left of *Sh*, and only in plants that also have *Ac*. If *Ac* is absent, no such mutations appear. *Ac*-controlled events, therefore, in each of these two cases where *Ds* is near and to the left of *Sh*, are responsible for this high frequency of mutation to *sh*. When such a mutation occurs, *Ds* is not always lost to the chromatid; it is sometimes still present between *C* and *Bz*. Some of these "mutations" may prove to be newly arisen mutable *sh* loci, but the tests for this mutability have not been concluded.

Knowledge gained from the cases reviewed has led to the conclusion that the appearance of *Ds* at or close to the locus of a known genetic factor can give rise to frequent changes in the action of the factor. The initial change is to an action resembling that of the known recessive allele. In the cases of $c^{m\text{-}1}$, $bz^{m\text{-}1}$, and $bz^{m\text{-}2}$, a subsequent alteration produces a return to apparently full dominant expression of the factor. This common type of mutational expression in these cases is of considerable significance. Its importance will become evident in the discussion of the behavior of other newly arisen mutable loci.

Origin and Behavior of One Class of Autonomous Mutable Loci

The behavior of some other types of mutable loci may now be considered for the additional and important knowledge they have contributed to an understanding of the basic processes involved. One of them is called mutable luteus. The luteus character is distinguished by a yellowish chlorophyll expression. This mutable luteus first appeared in the progeny derived from self-pollination of one of the original plants that had undergone the breakage-fusion-bridge cycle in early development. It resulted from some alteration at a normal locus, but the position of the locus in the chromosome complement was not known. This mutable locus is characterized, first, by its autonomous behavior. It required no recognizable, separate activator factor in order to undergo the mutation phenomenon. The mutations are registered in sectors of a plant as changes in the amount of chlorophyll that is produced. Alleles arise from germinal mutations. They are characterized by various quantitative grades of chlorophyll expression. These alleles, in turn, need not be stable; some of them may mutate to give higher or lower levels of chlorophyll expression. Even an allele apparently producing the full dominant expression may be unstable for it may mutate to or towards the lowest expression, which is luteus.

In studying one aspect of the behavior of this mutable luteus locus, a number of sister plants in one culture were all self-pollinated. On a resulting ear of one, and only one, of these plants, the presence of a new mutable locus was revealed. The mutability, registered in some of the kernels on this ear, involved a factor associated with the formation of pigment in the aleurone layer. Colorless kernels were present in which mutations to color occurred. None of the ears produced by the sister plants showed the presence of such a mutable factor; all kernels on these ears had the full aleurone color. Further study of the plants derived from the variegated kernels on the aberrant ear showed that a mutable condition had arisen at a previously known locus, the A_2 locus in chromosome 5 (A_2, aleurone and plant color; a_2, recessive allele colorless aleurone, altered plant color). Before the appearance of the mutable condition in this one plant, the A_2 locus had given the normal dominant expression in both parent plants. It had shown no indication whatsoever of any instability. One parent had contributed mutable luteus to this culture. The mutable luteus locus, however, was not linked with the A_2 locus. It should be emphasized that this newly arisen mutable a_2 behaved in many respects like mutable luteus. It was autonomous; and quantitative alleles were produced, some of which, in turn, were mutable.

Another mutable locus arose in a culture derived from one having mutable luteus. It first appeared as a single aberrant kernel on one ear. This ear was produced from a cross in which a plant carrying mutable luteus was used as a female parent. The male parent was homozygous for the stable recessive a_1. (A_1, aleurone and plant color, located in chromosome 3; a_1, recessive allele, colorless aleurone and altered anthocyanin pigments in the plant.) This single aberrant kernel exhibited variegation. Sectors of colored aleurone appeared in an otherwise colorless kernel. Tests were initiated with the plant derived from this kernel, and continued with the subsequent progeny. From these tests it was learned that the aberrant kernel carried a newly arisen mutable a_1 locus, designated $a_1^{m\text{-}1}$, whose general behavior resembled that shown by mutable luteus. It was autonomous; it produced a series of alleles showing various grades of quantitative expression of anthocyanin, in both the aleurone and the plant; and of these alleles, in turn, some were unstable, mutating to give higher or lower levels of quantitative expression of the anthocyanin pigments in both aleurone and plant. Other important aspects of the mutation phenomenon at this locus will be considered later.

In the discussion of the *Ac*-controlled mutable loci, $c^{m\text{-}1}$, $bz^{m\text{-}1}$ and $bz^{m\text{-}2}$ that arose in *Ds*-*Ac* carrying plants, it was emphasized that the types of mutational response were similar. Here also, the mutational expressions of the two mutable loci that have arisen in the mutable luteus stocks are much alike, and they resemble that shown by mutable luteus itself. One further example of related mutable loci will be given. It also shows the similarity of behavior of the newly arisen mutable locus to the one already present in the plant. The direction this discussion is taking may now be apparent. It is towards the conclusion that the type of mutation occurring at a locus is a function of the type of chromatin material that is present at the locus or is transposed to it, and does not involve changes in the components of the genes themselves. Rather, it is this chromatin that functions to control how the genic material may operate in the nuclear system. With this in mind, a third example of related origins and behaviors of mutable loci may now be considered.

ORIGIN AND BEHAVIOR OF $c^{m\text{-}2}$ AND $wx^{m\text{-}1}$: TWO RELATED MUTABLE LOCI

The progeny derived from self-pollination of another one of the original plants that had undergone the breakage-fusion-bridge cycle in early development was grown, and a number of these plants again self-pollinated. On a resulting ear of one of these plants, a new mutable locus was recognized. The factor involved was again associated with the production of pigment in the aleurone layer. Some of the kernels on this ear showed colored areas in a colorless background. Beginning with the plants derived from these kernels, a study was made of the condition responsible for the variegation. This proved to be due to another new mutable locus, and involved the previously discussed *C* locus in chromosome 9. This locus in the parent plant and in the sister plants of the culture, gave the normal dominant *C* expression. The new mutable condition was designated $c^{m\text{-}2}$, because it was the second case that appeared in this study. The types of mutation that arise from events at $c^{m\text{-}2}$ are strikingly different from those shown by $c^{m\text{-}1}$. A series of alleles, as expressed by quantitative grades of pigment formation associated with the production of at least two different precursor-type diffusible substances, is produced by mutations at $c^{m\text{-}2}$. The intermediate alleles are not always stable, for some of them, in turn, can mutate to alleles showing higher or lower grades of color expression.

In the course of the study of $c^{m\text{-}2}$, a number of crosses were made, using pollen of plants homozygous for $c^{m\text{-}2}$ and carrying a normal dominant *Wx* factor in each chromosome 9, on silks of plants carrying a stable recessive *c* and a stable recessive *wx* in both chromosomes 9 (see Fig. 2). A single aberrant kernel appeared on one of the ears resulting from this type of cross. It showed mutations to *C* and of the $c^{m\text{-}2}$ type, and, in addition, mutations from the *wx* to and towards the *Wx* phenotype. A plant was obtained from the kernel and a study commenced to determine the nature of this instability expression. It proved to be a new mutable *wx*, and was designated $wx^{m\text{-}1}$. The tests showed that it had arisen in the male parent plant, which carried $c^{m\text{-}2}$ and normal dominant *Wx* in each chromosome 9. It was present, however, in only one of the many tested male gametes of this plant. The pattern of mutational behavior of $wx^{m\text{-}1}$ strikingly resembled that shown by $c^{m\text{-}2}$. A series of quantitative alleles was produced by mutations of $wx^{m\text{-}1}$ as registered by the amount of amylose starch produced. These alleles, in turn, could mutate to give greater or lesser amounts of amylose starch. Another endosperm character also was affected by some of the mutations of $wx^{m\text{-}1}$. This was expressed by an altered growth of the endosperm tissue, and accompanied some but not all of the mutations to the intermediate alleles, appearing particularly often in association with a mutation to one of the lower alleles. This accompanying mutation behaved as a dominant or a semidominant.

It is of particular significance, in comparison of the behavior of the series of mutable loci $c^{m\text{-}1}$, $bz^{m\text{-}1}$, and $bz^{m\text{-}2}$ with that of the series $c^{m\text{-}2}$ and $wx^{m\text{-}1}$, that the members of both series are controlled by the very same *Ac* factor—wherever it may be located—and in precisely the same manner, with respect to time and place of occurrence of mutations. When *Ac* is absent, no mutations occur in either series. In the latter series, mutations of the intermediate alleles also occur, but only when *Ac* is present. Because of this, it has been possible to isolate a series of quantitative alleles of *C*, and also a series of quantitative alleles of *Wx*, that are stable. Stability is maintained when *Ac* is removed. The percentages of amylose starch in the endosperm, produced by a number of the alleles arising from mutations at $wx^{m\text{-}1}$ and freed of *Ac*, have been determined in terms of single, double, and triple doses of the particular allele. (Note: The writer is grateful to Dr. G. F. Sprague and to Dr. B. Brimhall, of Iowa State College, and also to Dr. C. O. Beckmann, of Columbia University, and Dr. R. Sager, of the Rockefeller Institute for Medical Research, for their chemical analyses of the amylose content produced by some of these mutants.) The preliminary tests suggest that alleles, falling into an almost continuous series with respect to the quantity of amylose starch they produce, may be obtained. It should be mentioned here that, as with *C* and *Bz*, previous investigation had shown that a tissue homozygous-deficient for the *Wx* locus will give the known recessive *wx* expression (McClintock, unpublished).

In order to compare the action of *Ac* on the members of these two series of *Ac* controlled mutable loci, crosses were made combining several of them in a single plant so that they might be present together in the nuclei of a tissue. By this means, it was possible to determine that the mutations at these various mutable loci arise as a function of the state and dose of *Ac*, irrespective of which mutable locus is involved or how many such loci from the same series or from the two different series are present in the nuclei of an individual plant.

In further comparison of the behavior of the different *Ac*-controlled mutable loci, one very significant correlation may now be given. It is known that all these loci show one other common characteristic. At all of them, some chromosome-break-inducing events occur, but only when *Ac* is also present in the nucleus and only at those times in the development of a tissue where mutations leading to changes in expression of the respective phenotypic character are also occurring. Again, it has been determined that such breaks may occur at the locus of *Ac* itself. The conclusion that the mutation-producing events in these two series of related mutable loci, and also at *Ac* itself, are associated with such a chromosome-break-inducing mechanism is difficult to avoid. This relationship will be explored after consideration has been given to a comparison of the types of mutability that may arise at any one known locus in a chromosome.

The descriptions of mutable loci given so far in this discussion have shown that the type of mutability and the mode of its control are not alike for all. Nevertheless, there appear to be classes of mutable loci, the members of which show similar types of changes in phenotypic expression—that is, of mutations—and similar types of control of these mutations. It is now necessary to indicate the extent to which various types of mutability expression may arise at any one particular locus. For this purpose the *Wx*, the *C*, and the A_1 loci will be chosen as examples.

COMPARISONS OF TYPES OF MUTABILITY ARISING AT ANY ONE LOCUS

a. *The Wx locus*

Six independent mutable conditions are known for the *Wx* locus. Five of them have arisen during the present study. One, $wx^{m\text{-}1}$, has been considered above. It is *Ac*-controlled, and produces quantitative alleles that may be unstable when *Ac* is present but are stable when *Ac* is absent from the nucleus. Mutations at this locus also give rise to an endosperm-growth-altering factor that is dominant in expression. The second mutable *wx*, $wx^{m\text{-}2}$, arose from a previously stable recessive *wx* carried in genetic stocks for many years. This mutable condition first appeared in a chromosome 9 in which a complex chromosomal rearrangement was present. Its mutations are expressed by different quantitative grades of the *Wx* phenotype. In the endosperm tissues, the sector produced by a cell in which a mutation has occurred is always markedly distorted in growth-type. The third mutable condition at this locus, $wx^{m\text{-}3}$, originated in a plant carrying a normal dominant *Wx* and also several mutable loci. It is autonomous, in that no separate activator factor is required for mutations to be expressed. It almost always mutates to give the full *Wx* phenotypic expression. The derived mutant giving the dominant expression is also mutable, for it produces mutants giving the full recessive expression, that is, *wx*. This recessive, in turn, may mutate again to give the full dominant expression. No altered growth conditions in the endosperm tissue accompany any of these mutations. The fourth case was recognized by a sudden change in the behavior of a previously normal *Wx* locus. It shows mutations producing various grades of quantitative expression between the full dominant and the full recessive. It is autonomous, and no alterations in growth conditions accompany the mutations. Another case, somewhat similar to the last, has recently arisen. It produces alleles giving various lowered expressions of the *Wx* phenotype, and appears to be autonomous although the information on its behavior is too incomplete to allow a full description. The sixth case is one that has been investigated by Sager (1951). It is autonomous, and gives quantitative grades of expression in the endosperm; but the germinal mutations that have been studied all give full or nearly full *Wx* expression, and the mutants are stable. No altered growth conditions appear to accompany these mutations.

Genetic analyses have indicated that all these various mutational changes occur at this one locus in chromosome 9, and yet all show a different kind of mutational behavior. It is evident that each arose in association with a particular type of alteration at the locus, and that different mutation-controlling mechanisms can be involved.

This is especially well illustrated by a comparison of the types of mutations produced by wx^{m-1} and wx^{m-3} and of their controlling mechanisms.

b. The C locus

The contrasts in the kinds of phenotypic expression produced by mutations at c^{m-1} and c^{m-2} have been discussed above. Although several other independent expressions of mutability at this *C* locus have also arisen, the study of them is too incomplete to allow detailed comparisons to be made. With respect to this locus in chromosome 9 it is necessary to mention, however, that in the cultures having mutable loci a heritable factor carried by a chromosome other than 9 has appeared on several occasions. The presence of such a factor results in the production of pigment in the aleurone when the endosperm is homozygous for the well-investigated stable recessive *c*, used for many years in genetic investigations. The pattern of pigment formation differs markedly from that produced by mutations at c^{m-1} or c^{m-2}. It resembles that associated with the factor *Bh* (Blotch), previously studied by R. A. Emerson (1921) and Rhoades (1945b). To complete the discussion of the series at this locus, it may be mentioned that a mutable condition has also arisen involving the expression of *I*, an allele of *C*. Changes in the degree of inhibitory action of *I* occur as a consequence of such mutations.

c. The A_1 locus

A study of changes at the locus of A_1 has contributed some very important information regarding the origin and behavior of mutable loci. For a number of years, a type of control of mutability of the recessive, a_1, has been known. A dominant factor, called Dotted (*Dt*), provokes mutability at the a_1 locus (Rhoades, 1936, 1938, 1941, 1945*a*). In many respects, *Dt* is comparable to *Ac*. It is an activator, for it produces mutations at a_1, just as *Ac* produces mutations at c^{m-1}, c^{m-2}, bz^{m-1}, bz^{m-2}, and wx^{m-1}. Moreover, when *Dt* is absent no mutations occur at a_1; the a_1 locus then gives a stable recessive phenotype. In the presence of *Dt*, mutations occur at a_1 to give mainly the higher alleles of the A_1 phenotypic expression. The time of occurrence of visible mutations at a_1 is usually late in the development of the plant or the endosperm tissues; and they occur in only some of the cells. This results in the presence of dots of the A_1 phenotype. The *Dt* factor has been located by Rhoades in the knob region terminating the short arm of chromosome 9. *Dt* and *Ac* appear not to be the same activator, as plants and endosperm tissues that are homozygous for the recessive a_1 have not shown the dotted-type mutations to A_1 in the presence of *Ac*.

In the early period of this investigation of newly arisen mutable loci, the unexpected appearance of modifications in the knobs, and in the other chromosome elements previously mentioned, of plants that had undergone the breakage-fusion-bridge cycle in their early development, suggested that disturbances in these elements might have been responsible for the initial burst of mutable loci, including the origin of *Ac* itself. It was suspected, therefore, that this cycle might induce alterations in the heterochromatic elements that could initiate a *Dt* factor as they may have originated the *Ac* factor. Once initiated, this factor would activate a_1 to undergo alterations in somatic cells leading to A_1-type expression. The most direct way to induce changes in the heterochromatic elements was considered to be the breakage-fusion-bridge cycle itself. By subjecting tissues to this cycle during their developmental periods, and then examining the matured tissue, this hypothesis could be tested. A preliminary experiment, designed to test for production of mutations of a_1 to A_1 by the breakage-fusion-bridge cycle as an inductor, was performed in 1946. The experiment was repeated in 1950 on a much larger scale. Because this experiment was of particular significance in revealing the mode of origin of mutable conditions, and because it provided evidence about the relation of chromosome organization to genic expression and its control, the details will be given.

The silks of plants homozygous for a_1 and carrying no *Dt* factor received pollen from plants of similar constitution with respect to a_1 and *Dt*. The pollen parents carried one chromosome 9 with a duplicated segment of the short arm. The homologous chromosome 9 was deficient for a terminal segment of the short arm. Newly broken ends of chromosome 9 were produced in some meiotic cells, as diagrammed in Figure 1. Pollen grains of these plants carried either: (1) a deficient chromosome 9, which did not function in pollen-tube growth, (2) a chromosome 9 with a full duplication of the short arm—that is, the homologous chromosome 9, or (3) a chromosome 9 with a newly broken end. Among this last type, duplications or deficiencies of the short arm were present. Those carrying an extensive deficiency were nonfunctional but those carrying a relatively short duplication were better able to compete in functioning than those carrying the full duplication of the short arm. Thus the majority of the functioning pollen grains of these plants carried a newly broken end of the short arm of chromosome 9 in their nuclei. These chromosomes had undergone the breakage-fusion-bridge cycle since the meiotic anaphase and continued to do so after being incorporated into the primary endosperm nucleus. Either before fertilization or during the development of the kernel, the breakage-fusion-bridge cycle might produce alterations in heterochromatic elements and some of them might include an alteration that would recreate the condition associated with *Dt* action. Mutations at the a_1 locus to give the A_1 phenotype could subsequently appear in the descendant cells. If this should occur early in development of the endosperm, a sector with dots of the A_1 phenotype would be produced. Examination of 95 ears resulting from the preliminary test conducted in 1946 revealed A_1 dots on 15 kernels from 14 different ears. In five of these kernels, more than one A_1 dot was present, and in one restricted region of the kernel in each case. The number of kernels with mutations to A_1 in this trial experiment was lower than anticipated, and the experiment was not expanded the following year. Later, as the probable relation between the origins of mutability and the alterations induced in the knobs or other chromosome elements by the breakage-fusion-bridge cycle became more clearly apparent, the same experiment was conducted in 1950 on a much larger scale. The results were rewarding, for now many kernels were obtained that had one or more A_1 dots. One hundred and twenty such kernels appeared in this second trial, and 24 of them had more than one A_1 spot. One of these kernels had 84 A_1 spots, distributed rather evenly over the kernel. In the other 23, the spots were not distributed at random over the aleurone layer but were restricted to well-defined sectors. In none of these kernels did any large areas of the A_1 phenotype appear. In all cases, the time of mutation, the pattern of mutation, and the type of mutation were much like those produced when the known *Dt* factor is present in endosperms homozygous for a_1. It was obvious that in each case the initial alteration had occurred in the ancestor cell that produced the dotted sector. This initial event was responsible for mutations that occurred at a_1 in some cells during the subsequent development of the endosperm. The observed mutations at a_1, therefore, were not produced directly by the breakage-fusion-bridge cycle but arose secondarily, as a consequence of an event that altered some particular component in the nucleus. It was the alteration of this component that was responsible for the subsequent mutations at the a_1 locus. And this initial alteration was one that imitated the effect produced when the known *Dt* factor is present. It is difficult to avoid the conclusion that a new *Dt*-like factor has been produced in each such case, and that it was created by some event associated with the breakage-fusion-bridge cycle. Unfortunately, the plant grown from the one kernel having 84 dots distributed over the whole aleurone layer did not show any mutations to A_1, nor did mutations appear in the kernels when this plant was crossed to plants homozygous for a_1. The effective alteration probably was present in only one of the two sperms carried in the pollen grain. Because the break in chromosome 9 that initiated the breakage-fusion-bridge cycle was produced at the meiotic anaphase, the event giving rise to the *Dt*-like factor would have had to occur in the subsequent microspore division in order to be incorporated in the two sperm nuclei. An even larger experiment of this same type must be conducted if such a case is to be obtained. It should be mentioned in this connection that the size of the sectors within which A_1 spots appeared graded from large to small, the smaller sectors being most frequent. Also, about three-fifths of the kernels showing mutations to A_1 had only one A_1 spot. These frequencies are to be expected if the creation of a *Dt*-like factor is a consequence of an event, associated with the breakage-fusion-bridge cycle, that has a probability of occurrence in a limited number of mitotic cycles. In order to indicate why the dotted pattern of mutations to A_1 is to be anticipated, rather than any other, it is necessary to review the origin of the previously discovered $Dt - a_1$ mutable condition.

The $Dt - a_1$ mutable condition first appeared on one ear after self-pollination of a plant belonging to the commercial variety known as Black Mexican Sweet Corn. This variety is homozygous for A_1. The recessive a_1 in this case represented a new mutation from A_1, and was associated with the appearance of *Dt*. The original a_1 mutant, known for many years and used in genetic studies, had originally been found to be present in a commercial variety of maize. Both a_1 mutants responded in much the same manner when *Dt* was present in the nucleus. In both cases, the dotted mutation pattern was produced in the presence of *Dt*. The states of the two a_1 mutants thus appeared to be alike. This suggests that the older a_1 mutant may have been produced by a mechanism similar to that responsible for the origin of the newer a_1 mutant. A *Dt* factor may have arisen at the same time but subsequently been lost from the commercial variety during its propagation, leaving an apparently stable a_1 mutant. The change at *C* that produced $c^{m\text{-}1}$ would have behaved quite comparably had *Ac* been absent from the nuclei in the initial gamete carrying $c^{m\text{-}1}$, or had it been removed by crossing before the change at this locus had been detected. If the mutation had been discovered several generations after its origin, and if *Ac* had been removed by a previous cross, it would have appeared to be a newly arisen, stable, recessive *c*. Only after an incidental cross to a plant carrying *Ac* would its potential mutability have been revealed. It is possible, therefore, that many known recessives may prove to be potentially mutable.

The essential similarity of the $Dt - a_1$ system to the $Ac - c^{m\text{-}1}$, etc. system is also expressed in the changes in state of a_1 that may occur in the presence of *Dt*. Such changes in state of a_1 have recently been described by Nuffer (1951). They are recognized individually by marked departures in frequency of visible mutations, in types of mutation and in time of occurrence of these mutations. The types of different phenotypic expression produced by mutations at altered states of a_1 are much the same as those produced by a mutable a_1 locus that has appeared in the Cold Spring Harbor cultures. This new mutable a_1 locus, called $a_1^{m\text{-}1}$, differs from the mutable a_1 studied by Nuffer in that it is autonomous and does not require *Dt* for mutability to be expressed. In this respect, mutability at the A_1 locus behaves like that at the *C* and the *Wx* loci, for both autonomous and activator-controlled mutable conditions may arise.

The origin of $a_1^{m\text{-}1}$, in a culture carrying mutable luteus, was described previously. It is autonomous, and produces a series of quantitative alleles, many of which are unstable in that they may mutate to give higher or lower levels of quantitative expression. Difference in degree of quantitative expression is only one of the consequences of mutations occurring at $a_1^{m\text{-}1}$, however. The diversity of phenotypic changes arising from these mutations is so great that an adequate analysis of all the observed types is a large task. They are distinguished not only by quantitative but also by qualitative differences in the anthocyanin pigments formed. Diversity is shown in other respects. For example, some of the mutations giving pale aleurone color are related to changes involving the rate of a particular reaction responsible for pigment formation. Others appear to be related to the absolute amount of pigment that may be produced, regardless of a time factor. This becomes evident when comparisons are made of pigment-forming capacities in plants arising from kernels carrying different mutants of $a_1^{m\text{-}1}$, each producing a pale color in the aleurone of the kernel. In some cases, such plants are pale in their expression of anthocyanin color throughout their lives. Others are pale in anthocyanin color up to the time of anthesis, when growth of the plant terminates; but in the six or seven following weeks, as the plants mature their ears, the anthocyanin color gradually deepens, becoming intensely dark by the time the ears are mature. The kernels derived from both these two types of plants, however, may be equally pale. The fact that pigment forms late in the development of the kernel, and dehydration of the tissues occurs shortly thereafter, may explain this similarity in color of the kernels.

Other types of mutation occur at $a_1^{m\text{-}1}$. Some produce sectors of deep color that are rimmed by areas in which the color gradually fades off to colorless, as if a diffusible substance associated with pigment formation had been produced in excess in the mutant sector. The area of diffusion may be very extensive for some mutations, and only slight for others, whereas still other mutations give rise to no such diffusion areas at all. Some of the mutations that result in strong A_1 pigmentation are associated with failure of development or degeneration of some of the aleurone cells within the mutated sector.

Besides the mutational changes at $a_1^{m\text{-}1}$ that affect the type and amount of pigment formation, a number of other changes occur which affect the subsequent behavior of $a_1^{m\text{-}1}$. These alterations are termed changes in state, since they affect not only the times when pigment-forming mutations will occur at the locus in future plant and endosperm cells but also the kinds and frequencies of such mutations, and their distributions and their sequences in the development of the tissues.

Any interpretations that attempt to explain the primary action of a specific locus in a chromosome, and how this action may be changed, must take into consideration the facts just enumerated concerning the behavior of this $a_1^{m\text{-}1}$ locus. It is not reasonable to regard such changes in expression and action as being produced by changes in a single gene—that is, according to the usually accepted concepts of the gene that have been developed. The evidence suggests, rather, that the observed changes result either from alterations at a locus that has many individual components or from alterations at the locus affecting its relationship to other loci in the chromosome complement. If the latter is true, a combination of loci functions as an organized unit in the production of pigmentation. If such functional organizations exist within the nucleus —and it is reasonable to assume they do—then the large numbers of alleles known to arise at certain loci need not express altered genic action at the identified locus. Rather, any one alteration may affect the action of the organized nuclear unit as a whole. The mode of functioning of various other loci concerned may thereby be modified. In other words, the numerous different phenotypic expressions attributable to changes at one locus need not be related, in each case, to changes in the genic components at the locus, but rather to changes in the mechanism of association and interaction of a number of individual chromosome components with which the factor or factors at the locus are associated. According to this view, it is organized nuclear systems that function as units at any one time in development. In this connection it may be repeated that at a_1^{m-1}, and also at other mutable loci, many of the alterations observed represent changes in the potential for patterns of genic action during development (changes in state). Thus a pattern-controlling mechanism is being altered. If particular nuclear components are formed into organized functional nuclear units, the evidence would suggest that this may happen only at prescribed times in the development of an organism. In this event it may readily be seen that changes in pattern-controlling mechanisms would serve as a primary source of potential variability of genic expression without requiring any changes in the genes themselves.

A few more pertinent facts about the A_1 locus may now be mentioned. Mutability has arisen at this locus independently on a number of different occasions. Several cases have recently appeared in the stocks having *Ds* and *Ac*. Analysis of these cases has not proceeded to a stage where a complete description of their behavior may be given. Both Laughnan (1950) and Rhoades (1950) have found new cases of instability at this locus. Several have appeared in plants derived from kernels that had been aged for some time (Rhoades, 1950). Such aging is known to give rise to chromosomal aberrations as well as mutations; the observed instability may be an expression of one such structural change. Laughnan (1949, 1951) has shown that mutations at the A_1 locus may be associated with the mechanism of crossing over, suggesting again that mutations arise from structural changes at the locus. The crossover studies may elucidate the nature of some of these changes. Not only the cases described in this report but also others have produced evidence converging in support of a hypothesis that mutations originate from structural alterations in chromosome elements. The evidence derived from a study of progressive changes in state of c^{m-1} has shown the close relation between a structural change at a locus and one that so often has been called a "gene mutation." This study will now be described.

SIGNIFICANCE OF CHANGES IN STATE OF A MUTABLE LOCUS: SELECTED EXAMPLES

The foregoing review of the very different types of phenotypic expression that may be produced as a consequence of mutations arising at any one locus, and of the relation between the origin of a mutable condition and the type of mutations expressed, clearly indicates the necessity for caution in attempting to interpret the mutation process as one associated with a "change in a gene." With respect to events occurring at *Ac*, c^{m-1}, c^{m-2}, bz^{m-1}, bz^{m-2}, and wx^{m-1}, it has been established that a mechanism capable of producing chromosomal breaks at the locus is associated with the mutation-producing process. It could be argued from the evidence so far presented that these cases fall into a special category, and that what they may indicate regarding the mechanism of mutation at these mutable loci may not be used to interpret the mutation process in general. Knowledge gained from a study of the changes in state of c^{m-1} has shown, however, that no line may be drawn between those events at a locus that produce detectable chromosomal alterations and those that give rise to mutations but produce no readily detectable chromosomal alterations.

The origin of c^{m-1} by a transposition of *Ds* to the normal *C* locus has been discussed above. The state of *Ds*, when first transposed to the *C* locus, was one that produced many detectable chromosome breaks at this locus and few mutations to *C*. When plants having this state of c^{m-1} were crossed to plants that were homozygous for the stable recessive *c*, the majority of the resulting kernels showed this relationship. Some of these kernels, however, had sectors with higher rates of mutation to *C* and lower rates of detectable chromosome breaks. In some sectors, no chromosome breaks were evident; but often a high frequency of mutation to *C* had occurred. There were also a few kernels on these ears that had this pattern throughout the kernel. When found, such kernels were selected from the ears (as well as others that showed changed mutational patterns). The plants grown from them were again crossed to plants homozygous for the stable recessive *c*. The kernels on the resulting ears now showed the types and frequencies of the different detectable events at $c^{m\text{-}1}$ that had been observed in the kernel from which the plant had arisen. It was possible to determine in this manner that a heritable change had occurred at $c^{m\text{-}1}$. It was this change which was responsible for the altered frequencies of expression of the detectable events that subsequently occurred at this locus. That the altered response, in each isolated case, arose from a change at $c^{m\text{-}1}$ and was not produced by a change at *Ac*, was determined by testing the responses of these $c^{m\text{-}1}$ isolates with different isolates of *Ac*, each having a known type of action. *Ac* controlled, in each test, the time and place of the event occurring at $c^{m\text{-}1}$, but not its type. Because, in each case, the change in behavior of $c^{m\text{-}1}$ was heritable, it must have arisen by an event that produced an alteration at this locus. It is this heritable altered condition that has been termed an altered state of the locus.

The various altered states of $c^{m\text{-}1}$, as previously mentioned, arise only when *Ac* is also present in the nuclei, and only at the times in development when mutations or chromosome breaks may occur at $c^{m\text{-}1}$. For our purposes, the most instructive of the changed states are those giving reciprocal frequencies of chromosome breaks and mutations to *C*. A series of isolates, each showing a particular relation between these two events at $c^{m\text{-}1}$, has been studied. The isolates ranged from those showing no mutations to *C* or only a very occasional one, but having a very high frequency of detectable chromosome breaks, to those showing a high frequency of mutations to *C* and no detectable breakage events or only an occasional one. The states of $c^{m\text{-}1}$ that give high frequencies of chromosome breaks are unstable, for other altered states may be produced as a consequence of events at the locus, but only, as emphasized above, when *Ac* is also present in the nucleus. A particular state of $c^{m\text{-}1}$ remains constant if maintained in plants having no *Ac*. The state of $c^{m\text{-}1}$ giving no detectable chromosome breaks, and a correspondingly high frequency of mutations to *C*, is very stable with respect to the absence of breakage events. Had the state of $c^{m\text{-}1}$ been of this type when it first arose, there would have been no opportunity to discover that the chromosome-break-producing mechanism and the mutation-producing mechanism were related. If chromosome breaks are not exhibited by a mutable locus, therefore, it cannot be argued that because of this the basic mechanism producing the mutations must be different from that known to operate at $c^{m\text{-}1}$ and at the other *Ac*-controlled mutable loci. The evidence obtained from this study of the origin and subsequent behavior of altered states of $c^{m\text{-}1}$ argues, rather, for similarities if not identities in the basic mechanism.

Another type of change in state of $c^{m\text{-}1}$ should be mentioned, although it occurs infrequently. It is detected by a much altered expression of the mutation at this locus that affects aleurone color. The frequency of origin of such states of $c^{m\text{-}1}$ is so very low as to suggest that they represent entirely new modifications at this locus, comparable to the original inception of a mutable condition at any locus. They may well represent just such new inceptions, for these are to be expected in view of the fact that very different types of mutational behavior are exhibited at this same locus by $c^{m\text{-}1}$ and $c^{m\text{-}2}$, both of which are *Ac*-controlled.

All the *Ac*-controlled mutable loci exhibit changes in state. These are characterized by changed relative frequencies of the different recognizable types of mutations that occur. In other words, as we have already seen, the types of mutation produced in *Ac*-controlled mutable loci are related to the state of the mutable locus itself. The time and place during development of occurrence of mutations, on the other hand, are controlled by the state and dose of *Ac*; and therefore alterations in them are related to changes in state of *Ac* rather than to changes in state of the mutable locus. The changes in state of *Ac* have been described. The autonomous mutable loci also undergo changes in state, as described previously for the $a_1^{m\text{-}1}$ locus. In this group, however, the controller of the time and place of occurrence of mutations is a component of the locus itself. Consequently, the changes of state that arise at these loci are reflected in changes in the control of time and place of mutations as well as in the type or types of mutation that may occur, and also in the time and place of occurrence of each such mutation if several types are produced. Thus there is a much more diverse group of altered states associated with changes at any particular locus in the autonomous group. The general similarities between the autonomous and the *Ac*-controlled mutable loci are nevertheless striking.

EXTENT OF INSTABILITY OF GENIC EXPRESSION IN MAIZE

The discussion so far has mentioned a number of different mutable conditions at known loci in maize. In order to show that the phenomenon is much more prevalent than the particular cases described would indicate, some additional observations of changes in genic expression may be discussed briefly. Not only have *C*, *I*, *Bz*, *Wx*, A_1, and A_2 become mutable during the course of these studies, but instability of other known dominants has been noted. These are *R*, *Pr*, *Yg*, *Pyd*, *Y*, and possibly *B* and *Pl*, although the evidence for the last two is observational and not genetic. Some previously unknown dominant loci have also become unstable. These are associated with various chlorophyll-determining factors, endosperm-starch-controlling factors, aleurone-color factors, growth-controlling factors, etc. Instability has arisen at the loci of some of the known recessives such as *wx*, *yg*, and the special case a_1 described previously. Instability at the loci of the recessives *y* and *p* also appears to have arisen on several independent occasions. Genetic tests were not made, however, to determine the association of the instability with the known loci. Instability arising at recessive loci is recognized by mutations to or towards the expression of the dominant allele. It may be concluded, therefore, that many of the known recessive alleles are potentially capable of expressing action that is characteristic of the dominant alleles.

The expression of instability of various factors in maize is probably far more common than has been suspected in the past. Until recently, only a few such cases had been reported in the literature. One of the earliest recognized was that occurring at the *p* locus (pericarp and cob color, chromosome 1), studied by R. A. Emerson and his students (for literature citations, see Demerec, 1935) and recently being studied by Brink and his students (1951, and personal communication) and by Tavcar (personal communication). Reported cases of instability at the a_1 and the *wx* loci have been mentioned previously in this discussion. In addition, Rhoades (1947) has studied instability at the *bt* locus (brittle endosperm, chromosome 5) and has reported two cases involving chlorophyll characters that appeared in the progeny of irradiated seed (Rhoades and Dempsey, 1950). Fogel (1950) has been investigating instability at the *R* locus (aleurone, plant, and pericarp color, chromosome 10); and Mangelsdorf (1948) has reported instability at the *Tu* locus (Tunicate, chromosome 4). It is believed that critical examination will uncover many such cases in maize, and that they will involve many different loci.

MUTABLE LOCI AND THE CONCEPT OF THE GENE

It will be noted that use of the term gene has been avoided in the foregoing discussion of instability. This does not imply a denial of the existence within chromosomes of units or elements having specific functions. The evidence for such units seems clear. The gene concept stems from studies of mutation. That heritable changes affecting a particular reaction, or the development of a particular character, in an organism arise repeatedly and are associated with a change of some kind occurring at one specific locus or within one specific region of a chromosome, has been established. This knowledge has been responsible for the development of a concept requiring unitary determiners. It cannot be denied, in the face of such evidence, that certain loci or regions in the chromosomes are associated in some manner with certain cellular reactions or with the development of particular phenotypic characters. This is not the major questionable aspect of current gene concepts. The principal questions relate to the mode of operation of the components at these loci, and the nature of the alterations that affect their constitution and their action. Within the organized nucleus, the modes of operation of units in the chromosomes, of whatever dimensions these may be, and the types of change that may result in specific alterations in their mode of action, are so little understood that no truly adequate concept of the gene can be developed until more has been discovered about the function of the various nuclear components. The author agrees with Goldschmidt that it is not possible to arrive at any clear understanding of the nature of a gene, or the nature of a change in a gene, from mutational evidence alone. At present, the most we know about any "gene mutation" is that a heritable change of some nature has occurred at a particular locus in a chromosome, and that any one locus is somehow concerned with a certain chemical reaction, or with a certain restricted phenotypic expression, or even with the control of a complex pattern of differentiation in the development of a tissue or organ. The various types of known mutation, each showing unitary inheritance, obviously reflect various levels of control of reactions and reaction paths. It is necessary to consider these various widely different levels of unitary control and how they may operate in the working nucleus, and also to consider the nature of the changes that can affect their operation. It is with the nature of such heritable changes, the conditions that induce them, and their consequences, that this report has been concerned. Various levels of unitary control, as witnessed by inheritance behavior, are evident from the study. That genes are present in the chromosomes and that they function to produce a specific type of reactive substance will be assumed in this discussion, even though such a restricted assumption may prove to be untenable. The knowledge gained from the study of mutable loci focuses attention on the components in the nucleus that function to control the action of the genes in the course of development. It is hoped that the evidence may serve to clarify some aspects of gene action and its control. Some of the interpretations of the author, based on this evidence, have been stated or implied at various points in the previous discussion, and may now be summarized.

The primary thesis states that instability arises from alterations that do not directly alter the genes themselves, but affect the functioning of the genic components at or near the locus of alteration. The particular class to which a mutable locus belongs is related to the particular kind of chromatin substance that is present at or near the genic component in the chromosome. It is this material and the changes that occur to it that control the types and the rates of action of the genic components. Thus the basic mechanism responsible for a change at a mutable locus is considered to be one that is associated with a structural alteration of the chromatin materials at the locus. The mechanism that brings about these changes is related to the mitotic cycle; and it may involve alterations of both sister chromatids at the given locus. Some of these alterations may immediately result in the expression of an altered phenotype, a "gene mutation." Others produce modifications controlling the type of events that will occur at the locus in future cell and plant generations. Still others produce changes of a more extensive type, such as duplications and deficiencies of segments of chromatin in the vicinity of the locus. With regard to these conclusions, the evidence presented in the discussion of changes in state of $c^{m\text{-}1}$ may be recalled.

BEHAVIOR AND ACTION OF *Ac* IN CELL AND PLANT GENERATIONS

The interpretations given above deal with the organization and the kinds of events that occur at genetically detectable loci in the chromosomes. The next level of consideration deals with mode of operation of the nucleus in controlling the course of events during development. Do these studies suggest a mode of operation, or at least one component in the operative system? It is believed that the behavior of *Ac* may be of importance in such a consideration. With reference to $c^{m\text{-}1}$, for example, it has been shown that mutations to *C* occur at particular times and places in development, under the control of the particular state and dose of *Ac*. It has also been shown that *Ac* is unstable, for changes arise affecting its dosage action, and changes also occur in its location in the chromosome complement. It has not been explained, however, that the time of occurrence of these changes at *Ac* during development is also a function of the particular state and dose of *Ac* that is present in a tissue. With any particular *Ac*, and any particular dose of this *Ac*, or with any combination of different isolates of *Ac*, the time when these events will occur to *Ac* is a function of the single or combined action of the *Ac* factors. An example of this may now be given.

One plant having an *Ac* factor at the same location in each homologue of a pair of chromosomes, and carrying *I*, *Sh*, *Bz*, *Wx*, and *Ds* (standard location) in both chromosomes 9, was crossed to a number of plants homozygous for *C*, *sh*, *bz*, and *wx* and carrying no *Ds* or *Ac*. On all the many ears resulting from these crosses, approximately 90 per cent of the kernels were sectorial with reference to the time of occurrence of *Ds* events. Pollen from this same plant was placed on silks of plants having other combinations of markers in chromosome 9, and similar types of sectorial kernels appeared. The sectoring was produced by segregations, occurring in the earliest nuclear divisions of the endosperm, that involved the controller of the time of occurrence of *Ds* events, that is, the Activator. The action of *Ac* in these different sectors resembled that occurring either (1) when no *Ac* is present, (2) when a sharply decreased dose of *Ac* is present, or (3) when a sharply increased dose of *Ac* is present. Illustrations that will make this relationship clear are shown in the photographs of Figures 17 to 22. In a definite fraction of the cases, a chromosome break at *Ds* was associated with segregations of *Ac* action. The mechanism responsible for this precise, somatically occurring segregation for *Ac* action was very likely the same as that responsible for the origins of germinal changes in action of *Ac*, as well as changes in its position in the chromosome complement. Such changes have been mentioned earlier in this report. It can be seen that if this same type of segregation occurred within some cells early enough in the development of a plant to be incorporated in a microspore or megaspore nucleus, the altered state or location of *Ac*, or both, would be recoverable in the plant that subsequently resulted from the functioning of the male or female gametophyte arising from such a spore. Such an early-timed event would allow for isolation and subsequent study of the transpositions and changes in state of *Ac*.

A study was initiated to determine the nature of the changes that occur at *Ac* by an analysis of the *Ac* constitutions in the gametes of plants having an *Ac* factor at the same location in each member of one pair of chromosomes, that is, in plants homozygous for *Ac* in allelic positions. In these plants the *Ac* factors were alike in state, and the homozygous condition was produced by self-pollination of plants having one *Ac* factor. Both chromosomes 9 in these plants carried the stable factors *c*, *sh*, and *Wx*. No *Ds* was present in these chromosomes. To test for *Ac* inheritance, these plants were crossed by plants having no *Ac* but carrying *C*, *Sh*, *wx*, and *Ds* in both chromosomes 9 (standard locations of *Ds* and identical states). If no changes had occurred to *Ac*, all the kernels on the resulting ears should be variegated. Similar variegation patterns, produced by sectors showing the *c*, *sh*, and *Wx* phenotype, should be present, because *Ac* would initiate chromosome breaks at *Ds* in the *C*, *Sh*, *wx*, *Ds*-carrying chromosome 9 contributed by the male parent. With the exception of a few kernels, just such conditions were realized on these ears. A photograph of one such ear appears in Figure 23. It will be noted that the majority of kernels on this ear show very similar patterns of variegation. A few kernels that differ from the majority are completely colored, with no colorless sectors of any size. In them, no *Ds* breaks at all occurred. A few other atypical kernels show an altered timing of the breakage events at *Ds*. In them such events occurred either much earlier or very much later in development of the endosperm.

A study was made of the *Ac* constitution in plants derived from selections of all the different types of kernels appearing on such ears. In the plants coming from the kernels showing altered variegation patterns, it was necessary to determine the subsequent behavior of *Ac*; and in the plants coming from the nonvariegated kernels it was necessary to determine the presence or absence of *Ac*, and the presence or absence of *Ds* in the *C*, *Sh*, *wx*-carrying chromosome. The results of this study may be summarized. In the plants derived from the majority class of kernels, a single *Ac* factor was present. Its state was similar to that present in the parent plant (more than 25 cases studied). The *Ac* constitution in the plants derived from the nonvariegated kernels was most instructive. In 19 plants, no *Ac* was present. In 17 plants, two nonlinked *Ac* factors were present. In six plants, an *Ac* factor inherited as a single unit was present, but it gave a dose action equivalent to two doses of the *Ac* factor in the parent plant. In the plants derived from kernels that showed very late-occurring *Ds* events, either two nonlinked *Ac* factors were present (5 cases), or a single *Ac* factor was present giving a dose action greater than that of the *Ac* factor in the parent plant (3 cases).

FIGS. 17 to 22. Photographs of kernels illustrating the somatic segregations of *Ac* that may occur very early in the development of a kernel. These kernels arose from the cross of plants (♀) carrying *C*, *bz* and no *Ds* in each chromosome 9 and having no *Ac* factor, by plants carrying *I*, *Bz*, and *Ds* (standard location) in chromosome 9 and also carrying *Ac*. For phenotypes expected from breaks at *Ds*, see descriptions accompanying Figures 10 to 15. In Figures 17 and 18 there are 4 large sectors in each kernel: one is *C bz* (above in Figure 17, to right in Figure 18), one is *I*, non-variegated (to right in Figure 17, upper left in Figure 18), one is characterized by late occurring *Ds* breaks, producing speckles of the *C bz* genotype (left in Figure 17, lower left in Figure 18), and one shows that numerous *Ds* breaks occurred earlier in the development of the kernel (lower segment in Figure 17, middle segment in Figure 18). The kernel in Figure 19 has 3 sectors: one that is *C bz*, one with few specks of the *C bz* genotype and one with many specks of the *C bz* genotypes. Figure 20 shows a kernel with 3 sectors: one that is *C bz*, one that is wholly *I Bz* and one having many specks of the *C bz* genotype. Figure 21 shows a kernel with two sectors: one with many specks of the *C bz* genotype and one with few such specks. Figure 22 shows a kernel with five sectors: a large *C bz* sector (lower left), a large sector having many specks of *C bz* (upper), a large sector showing many larger *C bz* areas (upper right), a small sector with few *C bz* specks (middle), and a sector of *I Bz* with no *C bz* specks (lower right).

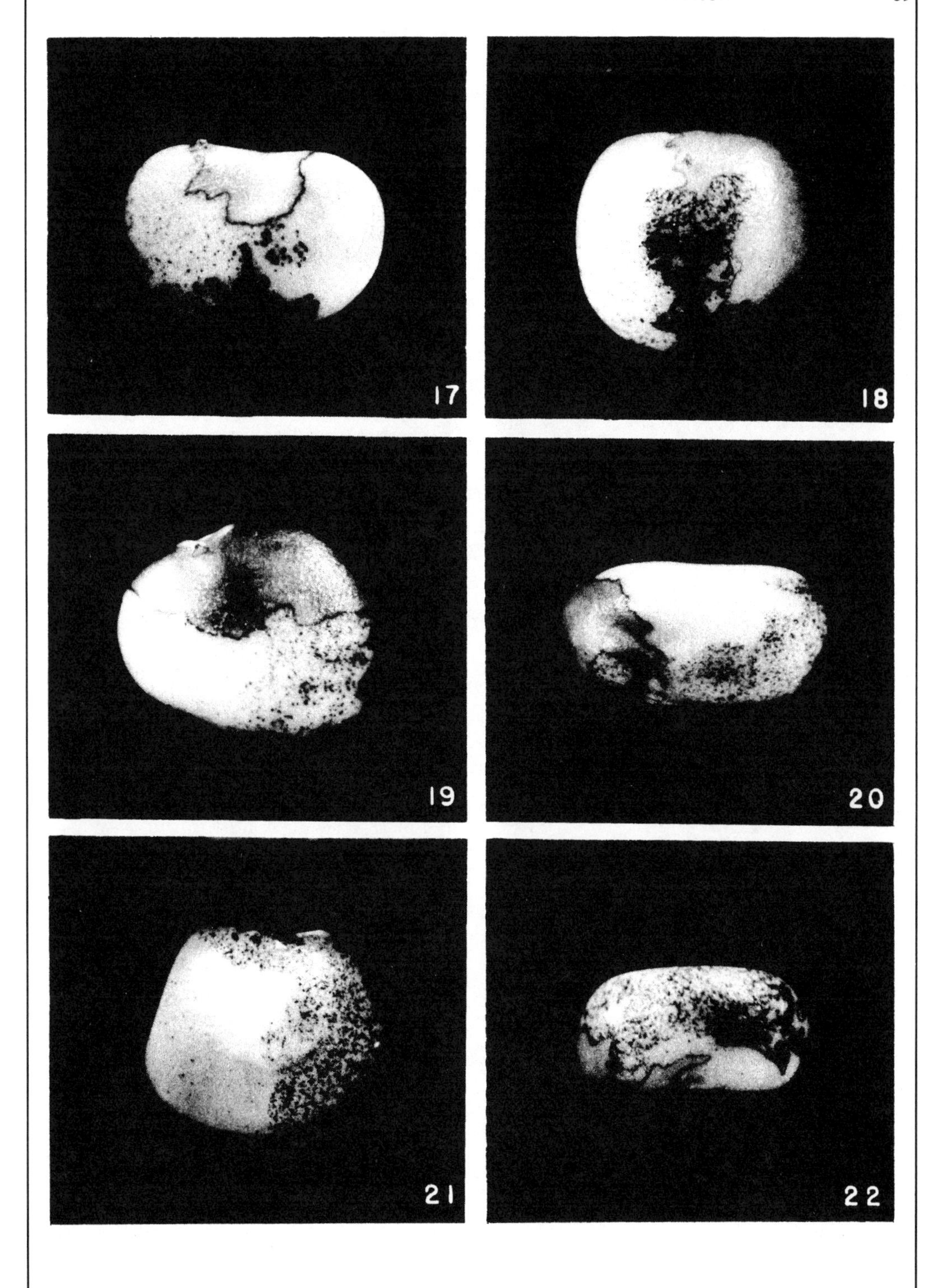
17
18
19
20
21
22

From this analysis it is clear that all the aberrant kernels on ears of the type shown in Figure 23 were produced because of some alteration of *Ac* that had occurred in cells of the parent plant. The reason that no *Ds* breaks were detected in some of these kernels is related either to the absence of *Ac* in the endosperm or to the presence of a marked increase in the dose of *Ac*. It will be recalled that the female parent contributed two gametophytic nuclei to the primary endosperm nucleus. If each nucleus carried two *Ac* factors, or a single *Ac* factor with a double-dose action, the endosperm would have either four *Ac* factors, or two *Ac* factors equivalent in action to four *Ac* factors. In such kernels, the high dose of *Ac* so delayed the time of occurrence of *Ds* breaks that none took place before the endosperm growth had been completed.

In order to verify the analyses of *Ac* constitution in some of these cases, tests were continued for another generation. For example, if a plant contains two nonlinked *Ac* factors, the gametic ratios approach 1 two-*Ac* : 2 one-*Ac* : 1 no-*Ac*—that is, a three-to-one ratio for the presence of *Ac*. On ears derived from crosses in which such plants are used as male parents, the kernels with one or with two doses of *Ac* may be distinguished because of clearly seen differences in the time of response of *Ds* to *Ac* doses. Therefore, some of the kernels considered to have two *Ac* factors and others considered to have only one *Ac* factor were selected from the test ears. The plants grown from them were again tested for gametic ratio of *Ac*. In each case, verification was realized. The gametic ratios produced by the latter plants approached 1 one-*Ac* : 1 no-*Ac*, whereas those produced by the former approximated 3 with one or two *Ac* factors to 1 with no *Ac*.

FIG. 23. Photograph of an ear derived from a plant having two identical *Ac* factors located at allelic positions in an homologous pair of chromosomes. This plant carried *c* in each chromosome 9. The ♂ parent, having no *Ac*, introduced a chromosome 9 with *C* and *Ds* (standard location). The majority of the kernels are similarly variegated for sectors of the *c* genotype due to breaks at *Ds* that occurred in the *C Ds* chromosome during the development of the kernels. Note the few fully colored kernels in which *c* sectors are absent, and also the several kernels that show large sectors of the *c* genotype.

The above-described series of tests, and still others that have been concerned with the time and type of changes occurring at *Ac*, have made it possible to understand the nature of its inheritance patterns. It has been found that, with any particular state or dose of *Ac*, the time of occurrence of changes of *Ac* is controlled by *Ac* itself. If, with a particular *Ac* state, the time of such changes is delayed until late in the development of the endosperm, then all the kernels should show this same late timing. This is known to occur with many of the isolates of *Ac*. As the photograph in Figure 23 has shown, however, a few aberrant kernels may be present on some of these ears. Some internal or external alteration in environmental conditions may have caused these few early-occurring changes at *Ac*. No attempts have been made, however, to study conditions that might alter the time of such changes.

If these tests for determining the inheritance behavior of *Ac* had not been made, considerable confusion might have arisen. This would certainly have been true had states of *Ac* giving relatively early changes been used in the initial inheritance studies. It must be stated that just such a situation has been observed. States of *Ac* giving aberrant gametic ratios have arisen. It is now realized that this is to be anticipated. It has been determined that the reason for the difference in patterns of inheritance between an *Ac* isolate that gives clear-cut mendelian gametic ratios and one of its modified derivatives, that gives aberrant gametic ratios, is related to the time in the development of the sporogenous or gametophytic cells at which such changes in *Ac* arise. With reference to the gametic constitutions that will be produced, the time when these changes occur is most critical. If they occur in somatic divisions before the meiotic mitoses, or in the male or female gametophytes, an apparently unorthodox inheritance pattern for *Ac* will result. If they occur late, that is, in the endosperm tissues—which act in this connection like a continuation of the development of the gametophyte—then no such confused pattern of inheritance will arise. The gametic constitutions will then closely approximate those predicted for mendelizing units. In the study of *Ac* inheritance, it was necessary to make selections for these latter states of *Ac*. A few exceptions with regard to the time of changes at *Ac* may occur in some cells, even with such selected states of *Ac*. It was the analysis of *Ac* constitutions in plants derived, in cell lineage, from those cells in which such exceptional timing of changes at *Ac* had occurred, that provided the information leading to appreciation of the somatic origins of altered states and locations of *Ac*.

Confused patterns of inheritance behavior of mutable loci have been described in the literature many times. The ratios obtained have often been so irregular that no satisfactory formulation of the nature of the inheritance patterns could be derived. This would be just as true of some of the autonomous mutable loci in maize if attention had not been given to altered states and their behavior. Two examples may illustrate this. Both $a_1^{m\text{-}1}$ and $a_2^{m\text{-}1}$, when first discovered, produced many mutations and changes in state very early in the development of the plant. The plant, therefore, was sectorial for the altered conditions at these mutable loci. The sectors were present in the tassel. When pollen was collected without reference to the sectors present, and placed on the silks of plants carrying the stable a_1 or a_2 alleles, the kinds of kernels appearing on the resulting ears, and their frequencies, were not readily analyzable in terms of mendelian ratios. No such difficulty arises, however, when similar tests are made for gametic ratios in plants derived from those kernels on the original test ears that show only very late-occurring mutations. The inheritance pattern is now of the obvious mendelian type, for mutations and changes in state are mainly delayed until after meiosis and gamete formation. As with *Ac*, the selection of states of autonomous mutable loci that produce very late-occurring mutations makes it possible to examine the inheritance behavior of such loci, freed from the apparent confusion resulting from early-occurring modifications at the locus, which can distort the expected mendelian ratios.

Is the Behavior of *Ac* a Reflection of a Mechanism of Differentiation?

We now return to the original question. What is the significance of the somatically occurring changes at *Ac*, and the changes in state that occur at the autonomous mutable loci? Do they suggest the presence of nuclear factors that serve to control when and where certain decisive events will occur in the nucleus? With regard to *Ac*, it is known that the events leading to its loss, to increase or decrease of its dosage action, or to other changes involving its action or position in the chromosome complement, are related; and that they appear as the consequence of a mitotic event, controlled in time of occurrence by the state and dose of *Ac* itself. Sister nuclei are formed that differ with respect to *Ac* constitution, as the photographs of Figures 17 to 22 illustrate. Because of this somatically occurring event involving *Ac*, the *Ac*-controlled mutable loci will differ markedly as to the time when mutational events will occur at them, or as to whether or not any such events will occur at all in the cells arising from the sister cells. This precise timing of somatic segregations effects a form of differentiation, for it brings about changes in the control of occurrence and time of occurrence of genic action at other loci, and does so differentially in the progeny of two sister cells. This likewise applies to the autonomous mutable loci; but in these cases the controller of the time and place of appearance of genic activity is a component of the locus itself.

The process of differentiation is basically one involving patterns of action arising in sequential steps during development and affecting the types of activities of definitive cells. The ultimate expression of component parts of an organism represents the consequence of segregation mechanisms involving the various cellular components. The part played by any one component of the cell in this segregation system can not be divorced from that played by any other component. It is possible, however, to attempt to examine the various components in order to determine their respective relationships and the sequential events that involve them. Embryological studies have contributed much to our knowledge of the segregation of cytoplasmic components. The segregation of nuclear components is less well understood, although some outstanding examples are known. These examples show segregations or losses of obvious components of the nucleus—that is, of whole chromosomes or easily seen parts of chromosomes. The segregation or loss of smaller components, not readily visible on microscopic examination, may well be one of the mechanisms responsible for the nuclear aspects of control of the differentiation process. The phenomenon of variegation, as described here and observed in many other organisms, may be a reflection of such a segregation mechanism—exposed to view be- the timing of events leading to a specific type of genic action is "out-of-phase" in the developmental path. Variegation may represent merely an example of the usual process of differentiation that takes place at an abnormal time in development. Viewed in this way, it is possible to formulate an interpretation of the part played by the nuclear components in controlling the course of differentiation.

This interpretation considers that the nucleus is organized into definite units of action, and that the potentials for types of genic action in any one kind of cell differ from the potentials in another kind of cell. In other words, the functional capacities of the nuclei in different tissues or in different cells of a tissue are not alike. The differences are expressions of nonequivalence of nuclear components. This nonequivalence arises from events that occur during mitotic cycles. The differential mitotic segregations are of several types. Some involve controlling components, such as *Ac*, and produce sister nuclei that are no longer alike with respect to these components. As a result, the progeny of two such sister cells are not alike with respect to the types of genic action that will occur. Differential mitoses also produce the alterations that allow particular genes to be reactive. Other genes, although present, may remain inactive. This inactivity or suppression is considered to occur because the genes are "covered" by other nongenic chromatin materials. Genic activity may be possible only when a physical change in this covering material allows the reactive components of the gene to be "exposed" and thus capable of functioning.

A mechanism of differentiation that requires differences in nuclear composition in the various cells of an organism finds considerable support in the literature. The most conspicuous example is in *Sciara*, where a thorough cytological and genetical analysis has been made. (For reviews, see Metz, 1938, and Berry, 1941.) It is known, in this organism, at just what stage of development differences in nuclear composition will arise; and, with regard to the X chromosome, it is known what element in the chromosome controls the differential behavior. This element is at or near the centromere of the chromosome (Crouse, 1943). Furthermore, differential segregations of the B-type or accessory chromosome have been found to occur in a number of plants. (For reviews of literature to 1949, see Müntzing, 1949.) Numerous other examples are known of differential segregation involving whole chromosome complements, certain types of chromosomes of a complement, or, occasionally, a certain component of a chromosome. (For literature citations see Melander, 1950; White, 1945, 1950; Berry, 1941.) Whether the differential segregations involve whole complements of chromosomes, individual chromosomes, individual parts of chromosomes that can be seen, or submicroscopic parts of chromosomes, may well be a matter of degree rather than type. Certainly, the evidence for differential segregation is not wholly negative.

With regard to mechanisms associated with differentiation and genic action, an additional factor may be mentioned. The part played by the doses of component elements in the chromosomes appears to be of considerable importance. First, a number of genetic factors associated with known loci produce measurable quantitative effects that are related to dose: the higher the dose the greater the effect. Such dosage actions, probably reflecting rates of reaction, are familiar to all geneticists, and some of them have been reviewed in this study of mutable loci. Dosage controls of the *Ac* type, affecting the time of action of certain other factors carried by the chromosomes, has been less well appreciated. A third type of dosage action has made itself evident in these studies. In some aspects, however, it resembles the action of different doses of *Ac*. In the study of the autonomous $a_1^{m\text{-}1}$ mutable locus, a number of mutants appeared, particularly on self-pollinated ears, showing a pale aleurone color. Study of the behavior of these pale mutants has revealed the following. Some of them produce pale-colored aleurone in one, two, and three doses and give no evidence of instability in the expression of the phenotype. (One and two doses are obtained by combinations of the pale-mutant allele with the stable a_1 allele.) That this stable expression may be deceptive is shown by the dosage effects of other similarly appearing pale-producing isolates derived from mutations at $a_1^{m\text{-}1}$. These may give pale aleurone color, and no indication of instability, in three and two doses; but with one dose something unexpected occurs. The kernels show a colorless aleurone in which mutations to deep aleurone color appear. Still other isolates give pale color in three doses, but in one or two doses produce the colorless background with deep-colored mutant areas. In these cases, it is clear that some of the mutations at a_1^{m-1} giving pale color and appearing to be stable are stable only because of some dosage action produced by a mutation of the original a_1^{m-1} to the pale-producing type.

The study of dose-provoked actions in the pale mutants mentioned above and those of *Ac* have given some indication of the importance of dosage action in affecting genic expression. The original isolate of a_1^{m-1} did not give evidence of such striking dosage action. When present in one, two, or three doses, it gave rise to colorless kernels in which mutations, mainly to a deep color and occasionally to a pale color, appeared. The graded series of dosage action exhibited by the various pale mutants derived from a_1^{m-1} is very much the same as the graded series exhibited by the various isolates of *Ac*. In these cases, it appears as if each isolate is composed of a specific number of reactive subunits and that the dosage expressions are related to the total number of such units that are present in the nucleus. Although these graded dosage effects may be visualized on a numerical basis, it is not claimed that such an interpretation is necessarily the correct one. The large differences in dosage expression exhibited by the various isolates of *Ac*, and also the various isolates of the pale mutants derived from a_1^{m-1}, nevertheless appear to follow such a scheme.

Why different doses of components of the chromosomes function as they do in controlling developmental processes takes us to another level of analysis that is not under consideration here. A relation to rates of particular reactions can be suspected. It is tempting to consider that changed environmental conditions may well alter otherwise-established rates of reaction, and thus initiate alterations in the nuclear components at predictable times, leading to strikingly modified phenotypic expression. Just such effects have been observed by students of developmental genetics. They have shown that alterations of environmental conditions at particular times in development can lead to predictable changes in the subsequent paths of differentiation.

Consideration of the Chromosome Elements Responsible for Initiating Instability

It will be recalled that this study of the origin and behavior of mutable loci was undertaken because a large number of newly arisen mutable loci appeared in the progeny of plants in which an unusual sequence of chromosomal events had occurred—that is, the breakage-fusion-bridge cycle. Striking similarities in the patterns of behavior of these mutable loci were immediately noticed. It was the pattern of behavior, rather than the change in expression of the particular phenotypic character, that was obviously of importance. This pattern, revealed in all cases, stemmed from an event occurring at mitosis, which altered the time and frequency of mutations that would subsequently occur in the cells derived from those in which this event occurred. It was noticed that sister nuclei could differ in these respects—and sometimes reciprocally, as if the mitotic event had resulted in an increase in one nucleus of a component controlling the mutation time or frequency, and a decrease of this component in the sister nucleus. It was also noted that the change in phenotypic expression—that is, the mutation—likewise resulted from a mitotic event; and that the mutation itself and changes of the controller of the mutation process could result from the same mitotic event: one cell showing the mutation, the sister cell showing an altered condition with respect to control of future mutations in the cells derived from it.

Further, it may be recalled that the mechanism which resulted in the appearance of newly arisen mutable loci—that is, the breakage-fusion-bridge cycle involving chromosome 9—gave rise to numerous obvious alterations of the heterochromatic materials, in other chromosomes of the complement as well as in chromosome 9. It was also demonstrated that the effect of a known activator, *Dt*, located in the heterochromatin of the chromosome-9 short arm, and producing a very definite pattern of mutations of the otherwise stable a_1 locus in chromosome 3, could be recreated independently and on a number of different occasions in cells of a tissue in which the breakage-fusion-bridge cycle was in action. The combined observations and experiments point to elements in the heterochromatin as being the ones concerned with differential control of the times at which certain genes may become reactive. It is believed that somatic segregations of components of these elements may initiate the process of nuclear control of differentiation.

On the basis of these interpretations and those given in the previous section, it becomes apparent why a large number of newly arisen mutable loci appeared in the self-pollinated progeny of plants that had undergone the chromosome type of breakage-fusion-bridge cycle. This cycle induced alterations in the heterochromatin. These alterations changed the organization of the heterochromatic chromosome constituents and probably also, in many cases, the doses of their component elements. Changes were induced in these heterochromatic elements at times other than those at which they would normally occur during differentiation. This resulted in changes in the times in development when their action on specific chromatin material, associated with genic components of the chromosome, was expressed. The altered timing of their actions was consequently "out-of-phase" with respect to the timing that occurs during normal differentiation. This was made evident by the appearance of a "mutable locus." The "mutable locus" is thus a consequence of the alteration of an element of the heterochromatin produced by the breakage-fusion-bridge cycle. Once such an "out-of-phase" condition arises, others may subsequently appear because of the physical changes in the chromatin that occur at the mutable locus, leading, at times, to transpositions of this chromatin to new locations, as described earlier. In their new locations, these transposed chromatin elements continue their specific control of types of genic action but now affect the action of the genic components at the new locations.

Relation of "Mutable Loci" to "Position-effect" Expressions in Drosophila and Oenothera

In a previous publication (McClintock, 1950) the author has suggested that the position-effect variegations in *Drosophila melanogaster* and the variegations observed in many other organisms, including those associated with the mutable loci here described. are essentially the same. An adequate discussion of the interrelations would require more space than can appropriately be given here. Attention will be drawn, therefore, only to a few relevant facts, which may serve to indicate why this conclusion has been reached. In the first place, a number of different types of position-effect expression are found in *Drosophila* (for review and literature citations, see Lewis, 1950). In maize, comparable types of instability expression have appeared. In *Drosophila,* some of the variegations appear to result from loss of segments of chromosomes. This applies to those cases where the expression of the dominant markers, carried by the chromosome showing the "position-effect" phenomenon, is absent in some sectors of the organism. The extent of the deficiency varies, but it includes in each case the region adjacent to the heterochromatic segment with which many of the variegation types of position-effect expression are known to be associated. It may be recalled that such deficiencies are produced in maize when *Ds* is present.

That heterochromatic elements of the chromosomes of *Drosophila* undergo breakage events in somatic cells is suggested by the study of "somatic crossing over" in this organism (Stern, 1936). The appearance of the abnormally timed exchanges between chromosomes is conditioned by the presence of certain Minute factors, for example, *M(1)n*, much as the occurrence of structural aberrations at certain loci in maize (i.e., *Ds*, wherever it may be) is dependent on the presence of *Ac*.

Of particular significance for comparative purposes is the study of Griffen and Stone (1941) on the induction of changes in the position-effect expression in *Drosophila* of the white-eye variegation, w^{m5}. The w^{m5} case arose through an X-ray-induced translocation of the segment of the left end of the X chromosome at 3C2 (the w^+ locus) to the heterochromatic region of chromosome 4. Males carrying w^{m5} were X-rayed, and the progeny examined for changes in the variegation expression of the eye mottling. Many such changes were found. Studies of these cases were continued in order to determine the nature of the events associated with the changes. In all cases, the new modification in the phenotypic expression of the w^+ locus was found to be associated with a translocation, which placed the segment of the left end of the X chromosome, from 3C2 to the end of the arm, at a new location. In many cases, the new position was to a euchromatic region of another chromosome, and yet variegation persisted. Some of the new positions, however, gave rise to apparent "reversions" to a wild-type expression. Individuals having these "reversions" were X-rayed, and variegation types again appeared in the progeny. Here also the variegation was shown to be associated with a translocation involving the left end of the X chromosome at 3C2, from the location in the "reversion" stock to a new location—again, sometimes a euchromatic region. It may be suspected that the maintenance of variegation potentialities in all these cases was associated with the presence of a segment of heterochromatin of chromosome 4 that remained adjacent to the w^+ locus when the successive translocations occurred. This would not readily be detected in the salivary chromosomes. The presence of such "inserted" heterochromatin could be responsible for the continued expression of variegation at the w^+ locus in repeated translocations. If such was the case, then the resemblance to the maize cases, described in this report, is obvious. The appearance of "reversions," and the subsequent appearance of variegation after X-radiation of individuals carrying such "reversions," might seem to present a contradiction. On the basis of an analysis of the cases described by Griffen and Stone, the writer believes that no contradiction is involved. This analysis has suggested that the timing of variegation-producing events during development is, in part, a function of the relative distance of the translocated segment—i.e., the left end of the X chromosome—from the centromere of the chromosome that carries it: the farther removed the segment is from the centromere, the later in the development the variegation-producing events will occur. In the "reversions," this segment has been placed close to the end of one arm of a chromosome. The reappearance of variegation occurs when the segment is translocated to a position closer to a centromere. Another factor is also associated with the timing of the variegation-producing events. This is the Y chromosome. When the Y is absent, the areas of altered phenotype are larger than when it is present, indicating an earlier timing of the variegation-producing events. It may be noted in this connection that some of the cases of "reversions" are only apparent reversions. In XY constitutions they appear to give a stable wild-type expression but in XO constitutions, the eyes show a light speckling of the altered phenotype. With the latter constitution variegation occurs, but only very late in the development of the eye. The similarity of this effect of the Y chromosome to that of dosage action of *Ac* is apparent in these cases as well as in many others in *Drosophila* that have been examined.

In a recent report, Hinton and Goodsmith (1950) gave an analysis of induced changes at the bw^D (Dominant brown eye) locus in chromosome 2 of *Drosophila.* This case was considered to be a stable-type position effect. It arose originally through the insertion of an extra band next to the salivary-chromosome band where bw^+ is located. Males carrying bw^D were irradiated and crossed to wild-type females. The offspring (9,757 individuals) were examined for changes in the bw^D expression. Twenty-one individuals showing the wild-type expression appeared in the F_1, and progeny was obtained from one-third of them. A study of the inheritance behavior of each modification was undertaken, and a study was also made of the salivary-gland chromosomes. From these studies it was clear that the modifications arose from changes that occurred in the vicinity of the bw^D locus and involved the inserted band. In four cases, restoration of the wild-type expression followed removal of this band. In two cases, it followed separation of this band from the bw^+ band by translocations. In one case, no obvious change in the salivary chromosomes was noted, but nevertheless a change in phenotypic expression had occurred. Of considerable importance, also, was the appearance, in some of these cases, of somatic instability of expression of the bw^+ phenotype. Variegation began to appear. It had never been observed in the brown-Dominant stock itself. It may be noted that the changes in the bw^D expression are associated with types of chromosomal alterations which are much the same as those proposed to account for changes in phenotypic expression at some mutable loci in maize.

A further resemblance between *Drosophila* and maize will be mentioned. In *Drosophila,* many of the translocations, inversions, and duplications are believed to be associated with the formation of dominant-lethal effects. In maize, a number of dominant lethals have arisen from transpositions of *Ds*. Some produce defective growth of the endosperm and embryo; others affect the development of the embryo but not the endosperm; and still others affect the capacity of the embryo to germinate, without affecting its morphological characters. Over half the newly arisen transpositions of *Ds* that are of this latter type have not produced viable plants, owing to lack of germination of the embryos in the kernels.

There are similarities between the maize cases and a case in *Drosophila pseudoobscura* described by Mampell (1943, 1945, 1946). In this *Drosophila* case, a heterochromatic element appeared to be associated with the initiation of instability at another locus, which in turn led to changes in chromosome organization and to numerous changes in genic action at various loci in the chromosome complement. These changes were expressed both somatically and germinally.

The position-effect behavior reported in *Oenothera* (Catcheside, 1939, 1947a, b) is much like that of *Ds*. The chromosomal events responsible for the observed types of change in phenotypic expression may be the same in the two organisms. In *Oenothera* as in maize, gross changes in chromosome constitution arise, such as duplications and deficiencies of segments of the chromosome involved. Similar cases in other organisms undoubtedly exist. It is probable, however, that the lack of a critical mode of detection of a chromosome breakage mechanism has been responsible for the apparent delay in reporting such cases in connection with studies of somatic variegation and mutable loci. Also, because changes in state occur that involve reduction in the frequency of chromosome breaks, and because such breaks lead to lethal gametes, it is probable that states of a mutable locus producing some detectable breaks are rapidly eliminated from a population, leaving a state of the mutable locus that produces few or no such events to be propagated.

It has been argued that the variegation types of position effect in *Drosophila* usually do not give rise to germinal mutations, and that they belong, therefore, to a separate category of instability expression. Since some variegation position effects do give rise to germinal changes, this argument could in any case be only partially applicable. However, whether or not germinal changes arise is not considered relevant in the interpretation developed here. The time and place of occurrence of such changes is related to controls, existing in the nucleus. The differentiation mechanism described above should effect controls that would exclude the germ lines from undergoing many changes, but should allow numerous alterations in the soma that would lead to altered patterns of genic expression. Whether or not a particular somatically expressed pattern of genic action—for example, the distribution of pigment—arises from mutations at a "mutable locus" or from the action of a particular "stable" allele of the locus cannot be decided by using the criterion of presence or absence of germinal mutations. The important consideration is when, where, and how the patterns of genic action are controlled and eventually expressed.

The combined evidence from many sources suggests that one should look first to the conspicuous heterochromatic elements in the chromosomes in search of the controlling systems associated with initiation of differential genic action in the various cells of an organism; and secondarily to other such elements, which are believed to be present along the chromosomes and to be either initially or subsequently involved in the events leading to differential genic action. Evidence, derived from *Drosophila* experimentation, of the influences of various known modifiers on expression of phenotypic characters has led Goldschmidt (1949, 1951) to conclusions that are essentially similar to those given here.

The conclusions and speculations on nuclear, chromosomal, and genic organization and behavior included in this report are an outgrowth of studies of the instability phenomenon in maize. They are presented here for whatever value they may have in giving focus to thoughts regarding the basic genetic problems concerned with nuclear organization and genic functioning. Until these problems find some adequate solution, our understanding and our experimental approach to many phenomena will remain obscured.

REFERENCES

BERRY, R. O., 1941, Chromosome behavior in the germ cells and development of the gonads in *Sciara ocellaris*. J. Morph. *68:* 547–576.

CATCHESIDE, D. G., 1939, A position effect in *Oenothera*. J. Genet. *38*: 345–352.

1947a, The *P*-locus position effect in *Oenothera*. J. Genet. *48*: 31–42.

1947b, A duplication and a deficiency in *Oenothera*. J. Genet. *48*: 99–110.

CROUSE, HELEN V., 1943, Translocations in *Sciara;* their bearing on chromosome behavior and sex determination. Res. Bull. Mo. Agric. Exp. Sta. *379*: 1–75.

DEMEREC, M., 1935, Mutable genes. Bot. Rev. *1*: 233–248.

EMERSON, R. A., 1921, Genetic evidence of aberrant chromosome behavior in maize. Amer. J. Bot. *8*: 411–424.

FOGEL, S., 1950, A mutable gene at the *R* locus in maize. Rec. Genet. Soc. Amer. *19*: 105.

GOLDSCHMIDT, R. B., 1949, Heterochromatic heredity. Hereditas, Suppl. *5*: 244–255.

GOLDSCHMIDT, R. B., HANNAH, A., and PITERNICK, L. K., 1951, The podoptera effect in *Drosophila melanogaster*. Univ. Calif. Publ. Zool. *55*: 67–294.

GRIFFEN, A. B., and STONE, W. S., 1941, The w^{m5} and its derivatives. Univ. Texas Publ. No. 4032: 190–200.

HINTON, T., and GOODSMITH, W., 1950, An analysis of phenotypic reversion at the brown locus in *Drosophila*. J. Exp. Zool. *114*: 103–114.

LAUGHNAN, JOHN R., 1949, The action of allelic forms of the gene *A* in maize. II. The relation of crossing over to mutations of A^b. Proc. Nat. Acad. Sci. Wash. *35*: 167–178.

1950, Maize Genetics Coöperative News Letter *24*: 51–52.

1951, Maize Genetics Coöperative News Letter *25*: 28–29.

LEWIS, E. B., 1950, The phenomenon of position effect. Advances in Genetics *3*: 73–115.

MAMPELL, K., 1943, High mutation frequency in *Drosophila pseudoöbscura*, Race B. Proc. Nat. Acad. Sci. Wash. *29*: 137–144.

1945, Analysis of a mutator. Genetics *30*: 496–505.

1946, Genic and non-genic transmission of mutator activity. Genetics *31*: 589–597.

MANGELSDORF, P. C., 1948, Maize Genetics Coöperative News Letter *22*: 21.

McCLINTOCK, B., 1941, The stability of broken ends of chromosomes in *Zea Mays*. Genetics *26*: 234–282.

1942, The fusion of broken ends of chromosomes following nuclear fusion. Proc. Nat. Acad. Sci. Wash. *28*: 458–463.
1950, The origin and behavior of mutable loci in maize. Proc. Nat. Acad. Sci. Wash. *36*: 344–355.
MELANDER, Y., 1950, Accessory chromosomes in animals, especially in *Polycelis tenus*. Hereditas *36*: 19–38.
METZ, C. W., 1938, Chromosome behavior, inheritance and sex determination in *Sciara*. Amer. Nat. *72*: 485–520.
MÜNTZING, A., 1949, Accessory chromosomes in *Secale* and *Poa*. Proc. Eighth Intern. Congr. Genetics. (Hereditas, Suppl. Vol.).
NUFFER, M. GERALD, 1951, Maize Genetics Coöperative News Letter *25*: 38–39.
RHOADES, M. M., 1936, The effect of varying gene dosage on aleurone color in maize. J. Genet. *33*: 347–354.
1938, Effect of the *Dt* gene on the mutability of the a_1 allele in maize. Genetics *23*: 377–395.
1941, The genetic control of mutability in maize. Cold Spring Harb. Symposium Quant. Biol. *9*: 138–144.
1945a, On the genetic control of mutability in maize. Proc. Nat. Acad. Sci. Wash. *31*: 91–95.
1945b, Maize Genetics Coöperative News Letter *20*: 14.
1947, Maize Genetics Coöperative News Letter *21*: 3.
1950, Maize Genetics Coöperative News Letter *24*: 49.
RHOADES, M. M., and DEMPSEY, E., 1950, Maize Genetics Coöperative News Letter *24*: 50.
SAGER, R., 1951, On the mutability of the waxy locus in maize. Genetics (in press).
STERN, C., 1936, Somatic crossing over and segregation in *Drosophila melanogaster*. Genetics *21*: 625–730.
WHITE, M. J. D., 1945, Animal Cytology and Evolution. Cambridge Univ. Press.
1950, Cytological studies on gall midges (Cecidomyidae). Univ. Texas Pub. No. 5007: 1–80.

Insertion by Phages and Transposons

ALLAN CAMPBELL
Stanford University, Stanford, California 94305

One scientist can influence another scientist's career in at least two ways: through direct, personal interaction or through laying the experimental and conceptual foundations on which the second scientist builds. Barbara McClintock has been a personal friend of mine for many years. The interest she took in my work from its early stages and the insight she has provided have been invaluable to me. Here, I touch on a few such personal interactions, but I concentrate on the relationship between her experimental work and my own.

My first serious introduction to Barbara's elegant work on maize cytogenetics (including, but by no means restricted to, the transposon studies) came from a cytogenetics course that I took from Marcus Rhoades at the University of Illinois in 1951. At the time, her reputation among the geneticists I knew verged on the legendary. There was a common opinion that she was always right and did everything better than anyone else could. While I was taking Rhoades' course, Barbara visited Urbana to give a seminar. She also dropped in on the cytogenetics laboratory, where she looked through our microscopes at our chromosome preparations. My own staining technique was messy (to put it mildly), and I was apprehensive about showing her the cell that happened to be under the microscope at the time—a translocation heterozygote that I was struggling to classify. She glanced at it, then asked Rhoades if this was not one of her stocks (it was) and went on.

The incident impressed me because up to that point I had imagined that the main skill of cytologists lay in making good preparations, which mine certainly was not. Several years later, I mentioned this to Barbara. Her opinion was definite. In looking at chromosomes, what mattered most was orientation. As you focused up and down, you needed to be aware at all times of what plane you were viewing and how it related to the one you had just seen. This required concentration. Once in a while, she herself found her concentration fading and stopped briefly to recover. It was that simple (if you were Barbara McClintock, of course).

Prophage Insertion

Most of my career has been spent examining various aspects of the interactions between bacteriophages and their bacterial hosts. In the early stages, this consisted of investigating the manner in which prophages were associated with bacterial chromosomes. There were three dominant influences: classical cytogenetics, as I had learned in Rhoades' course; fine structure genetics, as developed by Seymour Benzer; and the work on lysogeny initiated by André Lwoff and extended by Salvador Luria and Giuseppe Bertani at Illinois.

The Luria-Lwoff philosophy was summed up by Lwoff (1953) in his statement that lysogeny stands at the crossroads between infection and normal heredity, the prophage behaving, on the one hand, as part of the life cycle of an infectious element and, on the other hand, as a host gene. My own contribution depended on following through on the logical implications of that statement, which I may well have taken more literally than Lwoff intended. In Lwoff's department at the Institut Pasteur, François Jacob and Elie Wollman had studied the relationship between prophage and bacterial chromosome and had concluded that prophages were probably attached laterally to the bacterial chromosome, rather than being inserted into its continuity. Many bacterial geneticists, such as William Hayes (1964), considered it self-evident that specialized transducing phages like λ*gal* were extrachromosomal, because they were readily lost from bacterial lineages.

These positions seemed at odds with Lwoff's concept of the prophage as being like a host gene, and especially with a corollary to it that Luria frequently emphasized: Once a cell has been lysogenized, we have no way of knowing that the prophage came from an external source. Therefore, many of the things we classify as genes may really be prophages or derived from them. At least there is no operational criterion to distinguish a prophage from a gene (or genes) once the prophage is in place.

The question of prophage integration then became part of a larger issue: the structure of the chromosome itself. In 1960, there were many advocates of a chromosome whose central backbone was decorated with lateral appendages; indeed, the "classical" picture of genes as beads on a chromosomal string (perhaps taken more seriously by textbook writers than by working geneticists) implied that genes were added to a chromosomal backbone rather than being part of it. However, the fine structure genetic results that I trusted most—such as Seymour Benzer's with T4 phage and Edward Lewis's and Melvin Green's in *Drosophila*—indicated that the internal linear structure of the gene was part of the backbone itself. If that conclusion were general, it seemed that the only way a prophage could become permanently associated with a chromosomal site was by splicing the prophage into the continuity of the chromosome.

I cannot claim any great originality for that sentiment, which was probably shared by many geneticists. Most of them found Jacob and Wollman's experiments on the subject more convincing than I did. I was frequently told that my ideas were beautiful but wrong; a common position spawned by the misapprehension that experimental facts, however complex or ambiguous, should always take precedence over ideas, however simple and clarifying. A simple idea should be abandoned only when the facts are compelling, rather than merely indicative.

At any rate, my formulation of a model for the integration of λ prophage came about in the course of writing a review on lysogeny, in which I went over the available data on the relationship between host and prophage chromosomes. (Both François Jacob and Barbara McClintock read the manuscript and provided encouragement and constructive criticism.) One result that challenged interpretation had been reported by Calef and Licciardello (1960). They constructed a recombinational map of the lysogenic chromosome from bacterial crosses and showed that their results fit linear insertion of λ prophage into the bacterial chromosome, but with a change in gene order between phage and prophage. Much of Jacob and Wollman's argument for lateral attachment of λ depended on the assumption that the internal gene order of the prophage should be the same as that of the phage. In contemplating the possible ways that Calef and Licciardello's result might come about, I noted that if the ends of the phage chromosome were to become joined together within the cell, then a reciprocal crossover between the phage and the bacterial chromosome would insert the prophage with the observed gene order.

Put that way, this seems like an enormously ad hoc explanation for Calef and Licciardello's result, which hardly demonstrated the proposed cyclic permutation of gene order. There were three prophage markers in their cross, and the evidence for permutation was only that the wrong one was in the middle. Given that result, their argument for intercalation into the bacterial chromosome (based on exchange of linked flanking markers) was fairly strong. Calef pointed out that earlier results of Jacob and Wollman (1957) could be read in the same way. Jacob and Wollman had assumed that the gene order must be the same in the prophage as it is in the phage and proceeded to conclude that linear insertion was unlikely.

Most ideas designed to explain a few specific findings are rejected by their creators in short order. Certainly most of mine have been. This is because they generally generate as many problems as they solve and imply no strong testable predictions. Cyclic permutation during prophage insertion, however, seemed to explain several results and to make some predictions testable with the methodology available at the time.

Some experimental facts came from my own work on the genetic content of specialized transducing phages. Morse, Lederberg, and Lederberg (1956) had found that, on coming out of the chromosome, λ can occasionally pick up the *gal* genes, which lie adjacent to it; Werner Arber and I had both found that the λ*gal* particles were "defective" as phages because they had lost some phage genes along with acquiring these bacterial genes. Jacob and Wollman interpreted the result as indicating homology between λ and bacterial DNA, allowing recombinational replacement of phage genes by bacterial genes. However, that hypothesis did not really fit my observations. Each λ*gal* phage that I studied had lost a segment of phage DNA characteristic for that particular λ*gal*, and this property was stably reproduced in its subsequent reproduction, just as rare rearrangements like translocations or inversions are stable once they have occurred.

If I considered all of the λ*gal* phages that I had studied, the properties of the whole collection were readily explainable on the cyclic permutation model by the assumption that every λ*gal* resulted from rare breakage and joining between bacterial and phage genes that excised a connected segment of the lysogenic chromosome—a unique segment for each λ*gal*. This model made the strong predictions that, since phage 434 transduced the same genes that λ did, the two phages must insert at the same site on the bacterial chromosome and that any bacterial gene between *gal* and λ must be present in every λ*gal* phage. There were reports in the literature (which eventually proved incorrect) that contradicted both of these predictions. Had they stood up, I would have abandoned the model.

My model has sometimes been designated by other authors as "purely theoretical." I do not object to that description, but it may be misleading. If I compare the array of data on gene contents of λ*gal* phages with Benzer's deletion mapping of the T4*r*II genes, the argument seems as strong in one case as the other that the results generate a unique linear gene order that (in the λ case) reveals the cyclic permutation of prophage with respect to phage. One must of course assume without independent proof that each λ*gal* is generated by a rare event that creates one novel joint between phage and host DNA, but deletion mapping requires the equivalent assumption that each deletion removes a connected segment of the genetic map. The difference is more one of style of presentation than of theory versus experiment. I have always preferred to keep experimental facts distinct from their interpretations, even though this is seldom the most effective manner of communicating the significance of the former or the strength of the latter.

One corollary to the postulated structure of λ*gal* was that, in cells lysogenic for a λ*gal*, the prophage part of the λ*gal* is located at the junction between the endogenous *gal* genes and the introduced *gal* genes and that the haploidization of *gal* noticed earlier by Morse, Lederberg, and Lederberg resulted from internal recombination that looped out one copy of *gal* along with the phage genes. Recombination between tandem duplications was not a process that most prokaryotic geneticists of the time considered relevant to their material, and it was some years (even after the basic model was widely accepted) before the full disappearance of an older notion that anything that was lost easily must be extrachromosomal.

Insertion of Transposons

All of the general arguments about why prophages ought to be inserted into chromosomes, rather than attached laterally, applied with equal force to the transposable elements that Barbara McClintock had found in maize. I discussed the question once with Barbara in the 1960s. She seemed to understand my interest in the question but considered it unanswerable by the methods then available. She was less ready than I to equate the simple linearity of bacteriophage genomes with the structure of eukaryotic chromosomes, which she thought were much richer in complexity. I had broached the question to her because of a meeting volume I had read, where Boris Ephrussi had suggested that transposable elements might increase the genetic distance between flanking markers. Barbara rejected the idea that an element would need to be inserted in (or even map within) a genetic region to affect its recombination frequency, and I concurred. I regarded such a "stretching" experiment (which had also been done with λ) as providing rather weak evidence for insertion, compared to crosses that ordered the internal structures of the element with flanking markers; but the latter required genetic markers within the element, which were unavailable for the maize elements.

Eventually, maize transposons were cloned and junction fragments were sequenced, demonstrating insertion directly. Long before that, however, insertion was assumed (see, e.g., Fincham and Sastry 1974). The assumption seems to have been based partly on analogies between maize transposons and bacterial insertion sequences (not altogether compelling in 1974), but more on the mounting evidence that chromosomes really are long linear DNA molecules. Given a dominant paradigm with no room for lateral appendages, few people would raise (or perhaps even understand) the question of whether transposons might be attached laterally. There is substantial contrast between this perspective and Barbara McClintock's earlier treatment of "controlling elements" as objects quite different from ordinary genes and not necessarily governed by any of the same rules.

Transposons and Viruses

Despite the many parallels between prophages and transposons, Barbara was always reluctant to discuss the possibility that the two had much in common. I never tried to elicit her opinions on this in depth. If she seemed to find a topic distasteful, I generally just dropped it. Once I mentioned Kyoshi Mizuuchi's beautiful work on transposition mechanisms with phage Mu, but she quickly truncated the conversation. Virus replication should not be discussed along with transposition. She saw a close relationship between her work and the G loop inversion of Mu, along with its biological consequences, but Mu replication did not interest her.

We have therefore never communicated effectively on the parallels between phages and transposons. I gradually realized the roots of the problem but never succeeded in overcoming it. Soon after her discovery of transposons, McClintock had been challenged with the hypothesis that the maize stocks in which they were discovered were virus-infected. She tested the hypothesis, generating unequivocal evidence that the elements were present in every maize stock she examined but only manifested themselves after the genome had experienced some shock, as in the breakage–fusion–bridge cycle.

Her results clearly ruled out casual infection; but from my perspective, they barely addressed the question of whether the elements were viral or virus-related. I was philosophically committed to the assumption that proviruses are buried in most if not all genomes and even happy to entertain the possibility that some of them played constructive roles in the genetic programs of their hosts. A general operational distinction between endogenous and exogenous elements is to my mind impossible, and energy is more profitably directed at evaluating the selective forces that maintain both transposons and viruses over evolutionary time.

Reprise

I have tried to indicate some of the intersections between McClintock's work and my own and to describe a few of our personal contacts. I have not yet done justice to the impact of her career on my own. I think the major impact has been through her work and would have been felt by me even had we never met. I have described above some parallels between prophage insertion and transposition, but I am not sure whether that is even the most important point. My initial work on prophage integration and related problems (such as tandem duplications in bacteria) conceptually amounted to little more than applying the principles of classical cytogenetics to prokaryotic genomes. McClintock was one of the prime architects of classical cytogenetics, long before transposons attracted her interest. When today I read her papers from the 1930s (on ring chromosomes, for example), I see many steps of my own thought processes anticipated in them.

On the more personal side, I stress the value I have received from her insights, her interest, and most of all, her enthusiasm for science. While preparing this article, I came across an interview with my Stanford colleague Carl Djerassi, where he implies that the portrait of a competitive elitist scientist in his best-selling novel *Cantor's Dilemma* is not atypical: "A scientist's drive, his self-esteem are really based on a very simple desire: recognition by one's peers. That recognition is bestowed only for originality, which, quite crassly, means that you must be first. No wonder that the push for priority is enormous."

To those of us who believe that the most important advances of the future will come from the creative minds of gifted individuals rather than from the collective efforts of many cogs in the wheels of Big Science—Human Genome-type projects—Barbara McClintock provides an inspiring example of the antithesis of Djerassi's elitist scientist. Throughout her career, she has shown that true originality has no need to fear competition, that a first-rate intellect can set its own criteria for self-esteem, and that the joy of discovery can be given precedence over all else.

References

Calef, E. and G. Licciardello. 1960. Recombination experiments on prophage-host relationships. *Virology 12:* 81.

Fincham, J.R.S. and G.R.K. Sastray. 1974. Controlling elements in maize. *Annu. Rev. Genet. 8:* 15.

Hayes, W. 1964. *The genetics of bacteria and their viruses,* 1st edition. John Wiley, New York.

Jacob, F. and E. Wollman. 1957. Genetic aspects of lysogeny. In *The chemical basis of heredity* (ed. W.D. McElroy and B. Glass), p. 468. Johns Hopkins Press, Baltimore.

Lwoff, A. 1953. Lysogeny. *Bacteriol. Rev. 17:* 269.

Morse, M.L., E.M. Lederberg, and J. Lederberg. 1956. Transduction in *Escherichia coli* K12. *Genetics 41:* 142.

Cold Spring Harbor 1944-1955: A Minimemoir

EVELYN M. WITKIN
Waksman Institute of Microbiology, Rutgers, The State University of New Jersey
Piscataway, New Jersey 08854

My first significant encounter with Barbara McClintock took place in the Victorian "Dormitory" of the Carnegie Institution's Department of Genetics at Cold Spring Harbor in June of 1944. Barbara then lived on the main floor, and I had just been installed, for the summer, in an attic room. Barbara was a member of the staff, a very famous scientist whose classical work in cytogenetics I had studied in my genetics courses and for whom I felt overwhelming awe and admiration. I was then a graduate student, based at Columbia University with Professor Theodosius Dobzhansky, who had encouraged me to begin my dissertation research at Cold Spring Harbor, with Salvador Luria and Milislav Demerec, in the new field of bacterial genetics.

At the time, I wrote almost daily to my soldier husband, who saved my letters. I have read some of them recently, and one letter, dated June 19, 1944, serves to reinforce my memory of that first long conversation with Barbara. I wrote, "I had a rare and exciting experience tonight. You've heard of Dr. Barbara McClintock, the maize geneticist Professor Rhoades always raves about? Tonight I had a long talk with her—for about two hours—and a more fascinating two hours I have never spent! I wandered into the living room after dinner, and she was alone there, smoking a cigarette. We started to talk, and before long the conversation really took off..." The letter goes on for several pages, describing the highlights of the discussion and recounting some of Barbara's comments on various philosophical, scientific, and political questions. After almost half a century, the letter recalls vividly my sense of having met a truly extraordinary person and of having undergone a memorable experience. That feeling has been renewed in every encounter I have had with Barbara during all the years I have been fortunate enough to have had her friendship.

During that summer of 1944, I took to visiting Barbara's laboratory often and listening to her enthusiastic explication of her work with maize. She was at that time studying some of the mutagenic consequences of the breakage–fusion–bridge cycle and was already observing some of the intriguing patterns of variegation that led to her discovery of transposable elements. I found her material fascinating, and although my knowledge of maize genetics was limited, Barbara's briefings filled some of the gaps. I came back to Cold Spring Harbor the following summer to resume my predoctoral research, this time to stay for 10 years. Barbara continued to put up with my frequent visits to her laboratory, as I became increasingly captivated by the story unfolding there.

Cold Spring Harbor in the years between 1945 and 1955 was an intensely exciting place for a young person obsessed with the gene. The people one encountered in the laboratories or on the beach included Salvador Luria, Max Delbrück, Alfred Hershey, Joshua Lederberg, James Watson, François Jacob, Leo Szilard, and Jacques Monod, to mention only a few of the scientific giants who worked, spoke, or visited there. It was a decade of great discoveries that revolutionized biology, and Cold Spring Harbor was its center. To have been there during those years was indescribably thrilling.

For me, perhaps the greatest thrill was the opportunity to follow closely Barbara McClintock's discovery of transposable elements and her elucidation of their role in controlling gene expression. I was able to look over her shoulder often as her careful experiments began to tell her what the patterns of spots on her maize kernels meant and to follow her patient explanations as she increasingly understood how the frequency and timing of the responsible mutagenic events were controlled. I recall, in particular, some of the times I spent in her laboratory when she had just learned of something especially exciting, for instance, when she first recognized the interaction of *Ds* and *Ac* or when she found that the timing of *Ds* mutations depends on the dosage of *Ac*. Although I could not always follow the intricacy of the crosses behind these results, Barbara's detailed explanations and the striking visual evidence provided by her material made her conclusions compellingly convincing. I remember clearly her first use of the term "transposition" when she found unmistakable genetic and cytological evidence that both *Ds* and *Ac* can move from one chromosomal position to another. How startling that was in 1948 or 1949! It was more than two decades before transposons were rediscovered in bacteria, in the more hospitable context of the maturing molecular phase of genetics.

What made Barbara's achievement so extraordinary was that it was accomplished in a context not nearly ready to incorporate or accommodate it. She started with curiosity about how broken chromosomes and mutable genes are related and proceeded to wrest the answers from an unlikely organism at an unripe stage of genetics by the sheer force of powerful scientific thinking, brilliant experimentation, persistent hard work, and total command of her material. My favorite poet, Robert Browning, wrote "A fact looks to the eye as the eye likes the look." Barbara had trained her eye to like the look of any fact, as long as she could be sure of its truth.

Barbara's absorption in her work has never prevented her from maintaining her intense interest in other people's work, in world events, in friendship, in literature, and in music, nor has it ever interfered with her ability to see humor in much of the passing scene. My old letter documents how fascinating I found her conversation in 1944. This, if anything, is even more true in 1991.

Annals of Mobile DNA Elements in Drosophila: The Impact and Influence of Barbara McClintock

MEL GREEN
University of California, Davis, California 95616

It is with unlimited pleasure and ongoing affection that I join my colleagues in extending best wishes to Barbara McClintock on the occasion of her 90th birthday. It has taken me many years to publicly acknowledge Barbara's birthday. Some 15 years ago at a symposium of the Society of Developmental Biology convened in Raleigh, North Carolina, I tried to publicly acknowledge her birthday. However, politely but firmly I was instructed not to make a public announcement. At long last, I have had at least one last word! We honor her with birthday greetings publicly.

In the remarks to follow, I have two objectives in mind. The first is to examine the purported delay in recognizing Barbara's seminal discovery of mobile DNA elements. The second is to emphasize the pervasive influence the maize mobile elements discovery had upon the discovery of mobile elements in *Drosophila melanogaster*.

There is a widely extant viewpoint that Barbara's research was much unappreciated and appropriate recognition was too long delayed. This is believed to be the case especially for the discovery of mobile DNA elements. Here, I wish to enter my personal comments because I believe this viewpoint to be a half-truth. In order to comment, it is necessary to describe the state of genetics during the period of 1945–1970.

For the period beginning with the rediscovery of Mendelism to the end of World War II, the dominance of eukaryote genetics was obvious. During this time, the arithmetic of eukaryote genetics was brilliantly worked out. Corn (maize) and *D. melanogaster* were the primary experimental organisms of choice. Combining genetics with cytology, the gene could be delimited to a precise place on the chromosome. Barbara McClintock's contributions here are noteworthy. Her proof in 1931, coincidental with that of Curt Stern in *Drosophila*, that crossing over involved the physical exchange between homologous chromosomes was classic. Despite the elegance of the cytogenetic experimentation, the precise chemical nature of the gene defied definition and remained essentially a black box. The advent of prokaryote genetics with the development of bacterial and phage genetics culminating in the determination of DNA as the genetic material established the temporary hegemony of prokaryote genetics. The comparative genetic simplicity of the prokaryotes, their generation time measured in minutes rather than weeks or months, and their progeny numbered in hundreds of thousands, even millions, served to attract a group of researchers to genetic questions whose solution was practically unimaginable with eukaryote organisms. Many of the researchers were physicists or chemists with little or no biological background, and accordingly unencumbered by Mendelian genetics. They brought to genetics a fresh, even refreshing, point of view, an esprit, a cockiness, and often intellectual arrogance predicated on the belief that with prokaryotes, the outstanding issues of genetics would be solved in their hands. Here, I cannot help but relate a personal experience that illustrates the point.

At the 6th International Congress held in Bellagio, Italy during August of 1953, I gave a paper on genetic fine structure in *D. melanogaster*, at the time a topic of some importance because intragenic crossing over posed important questions on the then conceived organization of the gene. After my presentation, I got into a conversation with the late Harriet Ephrussi Taylor, an elegant lady and an outstanding student of bacterial (pneumococcus) transformation. What she said to me was essentially the following: "You are wasting your time with *Drosophila* (eukaryote) genetics. *Drosophila* is a dead organism. The action is with prokaryote genetics where the fundamental questions will be answered." I could only reply feebly that *Drosophila* is all I know and that, aesthetically, the adult *Drosophila* flies are vastly superior in phenotype to a bacterial colony. But the take home lesson was clear: By and large, prokaryotes will provide the answers and eukaryote genetics is inconsequential and can be ignored.

It is into this intellectual climate that Barbara McClintock introduced her concept of controlling elements presently included under the rubric mobile DNA elements. The idea was revolutionary but of only passing interest to the dominant body of prokaryote geneticists. Why? Here I can only conjecture. A number of factors come to mind. In 1951 when Barbara presented her detailed paper on controlling elements at the Cold Spring Harbor Symposium, DNA as the genetic material had only recently been accepted by the community of biologists. Among the properties of the genetic material, stability was a necessary requirement. The concept of controlling elements included the revolutionary idea that a portion of the genetic material was indeed unstable and moved from site to site in the genome. Such a concept clearly contradicted the idea of the stable genetic material. Furthermore, since the controlling elements concept derived from maize, a domesticated eukaryote organism for which there was not even a wild type, the whole idea could be of little general application or interest. What is all too often the fate of revolutionary ideas that contradict the accepted orthodoxy? Such ideas are either ignobly dismissed or "swept under the rug." Certainly, for prokaryotes, the controlling element concept had little heuristic merit or value.

Despite the aforementioned reservations, there existed a group of geneticists who could not dismiss the controlling element concept out of hand. Among eukaryote geneticists, especially the maize and *Drosophila* researchers, Barbara McClintock was so highly respected for her extraordinary insights and elegant experimentation that however revolutionary the controlling element concept might be, it could not be ignored. As noted above, her classic cytogenetic experiments proving that crossing over must involve a physical exchange between homologous chromosomes, plus her demonstration of the breakage–fusion–bridge cycle of chromosome behavior, and her study of the fate of ring chromosomes in maize represent a few textbook examples of her enormous prowess as an experimenter and thinker. For many, she was the premier cytogeneticist of the century! Like it or not, Barbara was rarely wrong. However revolutionary controlling elements might be, the idea must be given serious consideration. To illustrate the enormous faith and confidence we had in Barbara, I relate here yet another relevant anecdote. Shortly after the famous 1951 Cold Spring Harbor Symposium, a small group of interested geneticists, myself included, gathered in the office of Alfred H. Sturtevant at the California Institute of Technology to discuss the Symposium presentations. Sturtevant, to my mind the greatest all-around geneticist of this century, whose incisive mind was widely recognized, had attended the Symposium. During the discussion, someone put the question to Sturtevant: "Sturt, what did McClintock have to say?" Sturt sucked on his ever-present pipe, looked briefly at the ceiling, and then replied. "I didn't understand one word she said, but if she says it is so, it must be so!" Such was the reputation and confidence in Barbara widespread among the maize and *Drosophila* geneticists of the period. Her ideas were, for sure, not ignored.

I am certain the discovery of mobile elements in *D. melanogaster* would not have been made at that time without the prior discovery of maize controlling elements. In the case of *Drosophila*, the discovery was a fulfillment in part of Pasteur's dictum that "chance only favors the prepared mind," and in part of the already described maize elements. Elsewhere, I have discussed the *Drosophila* case in detail. Here, I consider only the critical facts.

Barbara and I, independently, share an experience in common. We served briefly on the faculty of the University of Missouri, Columbia. We survived working in Lewis J. Stadler's cornfield, a daunting experience for anyone who has not experienced the rigors of the summer in a Missouri cornfield. We had our separate problems with the University administration, a topic that merits no further discussion. Most important, we were able to interact with Lew Stadler, one of the keenest brains of his time. Independently, Stadler had discovered X-ray mutagenesis in plants. He had a long-standing interest in the X-ray mutation, specifically the nature of the mutational lesion. Because the overwhelming majority of putative X-ray-induced mutations were not transmitted by the plant (barley or maize) gametophyte, he concluded that X-ray mutations were not sensu strictu gene mutations but deletions. Here, he differed with Hermann J. Muller with whom he should have shared the Nobel prize for the discovery of X-ray mutagenesis. One published set of data puzzled and perplexed Stadler. This was the report by Nicolai Timoféeff-Ressovsky that certain mutations at the white eye color locus in *D. melanogaster* could be reverted to wild type by X-irradiation. I recall only too vividly the frequent discussions of this report that Stadler and I had. Stadler, the consummate skeptic, could not rationalize X-rays making gene reversions. If X-rays made deletions, how could such deletions generate a reversion? There was no information in hand then (1946–1950) to explain reversions of a gene via a deletion. The frequent discussions with Stadler made a lasting impression, and for many months I pondered the problem: How can one check Timoféeff's results? Timoféeff's fly stocks were lost during World War II, so his mutants were not available. After long consideration, I concluded there was only one unambiguous way of solving the riddle of Timoféeff's results. Gene mutations that had spontaneously reverted were needed. If they were capable of spontaneous reversion, it should be comparatively easy to determine whether or not X-rays significantly increased their reversion frequency. Unfortunately, the number of examples in *Drosophila* of spontaneous reversion of mutant genes to wild type was small. This is because the reversion event is rare. However, over a period of years, a number of X-linked mutations were collected, each of which had undergone a bona fide mutation to wild type. Among this group, the mutation at the white eye locus, white-ivory (w^i) was noteworthy. In a short communication, E.B. Lewis reported that w^i reverted to wild type at a low frequency both somatically and germinally. Furthermore, the frequency of somatic w^i reversions could be increased if larvae were X-rayed. Since Lewis did not pursue this line of research further, I decided that w^i was a prime candidate to test whether or not X-rays could produce gene reversions. Before the X-ray analysis, a detailed study of spontaneous w^i mutation was undertaken by J.B. Bowman as part of a Ph.D dissertation done under my guidance. Together, Bowman and I initiated a study of X-ray reversion of w^i. When Bowman completed his dissertation and left to take a teaching position, I continued the study of X-ray induction of w^i reversions. To increase the efficiency in detecting X-ray-induced reversion of w^i, I decided to study reversions in attached-X females homozygous for w^i. There were two advantages to this design. Because attached-X females possess two X chromosomes attached to one centromere, they are transmitted together to the next generation. After irradiating attached-X females, only their female progeny need be scored for reversions, and simultaneously, two irradiated w^i genes are scored. In one X-ray experiment, a single female was found whose eye color was intermediate between that of w^i and wild type. This I assumed was caused by the induction of a partial reversion, a class of reversions that had been recovered on rare occasions before. I decided to save this partial reversion for later study and proceeded to make a stock in which the attached-X females were homozygous for the partial reversion. Making such a genetic stock was simple, but before sequestering the stock in my stock collection for future study, I decided to check the flies to see that all was in order. To my surprise and chagrin, all was not in order. Among the expected homozygous partial reversion females, I found females of a wild-type eye color and some with the w^i eye color. Contamination could be ruled out since linked marker genes were present in all females as expected. It took no great insight to decide that this was no ordinary partial reversion of w^i. Somehow, subsequent to X-irradiation, a derivative of w^i occurred that was mutationally highly unstable. Such mutations were not new and had been described in *Drosophila virilis* by Milislav Demerec during the period 1927–1939. Unfortunately, the genetic analysis of these mutations provided no explanation for their mutation instability. In part because no prior convincing mutationally unstable gene had been described in *D. melanogaster* and in part because *D. melanogaster* is a better "cytogenetic" species than *D. virilis*, I undertook the cytogenetic dissection of the partial reversion of w^i, which was named white-crimson (w^c). The objective was simple: determine why w^c is mutationally unstable. I shall not discuss here the details of the analysis except to note that there were noteworthy parallelisms with Barbara's controlling element instabilities. But one salient feature of the w^c story was crucial: the finding that in addition to mutation back to wild type and to the original w^i, w^c generated deletions. Deletions of the *w* gene in *D. melanogaster* were nothing new, but those generated by w^c were different from those heretofore reported. They were fixed endpoint deletions, i.e., they began at the *w* locus and extended either to the left or to the right along the X chromosome but did not overlap the *w* locus. How could this be? Using the maize controlling element concept as a guide and arguing by analogy, I concluded that the w^c mutation, its inordinate mutational instability, and in particular the generated fixed endpoint deletions could be best explained by the insertion of a "foreign" agent akin to a maize controlling element into the w^c gene. Without the precedent of maize controlling elements, I am quite certain I would not have come up with a satisfactory explanation for w^c and its properties. (Parenthetically, with the advent of molecular biology, it was proved by Gerald Rubin and associates that w^c is associated with the insertion of a DNA foldback sequence. Thus, the controlling element explanation proved to be correct.)

Subsequently, I made another observation, one which at that time had not been observed in maize. The observation was that, mediated by the controlling element, the *w* gene could transpose from its normal location on the X chromosome to several different locations on the third chromosome, a *D. melanogaster* autosome. Although transposition of chromosome segments mediated by X-irradiation had already been reported in *Drosophila,* spontaneous gene transposition was a new genetic phenomenon. The w^c transposition story was published in *Genetics* in 1969 under the title "Controlling element mediated transpositions of the white gene in *Drosophila melanogaster*." At that time, copy machines were essentially unknown, so that audience response to publications was gauged by the number of reprint requests received. The response was at best indifferent and thus disappointing. As I recall, I ordered 100 reprints and I still have 80! (Of late I have given away a number of reprints, which for some colleagues are a collector's item.) I could not understand the almost total nonresponse because I had thought that spontaneous gene transposition especially mediated by a controlling element would be of great interest to the community of geneticists. During this time, I served on the National Science Foundation genetics panel and traveled to Washington, D.C. three times a year. I thought it might be useful to seek Barbara's counsel. So before one panel meeting, I telephoned to ask Barbara whether it would be convenient to stop by Cold Spring Harbor and talk. Being eminently polite, she replied to come along.

Accordingly, I visited Cold Spring Harbor and put the question to Barbara pretty much as follows. "What's wrong with the transposition paper? Why has the response been so poor because I do think this is a significant paper?" Her response was, and here I paraphrase, as follows. "Relax, there's nothing wrong with the transposition paper. People aren't ready for this yet. I stopped publishing in refereed journals in 1965 because there was no interest in the maize controlling elements." Here I do believe is the root cause why it took so long for the mainstream of geneticists, especially those working with prokaryotes, to appreciate the significance of controlling elements. When by 1970, insertion sequence (IS) elements were discovered in *Escherichia coli* and when the published literature was examined in earnest, it was quickly found that the seemingly unique IS elements were not unique and the principles of mobile DNA elements had already been described in maize and secondarily in *Drosophila*. Thus, the community of geneticists was now ready for controlling elements.

I am moved to end this narrative with yet another reminiscence and to disclose a long-held secret. Among a small group of geneticists, Barbara was affectionately known as Big Mac. This nickname had everything to do with her intellectual stature and nothing to do with physical stature. Furthermore, this nickname antedates by many years that dubious culinary concoction of the same name. It stems from the following situation. My personal friendship with Barbara began during the summer of 1948 which my wife and I spent at the Cold Spring Harbor Laboratory. At the time, I was on the faculty of the University of Missouri, Columbia. As was her habit, Barbara worked in the field performing those tasks common among corn geneticists: cutting back silks, bagging tassels, making pollinations, etc. When working in the field, Barbara customarily wore blue denim work clothes. There was nothing unusual about this outfit except for the buttons on the denim jacket she usually wore. The buttons were made of brass and embossed on their outer surface in large letters was the brand name, *Big Mac!* To those of us who as students and associates knew Barbara's research and the enormous impact her cytogenetic studies in maize had made on the field of genetics in the 1930s and 1940s, those buttons were most appropriate. For us she was *the* original *Big Mac!* It is therefore fitting to acknowledge my sincere thanks for her scientific contributions and guidance by repeating here the final paragraph of a review I wrote on *Drosophila* mobile DNA elements which appeared in the 1980 volume of the *Annual Review of Genetics*: "No review of transposable elements can be deemed prudent without explicitly acknowledging indebtedness to Barbara McClintock. Combining elegant experimentation with unsurpassed genetic intuition she deduced the existence of transposable genetic elements in maize more than three decades ago. She spelled out the basic ideas and events. The rest of us follow in her footsteps confirming in pro- and eukaryotes what she told us is so but what we have been unprepared or reluctant to believe."

The Mutable waxy and bronze1 Alleles of Maize

OLIVER E. NELSON
University of Wisconsin, Madison, Wisconsin 53706

I am delighted to have the opportunity to add my contribution to this volume dedicated to Barbara McClintock. Her discovery of the transposable elements in maize has been widely recognized as one of the seminal advances in the biological sciences in the twentieth century. What has been less extensively reported is her kindness to and great support of younger researchers who were interested in investigations bearing on any phase of maize genetics and particularly transposable elements. Since I was one beneficiary of this solicitude, it is this aspect of her contributions to genetics that I wish to discuss from a personal perspective using examples showing how her assistance aided our research, although I am well aware of other investigators who have benefited as much from her advice and generous sharing of unique genetic stocks.

I was only recently launched on an independent career when Barbara McClintock's report on the origin of mutable loci in maize was published (McClintock 1950). Although I was aware that many of the elder statesmen of genetics were more than slightly skeptical that there could be genetic entities capable of moving from site to site in the genome, I found the report tremendously stimulating and followed further reports avidly. My copy of the 1951 paper "Chromosome organization and genic expression" (McClintock 1952) is well-thumbed and extensively underlined as is a reprint of the 1953 paper "Induction of instability at selected loci in maize" (McClintock 1953). However, my primary assignment at that time was as a maize breeder, and my forays into more basic research involved investigations of unilateral cross-sterility and developmental mutants such as *Polytypic Ear* that turned up in my breeding plots. There was little opportunity to expand my research further.

In 1955 after returning from a sabbatical year at the Biochemical Institute of the University of Stockholm, my assignment was changed so that I no longer had responsibility for a breeding program, and my research horizons could be expanded. My first initiative was to attempt to test whether intragenic recombination occurred in maize, one of the classic organisms for genetic research. Recombination within genes, which was first unambiguously shown in prokaryotic organisms, had also been documented in eukaryotic organisms such as *Aspergillus* and *Neurospora* but not in the higher eukaryotes with which genetic theory had been developed during the preceding half century. The requirement for very large populations to detect the events that would occur, if they did, at exceedingly low frequencies led to the realization that the *waxy* locus would be the most suitable locus for such an investigation (Nelson 1957). The reason was that Brink and MacGillivray as well as Demerec had reported in the same volume of the *American Journal of Botany* in 1924 that the pollen produced by a *Wx/wx* maize plant is 50% *Wx* phenotypically (stains black with a KI/I_2 solution), whereas the remainder of the pollen stains light brown and is *wx* in phenotype. These observations made it clear that the phenotype of a haploid pollen grain is determined by its own genotype. Thus, if two independently occurring mutations affected different regions of the *wx* locus and recombination occurred between the mutant sites, one product should be a *Wx* pollen grain that should be detectable amid the myriad of *wx* pollen grains under appropriate conditions. Furthermore, the entire population of pollen grains produced by a plant is theoretically available for testing, and a vigorous plant may produce in excess of 20 million. Detailed instructions for collecting pollen and making slides with approximately 5×10^4 microspores have been provided previously (Nelson 1968).

The first set of crosses in all possible combinations between *wx* mutants of independent origin collected from maize geneticists in whose plots these had occurred made it clear that each heteroallelic cross had a characteristic frequency of *Wx* microspores ranging from a frequency that was not higher than that of the homoallelic parents to about 1×10^{-3}. In addition, for heteroallelic plants with a frequency of *Wx* microspores above the homoallelic parents, the distribution of *Wx* microspores per anther fitted a Poisson distribution as would be expected if the events producing them were independent and occurred at meiosis. In addition, it could be shown with the heteroallelic combination (90 x C) that gave the highest frequency of *Wx* microspores in the pollen assay that the frequency could be validated in a conventional test using seeds instead of pollen as the units of genetic observation. Since the assortment of flanking markers could now be ascertained for the *Wx* gametes, it was possible to order the mutant sites within the locus as *bz-wx-C, wx-90-v1* (Nelson 1959, 1962).

The Location of the Mutable *wx* Alleles

As soon as I was convinced that it was possible to observe intragenic recombination with this system, one goal was to ascertain whether the mutable *wx* alleles that had been isolated by Barbara McClintock mapped at different locations within the locus or were concentrated at one site (McClintock 1962). The mutable alleles that belonged to two element systems such as *Ac, Ds* or *dSpm, Spm* could be used if the autonomous member of the system (*Ac* or *Spm*) were absent from the genome. Accordingly, I wrote to Barbara requesting seed of the mutable *wx* alleles of this type, and I received a generous and helpful response, which I was to learn was typical of Barbara's reply to requests. Although it appeared at first that several of the mutants that I requested had been discarded, they could be resurrected from older stocks. I quote here an excerpt from her letter of April 30, 1960. I hope that I will be forgiven for quoting without permission.

> Under separate cover, I am sending seed of three independently arising cases of mutable wx. I found some seed of wx-m1 that came from an ear harvested in 1948. It had been kept for demonstration purposes. It is being sent to you just in

> case you did not receive wx-m1 from Sprague, and with the hope that a few of the kernels may germinate. In addition, I'm sending four ears in which wx–m6 is segregating. You will recognize readily the kernels that carry it. These ears were harvested in 1953, but I believe that some of the kernels will germinate. wx-m6 is Ac-controlled. The rate of mutation to higher alleles of Wx is not high. I suspect that many of the mutations give rise to stable wx alleles, but I did not make tests of this. Two of the ears are segregating kernels carrying the rearranged chromosome 9, described in an early paper of mine that I am also sending you. You may find this rearrangement useful in some of your studies as it cuts down crossing over in the short arm of chromosome 9 to a marked degree. Single crossovers give rise to bridges at meiosis and you will see the consequences of this in some of the kernels on these ears that exhibit the breakage-fusion-bridge cycle.

In a similar vein, a letter of March 16, 1966 offered help that was far beyond what might have been expected in response to a request for seed of a desired genotype.

> Some time ago you told me that you were unable to isolate the original *wx-m6* and that you were using a mutant of it that no longer responded to *Ac*. After some delay, I decided that I should isolate it. This past summer, I obtained some stocks of it without *Ac* but capable of responding to *Ac*. Only one plant was homozygous for *wx-m6* and its self-pollinated ear was poor. There were plenty of *wx-m6/wx* plants that produced good ears. In order to speed isolation, I used a state of *a1-m3* that gives a light to medium pigmentation in the aleurone layer in the absence of *Ac* and gives good mutant expression in its presence. In case you might wish to use the original *wx-m6*, capable of responding to *Ac*, I am enclosing some seed. If you don't wish to use the seed, just throw it away.

The rearranged 9 stock did indeed prove to be useful in future studies since I was able to use it together with two other rearrangements, Tp 9 and Inv 9a, to demonstrate that the *Wx* gametes from the heteroallelic F_1, 90 x C, that had the flanking markers, *bz* and *v1*, which entered the cross with the *wx*-C parent, did not result from double crossovers, one within the *wx* locus and one between the locus and a flanking marker (Nelson 1975). For example, heterozygosity for rearranged 9 drastically reduces crossing over between *wx-90* and *wx-C*, and all of the *Wx* gametes have the flanking markers, *bz* and *v1*, contributed by the *wx*-C parent. The *bz Wx v1* gametes, which comprise 31% of the *Wx* gametes when no rearrangement is involved, presumably arise via gene conversion, although that cannot be directly tested.

The results of the recombination tests involving five mutable *wx* alleles and 19 stable *wx* alleles showed unambiguously that the mutable alleles could recombine with each other and with many of the stable alleles with which they were crossed (Nelson 1968). The alleles that were crossed in all possible combinations could be arranged in linear order within the locus using the method of overlapping deletions (Benzer 1959), where the datum used for any cross is whether or not the two alleles recombined. This was possible since many of the alleles behaved in recombination tests as though they possessed physical size. The physical size as well as the order of the alleles within the locus was later validated by Wessler and Varagona (1985) using molecular techniques. The alleles could not be mapped using the observed crossover frequencies presumably owing to the different *Wx* alleles in which the mutations occurred and the fact that many of the mutant alleles were the result of large insertions or deletions.

The Mutable bronze1 Alleles

During research on the *wx* locus, there emerged a second area in which Barbara McClintock was a major resource for both genetic stocks and advice. In the biological era B.C., which in this context means Before Cloning, it was clear that the association of a transposable element with a functional gene could suppress or attenuate the function of that gene, but there was no information as to how the product of that altered gene was affected. An attempt to fill this particular void became a major goal, and this required a locus whose product was known and at which a number of mutable alleles had been isolated. We knew the product of the functional alleles at the *wx* locus, which is a nucleoside diphosphate glucose:starch glucosyltransferase that transfers glucose to the nonreducing end of starch molecules (Nelson and Rines 1962), and there were certainly mutable alleles available. However, the *Wx* enzyme is very tightly bound to the starch granules and could not be released in an active form, making the locus unsuitable for studies of this type. In its stead, the *bronze1* (*bz1*) locus was selected as a likely candidate. McClintock (1956, 1965) had isolated several mutable *bz1* alleles. Larson and Coe (1968) had suggested in an abstract that the functional alleles encode an enzyme that glucosylates flavonoid molecules. When Hugo Dooner joined the laboratory in the early 1970s, he undertook to examine this suggestion using endosperms as the experimental material and showed that *Bz1* alleles are the structural gene for a UDPglucose:flavonoid glucosyltransferase that transfers glucose to the 3-OH on anthocyanidin molecules (Dooner and Nelson 1977a). Larson and Coe (1977) provided similar evidence.

With this information as a background, it was possible to assess the results of *Ds* insertions in *Bz1* alleles using a group of mutable *bz1* alleles, *bz1-m1*, *bz1-m2(DI)*, *bz1-m2(DII)*, *bz1-m4*, and *Bz1-wm*, that had come originally from Barbara McClintock. Enzyme activity was assayed in extracts of developing or mature endosperms of the mutables when no *Ac* was present in the genome (Dooner and Nelson 1977b). The activity is as high in the mature endosperms of *Bz1/Bz1* seeds as at any time during development. Three different responses to the *Ds* insertion could be noted. No enzyme activity was detectable in any of the alleles when assayed at maturity, nor in many of the alleles during development. However, *Bz1-wm* in one copy per endosperm conditions the production of considerable anthocyanin pigmentation, and homozygous *bz1-m4* seeds show a flush of pigment. The developmental profiles of enzyme activity in these mutants showed that each had measurable enzyme activity early in endosperm development, but enzyme activity started to decrease at between 26 and 30 days postpollination and was unmeasurable at maturity. Investigations subsequently demonstrated that the *Bz1-wm* enzyme was extremely heat-labile as compared to a nonmutant enzyme and apparently equally sensitive to some concomitant of seed maturation. The *bz1-m4* allele conditions a change in tissue specificity. In nonmutant endosperms and in the other mutants that retain any enzymatic activity, all enzymatic activity is in the aleurone layer of the endosperm. In *bz1-m4* endosperms, the bulk of the enzyme activity is found in the subaleurone layers (Dooner 1981), confirming the conjecture of McClintock that a controlling element could have effects on gene action other than reducing or eliminating the gene product.

The isolation of a number of stable *Bz1'* alleles from *bz1-m2(DI)* in the presence of *Ac* offered the opportunity to test whether the revertant alleles were equivalent in function to the *Bz1* alleles from which *bz1-m2* was derived (Dooner and Nelson 1979). Of the 15 *Bz1* ' alleles (the ' being used to indicate that they have been derived from a mutable allele), 5 conditioned the production of an enzyme indistinguishable from that of the progenitor *Bz1* allele. The remaining 10 alleles, although they conditioned full pigmentation with one copy per endosperm, were a heterogeneous group with respect to enzyme activity present at maturity, but all were low relative to their *Bz1* progenitor. The enzymes produced by all were more heat-labile than nonmutant enzyme, had low amounts of cross-reacting material (CRM) at maturity, and lost all activity during gel electrophoresis. So it was apparent that excisions of a transposable element from the same location within a gene resulted in different alterations to the locus.

Although the availability of the collection of mutable *bz1* alleles from Barbara McClintock gave us the tools to investigate the biochemical consequences of insertions and excisions of *Ds* elements in the *bz1* locus, the lack of any *Spm*-controlled mutable alleles of *bz1* was a block to one research objective of long standing. Since learning of the *Spm*-controlled alleles like some states of *a1-m1*, in which what we now know is a defective *Spm* (*dSpm* is inserted in the locus, but a nonmutant phenotype is conditioned in the absence from the genome of an active *Spm* [McClintock 1957, 1965]), I had been fascinated by this particular aspect of the *Spm* system and wished to investigate it with mutants at the *bz* locus with which we had so much experience. The only avenue open was to attempt the isolation of a mutable *bz* allele of the type described. The starting point was a chromosome 9 stock of Barbara's, *c1-m1 Shl Bz1 wx-m8*. The *c1-m1* had an active *Spm* associated with the *c1* locus, and the *wx* locus had a *dSpm* present so that the target *Bz* allele was flanked by transposable elements. The initial strategy was to pollinate the homozygous McClintock stock by a *C shl bz1 wx* tester stock and hope to find variegated kernels (purple sectors superimposed on a *bz* background) indicating that the *Bz1* allele had come under the control of the *Spm* system. The first attempt yielded no heritable variegated *bz1* kernels in spite of a large population, but in the second year, several *bz1* variegated kernels were found in a relatively small population when the McClintock stock was used as the male parent on the *bz1* tester stock. One of these variegated kernels proved to be the sort of *Spm*-controlled mutable alleles we were seeking (i.e., a nonmutant or fully colored phenotype in the absence of an active *Spm* and heavily variegated when *Spm* was present), and it was designated *bz1-m13*.

The *bz1-m13* allele has interesting characteristics (Nelson and Klein 1984). The rate of gametic change of *bz1-m13* to stable *Bz1* ' or *bz1* ' gametes in plants carrying an active *Spm* is high, between 50% and 75%. In the same plants, 0.2 to 0.4 of the gametes have a heritable change to a mutable allele of a phenotype different from that of *bz1-m13* (smaller or fewer colored sectors or both). In the McClintock terminology, these represent "changes of state." Although the *bz1-m13* allele conditions a nonmutant phenotype in one copy per endosperm when *Spm* is absent and the developmental profile of pigment production is indistinguishable from that conditioned by a *Bz1* allele, the enzymatic activity produced by a *bz1-m13* allele is only about 10% of that of the *Bz1* alleles (Klein and Nelson 1983).

Our opportunity to investigate the mutable alleles of the *bz* locus from a molecular basis came when Nina Fedoroff invited my laboratory to join her in an attempt to clone the *bz1* locus using the McClintock allele, *bz1-m2*, which has a copy of *Ac* inserted in a *Bz1* allele, as the starting point. The Fedoroff laboratory had previously cloned *Ac*, so various subclones of that transposable element could be used as probes. Douglas Furtek, who was then a graduate student at Wisconsin, went to Baltimore in response to this invitation. In an intensive 3 weeks of work, they identified a *bz*-specific sequence (Fedoroff et al. 1984), and this was the first instance of using a transposon tag to clone a gene in plants. Furtek then used the *bz*-specific sequence to identify the genomic clones of two *Bz* alleles before sequencing them. The sequencing of the genomic clones in comparison with a cDNA clone of a third *Bz* allele revealed the surprising variation within the coding sequences of the three wild-type alleles, as well as the fact that the *bz1* locus is the simplest example of a mosaic gene with a single intron interrupting the coding sequence (Furtek et al. 1988).

The first application of molecular techniques to *Spm*-controlled *bz* alleles compared the structure of *bz-m13* with that of six changes of state (CS) derived from it in the presence of *Spm*. The *bz-m13* allele has a 2.2-kb *dSpm* inserted in the second exon of the gene. All of the derivatives have *dSpm* inserted at the same location, but in four CS, large deletions had occurred. The remaining two CS could not be distinguished from *bz-m13* by restriction mapping (Schiefelbein et al. 1985). These results were a striking confirmation of Barbara McClintock's hypothesis (1958) that changes in state could be caused by a change in the composition of the transposable element, which remained in the same association with the gene. CS6, which was one of the two CS that could not be distinguished from *bz1-m13* by restriction mapping, was subsequently shown to have suffered a 2-bp deletion in the 13-bp terminal repeat on the 5′ side of the *dSpm*, as it is oriented in the *bz1* locus (Schiefelbein et al. 1988b). CS1, the remaining CS, has a *dSpm* that is identical in sequence to that in *bz1-m13*, but the element is extensively methylated at positions that are not methylated in the *bz1-m13 dSpm* (V. Raboy, pers. comm.).

The examination of enzymatic activities in the husks of plants homozygous for the various CS alleles with no *Spm* present, in comparison with the husks of *bz1-m13* plants, yielded a surprise. The enzymatic activity of CS9 was approximately fivefold higher (or 50% of the *Bz* allele with which it was compared) than *bz1-m13* from which it was derived by a deletion. Owing to the greater enzymatic activity it conditioned, CS9 was selected as the *Spm*-controlled allele with which initially to attempt to assess how the genetic system dealt with a transposable element insertion in an exon to yield substantial activity from the product of that gene. The sequence of a full-length cDNA derived from the mRNA produced by CS9 and Northern blots of the other poly(A) RNAs hybridizing to *Bz* subclones yielded the answer (Kim et al. 1987). The transposable element insert is transcribed with the gene. The processing of this hybrid premessage involves the use of the donor splice site of the intron but a skip of the intron's acceptor splice site. The acceptor splice site used is 2 bp from the 3′ side of the *dSpm* inserted in the gene so that processing of the primary transcript removes the intron, the 38 bases of *Bz* sequence between the intron and the *dSpm*, and all of the *dSpm* except the last 2 bases. So although 38 bases of *Bz* sequence are removed, the two terminal bases of the *dSpm* remain, as does the 3-base duplication of host sequence created on the insertion of the *dSpm*. There is a net loss of 33 bases, but the reading frame is preserved. The protein product lacks 11 amino acids present in a *Bz* product, but the specific activity appears to be similar to that of a *Bz* enzyme, albeit more heat-labile.

It was then possible to learn why the CS9 allele conditioned so much higher enzymatic activity than the *bz1-m13* allele from which it was derived by a large deletion in the 2.2-kb *dSpm* present in *bz1-m13*. The deletion removed an internal acceptor splice site in the *dSpm* that is used in a substantial proportion of the processing events of the *bz1-m13* primary transcript. The use of the internal site yields an mRNA containing a long stretch of *dSpm* sequence that has a translation stop codon and produces no enzymatically active protein. The use of the terminal acceptor splice site gives the same message produced by CS9 but in much lower quantity. Even the enzymatic activity conditioned by *bz1-m13* in the absence of an *Spm* is more than sufficient to produce a nonmutant phenotype. When an *Spm* is present, no gene product is synthesized until excisions of the *dSpm* give revertant alleles with some enzymatic activity. Raboy et al. (1989) have reported on the premessage processing products and the enzymatic activities of the other CS alleles from *bz1-m13*.

To return to the McClintock mutable *bz1* alleles of the *Ac/Ds* system, Schiefelbein et al. (1988a) showed that the *Bz-wm* allele (McClintock 1962), which has weak pigment-synthesizing capacity, differs from the *Bz* alleles previously sequenced by two changes referable to transposable element activity. One is a 3-bp "footprint" in the second exon left after the *Ac* present at that site in *bz1-m2* excised. The footprint is the remnant of the 8-bp host sequence duplication created when the *Ac* inserted. The excision of that *Ac* created a revertant *Bz1* allele that subsequently again came under control of the *Ac/Ds* system. McClintock had shown this to result from the association of a *Ds* with the revertant *Bz'* allele. This *Ds* proved to be a 406-bp *Ds* inserted in the 5'-untranscribed region of the gene, creating a promoter mutation that is responsible for the reduced amount of message produced, and this is the second alteration produced in the gene. We believe that the footprint left in the coding sequence is the cause of the heat lability of the *Bz1-wm* protein noted by Dooner and Nelson (1977b). Sullivan et al. (1989) have shown with *Bz1-wm* and three *Bz1'* revertants that sequence changes in this region (–63 relative to the transcription start site) have tissue-specific effects in that enzymatic activity relative to *Bz* alleles is much more reduced in the aleurone and husk than in the pollen and coleoptile.

In this discussion of research activities that have a direct connection to Barbara McClintock, I have quoted several times from her letters to demonstrate the lengths to which she had gone to be helpful. Such examples could have been multiplied many times over with quotations from letters explaining the derivation of stocks that I was receiving and the beautifully detailed photographs showing what I should look for. For these and for perceptive comments on manuscripts over the years, I am extremely grateful. Her generous assistance made some of our research possible and facilitated a great deal of the remainder. I would like to think that my laboratory has repaid a portion of the debt by supplying some of the biochemical and molecular explanations for the behavior of the fascinating systems that she first called to the attention of the scientific community.

References

Benzer, S. 1959. On the topology of the genetic fine structure. *Proc. Natl. Acad. Sci. 45:* 1607.
Brink, R.A. and J.H. MacGillivray. 1924. Segregation for the *waxy* character in maize pollen and differential development of the male gametophyte. *Am. J. Bot. 11:* 465.
Demerec, M. 1924. A case of pollen dimorphism in maize. *Am. J. Bot. 11:* 461.
Dooner, H.K. 1981. Regulation of the enzyme UFGT by the controlling element *Ds* in *bz-m4*, an unstable mutant in maize. *Cold Spring Harbor Symp. Quant. Biol. 45:* 457.
Dooner, H.K. and O.E. Nelson. 1977a. Genetic control of UDPglucose:flavonol 3-O-glucosyltransferase in the endosperm of maize. *Biochem. Genet. 15:* 509.
———. 1977b. Controlling element-induced alterations in UDPglucose:flavonoid glucosyltransferase, the enzyme specified by the *bronze* locus in maize. *Proc. Natl. Acad. Sci. 74:* 5623.
———. 1979. Heterogeneous flavonoid glucosyl-transferases in purple derivatives from a controlling element-suppressed *bronze* mutant in maize. *Proc. Natl. Acad. Sci. 76:* 2369.
Fedoroff, N.V., D.B. Furtek, and O.E. Nelson. 1984. Cloning of the bronze locus in maize by a simple and generalizable procedure using the transposable controlling element *Activator. Proc. Natl. Acad. Sci. 81:* 3825.
Furtek, D.B., J.W. Schiefelbein, F. Johnston, and O.E. Nelson. 1988. Sequence comparisons of three wild-type *Bronze1* alleles from *Zea mays. Plant Mol. Biol. 11:* 473.
Klein, A.S. and O.E. Nelson. 1983. Biochemical consequences of the insertion of a Suppressor-mutator (*Spm*) receptor at the bronze-1 locus in maize. *Proc. Natl. Acad. Sci. 80:* 7591.
Kim, H.Y., J.W. Schiefelbein, V. Raboy, D.B. Furtek, and O.E. Nelson. 1987. RNA splicing permits expression of a maize gene with a defective Suppressor-mutator transposable element insertion in an exon. *Proc. Natl. Acad. Sci. 84:* 5683.
Larson, R.L. and E.H. Coe. 1968. Enzymatic action of the *Bz* anthocyanin factor in maize. *Proc. Int. Congr. Genet. 12(1):* 131 (Abstr.).
———. 1977. Gene dependent flavonoid glucosyltransferase in maize. *Biochem. Genet. 15:* 153.
McClintock, B. 1950. The origin and behavior of mutable loci in maize. *Proc. Natl. Acad. Sci. 36:* 344.
———. 1952. Chromosome organization and genic expression. *Cold Spring Harbor Symp. Quant. Biol. 16:* 13.
———. 1953. Induction of instability at selected loci in maize. *Genetics 38:* 579.
———. 1956. Mutation in maize. *Carnegie Inst. Washington Year Book 55:* 323.
———. 1957. Controlling elements and the gene. *Cold Spring Harbor Symp. Quant. Biol. 21:* 197.
———. 1958. The Suppressor-mutator system of control of gene action in maize. *Carnegie Inst. Washington Year Book 57:* 415.
———. 1962. Some parallels between gene control systems in maize and in bacteria. *Am. Nat. 85:* 265.
———. 1965. The control of gene action in maize. *Brookhaven Symp. Biol. 18:* 162.
Nelson, O.E. 1957. The feasibility of investigating "genetic fine structure" in higher plants. *Am. Nat. 81:* 331.
———. 1959. Intracistron recombination in the *Wx/wx* region in maize. *Science 130:* 794.
———. 1962. The *waxy* locus in maize. I. Intralocus recombination frequency estimates by pollen and by conventional analyses. *Genetics 47:* 737.
———. 1968. The *waxy* locus in maize. II. The location of the controlling element alleles. *Genetics 60:* 507.
———. 1975. The *waxy* locus in maize. III. Effect of structural heterozygosity on intragenic recombination and flanking assortment. *Genetics 79:* 31.
Nelson, O.E. and A.S. Klein. 1984. Characterization of an *Spm*-controlled bronze-mutable allele in maize. *Genetics 106:* 769.
Nelson, O.E.and H.W. Rines. 1962. The enzymatic deficiency in the waxy mutant of maize. *Biochem. Biophys. Res. Commun. 9:* 297.
Raboy, V., H.Y. Kim, J.W. Schiefelbein, and O.E. Nelson. 1989. Deletions in a *dSpm*

insert in a maize *bronze-1* allele alter RNA processing and gene expression. *Genetics 122:* 695.

Schiefelbein, J.W., D.B. Furtek, H.K. Dooner, and O.E. Nelson. 1988a. Two mutations in a maize *bronze-1* allele caused by transposable elements of the *Ac/Ds* family alter the quantity and quality of the gene product. *Genetics 120:* 767.

Schiefelbein, J.W., V. Raboy, N.V. Fedoroff, and O.E. Nelson. 1985. Deletions within a defective Suppressor-mutator element in maize affect the frequency and developmental timing of its excision from the bronze locus. *Proc. Natl. Acad. Sci. 82:* 4783.

Schiefelbein, J.W., V. Raboy, H.Y. Kim, and O.E. Nelson. 1988b. Molecular characterization of Suppressor-mutator (*Spm*)-induced alleles at the *bronze-1* locus in maize: The *bz1-m13* transposable elements. In *Plant transposable elements* (ed. O.E. Nelson), p. 261. Plenum Press, New York.

Sullivan, T.D., J.W. Schiefelbein, and O.E. Nelson. 1989. Tissue-specific effects of maize *Bronze-1* gene promoter mutations induced by *Ds1* insertion and excision. *Dev. Genet. 10:*412.

Wessler, S.R. and M. Varagona. 1985. Molecular basis of mutations at the *waxy* locus of maize: Correlation with the genetic fine structure map. *Proc. Natl. Acad. Sci. 82:* 4117.

Remembrances of Barbara McClintock

OSCAR L. MILLER, JR.
University of Virginia, Charlottesville, Virginia 22901

As a graduate student at North Carolina State College (it had not become a University yet) after WWII, I became aware of Barbara McClintock's work with the nucleolus of *Zea mays*. At that time, I was working with S.G. Stephens and Ernest Ball on cotton genetics. The question was, how could one introduce commercially valuable genes from wild diploid cotton species into the amphidiploid genome of cultivated cottons.

After I received my Masters Degree in 1950, my wife Mary Rose and I went to South Carolina to do some farming. After 6 years of real dirt farming, we decided it was time for me to go back to school. After putting in applications to several universities, I received a call from Professor Charles Burnham, in the Department of Agronomy and Plant Genetics on the Saint Paul campus of the University of Minnesota, offering me a research assistantship. So Mary Rose and I went to Minnesota. Mary Rose went on to become a fashion model and then a fashion director of a large department store in St. Paul. I consulted with Professor Burnham about a research project, and we decided that I should take a look at the cytology of the "asynaptic" gene in *Zea mays*. At this stage of my scientific career, I became very familiar with the observations Barbara McClintock had made on the nucleolus organizer of *Zea mays*. In fact, I actually saw Barbara's nucleolus organizer in my research (see Fig. 1).

While I was working on my project, Professor Burnham arranged for Barbara to give a seminar in our department. In preparation for her arrival, we were instructed to read all of her publications relative to mutable loci in maize, but especially her 1955 article on the *Ac*-induced instability at the *Bronze* locus on chromosome 9. I do not think any of us completely understood her observations, but I believe we all thought she was a fantastic scientist, as well as a delightful lady.

The next time I met Barbara was in December of 1965 at the International Symposium on the Nucleolus in Montevideo, Uruguay. She had been invited to be co-chairperson of the session on Nucleolar Genes. Crodowaldo Pavan chaired the first half of the session and Barbara, the second. The next to the last speaker was Max Birnstiel, who lucidly summarized his work with Wallace, Sirlin, and Fischberg on the localization of rDNA in the nucleolar

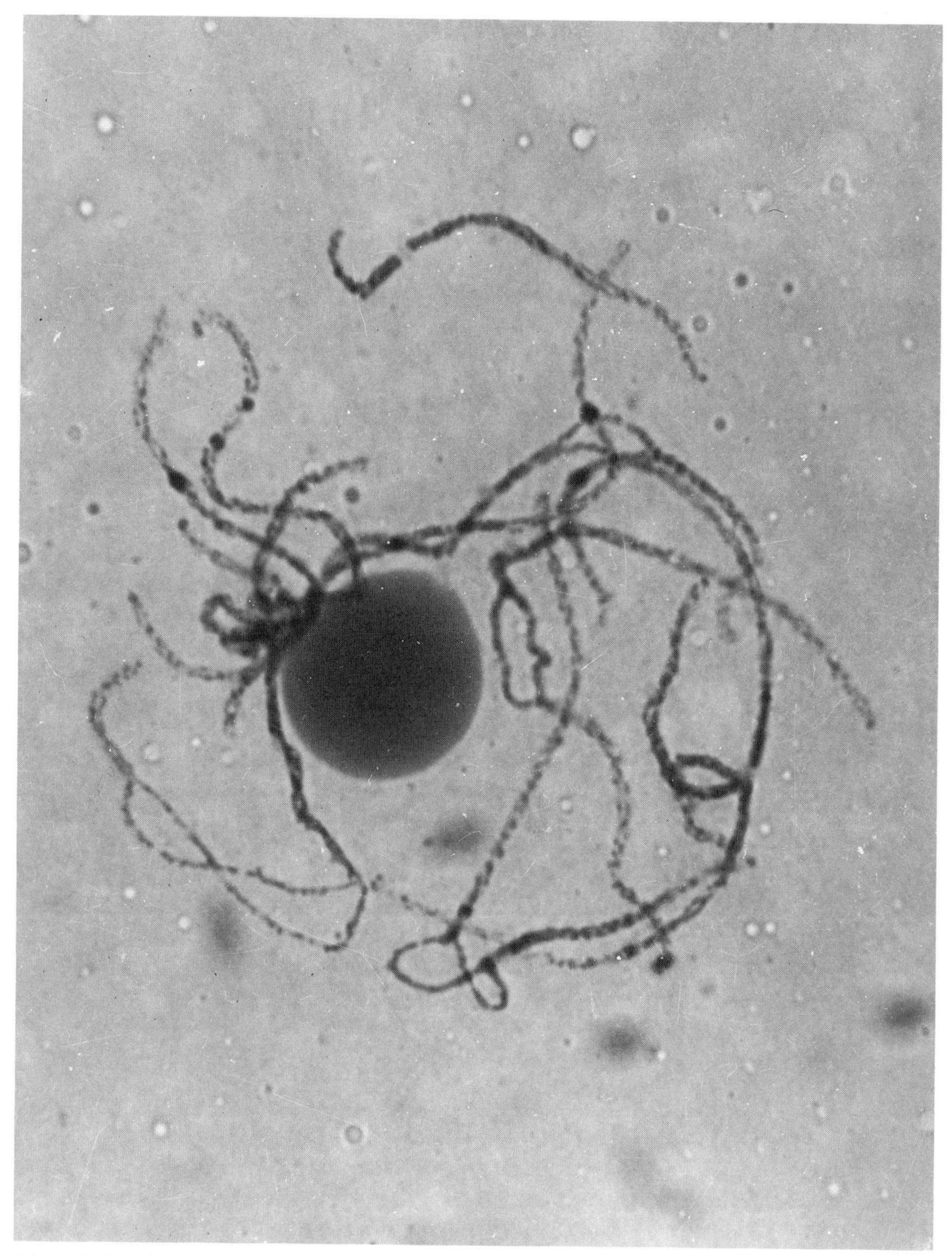

Figure 1 A pachytene stage exhibiting low asynapsis during male meiosis in *Zea mays* (Miller 1963).

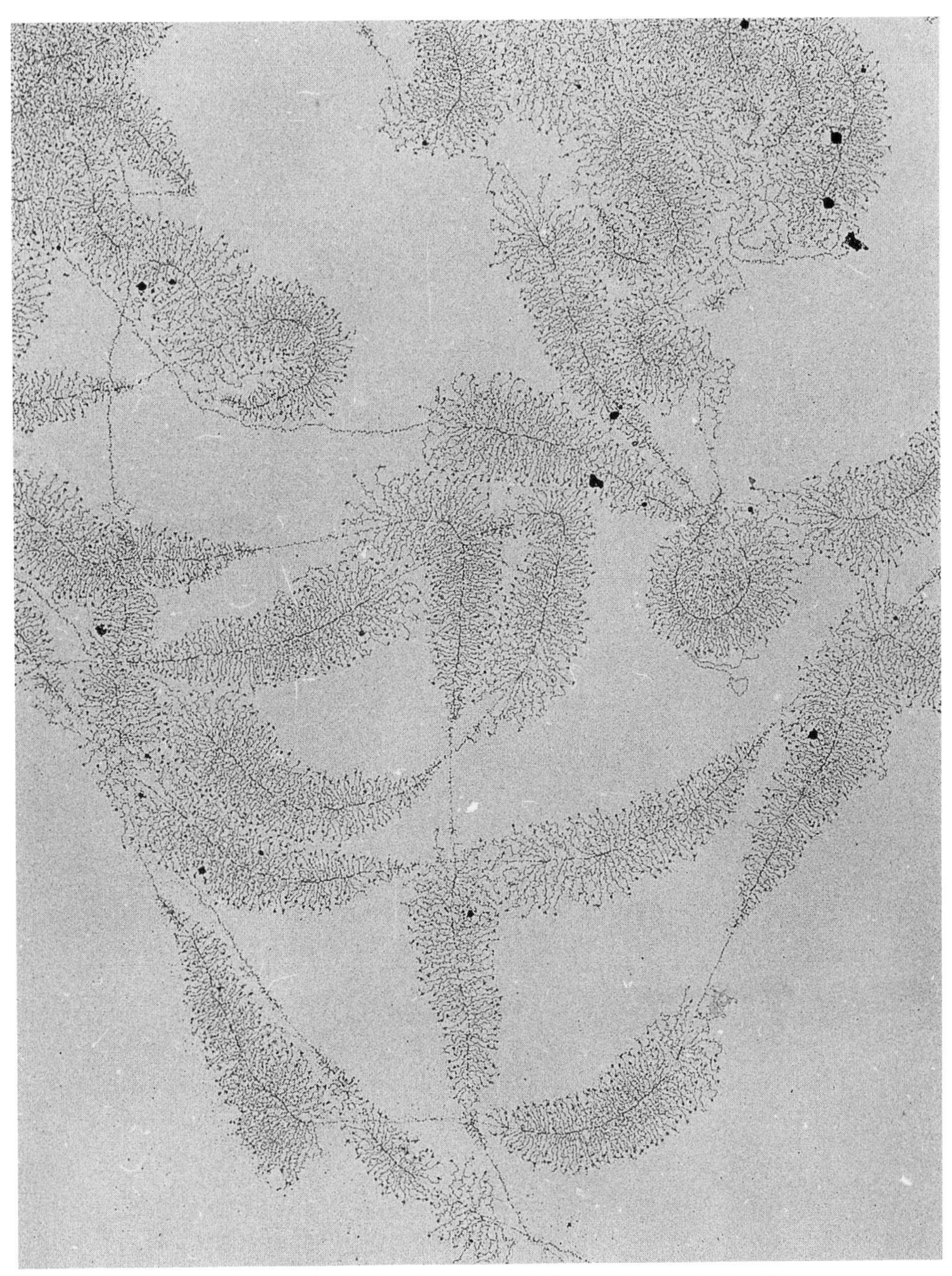

Figure 2 An electron micrograph of extrachromosomal ribosomal genes isolated from an oocyte of the common spotted newt, *Notophthalmus viridescens* (Miller and Beatty 1969 [copyright Wiley-Liss]).

organizer of *Xenopus laevis*. He stayed within his time! I was, as a member of the Organizing Committee, trying to control an orderly procession of the minutes given to each speaker, plus making certain that everything was being recorded for the future. Feruccio Ritossa was the next speaker, reporting on results from Sol Spiegelman's laboratory at Indiana (with Ken Atwood and Dan Lindsley).

Ritossa was presenting essentially the same results that Max had represented, but he and his colleagues did this experiment with *Drosophila*. Barbara is Chairperson. Ritossa has been assigned 15 minutes. Ritossa does not give a summary (the paper would not be published for several months), he gives a full paper. After 15 minutes, Barbara consults with me. I tell Barbara to tell Ritossa he is overtime. She consults with me two or three more times, and I tell Barbara to ask Ritossa to stop. Ritossa is hanging in. Barbara leans over to ask me, what to do? Barbara, take the microphone out of his hand! Then this petite lady goes over and tries to grab the microphone from my good friend, Ritossa. He resists. Although Barbara is persistent, by then, he has presented his entire paper! I was then running 30 minutes late, but it was a beautiful symposium.

I next saw Barbara when I was invited to give a seminar in 1969 at Cold Spring Harbor Laboratory. My seminar was on the Structure of Genetic Transcription, and an example of this work is shown in Figure 2. After the seminar, Barbara invited me to come to her apartment when we finished dinner to talk science. This was one of the most fantastic experiences of my life! Barbara and I had a brandy in her apartment while she talked about the scientific colleagues she had known in her life. It was very intriguing.

The next morning while I was waiting to depart from Cold Spring Harbor by limousine, Barbara met me at the pick-up site. I looked at her; she looked at me! She thanked me for coming to give the seminar. I looked at her, and said, will you be my "Kissing Cousin"? Barbara said, "Yes." We kissed, and I left.

The first course in Molecular Cytogenetics at Cold Spring Harbor was held in the summer of 1971. Charlie Thomas, Mick Callan, Herbert Macgregor, Mary Lou Pardue, Barbara Hamklo, and myself were the instructors. Barbara McClintock was kept informed of everything that was going on with regard to the course. During the course, she "entertained" most of our students, plus the "faculty," in her laboratory. What has impressed me about Barbara is her ability to take young people (graduate student, postdocs, etc.) and even us older people into her laboratory and to instill into all of us the excitement of the use of the human brain to find out something we had never known before!

One memorable experience at this course was that Barbara consented to be Mary Rose's and my dinner guest one evening in our apartment on the top floor of the Firehouse Apartments. Barbara kept us enthralled from 5:00 in the evening to 2:00 in the morning. The evening went all too quickly.

In 1983, Barbara was elected as a Nobel laureate. As soon as Mary Rose and I heard of this, we telegraphed Barbara a dozen red roses. The Lady comes out! The next day, I received a personal phone call from my "Kissing Cousin," thanking us for the roses (she must have had a hundred or so phone calls that day). One of the joys of Mary Rose's and my life was to watch Barbara on television walking down the stairs with her arm on top of the arm of the King of Sweden to go to receive her Nobel prize!

References

Miller, O.L., Jr. 1963. Cytological studies in asynaptic maize. *Genetics 48:* 1445.
Miller, O.L., Jr. and B.R. Beatty. 1969. Portrait of a gene. *J. Cell. Physiol.* (suppl. 1) *74:* 225.

The Nucleolar-organizing Element

JOSEPH G. GALL
Carnegie Institution, Baltimore, Maryland 21210

Out of the hundreds of papers we have each read, a half dozen or so stick in our minds because of their beautiful logic, their clarification of an otherwise obscure set of data, or simply their technical elegance. These are not always the best known works of their authors, and more likely than not, their appeal involves uniquely personal factors. For me, one of Barbara McClintock's early cytogenetic papers falls in this category—her analysis of the nucleolus of maize published in 1934 in the *Zeitschrift für Zellforschung und mikroskopische Anatomie* under the title, "The relation of a particular chromosomal element to the development of the nucleoli in *Zea mays*" (McClintock 1934). Today, the word cytogenetics is often used in the narrow sense of clinical cytogenetics, the mapping of human genes and the study of the relationship between chromosome abnormalities and disease. In its original and broader sense, however, cytogenetics encompassed all the ways in which the structure of chromosomes illuminates genetic analysis and vice versa: What happens when chromosomes are broken, rearranged, deleted, duplicated, or otherwise altered, and what can be learned about the movements or other functions of chromosomes by studying the centromere, the telomeres, the heterochromatic regions, and so forth? Barbara McClintock began as a cytogeneticist; indeed, she helped define the important problems in this field by her early studies on maize. Anyone familiar with her original papers on transposable elements knows that this work, too, began as a problem in cytogenetics, and remains so, even though the molecular aspects now receive the most attention.

The Nucleolus in Corn

McClintock's analysis of the nucleolus in maize concentrated on its behavior during male meiosis. She began by showing that the single nucleolus in the microsporocyte is attached to a specific locus on chromosome 6. This was a new finding for maize, but as she herself pointed out, "The association of a particular chromosome of a complement with the nucleolus is well established." Several earlier workers had noted this fact, and Heitz (1931a,b) had just published two extensive papers in which he demonstrated that the nucleolus in plants arises in telophase at the "secondary constriction" on a specific chromosome (the "primary constriction" in this earlier terminology being the centromere). McClintock carried Heitz's analysis further, however, to show that the association is not with the constriction itself, but with a deeply staining knob or mass of chromatin, which she named the *nucleolar-organizing body* or *element*. The bulk of her paper then dealt with an X-ray-induced translocation between chromosomes 6 and 9 in which the breakpoint in chromosome 6 passed through the nucleolar-organizing element. As a result, the translocated chromosomes (referred to as 6^9 and 9^6) each carried only part of the nucleolar-organizing element, the part on 6^9 being larger than that on 9^6. She studied the behavior of these chromosomes and their associated nucleoli in plants homozygous and heterozygous for the translocation. Her most important conclusion was that each part of the nucleolar-organizing element could organize an independent nucleolus; in other words, that the genetic locus responsible for the nucleolus was in some sense divisible. This finding presaged by over 30 years the discovery of repeated ribosomal RNA genes in the organizers of *Drosophila* (Ritossa and Spiegelman 1965) and *Xenopus* (Brown and Gurdon 1964; Wallace and Birnstiel 1966). Curiously, the two fragments of the nucleolar-organizing element produced nucleoli of different sizes, the small fragment giving rise to a large nucleolus and the large fragment giving rise to to a small nucleolus. That this involved some kind of competition was shown by the fact that when the large fragment was alone in a nucleus, it produced a normal-size nucleolus. The basis for such differences in the "strength" of nucleolus organizers is still not understood.

These findings required skill in setting up and analyzing the genetic crosses and, of course, skill in making the chromosome preparations that show the details of the nucleoli and their organizing elements. One reason so much new information came from McClintock's cytological analysis is that she used the acetocarmine squash technique recently introduced by John Belling. For many years, chromosome cytologists had worked primarily with sections of cells. Sections have many advantages, particularly when it is important to retain the spatial relationships between cells within a tissue. But because chromosomes are long and thin, they are cut into small pieces when cells are sectioned, making it difficult to reconstruct their original morphology. The squash technique overcame this problem by spreading chromosomes out in a single plane, where they could be observed unbroken. The photographs accompanying McClintock's nucleolus paper are of outstanding quality, illustrating the important morphological points with great clarity. To publish such diagrammatic examples, she must have rejected dozens of preparations of lesser quality.

After her work on the nucleolar-organizing element, McClintock studied other parts of the chromosomes, particularly the centromere (McClintock 1938) and the telomere (McClintock 1941, 1942). Again, she made use of chromosome aberrations. For the centromere, she analyzed a small ring chromosome derived from an internal deletion in chromosome 5, where one breakpoint passed through the centromere. As with the nucleolar-organizing element, her analysis showed clearly that the centromere is functionally divisible. The molecular basis for this divisibility remains uncertain, although it may well involve the repetitive satellite DNA sequences found at the centromere. For the telomere, she studied the behavior of broken ends of chromosomes produced by the "breakage–fusion–bridge" cycle set up in mitosis when newly broken chromosome ends fuse with one another. It was while analyzing chromosomes undergoing the breakage–fusion–bridge cycle that McClintock first came to the idea of transposable chromosomal elements. It is beyond the scope of this essay to trace the further development of her ideas on transposition; suffice it to say that they had their beginnings in an eminently classical cytogenetic analysis.

For many years, I tried without success to find a copy of the nucleolus paper for my own reprint collection. Sometime in the mid 1970s, I asked Barbara[1] if she had a copy to spare. Unfortunately, she did not, but she remembered my request and later found and gave me a reprint with the name Lester W. Sharp stamped on it. Sharp was Barbara's Ph.D. advisor at Cornell University. Although his name is no longer a household word, I recognized it as belonging to the author of a well-known textbook, *Introduction to Cytology*, the third and final edition of which was published in 1934. In the preface to the third edition, "Dr. Barbara McClintock" heads the short list of individuals who were credited with reading certain chapters and supplying new material. The book contains many photographs from Barbara's work, including several from the nucleolus paper, which (because of the dates involved) must have been given to Sharp before publication.

The reason I recognized Sharp's textbook is that it and Edmund B. Wilson's *The Cell in Development and Heredity* were the two books that introduced me to the wonders of the cell. As a teenager in the early 1940s, I had my own microscope and enough laboratory paraphernalia to make permanent slides; I also lived on a farm in Virginia where any and all types of plants and animals could be collected in the fields, ponds, woods, and barns. I had a number of amateur works on natural history, plus these two professional books on cells. I clearly remember reading and outlining the chapters in Sharp's text, trying to understand inversions and translocations in corn with the aid of Barbara McClintock's wonderful photographs and diagrams. Little did I know that within a few years I would meet her at a Cold Spring Harbor Symposium and again when she lectured at Yale, where I was a graduate student.

The Nucleolus in Amphibians

I am sure this early introduction to Barbara's work is one reason I find her nucleolus paper so appealing. Another is that my own research centered for a number of years on the nucleolus, particularly the nucleolus organizer. (McClintock's *nucleolar-organizing element* is now generally referred to as the *nucleolus organizer, NO,* or *NOR.*) My interest began with the long-standing puzzle of the multiple nucleoli in the amphibian oocyte nucleus or germinal vesicle (Gall 1954). Barbara's work suggested that the structures in the oocyte might not be true nucleoli, because there were so many of them—up to a thousand in one nucleus—and because they were not attached to the chromosomes like proper nucleoli. I decided to find the true nucleolus in the oocyte, thinking that this would be a good way to show that the extrachromosomal "nucleoli" were, in fact, some other kind of organelle. The nucleolus organizer could be identified on mitotic chromosomes from the position of the somatic nucleolus, and so the problem resolved itself into a search for the organizer on the lampbrush chromosomes. When I identified this locus, the puzzle seemed to be solved: attached to the organizer were one or two objects identical in all respects to the extrachromosomal nucleoli. I

[1]This essay is neither a full-fledged scientific review nor a purely personal account. Thus, I vacillate between "McClintock" when discussing the history of ideas and "Barbara" when recounting personal experiences. I have known and admired both for over 40 years.

concluded, incorrectly, that the multiple nucleoli arose from the nucleolus organizer by successive growth and detachment. During the several months of oocyte enlargement, there would be plenty of time to accumulate a thousand or so extrachromosomal nucleoli in this way. In fact, the true sequence of events had been deduced much earlier by Theophilus Painter and A.N. Taylor (1942). (Painter is sometimes credited with discovering the polytene chromosomes of *Drosophila*. It is more accurate to say that he was among the first to recognize their value for cytogenetic analysis. Polytene chromosomes had been described 50 years earlier by Balbiani [1881], but their usefulness for gene mapping had to await the application of the acetocarmine squash technique—the same technique that let Barbara McClintock find the nucleolus organizer in the meiotic chromosomes.) Painter and Taylor noticed that nuclei in the very earliest oocytes of the toad *Bufo* contain numerous Feulgen-positive granules that are completely separate from the meiotic chromosomes. These granules eventually migrate to the nuclear periphery, where they seem to connect the newly forming multiple nucleoli to the nuclear envelope. It was suggested that the granules are in fact nucleolus organizers that have somehow become polyploid without the chromosomes as a whole becoming polyploid. This remarkable conclusion was absolutely correct, although for a variety of reasons, it was ignored by nearly everyone in the field, myself included. It was only much later, after the discovery of rDNA and the demonstration that the nucleolus organizers of *Drosophila* and *Xenopus* contained rDNA, that the way was open to understand the situation in the oocyte more fully. It occurred to me that if Painter and Taylor were correct, one should find an unusually large amount of rDNA in "baby" *Xenopus* ovaries, which have a high proportion of oocytes to other cell types. I was able to demonstrate this by the newly developed technique of nucleic acid hybridization and also by finding an exceptionally high rate of rDNA synthesis in the early ovary (Gall 1968). Simultaneously, and quite independently, Don Brown and Igor Dawid came to the same conclusion by a different route (Brown and Dawid 1968). They extracted the DNA from some 10,000 hand-isolated *Xenopus* germinal vesicles (from mature oocytes) and analyzed it by ultracentrifugation and nucleic acid hybridization. They found a large excess of rDNA in the germinal vesicle, which they likewise ascribed to specific amplification of the nucleolus organizer sequences.

I have always thought it appropriate that the nucleolus organizer, one of the first objects of Barbara's attention, provided such an elegant example of a movable chromosome element, "jumping" out of the chromosome and replicating independently in a highly regulated way at a defined stage in the life cycle of the germ cell. Her insistence that elements of the chromosomes can move around semi-autonomously made it easy to accept this unusual behavior of the nucleolus organizer in amphibian oocytes.

Cold Spring Harbor Laboratory

During the mid to late 1970s, I taught in a summer course on Molecular Cytogenetics at Cold Spring Harbor organized by Mary Lou Pardue. Each year, a highlight for participants in this course was a lecture given by Barbara. In addition, Mary Lou and I had many opportunities to chat with Barbara personally. Not only did we learn how well she kept up with the latest advances in molecular genetics, but we were provided rare glimpses into the history of genetics and the personalities who helped shape it during the first part of this century. During this time, Barbara learned of my interest in microscopy and the history of microscopical research, and she insisted that I accept various items from her. Chief among these were many notes, drawings, and lantern slides of meiosis in the fungus *Neurospora*. In 1944, George Beadle had invited Barbara to Stanford to look at the chromosomes of *Neurospora*, an organism he and Edward Tatum made famous by their demonstration of the "one gene–one enzyme" hypothesis. During 10 weeks of intense research at Stanford, Barbara distinguished the seven haploid chromosomes of *Neurospora* by length and centromere positions, identified the nucleolus organizer on the second longest chromosome, worked out the chief events of meiosis, and showed that three X-ray-induced aberrations were classical reciprocal translocations (McClintock 1945). Over the next few years, she continued to study *Neurospora* along with her major interest in maize.

Barbara gave me books from her personal library, including a volume of Ernst Abbe's collected works, and others on microtechnique that came originally from John Belling. She also gave me a replica of Leeuwenhoek's microscope, several microscope accessories, and a cube of uranium glass, which is useful for demonstrating the path of light coming from a microscope condenser. Barbara said the cube was a gift from the Bausch and Lomb Optical Company for pointing out a bad case of chromatic aberration in their condenser! Finally, she insisted that I take care of her research microscope, with its set of beautiful apochromatic objectives, because "she no longer needed it and she wanted it to be in safe hands." When an appropriate archive of Barbara's scientific work is established, I will happily relinquish my custody of this historically important instrument.

I consider myself fortunate to have been influenced by Barbara McClintock's keen intellect from the beginning of my career and to have had so many opportunities to talk with her personally and benefit from her deep understanding of genes and chromosomes.

References

Balbiani, E.G. 1881. Sur la structure du noyau des cellules salivaires chez les larves de Chironomus. *Zool. Anz. 99; 100.*

Brown, D.D. and I.B. Dawid. 1968. Specific gene amplification in oocytes. *Science 160:* 272.

Brown, D.D. and J.B. Gurdon. 1964. Absence of ribosomal RNA synthesis in the anucleolate mutant of *Xenopus laevis*. *Proc. Natl. Acad. Sci. 51:* 139.

Gall, J.G. 1954. Lampbrush chromosomes from oocyte nuclei of the newt. *J. Morphol. 94:* 283.

———. 1968. Differential synthesis of the genes for ribosomal RNA during amphibian oogenesis. *Proc. Natl. Acad. Sci. 60:* 553.

Heitz, E. 1931a. Die Ursache der gesetzmässigen Zahl, Lage, Form und Grösse pflanzicher Nukleolen. *Planta 12:* 775.

———. 1931b. Nukleolen und Chromosomen in der Gattung *Vicia*. *Planta 15:* 495.

McClintock, B. 1934. The relation of a particular chromosomal element to the development of the nucleoli in *Zea mays*. *Z. Zellforsch. Mikrosk. Anat. 21:* 294.

———. 1938. The production of homozygous deficient tissues with mutant characteristics by means of the aberrant mitotic behavior of ring-shaped chromosomes. *Genetics 23:* 315.

———. 1941. The stability of broken ends of chromosomes in *Zea mays*. *Genetics 26:* 234.

———. 1942. The fusion of broken ends of chromosomes following nuclear fusion. *Proc. Natl. Acad. Sci. 11:* 458.

———. 1945: Neurospora. I. Preliminary observations of the chromosomes of *Neurospora crassa. Am. J. Bot. 32:* 671.

Painter, T.S. and A.N. Taylor. 1942. Nucleic acid storage in the toad's egg. *Proc. Natl. Acad. Sci. 28:* 311.

Ritossa, F.M. and S. Spiegelman. 1965. Localization of DNA complementary to ribosomal RNA in the nucleolus organizer region of *Drosophila melanogaster. Proc. Natl. Acad. Sci. 53:* 737.

Sharp, L.W. 1934. *Introduction to cytology,* 3rd edition. McGraw Hill, New York.

Wallace, H.R. and M.L. Birnstiel. 1966. Ribosomal cistrons and the nucleolar organizer. *Biochim. Biophys. Acta 114:* 296.

Wilson, E.B. 1925. *The cell in development and heredity,* 3rd edition. Macmillan, New York. (Reprinted [1987] in the series *Great books in experimental biology* [ed. J.A. Moore] by Garland, New York.)

Do Some "Parasitic" DNA Elements Earn an Honest Living?

MARY-LOU PARDUE
Massachusetts Institute of Technology, Cambridge, Massachusetts 02139

The invitation to take part in this celebration of Barbara McClintock comes at a time when I am, once again, rereading Barbara's papers because they have a new relevance to my work. My colleagues and I have recently found that the heterochromatin-specific DNA that we are studying includes a set of transposable elements. This set of elements is remarkable because, as I discuss below, our results strongly suggest that the elements we have identified have a role in chromosome structure. If the suggestion holds true, this set of transposable elements may be contributing members of the chromosome community, rather than the selfish, parasitic DNA that transposable elements are often thought to be.

This is only the most recent of several times when I have read and thought about Barbara's work because of its importance to the questions that I was thinking about at the moment. The questions have been different each time, but it has never been surprising to find that Barbara has been there before me. I was introduced to her work early in my study of biology. The work made a marked impression on me, partly because I was fascinated by genes and chromosomes and partly because Barbara's name was always brought up by professors who were encouraging me to go to graduate school. On the other hand, those professors were all men; Barbara was almost the only woman mentioned, and certainly the only woman mentioned consistently. Perhaps it is not surprising that it took me several years to be convinced that research careers really existed for women.

Early Times

Even though I was not convinced about a research career, I did know that I wanted to do research. I avoided graduate school by going to work in the Biology Division at the Oak Ridge National Laboratory. Looking back, I realize that decision gave me most of the benefits of a graduate education while also giving me a living wage. It also brought me to Barbara's work again. I went to work for Jack von Borstel who was at that time interested in determining the extent to which radiation-induced dominant lethals could be explained by the chromosome imbalance theory. The approach was to compare the radiation-induced lethality with the lethality from genetically contrived chromosome imbalance (von Borstel and Rekemeyer 1959). The project taught me a lot about chromosome mechanics, and, in particular, it introduced me to Barbara's papers on the breakage–fusion–bridge cycle (McClintock 1938, 1941, 1942).

My next encounter with Barbara's work came when I began doctoral research at Yale with Joe Gall. At that time, it had just become possible to study eukaryotic genes at the molecular level—as long as those genes were the genes for ribosomal RNAs. The limitation to rRNA genes may seem stringent to those who have now become accustomed to cloning almost any gene that strikes their fancy. Nevertheless, being able to use the rDNA genes opened up many questions about genes and associated chromosomal structures. It also gave a marvelous model system for developing the in situ hybridization technique (Gall and Pardue 1969; Pardue and Gall 1969). I spent a lot of time reading Barbara's wonderful cytogenetic analysis of the nucleolar region of the maize chromosome (McClintock 1934). Eventually, we were able to use in situ hybridization to show that the stalk attaching the satellite to the chromosome contained the rRNA genes (as did the dotted lines that Barbara drew through the nucleoli) (Pardue et al. 1970; Evans et al. 1974). Unfortunately, if the animal cells that we worked with actually had the nucleolar organizing body that Barbara identified (McClintock 1934), we were unable to detect it. I still wonder whether the nucleolar organizing body contains inactive rRNA genes or whether it contains other interesting components that remain to be discovered.

I finally met Barbara in the summer of 1971. At that time, I was a postdoctoral fellow at the University of Edinburgh and had gone to Cold Spring Harbor Laboratory to teach in the Molecular Cytogenetics course, a 3-week summer course initiated in 1971 to replace the phage course originated by Max Delbrück. I returned to Cold Spring Harbor to teach Molecular Cytogenetics or a successor course on *Drosophila* for the next 9 years. Each of those summers was a wonderful time of talking with and learning from Barbara. One of the many topics was heterochromatin. My interest in this mysterious material had intensified when I mapped mouse satellite DNA to pericentric heterochromatin during my thesis research. All of the satellite DNA appeared to be heterochromatin but not all heterochromatin was satellite DNA (Pardue and Gall 1970). I wanted very much to know what other sequences were in heterochromatin. Were they also satellite-type DNAs (very short tandemly repeated sequences) or were they very different kinds of DNAs? I also wanted to know how these sequences and the known satellite DNAs explained the varied properties that had been attributed to heterochromatin. Of course, I realized that the repeated DNAs in heterochromatin meant that any attack on the problem had to be made with finesse. For that reason, I wanted to know as much as possible about heterochromatin and considered myself extremely lucky to have the opportunities to learn from Barbara.

A New Class of Heterochromatic Sequences, HeT DNA

My search for interesting heterochromatin sequences was begun in a very relaxed way. *Drosophila* polytene chromosomes are the ideal subjects for such a study because the pericentric heterochromatin can be resolved into a tiny compact region of α-heterochromatin and the more diffuse surrounding β-heterochromatin (Heitz 1934). The satellite DNA sequences are in the α-heterochromatin and are very underreplicated (Gall et al. 1971). Thus, sequences hybridizing in situ specifically with β-heterochromatin should be the nonsatellite DNAs that I wanted. In situ hybridization is an early step in the characterization of most cloned DNAs from *Drosophila*. I simply watched for a cloned DNA fragment from any of the projects in my laboratory—or any other—that hybridized specifically with β-heterochromatin. Many of the clones did hybridize with β-heterochromatin but also hybridized with many euchromatic sites in patterns, suggesting that the sequences contained transposable elements. Those I did not consider bona fide heterochromatin, since transposable elements had not been seen to confer properties of heterochromatin on euchromatic regions. Eventually, Barbara Young, a postdoctoral fellow in my laboratory, found a clone fitting my criteria and named it λT-A (Young et al. 1983). This clone has served to define a complex family of heterochromatic DNA. To our surprise, the family includes at least one set of transposable elements, and, as discussed below, this set of elements may indeed be responsible for some of the properties of heterochromatin.

λT-A contained a 9.4-kb piece of *Drosophila melanogaster* DNA that hybridized only with β-heterochromatin and with the most distal band on each polytene chromosome. Telomeric regions can be seen to be heterochromatic in organisms with larger chromosomes. Cytological analysis is more difficult on smaller metaphase chromosomes, but they too might be expected to have telomeric heterochromatin. (The morphology of the most terminal band on each polytene chromosome cannot be unambiguously classified as either heterochromatin or euchromatin.) Our hybridization studies provide evidence that telomere regions share sequences with β-heterochromatin, supporting the hypothesis that *Drosophila* chromosomes have telomeric heterochromatin.

Because of its specific association with heterochromatic regions, the DNA in λT-A was named HeT DNA. Subfragments from the initial λT-A clone cross-hybridized with complex groups of middle repetitive DNA fragments in *Drosophila* DNA. In situ hybridization experiments, in which we can easily detect 40 base pairs of unique sequence in euchromatic regions of the polytene chromosomes, showed no hybridization of any HeT DNA sequences within the euchromatin of any of the more than 20 *D. melanogaster* strains analyzed. Thus, all sequences that have homology with λT-A are absolutely confined to heterochromatin (Valgeirsdottir et al. 1990). We now regard all sequences that have homology with the original λT-A probe as members of the HeT DNA family. This is a family that is related by chromosome location; however, since λT-A is a mosaic of several kinds of sequences, the HeT DNA family must contain several different sequence subfamilies.

Heterochromatin and Chromosome Ends

HeT DNA seems to be present in the most terminal regions on polytene chromosomes, even hybridizing across strands of DNA connecting ectopically paired chromosome tips. Still, from these cytological preparations, we cannot draw strong conclusions about the relationship of HeT DNA to the end of the chromosomal DNA because, even on the enlarged polytene chromosomes, resolution is not at the molecular level. In addition, the known underreplication of sequences in pericentric heterochromatin (Gall et al. 1971) suggests the possibility of similar underreplication of sequences in telomeric heterochromatin. Thus, there is still some question of whether the ends of polytene chromosomes represent true telomeres. Nevertheless, the apparent telomeric localization of HeT DNA raises questions about its relationship to the simple telomerase-generated G-rich repeats found at the ends of eukaryotic chromosomes (Zakian 1989). We and others have failed to find these simple repeats in *Drosophila*, although all of the techniques used to study telomere repeats in other organisms have been tried. The negative results cannot prove that *Drosophila* lacks telomerase and G-rich repeats, but they argue that, if the repeats exist, they are much less abundant than in other organisms.

It should be noted that, just internal to the simple telomerase-generated repeats, eukaryotic chromosomes have more complex sets of repeats (Zakian 1989). These complex repeats are called telomere-associated repeats; however, their constant association with chromosome ends raises the possibility that these repeats are responsible for some of the functions that cytologists have associated with telomeres. It seems likely that the telomere-associated sequences are actually part of the telomere if the telomere is defined as the element responsible for all of the functions that have been postulated by cytologists. These functions include mediation of telomere-telomere and telomere-lamina interactions. Such functions may be complex because some interactions appear to be cell-type-specific. Thus, it is possible that telomere-associated sequences play a role in chromosome pairing and/or in maintenance of the three-dimensional organization of chromosomes within the nucleus. The notable exceptions to the presence of telomere-associated repeats are the macronuclear chromosomes of protozoa. These macronuclear chromosomes are actually chromosome fragments. They do not take part in typical mitosis or meiosis, and it would not be surprising if their telomere structure were atypical.

The spectrum of organisms in which telomerase repeats have been found (protozoa, fungi, animals, and plants) argues that, if *Drosophila* truly lacks these repeats, it is the result of an evolutionary loss. The HeT DNA sequences that we see associated with *Drosophila* telomeres resemble telomere-associated sequences of other organisms in size and complexity. They may well be derived from the same ancestral elements as the telomere-associated sequences in other organisms. Perhaps *Drosophila* has dispensed with the telomerase-generated repeats and now makes do with only the telomere-associated sequence portion of the telomere. If so, understanding HeT DNA should give insights into telomere-associated sequences in general.

HeT DNA and "Healing" of Terminally Broken Chromosomes

What we have named the HeT DNA family appears to contain several sequence subfamilies. All subfamilies are confined to heterochromatin, but in situ hybridization detects some subfamilies only in telomere regions, some only in pericentric regions, and some in both regions (Valgeirsdottir et al. 1990). We have recently begun detailed characterization of one of these subfamilies, the HeT-A subfamily (H. Biessmann et al., in prep.). To our surprise, the HeT-A element appears to be a transposable element. HeT-A transposition has been detected only to the broken ends of chromosomes; it has been added to the ends of all of the known "healed" telomeres in

Drosophila. This specificity of transposition, its apparent confinement to telomeres in all unbroken chromosomes, plus features of its sequence (see below) suggest that HeT-A elements have a bona fide structural role in heterochromatin. If this is true, it has implications for our concepts of heterochromatin and transposable elements.

Our hybridization experiments have shown that the multiple copies of HeT DNA sequences are very well conserved, suggesting that sequences on different chromosomes are evolving together, exchanging information either by sequence movement or by gene conversion (Valgeirsdottir et al. 1990). Our first evidence that there was actual sequence movement came from Karen Traverse's analysis of a strain carrying the C(1)A chromosome (Traverse and Pardue 1988). C(1)A was a ring chromosome constructed in 1964 by joining the ends of two X chromosomes, thus deleting all telomeres (Lindsley and Grell 1968). Since 1964, it has been used as an attached-X chromosome in a number of stocks. Because it contains the complete genetic information of two X chromosomes, C(1)A provides the entire sex chromosome content needed for the female. Thus, it is passed through females in each generation and has never been associated with a normal homolog that could contribute telomeres by recombination. Sometime soon after it was made, the C(1)A ring opened, not at the site of original joining, but in the middle of the euchromatin (polytene region 13) of one of the X chromosomes. Although region 13 shows no homology with HeT DNA in unbroken chromosomes, both of the new telomeres on C(1)A have acquired HeT DNA.

The broken C(1)A chromosome has been stable as a linear attached-X for more than 20 years. The polytene band pattern gives no evidence that the chromosome ever entered the breakage–fusion–bridge cycle that Barbara discovered. However, the stability of C(1)A suggests that it has undergone the "healing" that Barbara described in broken chromosomes that passed into the sporocyte (McClintock 1941, 1942). We would like very much to know the molecular structure of the healed ends, but walking from currently cloned sites to the break in region 13 would be labor-intensive. It was fortunate for us that soon after we discovered the new C(1)A ends, a much more favorable set of healed telomeres appeared.

The new set of healed telomeres appeared in a collection of terminally deleted X chromosomes made by Jim Mason in 1983 (Mason et al. 1984). Searches for terminally deleted chromosomes in *Drosophila* have been carried out many times but have been uniformly unsuccessful (Roberts 1975). Jim has manipulated the genetics in ways that apparently circumvented mechanisms preventing the detection of terminal deletions in earlier studies. As in those earlier studies, chromosome breaks were produced by X-radiation; however, in Jim's experiments, the isolation of terminally deleted chromosomes was potentiated by radiation of *mu-2/mu-2* females, in which double-strand DNA breaks apparently fail to be repaired during oogenesis (J. Mason, pers. comm.). Broken X chromosomes were recovered and subsequently carried in males so that the broken ends could not be healed by recombination with an intact homolog. Although *mu-2* appears to be involved in establishing the chromosome breaks, it is not necessary for maintenance; the terminal deletions are carried in stocks that are wild type for *mu-2*. These males also have the y^2 *sc Y* chromosome which carries a duplication for the tip of X and should be able to cover the loss of any essential loci in the region. Jim selected for chromosomes broken in the *yellow* locus (very close to the end of the X chromosome) and recovered several lines (the RT lines) that he has maintained since 1983.

Jim Mason sent me some of the RT deletion stocks soon after they were established. I found no hybridization of HeT DNA to the newly broken ends on polytene chromosomes. That result could have been interpreted to mean that HeT DNA was not necessary at the telomere. On the other hand, it was also possible that the result meant that the broken ends did not have proper telomeres. The second possibility may seem overly optimistic in view of the evidence that broken ends are recombinogenic, producing either bridges or translocations. However, the second possibility later proved to be the correct one. When Harald Biessmann had cloned and sequenced the *yellow* gene, he and Jim Mason studied the structure of the ends of several of the RT chromosomes that had been broken in this gene (Biessmann and Mason 1988; Biessmann et al. 1990a). Most of the chromosomes had no DNA of any sort distal to the break in the *yellow* gene. The chromosomes simply ended at the break. The result was surprising because these chromosomes had been in culture for several years and appeared relatively stable, although phenotypic changes indicated progressive loss of the distal parts of the *yellow* gene. The sequence analyses showed that the broken ends were receding at approximately 75 nucleotides per generation, a rate consistent with the failure to replicate the DNA covered by the 5′ primer for DNA replication. Among the set of RT lines, two lines showed evidence of new DNA added to the broken end.

When Harald and Jim found that two of the RT lines appeared to have new DNA added to the broken end, I suggested that the chromosome might have acquired HeT DNA, as had the C(1)A chromosome. This began a three-way collaboration (Biessmann et al. 1990b) that has shown that a defined subclass of HeT DNA, the HeT-A element, has been involved in "healing" these broken chromosomes. The term "healed" is placed in quotation marks to emphasize that this may not be the complete reestablishment of the telomere. It is very likely that this is only the first step in a longer process. The possession of the HeT-A element gives the new telomere homology to the telomere regions of other chromosomes. It seems likely that this sets the stage for recombinations that add other sequences.

Our collaboration showed that the healing of the two RT chromosomes (in lines 475 and 394) had occurred in a very interesting and unexpected way. Each of the chromosomes had acquired approximately 3 kb of HeT DNA. Hybridization analysis showed that the two additions were colinear and very similar in sequence. Surprisingly, both of the additions had been joined to the broken ends (within the *yellow* sequence) by an oligo(A) segment. The 475 and the 394 *yellow* genes had been broken at different places and neither of the two sites of joining had any homology with the HeT DNA sequence. The sequence analyses provide strong evidence that the new sequences were not attached to the broken end by homologous recombination.

A Heterochromatic-specific Transposable Element, HeT-A

Further analyses have shown that the sequences added to the broken chromosome ends represent a specific element that we have named the HeT-A element (H. Biessmann, K. Valgeirsdottir, A. Lofsky, C. Chin, B. Ginther, R.W. Levis, and M.L. Pardue, in prep.). The name HeT-A was chosen on the assumption that HeT DNA will prove to contain additional families of elements. One constant feature of the HeT-A element is the oligo(A) segment (3–28 residues on the elements studied) that marks one end of the element. This oligo(A) segment is attached to the break in the *yellow* gene and is thus at the internal end of each of the elements on the healed chromosomes. What we know about the HeT-A elements at normal telomeres suggests that they have a similar polarity. Bob Levis has cloned a set of DNA fragments extending from a *P*-element insert partway into the 3R telomere region. This cloned DNA includes one HeT-A element and that element has the oligo(A) on the internal side. We do not know the orientation of any of the other elements that have been derived from normal telomeres. The λT-A insert appears to be derived from the telomere of a normal X chromosome. Although we have not been able to determine the insert's orientation on the chromosome, the two large HeT-A elements in λT-A are oriented in the same direction. These two elements give the λT-A insert a polarity and suggest that, in this case also, the oligo(A) may be on the internal end of the element. λT-A does have one very short (60 bp) element of opposite polarity; however, this element is flanked by 8-bp direct repeats. Most of the repeated bases lie within the element and include its 3-bp oligo(A) segment, suggesting that the tiny element has undergone an unusual rearrangement that perhaps inverted the sequence on the chromosome (Valgeirsdottir et al. 1990).

A second notable feature of the HeT-A elements is that, when elements are compared, sequence homology begins at the oligo(A) end and continues (with a few insertions or deletions) for varying distances up to nearly 2700 bp. These two characteristics, oligo(A) at one end and variable truncation of the other end, are very reminiscent of class II retroposons, a large class of nonviral elements that transpose through an RNA intermediate. For this class of retroposons, the oligo(A) segment indicates the 3′end and the variably truncated end is 5′. (The best known member of this class is the mammalian LINE element [Hutchison et al. 1989].) For elements of this class, it has been proposed (Bucheton 1990) that the 3′end of the polyadenylated RNA intermediate attaches to a break in chromosomal DNA. Reverse transcription is then primed by the 3′hydroxyl group of the other strand of the chromosomal DNA to make the first strand of the transposition insert. Truncation at the 5′end could be explained if the transcriptase does not always proceed to the 5′end of the RNA. We have not yet found an RNA transcript of HeT-A, but sequence analysis of the transpositions onto the broken *yellow* genes shows clearly that transposition has not been by homologous recombination. All of the HeT-A elements studied can be fit into the pattern of 3′oligo(A) and 5′variable truncation that characterizes the many class II retroposons. Our working hypothesis is that HeT-A elements are retroposons of the LINE class, although we cannot rule out transposition as a DNA molecule at this point.

Drosophila has a number of LINE-type retroposons. These retroposons include I elements (Finnegan 1989), DOC (Schneuwly et al. 1987), jockey (Mizrohki et al. 1988), F elements (DiNocera and Casari 1987), G elements (DiNocera et al. 1986), and the ribosomal DNA insertions, type I and type II (Jakubczak et al. 1990). HeT-A elements show one very significant difference from all of these other elements: HeT-A never appears at euchromatic sites unless the site has become the end of a broken chromosome. Most of the other LINE-type elements are distributed throughout the euchromatic regions of the chromosomes. There are two exceptions to this general pattern. The rDNA insertions show a very strong preference for insertion into a single site in the rDNA; G elements have been detected only in pericentric heterochromatin. It has been suggested that the positional restriction of the ribosomal insertions and the G elements is determined by sequence-specific endonucleases encoded by these elements (DiNocera 1988; Xiong and Eickbush 1988). We find no evidence that the HeT-A insertion is determined by such an endonuclease. The 3′-flanking sequences for some 19 HeT-A elements have been determined and each is different, giving no evidence for sequence-specific insertion.

The similarities between HeT-A elements and LINE-type retroposons suggest that HeT-A might encode proteins involved in its transposition; however, we have been unable to detect any open reading frames (ORFs) in HeT-A. Studies of retroposons have shown that many elements can be defective and no longer able to encode proteins. Our search for ORFs has been carried out with that in mind. The search has centered on the elements 473 and 394 since these transposed less than 3 years before they were sequenced. They have therefore had very little time for the sequence to decay since the time when they were at least capable of transposition, whether or not they were capable of providing the transposition proteins. We have found no significant ORFs, nor have we been able to find regions of similarity to any of the ORFs found in other retroposons. Several algorithms have been devised to look for ORFs in sequences that have been interrupted either by mutations or by sequencing errors. None of the algorithms that we have used have given evidence of large but degenerate ORFs in the HeT-A sequence.

Because the HeT-A elements studied appear to be variably truncated at one end, we may not have found a complete element yet. The complete element may have the typical ORFs. Nevertheless, whether or not there are additional coding sequences, the structure of HeT-A differs significantly from that of other retroposons. The known retroposons tend to have very little sequence outside the sequence of the ORFs (or the degenerate ORFs). In contrast, our HeT-A sequences show more than 2 kb at one end of the element (the 3′ end if HeT-A is indeed a retroposon), which gives no evidence of encoding an ORF but has more than 75% identity between all sequenced HeT-A elements. The high level of sequence conservation argues strongly that this rather long sequence has a function. A second novel feature of the HeT-A sequence suggests a possible function; there is a pattern of 80-bp repeats that are distinct in the central regions of these elements and in more degenerate form in other regions. Such repeats might play a role in chromatin structure of the telomere region. This feature, taken with our evidence about the sites in which HeT-A is found, strongly suggests that the HeT-A element is a bona fide structural element of the chromosome.

Perhaps There Are Other "Working Class" Transposable Elements

If the HeT-A element does in fact have a role in chromosome structure and its transposition serves a repair function, perhaps it does not encode its transposition proteins. Although other transposons do encode proteins, it is known that transposition elements can act in *trans* to mobilize defective elements. If HeT-A is a bona fide chromosomal component, it might be more economical for the cell to allow transposition of the element only when repair is needed. It seems likely that transposition could be controlled more efficiently if the necessary proteins were encoded by a master gene, rather than each individual element. This line of thought raises the possibility that the transposable elements that now populate euchromatin are sequences that managed to capture their own copies of master genes and escape the limits of heterochromatin.

The realization that bona fide chromosomal elements might have a relationship to transposable elements has again set me to reading Barbara's papers because she had already found a relationship between chromosome structure and such elements (McClintock 1978). Barbara's experiments showed that the breakage–fusion–bridge cycle mobilized transposable elements. The elements that she was detecting in those studies would be of the class that we postulate might be escapees from the restrictions that control movement within the genome. Such escapees would be the types of elements that now control their own transposition machinery. Perhaps those elements were also able to catch a ride on the cell's repair system when that system was activated by chromosome breakage.

When I was in Kathmandu several years ago, I bought a painting from a young Nepalese artist. When he learned I was a geneticist, the painter remarked that a lady geneticist had recently won the Nobel prize. I'm sure that Barbara's travels have not included Nepal; nevertheless, in one sense, she had also been there before me.

Acknowledgments

I have been fortunate in the colleagues who have shared in the work discussed here. They are Barbara Staller Young, Karen Lahey Traverse, Katrin Valgeirsdottir, Ariel Lofsky, Annalisa Pession, and Cynthia French at the Massachusetts Institute of Technology; Harald Biessmann at the University of California, Irvine; Jim Mason at NIEHS; and Bob Levis at the Hutchinson Cancer Center. The work in my laboratory has been supported by grants from the American Cancer Society and the National Science Foundation.

References

Biessmann, H. and J. Mason. 1988. Progressive loss of DNA sequences from terminal chromosome deficiencies in *Drosophila melanogaster. EMBO J 7:* 1081.

Biessmann, H., S.B. Carter, and J. Mason. 1990a. Chromosome ends in *Drosophila* without telomeric DNA sequences. *Proc. Natl. Acad. Sci. 87:* 1758.

Biessmann, H., J.M. Mason, K. Ferry, M. d'Hulst, K. Valgeirsdottir, K.L. Traverse, and M.L. Pardue. 1990b. Addition of telomere-associated HeT DNA sequences "heals" broken chromosome ends in *Drosophila. Cell 61:* 663.

Bucheton, A. 1990. I transposable elements and I-R hybrid dysgenesis in *Drosophila. Trends Genet. 6:* 16.

DiNocera, P.P. 1988. Close relationship between non-viral retroposons in *Drosophila melanogaster. Nucleic Acids Res. 16:* 4041.

DiNocera, P.P. and G. Casari. 1987. Related polypeptides are encoded by *Drosophila* F elements, I factors, and mammalian L1 sequences. *Proc. Natl. Acad. Sci. 84:* 5843.

DiNocera, P.P., F. Graziani, and G. Lavorgna. 1986. Genomic and structural organization of *Drosophila melanogaster* G elements. *Nucleic Acids Res. 14:* 675.

Evans, H.J., R.A. Buckland, and M.L. Pardue. 1974. Location of the genes coding for 18s and 28s ribosomal RNA in the human genome. *Chromosoma 48:* 405.

Finnegan, D.J. 1989. The I factor and I-R hybrid-dysgenesis in *Drosophila melanogaster.* In *Mobile DNA* (ed. D.E. Berg and M.M. Howe), p. 503. American Society for Microbiology, Washington, D.C.

Gall, J.G. and M.L. Pardue. 1969. Formation and detection of RNA-DNA hybrid molecules in cytological preparations. *Proc. Natl. Acad. Sci. 63:* 378.

Gall, J.G., E.H. Cohen, and M.L. Polan. 1971. Repetitive DNA sequences in *Drosophila. Chromosoma 33:* 319.

Heitz, E. 1934. Uber α- und β-Heterochromatin Sowie Konstanz und Bau der Chromomeren bei *Drosophila. Biol. Zentralbl. 54:* 588.

Hutchison, C. A., III, S.C. Hardies, D.D. Loeb, W.R. Shehee, and M.H. Edgell. 1989. LINEs and related retroposons: Long interspersed repeated sequences in the eucaryotic genome. In *Mobile DNA* (ed. D.E. Berg and M.M. Howe), p. 593. American Society for Microbiology, Washington, D.C.

Jakubczak, J.L., Y. Xiong, and T.H. Eickbush. 1990. Type I (RI) and type II (RII) ribosomal DNA insertions of *Drosophila melanogaster* are retrotransposable elements closely related to those of *Bombyx mori. J. Mol. Biol. 212:* 37.

Lindsley, D.L. and E.H. Grell. 1968. *Genetic Variations of* Drosophila melanogaster. Carnegie Institution of Washington. Washington D.C.

Mason, J.M., E. Strobel, and M.M. Green. 1984. *Mu-2:* Mutator gene in *Drosophila* that potentiates the induction of terminal deficiencies. *Proc. Natl. Acad. Sci. 81:* 6090.

McClintock, B. 1934. The relation of a particular chromosomal element to the development of the nucleoli in *Zea mays. Z. Zellforsch. 21:* 294.

———. 1938. The fusion of broken ends of sister half chromatids following chromatid breakage at meiotic anaphases. *Mo. Agric. Exp. Stn. Res. Bull. 290:* 1.

———. 1941. The stability of broken ends of chromosomes in *Zea mays. Genetics 26:* 234.

———. 1942. The fusion of broken ends of chromosomes following nuclear fusion. *Proc. Natl. Acad. Sci. 28:* 458.

———. 1978. Mechanisms that rapidly reorganize the genome. *Stadler Genet. Symp. 10:* 25.

Mizrohki, L.J., S.G. Georgieva, and Y.V. Ilyin. 1988. Jockey, a mobile *Drosophila* element similar to mammalian LINEs, is transcribed from the internal promoter by RNA polymerase II. *Cell 54:* 685.

Pardue, M.L. and J.G. Gall. 1969. Molecular hybridization of radioactive DNA to the DNA of cytological preparations. *Proc. Natl. Acad. Sci. 64:* 600.

———. 1970. Chromosomal localization of mouse satellite DNA. *Science 168:* 1356.

Pardue, M.L., S.A. Gerbi, R.A. Eckhardt, and J.G. Gall. 1970. Cytological localization of DNA complementary to ribosomal RNA in polytene chromosomes of *Diptera. Chromosoma 29:* 268.

Roberts, P. 1975. In support of the telomere concept. *Genetics 80:* 135.

Schneuwly, S., A. Kuriowa, and W.J. Gehring. 1987. Molecular analysis of the dominant homeotic Antennapedia phenotype. *EMBO J. 6:* 201.

Traverse, K.L. and M.L. Pardue. 1988. A spontaneously opened ring chromosome of *Drosophila* has acquired He-T DNA sequences on both new telomeres. *Proc. Natl. Acad. Sci. 85:* 8116.

Valgeirsdottir, K., K.L. Traverse, and M.L. Pardue. 1990. HeT DNA: A family of mosaic repeated sequences specific for heterochromatin in *Drosophila. Proc. Natl. Acad. Sci. 87:* 7998.

von Borstel, R.C. and M.L. Rekemeyer. 1959. Radiation-induced and genetically contrived dominant lethality in *Habrobracon* and *Drosophila. Genetics 44:* 1053.

Xiong, Y. and T.H. Eickbush. 1988. Functional expression of a sequence-specific endonuclease encoded by the retroposon R2Bm. *Cell 55:* 235.

Young, B.S., A. Pession, K.L. Traverse, C. French, and M.L. Pardue. 1983. Telomere regions in *Drosophila* share complex DNA sequences with pericentric heterochromatin. *Cell 34:* 85.

Zakian, V.A. 1989. Structure and function of telomeres. *Annu. Rev. Genet. 23:* 579.

The Plural of Heterochromatin

CHARLES D. LAIRD
University of Washington, Seattle, Washington 98195

As an Undergraduate Research Participant (URP) at Cold Spring Harbor Laboratory during the summer of 1960, I occasionally encountered Barbara McClintock on walks or while picking berries. A decade or so later, I had learned to ask more questions than I did that first summer. One of my most important questions was to ask whether McClintock would be able to help with the Molecular Cytogenetics course that Mary Lou Pardue and I, and others, taught at Cold Spring Harbor in the early 1970s. McClintock's acceptance of our invitation was tentative; she said that students probably preferred to learn about new molecular techniques than to hear an overview from her. Anyone who has seen McClintock interact with students—young and old—will understand that her reticence was unnecessary: Her contributions were among the truly memorable ones for our summer groups. And as they do for a few exceptional teachers, McClintock's contributions can resurface at unexpected times, often years and decades later.

Often our discussions with McClintock included a walk through local fields and woods. Each walk was a journey through biology. Old floral friends would first elicit exclamations of surprise and delight from McClintock and then help her to illustrate a concept of general importance. Queen Anne's Lace showed us variable pigmentation patterns that would come to mind years later as I worked on the human fragile-X syndrome. The coordinated flight pattern in a flock of birds raised questions of nonverbal communication. The collection of flower parts for chromosome preparations—or blackberries for consumption—was accompanied by discussions of the biology of each species.

After a walk, we would head back to the laboratory where we were treated to panoramas of chromosomes that included those of maize and *Neurospora,* often using the same slides whose chromosomes had appeared decades earlier in published micrographs (McClintock 1945). Looking through the microscope, we were told, is an important form of data collection; the brain may eventually sort out patterns if one is diligent and patient enough. And questioning enough, she would add. In particular, McClintock impressed upon us the importance of questioning the assumptions of cytogenetics. At the time of the Molecular Cytogenetics courses, I was a relatively new contributor to the cytogenetic literature (Laird 1971); the significance of McClintock's admonitions was not as apparent as it has since come to be. I suspect that cytogenetics is especially vulnerable to accumulating hidden assumptions because the discipline has such strong visual and logical components. It is often tempting to think that what we see can be interpreted definitively. A definitive interpretation, given a name, can easily mislead us until some lucky researcher or naive beginner ferrets out the hidden assumption.

And then, quickly, we would turn to new findings. One afternoon, McClintock brought up the recently described experiments of Rao and colleagues on precocious chromosome condensation in rodent cells (Sperling and Rao 1974). She wondered whether these observations had relevance to the "nucleic acid starvation" model proposed more than 30 years earlier to explain abnormal condensation of plant chromosomes (Darlington and La Cour 1940). Years later, this possible connection would resurface at a critical moment as we tried to understand the significance of "intercalary heterochromatin" in *Drosophila* (Lamb and Laird 1987). Our new understanding of this obscure phenomenon encouraged me to apply several concepts of cytogenetics to inherited human diseases. Cytogenetics has a rich tradition of enjoying and sharing information on chromosomes of any organism, as McClintock had demonstrated to us; my application to humans of concepts garnered from *Drosophila* and the mouse is squarely within this tradition.

My view of heterochromatin has broadened since the cytogenetics course. I now refer to heterochromatins in the plural to help remind me, and perhaps others, of the diversity of heterochromatic structures and circumstances that can be relevant to biological processes. In my laboratory, we have unexpectedly been confronted with cytogenetic phenomena that could involve, under present conventions of terminology, constitutive, facultative, and intercalary heterochromatins. My use of the plural refers to the variety of contexts in which my thoughts have unexpectedly turned to these topics. Projects that began with no overt connection have all been illuminated with various aspects of heterochromatin. I review here four projects from my laboratory that have provided glimpses of the diversity of heterochromatins.

Glimpse One: Function of a Single Ribosomal Gene Away from Its Home in Pericentric Heterochromatin

"Pericentric heterochromatin" describes the more intensely staining chromatin that flanks the centromeres of most eukaryotic chromosomes. Pericentric heterochromatin is usually enriched for highly repeated DNA sequences (Pardue and Gall 1970; Gall et al. 1971), for which there is little evidence of genetic function (Heitz 1934; Miklos and Cotsell 1990; see, however, Hareven et al. 1986; Hennig et al. 1989). It has also been termed "constitutive" heterochromatin to reflect this presumed genetic inertness (Chandra and Brown 1975). There are a few transcribable sequences of known genetic function that are imbedded in pericentric heterochromatin (Ritossa and Spiegelman 1965; Devlin et al. 1990), including, in *Drosophila melanogaster*, the ribosomal genes (Fig. 1) (Laird and Chooi 1976). If most pericentric heterochromatin is genetically inert, what functional reasons might there be for embedding in this heterochromatin a few transcribed genes?

The advent of germ-line transformation in *D. melanogaster* provided an important and long-awaited tool that could be used for this and other questions in genetic biology (Rubin and Spradling 1982; Spradling and Rubin 1982). Gary Karpen and I decided to use germ-line transformation to ask about transcriptional control of a *Drosophila* ribosomal gene. Midway through the construction of the appropriate plasmids, we realized that germ-line transformation with our control plasmid (Fig. 2) was itself the basis for a very

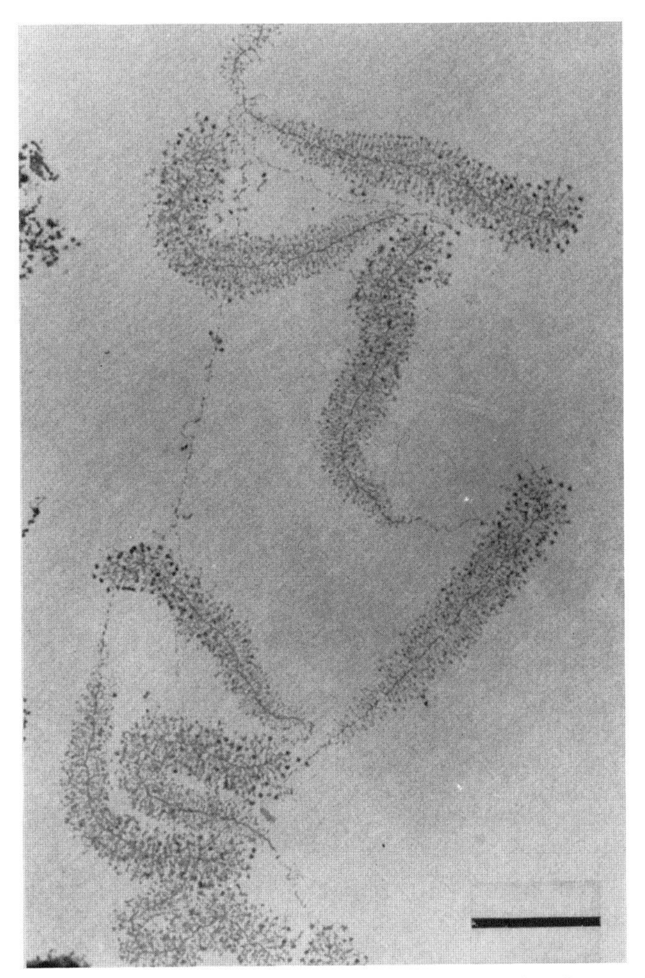

Figure 1 Interpretation of electron micrograph: ribosomal transcription units from embryos of *D. melanogaster*. The high level of transcriptional activity and the tandem repeat nature of the transcription units are characteristic of most ribosomal genes. The term "transcription unit" recognizes the relatively specific starts and stops of transcription that are visually and analytically apparent (Foe et al. 1976; Laird et al. 1976). This term is used interchangeably with "ribosomal gene," and refers to the encoding of the precursor RNA for the two large ribosomal RNA molecules. Bar, 1 μm. (Reprinted, with permission, from Laird and Chooi 1976.)

interesting experiment. We had been assuming that a ribosomal gene would function perfectly well in euchromatin. Although this assumption appears to be correct, it was initially unjustified because we had no way of knowing whether or not a single ribosomal gene would function away from its tandemly repeated neighbors and heterochromatic location. Since *P*-element-mediated transformation in *Drosophila* is usually random, rather than targeted to homologous genes, it was likely that a single ribosomal gene introduced by *P*-element transformation would integrate in the euchromatin arms, rather than in pericentric heterochromatin of the X or Y chromosomes. Integrating a single ribosomal gene into sites on the euchromatic arms allowed us to ask three interesting questions.

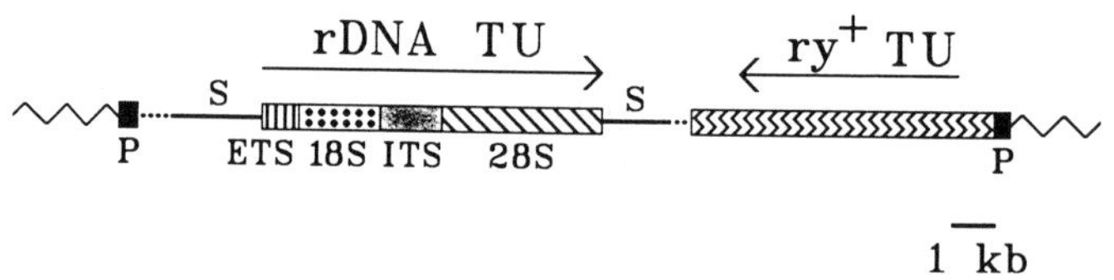

rDNA TU
ry⁺ TU
S S
P ETS 18S ITS 28S P
1 kb

Figure 2 Diagram of the *P*-element construct used for germ-line transformation of a single ribosomal gene into *D. melanogaster*. A single ribosomal transcription unit (TU), with flanking "nontranscribed spacers" (S), was inserted into the Carnegie 20 vector containing *P*-element ends and the *rosy*$^+$ transcription unit (ry$^+$ TU) (Rubin and Spradling 1982; Spradling and Rubin 1982). A single transformant at chromosome site 1A was obtained from injections of over 6000 embryos with this and other plasmids. The insert was mobilized to other chromosome sites by injecting embryos with the "wings-clipped" plasmid (Rubin and Spradling 1982; Spradling and Rubin 1982), which provided a source of transposase. (Reprinted, with permission, from Karpen et al. 1988.)

Can a single ribosomal gene be transcribed and make functional ribosomal RNA when it is removed from its normal location in heterochromatin?

In *Drosophila*, the answer appears to be "yes." Transcription of an ectopically located ribosomal gene was assessed in polytene chromosomes of salivary glands by in situ hybridization to nascent ribosomal RNA. Four sites were favorable for cytogenetic study. These were located on the X, 2nd, and 3rd chromosomes, and all four showed evidence of appropriate transcriptional activity in that the correct strand and not its complement was transcribed. Molecular characterization of DNA from the transformed lines indicated that the single rDNA gene had integrated without duplication (Karpen 1987 and pers. comm.; Karpen et al. 1988). We inferred that the transcriptional products were functional because the ectopically located ribosomal gene provided partial rescue from the deleterious effects of a *bobbed* mutation, which represents a partial deletion of the endogenous ribosomal genes (Karpen et al. 1988). Thus, ribosomal RNA can be synthesized and processed to functional RNA even when its template gene is removed from the context of its normal home in pericentric heterochromatin and from the tandem repeat organization that is typical of ribosomal genes.

It should be noted that the salivary gland chromosomes that we used for our assays of transcriptional activity provide a lateral repeat organization of the ectopic ribosomal gene that is generated during polytenization. Our experiments did not assess in diploid cells the transcriptional activity of a single, ectopically located ribosomal gene. There is, however, one activity of an ectopic ribosomal gene that has been assayed in nonpolytene cells. McKee and Karpen (1990) asked whether or not the single ribosomal gene located near the tip of the X chromosome could restore proper meiotic segregation of the X and Y chromosomes when the endogenous ribosomal genes had been deleted from one of the chromosomes. Partial recovery of normal disjunction was observed, indicating that in *D. melanogaster*, a ribosomal gene with flanking spacer sequences represents, or can substitute for, the DNA sequences involved in meiotic segregation of the X and Y chromosomes (McKee and Karpen 1990).

Is a single ribosomal gene sufficient to organize a nucleolus?

The molecular cytogenetic correlation between rDNA and the nucleolus (Ritossa and Spiegelman 1965; Brown and Dawid 1968) provided a molecular foundation for the proposal that ribosomal RNA is encoded, synthesized, and partially assembled into ribosomes at the site of the nucleolus. The concept of the nucleolar organizing region of a chromosome and the prediction of a repetitive nature of the underlying genetic information were developed to explain cytogenetic experiments involving chromosome breakage at the nucleolar organizer (McClintock 1934). We wondered whether or not the information in a ribosomal gene was sufficient to organize a nucleolus. In particular, could we use an ectopically integrated ribosomal gene to exclude in *D. melanogaster* a requirement for flanking pericentric heterochromatin in the formation of a nucleolus? We observed a small structure with morphological and molecular similarities to that of the main nucleolus of the salivary gland cells at the site of an ectopic ribosomal gene (Fig. 3). Our interpretation, based on cytological staining, the distribution of nascent RNA, and the detection of an antigen observed elsewhere only in the main nucleolus, is that this structure represents a mininucleolus.

If our interpretation is correct, we can conclude that a ribosomal gene can serve as a nucleolar organizer even when it lacks flanking heterochromatic sequences and the tandem repeat organization of its endogenous site. Further experiments are necessary to address the important question of whether, as seems likely, transcription of an ectopically located ribosomal gene is necessary to form a mininucleolus. Moreover, we do not know whether the two "nontranscribed spacer" regions that flank our ribosomal gene ("S" in Fig. 2) are necessary for transcription and nucleolus formation. Resolution of these two questions will be necessary to complete the dissection of the nucleolar organizing region begun by McClintock (1934).

If a single ribosomal gene can function outside of an endogenous nucleolar organizer, why are the multiple copies of the ribosomal genes normally clustered in centromeric or telomeric heterochromatin?

Karpen speculated that one possible reason for the clustering of ribosomal genes within heterochromatin observed in almost all organisms is that such clustering protects euchromatic genes from deleterious effects of the ribosomal genes (Karpen 1987). Since the ribosomal genes are transcriptionally active in most cell types and are usually transcribed at high rates, there may be special problems for transcription of other genes that find themselves in the neighborhood of a nucleolus. Competition for nucleoside triphosphates, for example, might lead to underexpression of a euchromatic gene that finds itself near a ribosomal gene that escaped from its normal site in heterochromatin. Karpen's speculation was based on the difficulty he experienced in obtaining *P*-element transformants containing a ribosomal gene. One cause of this difficulty may have been inhibition by the active ribosomal gene of expression of the marker gene, *rosy*$^+$. Preliminary experiments indicated that the level of transcripts from the transformed *rosy*$^+$ gene was indeed reduced in vivo when *rosy*$^+$ was adjacent to an ectopic ribosomal gene (Karpen 1987).

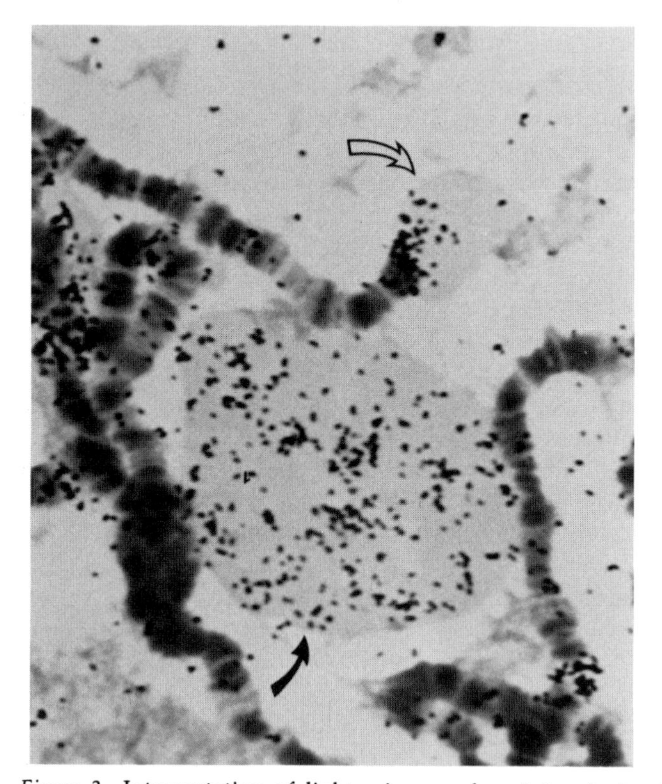

Figure 3 Interpretation of light micrograph: mininucleolus (open arrow) associated with site 1A of the polytene X chromosome of *D. melanogaster*. The large endogenous nucleolus is also shown (closed arrow). This preparation of salivary gland chromosomes is from a larva containing the single ribosomal gene integrated at site 1A; after heat shock, [^{3}H]uridine was added to salivary glands to label preferentially nascent ribosomal RNA. Autoradiography and Giemsa staining were carried out as described elsewhere (Karpen et al. 1988). (Reprinted, with permission, from Karpen et al. 1988.)

Karpen's suggestion that heterochromatin may have a protective effect on the expression of euchromatic genes highlights one way that heterochromatin could be beneficial without regard to a particular DNA sequence. The rapid evolutionary changes in the highly repeated sequences of some heterochromatic DNA (Southern 1970) may not reflect an absence of function of heterochromatin, but rather the relative independence of function and DNA sequence. The counterpoint to this suggestion is that heterochromatic sequences may also be important for the expression of the few coding sequences, such as *light*, that reside within it. These few genes may require the special context of heterochromatic DNA for either biochemical or positional information (Devlin et al. 1990).

Glimpse Two: Intercalary Heterochromatin

Kaufmann (1939) proposed the term "intercalary heterochromatin" to refer to several dozen unusual sites on the generally euchromatic arms of polytene chromosomes of *D. melanogaster*. These sites have some cytogenetic properties

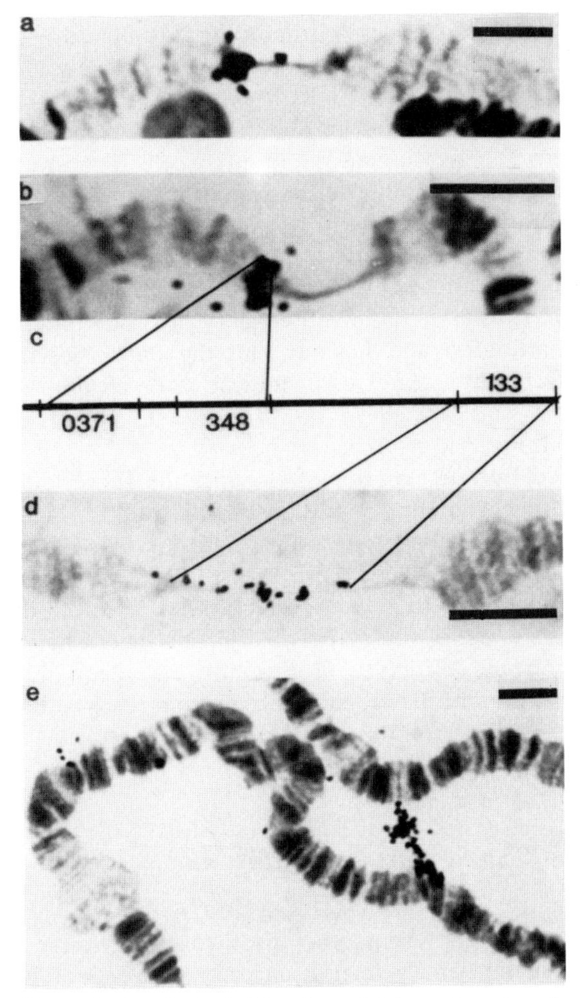

Figure 4 In situ hybridization of tritiated probes from site 11A to polytene chromosomes of *D. melanogaster*. Only probe 133 hybridizes to the constriction at 11A (*d*) and to ectopic fibers (*e*). This probe (but not the other two probes) shows reduced levels of polyteny (see Fig. 4 in Lamb and Laird 1987). Bar, 5 μm. (Reprinted, with permission, from Lamb and Laird 1987.)

in common with pericentric heterochromatin, including chromosome constrictions, fragility, and ectopic chromatin fibers. When molecular cytogenetic procedures became available, several laboratories noted that DNA sequences from two genetic loci, *Ultrabithorax* and the histone genes, and from chromosome site 11A had cytogenetic properties expected of intercalary heterochromatin (Fig. 4) (Pardue et al. 1977; Lifschytz 1983; Lamb and Laird 1987).

These observations led Mary Lamb and me to question the implication that intercalary heterochromatin represents DNA of constitutive heterochromatin that had been intercalated into euchromatin (Lamb and Laird 1987). Although some examples of such intercalation do exist (Cohen and Kaplan 1982), they may not be general. In particular, we proposed that the cytogenetic properties of intercalary heterochromatin arose from the property of late or delayed DNA replication of these sequences, a property that is shared with pericentric heterochromatin in *Drosophila* polytene cells (Lamb and Laird 1987). (Previously, Martin Hammond and I had shown that one highly repeated sequence from the pericentric heterochromatin is only delayed, not halted, in its replication pattern in salivary glands and nurse cells [Hammond and Laird 1985a,b].) This delayed replication also ties intercalary heterochromatin to the concept of facultative heterochromatin (Chandra and Brown 1975) in other organisms. For example, the inactive X chromosome in mammals appears primarily to be late-replicating. Our demonstration of delayed replication for several sequences at sites classified as intercalary heterochromatin supported a suggestion by Darlington and LeCour that the properties of intercalary heterochromatin (Kaufmann 1939) were indicative of chromosome changes induced by "nucleic acid starvation" as proposed for plant chromosomes (Darlington and La Cour 1940).

We were thus led by our experiments to question the implication of the term intercalary heterochromatin. Would the term facultative heterochromatin (Chandra and Brown 1975) be more appropriate for these *Drosophila* sites? Perhaps this designation would work for the *Ultrabithorax* locus, but it may not be appropriate for the histone locus, which is expressed in the same cells in which it manifests some of the cytogenetic properties of heterochromatin. There is obviously some difficulty in reaching an adequate terminology distinguishing the different heterochromatins from each other and from euchromatin. I return to this point in the final section.

Glimpse Three: The Fragile-X Syndrome, Imprinting, and Facultative Heterochromatin

The fragile-X syndrome represents the most common cause of inherited mental retardation in humans, affecting an estimated 1 per 1200 newborns (Brown 1990). This syndrome also represents the only known disease that is caused by, or closely associated with, a chromosomal fragile site. The term "fragile site" refers to a chromosomal site that is reproducibly broken under certain conditions of cell culture. Fragile sites are more often detected by failure of the chromosome to condense properly at that site (see Sutherland and Hecht 1985); in that property, they resemble a secondary constriction in condensed chromosomes that is characteristic of the nucleolar organizer (Heitz 1931; McClintock 1934).

The patterns of inheritance and expression of the fragile-X syndrome are complex in unusual and interesting ways. In particular, a large fraction of males who carry this X-linked mutation have no phenotypic manifestations of the syndrome; their daughters are never classified as mentally retarded, even though their granddaughters (and grandsons) frequently are so classified (Sherman et al. 1985; Sved and Laird 1990). I summarize here some of our work on this genetic disease. Our results on the fragile-X syndrome have heightened our interest in genomic imprinting and facultative heterochromatin as they relate to human disease; they have also led to an inference concerning human embryology (Laird et al. 1990).

General model of fragile sites

My colleagues and I proposed that chromosomal fragile sites represent *cis*-acting mutations to late replication of DNA at the fragile site (Laird et al. 1987). This proposal was based in part on our above-described observations on intercalary heterochromatin of *Drosophila* (Lamb and Laird 1987); it follows a suggestion that faulty DNA replication is involved in chromosome breakage at knobbed regions induced by B chromosomes of maize (Rhoades et al. 1967). Cytogenetic data have been reported that test one aspect of this model: The deleterious state of the fragile-X mutation appears to be later replicating in the cell cycle than the normal allele (Yu et al. 1990). Further experiments are necessary to quantify these replication patterns in more detail and to assess the relationship between late replication and the proposed imprinting event described in the following section.

Model of inheritance and expression of the fragile-X syndrome

On the basis of extensive cytogenetic, clinical, and epidemiological data that were available in the literature, I proposed that the fragile-X syndrome is a disorder of abnormal chromosome imprinting (Laird 1987). According to my model, the fragile-X mutation blocks the reactivation, prior to oogenesis, of a mutant X chromosome that was inactivated for dosage compensation in a female (Fig. 5). When inactivation and incomplete reactivation occur, there is a change in state of the mutant fragile-X allele: It becomes imprinted in the sense that a gene or genes at the fragile-X site fail to be inducible in some future cell and organismal generation. The use of the term "chromosome imprinting" follows the definition proposed by Crouse (1960) and applied to X-chromosome inactivation in female mammals by Chandra and Brown (1975).

The late-replicating, inactive chromosome in female mammals is considered to represent facultative heterochromatin: It is stably but temporarily heteropycnotic, representing the Barr body that is sometimes used to distinguish mammalian female cells from male cells (Barr and Bertram 1949). Some genes on the inactive X are so stably inactivated that not even removal of chromosomal proteins is sufficient to reactivate a previously inactive gene; removal of methylation of specific cytosine residues near the 5′ end of the gene appears to be necessary for gene activity (Gartler and Riggs 1983; Pfeifer et al. 1990). Thus, the proposed imprinted state of the mutant fragile-X allele may be considered as a small region of facultative heterochromatin that has persisted abnormally.

Many testable predictions are made by this proposal that the fragile-X syndrome represents a disorder of abnormal chromosomal imprinting. In particular, detailed molecular predictions are available because the proposal specifies that the cause of the imprinting is the failure to reactivate completely an inactive X chromosome. Three confirmed predictions that are especially relevant to this discussion of heterochromatin are that (1) abnormal patterns of hypermethylation of DNA would be found at the mutant fragile-X site in affected individuals (Bell et al. 1991; Heitz et al. 1991; Oberlé et al. 1991); (2) these changes will reflect a persistence of a pattern of hypermethylation similar to that established during X-chromosome inactivation (Bell et al. 1991;

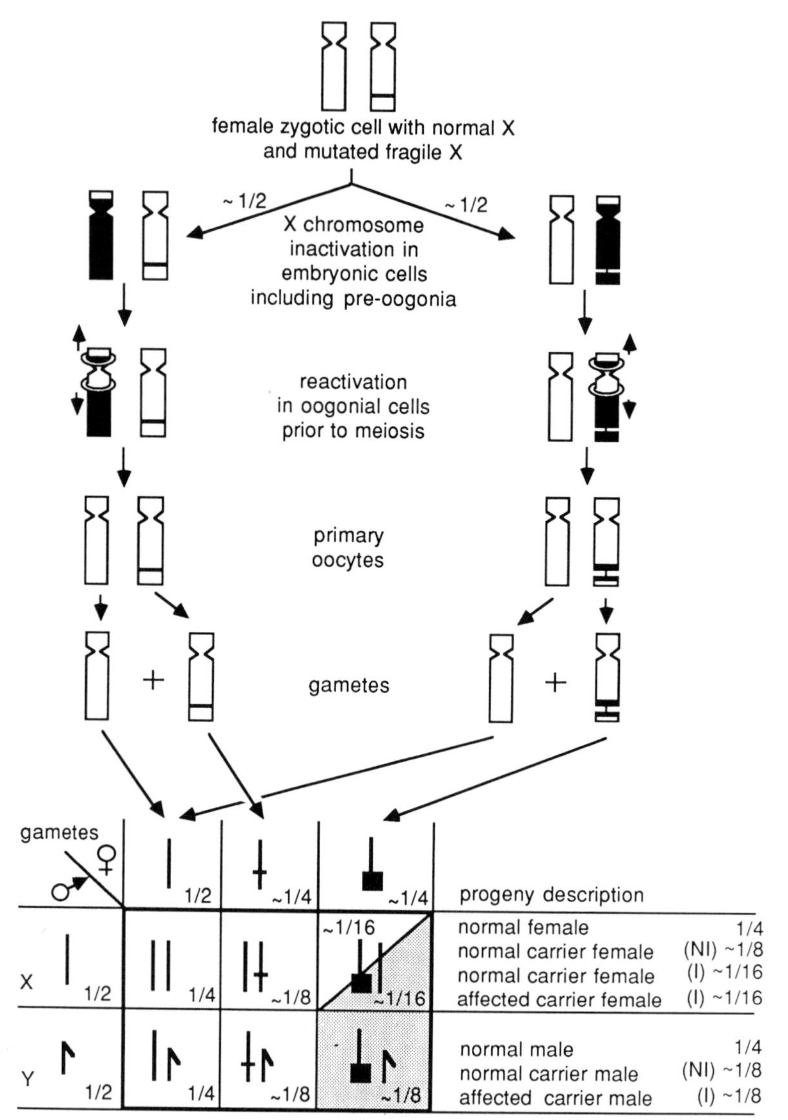

Figure 5 Proposed transitions in the mutant fragile-X allele during imprinting in females. According to this model, the involvement of X-chromosome inactivation and the failure to reactivate completely a mutant fragile-X chromosome lead to a change in state of the mutant allele from nonimprinted (NI) to imprinted (I). The predicted proportions of the various classes of gametes and progeny from females who imprint the mutant allele are indicated. These predictions refer to averages for all imprinting females; individual variation in imprinting among females may reflect a severe bottleneck in the number of oogonial progenitor cells after the time of X-chromosome inactivation (Laird et al. 1990). (Reprinted, with permission, from Laird 1987.)

Heitz et al. 1991); and (3) a gene (or genes) near the site of the mutation is silenced by the methylation rather than making an abnormal gene product (Pieretti et al. 1991).

The likely involvement in the fragile-X syndrome of an epigenetic modification of DNA raises the issue of epigenetic stability (Holliday 1987).

The fragile-X syndrome provides interesting data on instability of both genetic and epigenetic changes at this locus.

Genetic instability and the fragile-X mutation

The nature of the fragile-X mutation that leads to this apparent imprinting event is proposed to be a small expansion, which I call a primary expansion, in the number of CGG repeats within a candidate gene, the FMR-1 gene (Kremer et al. 1991; Yu et al. 1991). In addition, a secondary somatic expansion in the number of these CGG repeats is observed in DNA of affected individuals (Fu et al. 1991; Oberlé et al. 1991). These repeats are within the gene that is silenced by the hypermethylation of the CG island near its 5′ end (Verkerk et al. 1991).

One interpretation of these data, and the one that I favor, is that the imprinting event results from the primary expansion of CGG repeats; the secondary expansion of CGG repeats reflects somatic instability generated by the abnormally imprinted region. With this interpretation, it would seem that genetic instability—the primary expansion of CGG repeats—leads to epigenetic inactivation—imprinting—that is mediated by methylation of a CG island near a coding region. The imprinting in turn generates further genetic instability that is seen as a secondary expansion of CGG repeats. The involvement of impaired DNA replication for the mutant fragile site allele (Laird et al. 1987) is plausible because of the difficulty in replicating through the expanded number of CGG repeats by bacterial DNA polymerase (Fu et al. 1991). This difficulty may arise from the high guanine-cytosine content of the DNA and from the high concentration of methylated cytosines that are likely to be present in this region after imprinting.

Stability of the imprinted state

Viewing the molecular data from the perspective of our models, the fragile-X syndrome can be seen as a consequence of the persistence, in the next generation, of a small piece of facultative heterochromatin. Since the inactive state of facultative heterochromatin is potentially reversible (Chandra and Brown 1975), it is interesting to ask about the stability of the imprinted fragile-X state when it is passed through the germ lines of males and of females. In the fragile-X syndrome, the imprinted state was predicted from epidemiological data to be relatively stable when transmitted through females (Laird 1987). In a detailed assessment of carrier progeny of individual imprinted females, Peter Follette and I found that the stability was indeed high, but not complete; we estimated that an erasure frequency of 2% to 4% in females was consistent with cytogenetic and clinical data (Follette and Laird 1992). Molecular data have confirmed a similar low frequency in females of what we term imprint erasure (Fu et al. 1991; Rousseau et al. 1991).

The question of imprint stability through male meiosis is especially interesting because males, being the gender opposite to the imprinting gender, would be expected to erase the imprint if it were normal. Since the fragile-X imprint is expected to be an abnormal imprint, however, there is no basis for predicting imprint stability or erasure. The stability of the imprinted state through male meiosis is difficult to assess in a general way because only a few imprinted males have been reported to have children. On the basis of the cytogenetic and cognitive data from the few published examples of imprinted males who had daughters, I proposed that the imprinted state is efficiently erased when it passes through males (Laird 1991). Paradoxically, then, affected males may give rise to carrier daughters who are normal with respect to the fragile-X phenotype because of the change in state of the mutant allele. These daughters are at risk, however, for having mentally retarded children, which indicates that erasure probably does not affect the underlying mutation (Laird 1991). Thus, one rule of normal chromosome imprinting—that imprint erasure occurs in the parental gender opposite to the one that established the imprint—may also apply to the fragile-X syndrome. The predictions of imprint erasure in males are that the hypermethylation of the CG island next to the FMR-1 gene, the secondary expansion of the CGG repeats if they are caused by the persistent hypermethylation, and the inactivity of the FMR-1 gene on the mutant fragile-X chromosome will be reversed in daughters of fragile-X males. Molecular studies are in progress to test these predictions.

Somatic instability of the imprinted state is also observed in the fragile-X syndrome. About 25% of affected individuals exhibit, in their lymphocyte DNA, mosaicism for the hypermethylation and sometimes for the secondary expansion of the CGG repeats that are characteristic of the imprinted state (Rousseau et al. 1991). Cells of mosaic males exhibit some transcription of the FMR-1 gene (Pieretti et al. 1991), but it is not clear if the phenotype of these males is less severe than that of nonmosaic, imprinted males. Phenotypic mosaicism derived from novel junctions of heterochromatin and euchromatin, often referred to as position-effect variegation, has been observed in many organisms (see below). Some of the phenotypic variation in fragile-X syndrome and other human genetic diseases may arise from similar instabilities (Laird 1987).

Glimpse Four: Huntington's Disease, Dominant Position-effect Variegation, and the Concept of Epigene Conversion

Rearrangements between heterochromatin and euchromatin can put genes at risk for inactivation, a process sometimes described as position-effect variegation (Sturtevant 1925; McClintock 1952, 1953; Spofford 1976; Cattanach and Perez 1970; Devlin et al. 1990). Although most position effects are recessive, there is a very unusual case of a fully dominant position-effect variegation at the *brown* locus in *Drosophila* (Muller 1932). Steven Henikoff and his colleagues determined that the basis of dominant *brown* variegation is *trans*-inactivation, at the transcriptional level, of the wild-type *brown*$^{+}$ allele (Henikoff and Dreesen 1989).

I applied the concept of dominant position-effect variegation to Huntington's disease (HD) (Laird 1990). This application allows the two peculiar genetic properties of HD to be explained by one process: the fully dominant phenotype of the mutant *HD* allele is explicable by *trans*-inactivation of the *HD*$^{+}$ allele, mediated by the mutant allele (Fig. 6); the subtle differences in the age of onset of HD symptoms for carrier progeny of males and females can be accounted for by an X-linked modifier of position-effect variegation (Laird 1990). This model elaborates the concept of chromosome imprinting that had been offered (Erickson 1985) to explain variable expressivity of HD.

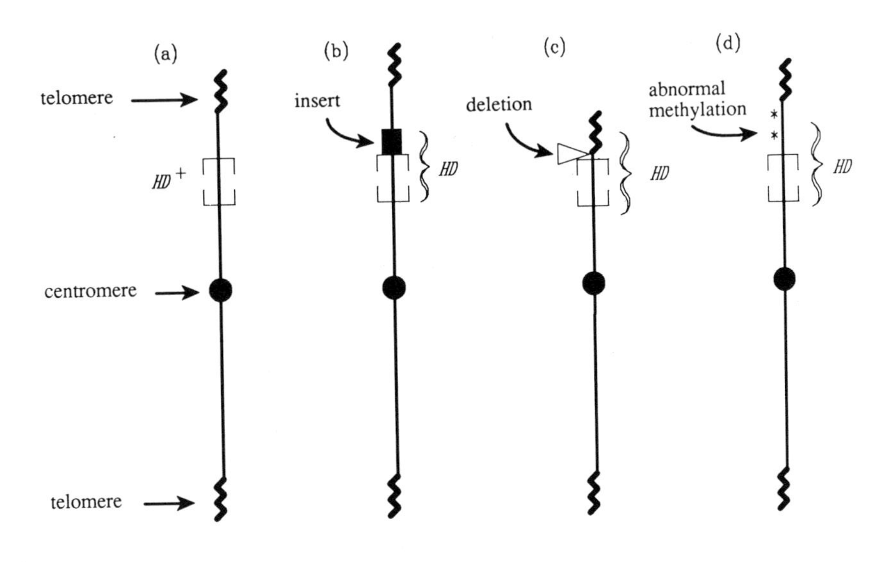

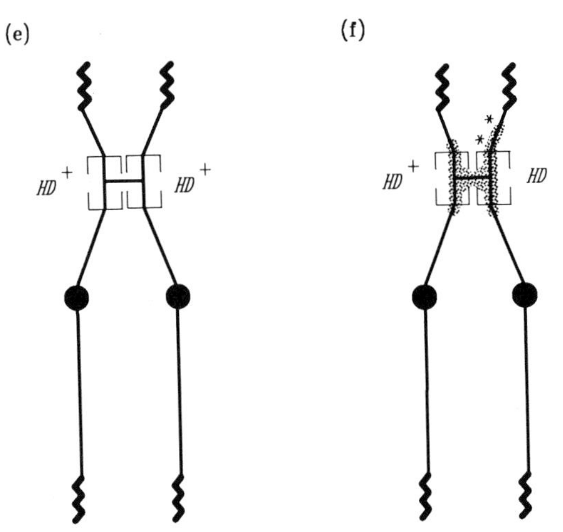

Figure 6 Proposed genetic basis of Huntington's disease. A chromosome alteration is predicted to lead to *cis*-inactivation of the Huntington's disease (HD^+) gene. Somatic pairing of homologs at the *HD* locus (*e*) leads to spreading of inactivation in *trans* to the HD^+ allele on the normal homolog (*f*). (Reprinted, with permission, from Laird 1990.)

In thinking about the detailed genetic and DNA linkage data for HD, Joy Sabl wondered whether or not the stable epigenetic inactivation of the normal locus could ever persist through meiosis. We knew that, in general, the proposed epigenetic inactivation of the normal locus did not persist through meiosis because segregation is approximately Mendelian and because DNA markers had been effective in arriving at a consistent position for the *HD* locus near the tip of the short arm of chromosome 4. But did the inactivation *occasionally* persist through meiosis?

This question turns out to be very interesting in light of the difficulties that human geneticists have had in mapping precisely the *HD* locus (Pritchard et al. 1991). We realized that if the proposed epigenetic inactivation of the normal allele persisted in a small percentage of meioses, then the inconsistent mapping data could be explained (Sabl and Laird 1992). We proposed the term "epigene conversion" for this concept, emphasizing the parallel of the proposed phenomenon to the process of gene conversion. In the case of epigene conversion, however, the mutant phenotype is proposed to exist without a corresponding mutant genotype, producing transient conversion in which the converted allele generally loses its imprinted state after another passage through meiosis.

Genetic and epigenetic interactions between alleles have been described in other organisms. The phenomenon of paramutation in maize involves a mutant allele that inactivates in *trans* a wild-type allele (Brink 1973). For the fungi *Neurospora crassa* (Selker 1990) and *Ascobolus immersus* (Faugeron et al. 1990), gene duplication leads to inactivation of both copies. The example in *Ascobolus* appears to be the most relevant to our proposal of epigene conversion in HD. In *Ascobolus*, the inactivated duplicate alleles are transmitted through meiosis in their inactive states, possibly as a consequence of hypermethylation. An inactivated allele can, however, become reactivated; in its reactivated state, the methylation pattern resembles that of the wild-type allele. If the proposal of epigene conversion in HD is correct, the *trans*-inactivation processes described for maize, *Drosophila*, and fungi may provide useful insights into this and other human diseases.

Concluding Remarks

During each of the projects described in these four glimpses, I have been surprised at the relevance and usefulness of classical cytogenetic observations on the heterochromatins. I should not have been surprised. The rich cytogenetic literature amply documents the enormous variety of genetic and epigenetic phenomena influenced by heterochromatin. The heterochromatins appear to be providing us with a glimpse of one limit of genetic and cytogenetic behavior; at the other limit are chromatins that facilitate stable gene activity. It is important to understand the molecular basis of chromatin at each of these limits as well as their interaction.

Approaching the phenomenon of genomic or chromosomal imprinting through concepts and details of the heterochromatins, for example, is likely to be productive. Facultative heterochromatin that persists through meiosis may uncover mutant alleles because of functional heterozygosity of a locus (Cattanach and Kirk 1985). Parental-source effects for mutant alleles are expected under these circumstances if the formation of facultative heterochromatin is gender-specific, as is often the case (Chandra and Brown 1975). Basic information on the genetic control of imprinting and related phenomena is becoming available in *Drosophila* (Locke et al. 1988; Eissenberg et al. 1990; Hayashi et al. 1990) and yeast (Pillus and Rine 1989; Klar 1990). This information is being used in the search for human modifier genes, some of which may affect the probability of forming heterochromatic structures (Singh et al. 1991). Products of modifier genes may influence the manifestation of human diseases caused by genomic imprinting (Sapienza 1989; Hall 1990; Laird 1990).

We are fortunate that the phenomenology of, and experimentation on, the heterochromatins have generally been described with accuracy and insight. Although periodic rediscovery, reinterpretation, and renaming of the heterochromatins will occur, frequent return to the observations of classical cytogenetics will almost certainly continue to be rewarding.

References

Barr, M.L. and E.G. Bertram. 1949. A morphological distinction between neurones of the male and female, and the behavior of the nucleolar satellite during accelerated nucleoprotein synthesis. *Nature 163:* 676.

Bell, M.V., M.C. Hirst, Y. Nakahori, R.N. MacKinnon, A. Roche, T.J. Flint, P.A. Jacobs, N. Tommerup, L. Tranebjaerg, U. Froster-Iskenius, B. Kerr, G. Turner, R.H. Lindenbaum, R. Winter, M. Pembrey, S. Thibodeau, and K.E. Davis. 1991. Physical mapping across the fragile X: Hypermethylation and clinical expression of the fragile X syndrome. *Cell 64:* 861.

Brink, R. 1973. Paramutation. *Annu. Rev. Genet. 7:* 129.

Brown, D.D. and I.B. Dawid. 1968. Specific gene amplification in oocytes. *Science 160:* 272.

Brown, W.T. 1990. Invited editorial. The fragile-X: Progress toward solving the puzzle. *Am. J. Hum. Genet. 47:* 175.

Cattanach, B.M. and M. Kirk. 1985. Differential activity of maternally and paternally derived chromosome regions in mice. *Nature 315:* 496.

Cattanach, B.M. and J.N. Perez. 1970. Parental influence on X-autosome translocation-induced variegation in the mouse. *Genet. Res. 15:* 43.

Chandra, H. and S. Brown. 1975. Chromosome imprinting and the mammalian X chromosome. *Nature 253:* 165.

Cohen, E. and G. Kaplan. 1982. Analysis of DNAs from two species of the virilis group of *Drosophila* and implications for satellite DNA evolution. *Chromosoma 87:* 519.

Crouse, H.V. 1960. The controlling element in sex chromosome behavior in Sciara. *Genetics 45:* 1429.

Darlington, C.D. and L. La Cour. 1940. Nucleic acid starvation of chromosomes in Trillium. *J. Genet. 40:* 185.

Devlin, R.H., B. Bingham, and B.T. Wakimoto. 1990. The organization and expression of the light gene, a heterochromatic gene of *Drosophila melanogaster. Genetics 125:* 129.

Eissenberg, J.C., T.C. James, D.M. Foster-Hartnett, T. Hartnett, V. Ngan, and S.C. Elgin. 1990. Mutation in a heterochromatin-specific chromosomal protein is associated with suppression of position-effect variegation in *Drosophila melanogaster. Proc. Natl. Acad. Sci. 87:* 9923.

Erickson, R.P. 1985. Chromosomal imprinting and the parent transmission specific variation in expressivity of Huntington disease. *Am. J. Hum. Genet. 37:* 827.

Faugeron, G., L. Rhounim, and J. Rossignol. 1990. How does the cell count the number of ectopic copies of a gene in the premeiotic inactivation process acting in *Ascobolus immersus? Genetics 124:* 585.

Foe, V.E., L.E. Wilkinson, and C.D. Laird. 1976. Comparative organization of active transcription units in *Oncopeltus fasciatus. Cell 9:* 131.

Follette, P.J. and C.D. Laird. 1992. Estimating the stability of the proposed imprinted state of the fragile-X mutation when transmitted by females. *Hum. Genet. 88:* 335.

Fu, Y.-H., D.P.A. Kuhl, A. Pizzuti, M. Pieretti, J.S. Sutcliffe, S. Richards, A.J.M.H. Verkerk, J.J.A. Holden, R.G. Fenwick, Jr., S.T. Warren, B.A. Oostra, D.L. Nelson, and C.T. Caskey. 1991. Variation of the CGG repeat at the fragile X site results in genetic instability: Resolution of the Sherman paradox. *Cell 67:* 1047.

Gall, J.G., E.H. Cohen, and M.L. Polan. 1971. Repetitive DNA sequences in *Drosophila. Chromosoma 33:* 319.

Gartler, S.M. and A.D. Riggs. 1983. Mammalian X-chromosome inactivation. *Annu. Rev. Genet. 17:* 155.

Hall, J.G. 1990. Genomic imprinting: Review and relevance to human diseases. *Am. J. Hum. Genet. 46:* 857.
Hammond, M.P. and C.D. Laird. 1985a. Control of DNA replication and spatial distribution of defined DNA sequences in salivary gland cells of *Drosophila melanogaster. Chromosoma 91:* 279.
———. 1985b. Chromosome structure and DNA replication in nurse and follicle cells of *Drosophila melanogaster. Chromosoma 91:* 267.
Hareven, D., M. Zuckerman, and E. Lifschytz. 1986. Origin and evolution of the transcribed repeated sequences of the Y chromosome lampbrush loops of *Drosophila hydei. Proc. Natl. Acad. Sci. 83:* 125.
Hayashi, S., A. Ruddell, D. Sinclair, and T. Grigliatti. 1990. Chromosomal structure is altered by mutations that suppress or enhance position effect variegation. *Chromosoma 99:* 391.
Heitz, D., F. Rousseau, D. Devys, S. Saccone, H. Abderrahim, D. Le Paslier, D. Cohen, A. Vincent, D. Toniolo, G. Della Valle, S. Johnson, D. Schlessinger, I. Oberlé, and J.L. Mandel. 1991. Isolation of sequences that span the fragile X and identification of a fragile X-related CpG island. *Science 251:* 1236.
Heitz, E. 1931. Nukleolen und Chromosomen in der Gattung Vicia. *Planta 15:* 495.
———. 1934. Über α-und β-Heterochromatin sowie Konstanz und Bau der Chromomeren bei *Drosophila. Biol. Zentralbl. 54:* 588.
Henikoff, S. and T.D. Dreesen. 1989. *Trans*-inactivation of the *Drosophila* brown gene. *Proc. Natl. Acad. Sci. 86:* 6704.
Hennig, W., R.C. Brand, J. Hackstein, R. Hochstenbach, H. Kremer, D.H. Lankenau, S. Lankenau, K. Miedema, and A. Potgens. 1989. Y chromosomal fertility genes of *Drosophila*: A new type of eukaryotic genes. *Genome 31:* 561.
Holliday, R. 1987. The inheritance of epigenetic defects. *Science 238:* 163.
Karpen, G.H. 1987. "The relationship between organization and function of ribosomal genes in *Drosophila melanogaster.*" Ph.D. thesis, University of Washington, Seattle.
Karpen, G.H., J.E. Schaefer, and C.D. Laird. 1988. A *Drosophila* rRNA gene located in euchromatin is active in transcription and nucleolus formation. *Genes Dev. 2:* 1745.
Kaufmann, B.P. 1939. Distribution of induced breaks along the X-chromosome of *Drosophila melanogaster. Proc. Natl. Acad. Sci. 25:* 571.
Klar, A.J. 1990. Regulation of fission yeast mating-type interconversion by chromosome imprinting. *Development* (suppl.), p. 3.
Kremer, E.J., M. Pritchard, M. Lynch, S. Yu, K. Holman, E. Baker, S.T. Warren, D. Schlessinger, G.R. Sutherland, and R.I. Richards. 1991. Mapping of DNA instability at the fragile X to a trinucleotide repeat sequence p(CCG)*n*. *Science 252:* 1711.
Laird, C.D. 1971. Chromatid structure: Relationship between DNA content and nucleotide sequence diversity. *Chromosoma 32:* 378.
———. 1987. Proposed mechanism of inheritance and expression of the human fragile-X syndrome of mental retardation. *Genetics 117:* 587.
———. 1990. Proposed genetic basis of Huntington's disease. *Trends Genet. 60:* 242.
———. 1991. Possible erasure of the imprint on a fragile-X chromosome when transmitted by a male. *Am. J. Med. Genet. 38:* 391.
Laird, C.D. and W.Y. Chooi. 1976. Morphology of transcription units in *Drosophila melanogaster. Chromosoma 58:* 169.
Laird, C.D., M.M. Lamb, and J.L. Thorne. 1990. Two progenitor cells for human oogonia inferred from pedigree data and the X-inactivation imprinting model of the fragile-X syndrome. *Am. J. Hum. Genet. 46:* 696.
Laird, C.D., L.E. Wilkinson, V.E. Foe, and W.Y. Chooi. 1976. Analysis of chromatin-associated fiber arrays. *Chromosoma 58:* 169.
Laird, C., E. Jaffe, G. Karpen, M. Lamb, and R. Nelson. 1987. Fragile sites in human chromosomes as regions of late-replicating DNA. *Trends Genet. 3:* 274.
Lamb, M.M. and C.D. Laird. 1987. Three euchromatic DNA sequences under-replicated in polytene chromosomes of *Drosophila* are localized in constrictions and ectopic fibers. *Chromosoma 95:* 227.
Lifschytz, E. 1983. Sequence replication and banding organization in the polytene chromosomes of *Drosophila melanogaster. J. Mol. Biol. 164:* 17.
Locke, J., M.A. Kotarski, and K.D. Tartof. 1988. Dosage-dependent modifiers of

position effect variegation in *Drosophila* and a mass action model that explains their effect. *Genetics 120:* 181.

McClintock, B. 1934. The relationship of a particular chromosomal element to the development of the nucleoli of *Zea mays. Z. Zellforsch. 21:* 294.

———. 1945. Neurospora. I. Preliminary observations of the chromosomes of *Neurospora crassa. Am. J. Bot. 32:* 671.

———. 1952. Chromosome organization and genic expression. *Cold Spring Harbor Symp. Quant. Biol. 16:* 13.

———. 1953. Induction of instability at selected loci in maize. *Genetics 38:* 579.

McKee, B.D. and G.H. Karpen. 1990. *Drosophila* ribosomal RNA genes function as an X-Y pairing site during male meiosis. *Cell 61:* 61.

Miklos, G.L.G. and J.N. Cotsell. 1990. Chromosome structure at interfaces between major chromatin types: Alpha- and beta-heterochromatin. *BioEssays 12:* 1.

Muller, H.J. 1932). Further studies on the nature and causes of gene mutations. *Proc. Int. Congr. Genet. 1:* 213.

Oberlé, I., F. Rousseau, D. Heitz, C. Kretz, D. Devys, A. Hanauer, J. Boué, M.F. Bertheas, and J.L. Mandel. 1991. Instability of a 550-base pair DNA segment and abnormal methylation in fragile X syndrome. *Science 252:* 1097.

Pardue, M.L. and J.G. Gall. 1970. Chromosomal localization of mouse satellite DNA. *Science 168:* 1356.

Pardue, M.L., L.H. Kedes, E.S. Weinberg, and M.L. Birnstiel. 1977. Localization of sequences coding for histone messenger RNA in the chromsomes of *Drosophila melanogaster. Chromosoma 63:* 135.

Pfeifer, G.P., S.D. Steigerwald, R.S. Hansen, S.M. Gartler, and A.D. Riggs. 1990. Polymerase chain reaction-aided genomic sequencing of an X chromosome-linked CpG island: Methylation patterns suggest clonal inheritance, CpG site autonomy, and an explanation of activity state stability. *Proc. Natl. Acad. Sci. 87:* 8252.

Pieretti, M., F. Zhang, Y.-H. Fu, S.T. Warren, B.A. Oostra, C.T. Caskey, and D.L. Nelson. 1991. Absence of expression of the FMR-1 gene in fragile X syndrome. *Cell 66:* 817.

Pillus, L. and J. Rine. 1989. Epigenetic inheritance of transcriptional states in *S. cerevisiae. Cell 59:* 637.

Pritchard, C., D.R. Cox, and R.M. Myers. 1991. Invited editorial: The end in sight for Huntington disease (erratum). *Am. J. Hum. Genet. 49:* 1106.

Rhoades, M.M., E. Dempsey, and A. Ghidoni. 1967. Chromosome elimination in maize induced by supernumerary B chromosomes. *Proc. Natl. Acad. Sci. 57:* 1626.

Ritossa, R.M. and S. Spiegelman. 1965. Localization of DNA complementary to ribosomal RNA in the nucleolus organizer region of *Drosophila melanogaster. Proc. Natl. Acad. Sci. 53:* 737.

Rousseau, F., D. Heitz, V. Biancalana, S. Blumenfeld, C. Kretz, J. Boué, N. Tommerup, C. Van Der Hagen, C. Blanchet-DeLozier, M.-F. Croquette, S. Gilgenkrantz, P. Jalbert, M.-A. Voelckel, I. Oberlé, and J.-L. Mandel. 1991. Direct diagnosis by DNA analysis of the fragile-X syndrome of mental retardation syndrome. *N. Engl. J. Med. 325:* 1673.

Rubin, G.M. and A.C. Spradling. 1982. Genetic transformation of *Drosophila* with transposable element vectors. *Science 218:* 348.

Sabl, J.F. and C.D. Laird. 1992. Epigene conversion: A proposal with implications for gene mapping in humans. *Science* (in press).

Sapienza, C. 1989. Genome imprinting and dominance modification. *Ann. N.Y. Acad. Sci. 564:* 24.

Selker, E.U. 1990. Premeiotic instability of repeated sequences in *Neurospora crassa. Annu. Rev. Genet. 24:* 579.

Sherman, S.L., P.A. Jacobs, N.E. Morton, U. Froster-Iskenius, P.N. Howard-Peebles, K.B. Nielsen, W.W. Partington, G.R. Sutherland, G. Turner, and M. Watson. 1985. Further segregation analysis of the fragile X syndrome with special reference to transmitting males. *Hum. Genet. 69:* 289.

Singh, P.B., J.R. Miller, J. Pearce, R. Kothary, R.D. Burton, R. Paro, T.C. James, and S.J. Gaunt. 1991. A sequence motif found in a *Drosophila* heterochromatin protein is conserved in animals and plants. *Nucleic Acids Res. 19:* 789.

Southern, E.M. 1970. Base sequence and evolution of guinea-pig α satellite DNA. *Nature 227:* 794.
Sperling, K. and P.N. Rao. 1974. Mammalian cell fusion. V. Replication behavior of heterochromatin as observed by premature chromosome condensation. *Chromosoma 45:* 121.
Spofford, J.B. 1976. Position effect variegation in *Drosophila*. In *The genetics and biology of* Drosophila (ed. M. Ashburner and E. Novitski), p. 955. Academic Press, London.
Spradling, A.C. and G.M. Rubin. 1982. Transposition of cloned P elements into *Drosophila* germ line chromosomes. *Science 218:* 341.
Sturtevant, A.H. 1925. The effects of unequal crossing over at the Bar locus in *Drosophila. Genetics 10:* 117.
Sutherland, G. and F. Hecht. 1985. *Fragile sites on human chromosomes.* Oxford University Press, New York.
Sved, J.A. and C.D. Laird. 1990. Population genetic consequences of the fragile-X syndrome, based on the X-inactivation imprinting model. *Am. J. Hum. Genet. 46:* 443.
Verkerk, A.J.M.H., M. Pieretti, J.S. Sutcliffe, Y.-H. Fu, D.P.A. Kuhl, A. Pizutti, O. Reiner, S. Richards, M.F. Victoria, F. Zhang, B.E. Eussen, G.-J.B. van Ommen, L.A.J. Blonden, G.J. Riggins, J.L. Chastain, C.B. Kunst, H. Galjaard, C.T. Caskey, D.L. Nelson, B.A. Oostra, and S.T. Warren. 1991. Identification of a gene (*FMR-1*) containing a CGG repeat coincident with a breakpoint cluster region exhibiting length variation in fragile X syndrome. *Cell 65:* 904.
Yu, S., M. Pritchard, E. Kremer, M. Lynch, J. Nancarrow, E. Baker, K. Holman, J.C. Mulley, S.T. Warren, D. Schlessinger, G.R. Sutherland, and R.I. Richards. 1991. The fragile X genotype is characterized by an unstable region of DNA. *Science 252:* 1179.
Yu, W.-D., S.L. Wenger, and M.W. Steele. 1990. X chromosome imprinting in fragile X syndrome. *Hum. Genet. 85:* 590.

A Tapestry of Transposition

ANNA MARIE SKALKA
Institute for Cancer Research, Fox Chase Cancer Center, Philadelphia, Pennsylvania 19111

When I first met Barbara McClintock, my own scientific career had not quite begun. Barbara had already accomplished the work that was ultimately to earn her a Nobel prize, and she was just at the beginning of a second, most rewarding phase in her career during which she would have the pleasure of watching the scientific world catch up with her. The occasion of our first meeting was a graduate course (1959–1960) at New York University Medical School where I was a student in Microbiology. Among the illustrious guest lecturers in this special course were François Jacob, Jacques Monod, and Barbara McClintock. The schedule was set up to allow the students (I recall no more than ten of us) an opportunity to spend a few private hours with these individuals, discussing their work and molecular aspects of genetics. These three guests made a lasting impression on me, as much for their unique personal qualities as for the scope of their intellectual powers.

I cannot say that I was able to follow all of what Barbara tried to communicate so enthusiastically. She seemed to think and speak largely in abstractions, a reflection, I presumed, of the classical geneticist's method of the deductive reasoning that was her primary tool. Like many budding molecular biologists of that time, I struggled to imagine what her observations might mean in biochemical terms. As will be described briefly in this short essay, most of my research efforts in the intervening years have been devoted essentially to that pursuit.

I had the chance to get to know Barbara much better and to enjoy many interactions with her during my 5 years as a postdoctoral associate with Alfred Hershey at the Carnegie Institution's Genetics Research Unit at Cold Spring Harbor. Barbara, whose laboratory was just downstairs, was a frequent and lively participant in our scientific discussions, to which she often contributed a unique perspective.

Beginning with Lambda

> Lambda is important mainly because it can recombine genetically with its host, mingling cellular and viral inheritance in ways that are fascinating to contemplate and, very likely, of practical importance to humans.
>
> A.D. Hershey, *Carnegie Institution Year Book* 1966

I became fascinated with bacteriophage λ when, still a graduate student, I took the Cold Spring Harbor Phage Course (1962) taught by Frank Stahl and the late George Streisinger. I was enthralled by Frank's description of Allan Campbell's then new model for λ DNA integration, which involved circularization of linear λ DNA followed by recombination with the host genome. Most intriguing to me were the profound functional ramifications of the relatively simple structural modification of joining DNA ends. In retrospect, this may not be so different from Barbara's early fascination with chromosomal abnormalities, including chromosomal rings, and her appreciation of their potential genetic significance. As I was to show later on, with Merv Smith, in Al Hershey's laboratory (Smith and Skalka 1966), and then as an independent investigator (Skalka 1971; Skalka et al. 1972; Enquist and Skalka 1973; Greenstein and Skalka 1975; Sogo et al. 1976), the ability to form circles had significant consequences for λ DNA replication as well as recombination. Our discovery of phage DNA molecules that were longer than genome length, which we later called "concatemers," provided some of the first evidence for the "rolling circle," or "printing press" model of replication as Merv and I had independently envisioned it. Concatemer formation is required for the packaging of phage DNA. We later found that when rolling circle replication is blocked, recombination can offer alternative pathways for the production of progeny phage (Skalka 1974). The interplay between different pathways of recombination and DNA synthesis is a theme that has recurred in our investigations over the years.

I had already begun to work with bacteriophage λ as a graduate student, but upon joining Hershey's laboratory at Cold Spring Harbor in 1964, I became a bona fide member of a special branch of the Phage Group, a "Lambdologist." The Phage Group's numerous contributions to the field of molecular biology are a matter of history, recounted by many of its luminaries in another volume of this sort, dedicated to Max Delbrück (Cairns et al. 1966). These were very exciting times; the Phage Group included some of the best and most creative minds in science. Furthermore, its members fostered, indeed demanded, an experimental rigor, intellectual honesty, and clarity of thinking that was a great challenge for a young scientist—one that I have always been grateful for.

Although we did not think about it quite that way at the time, it is now apparent that several other observations from our λ work added considerable flesh to Barbara's ideas about genetic transposition and its effects on regulation and genome evolution. One useful feature of λ, which we used early on to show that gene expression was subject to regional and temporal control (Skalka 1966), was our ability to separate left and right halves of its DNA, a consequence of λ's nonrandom distribution of nucleotides. Using Hg-CS_2SO_4 density equilibrium banding of DNA fragments of known average length, I was later able to distinguish several discrete segments in λ DNA (Skalka et al. 1968). As we noted at the time (extended by more recent work; Fig. 1), these segments correlated with different regions of function. We later found that λ is typical of a class of phages, all temperate, whose genomes appear to be structural mosaics of segments containing genes of related functions, which we speculated might be exchanged or inherited as units or "modules" (Skalka 1969; Skalka and Hanson 1972). We also proposed at that time that the evolutionary advantages of such exchanges might explain why genes of related functions are often clustered in viral genomes (Skalka et al. 1968). This, I believe, represented some of the first experimental support for

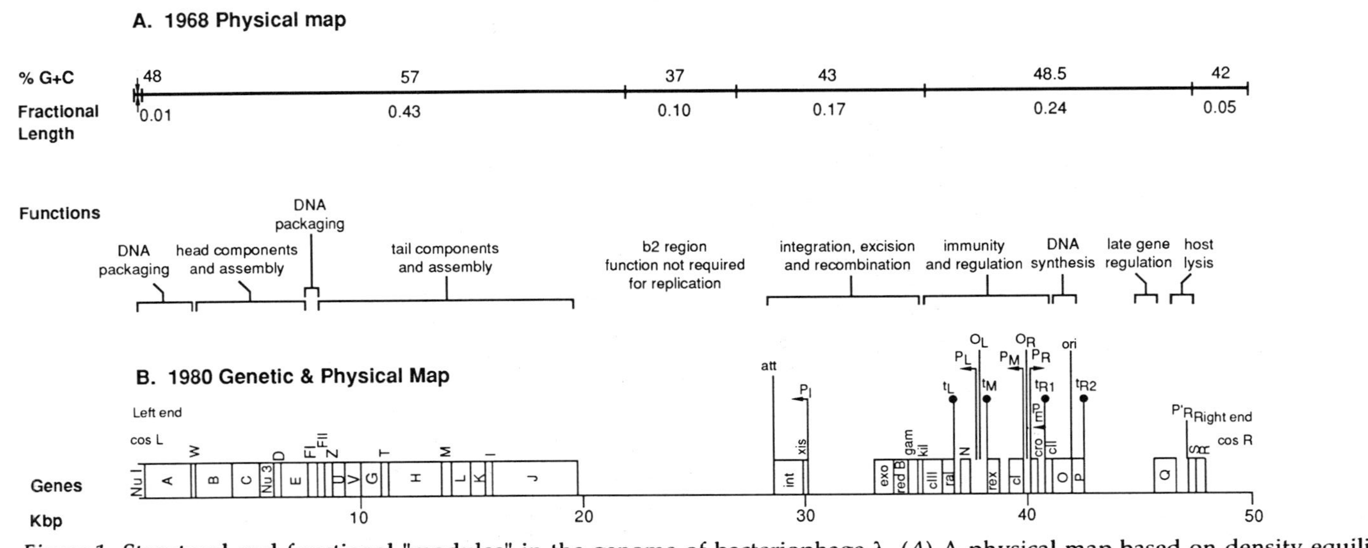

Figure 1 Structural and functional "modules" in the genome of bacteriophage λ. (*A*) A physical map based on density equilibrium analysis of λ DNA fragments. (Adapted from Skalka et al. 1968.) G + C content is indicated above the map and fractional lengths below. (*B*) A molecular map showing the location of genes above the line and kilobase pairs of DNA below the line. This map is adapted from a compilation in *Lambda II*, by Daniels et al. (1983), which includes the complete nucleotide sequence of λ DNA. Gene functions are indicated above the brackets, over map *B*.

the idea that not only phage chromosomes, but also clusters of genes within these chromosomes could be shuffled independently and that such shuffling could have important biological consequences. Such ideas are an integral part of current-day dogma.

Graduating to Retroviruses

By the early 1970s, the major pathways of λ replication, recombination, and gene regulation had been delineated. What was left was an elaboration of the molecular and biochemical details, a job not yet completed and one that continues to reward a number of investigators. At the same time, recombinant DNA technology loomed on the horizon and promised to revolutionize the study of molecular biology in eukaryotic cells. I began to look for a system in which I might use my knowledge of λ and the perspective of prokaryotic biology to explore this vast new frontier. And I was again beguiled by circles (Gianni et al. 1975; Guntaka et al. 1975).

Although we now know that the circular form of retroviral DNA is somewhat of a red herring, it seemed possible at the time that these circles, like those of λ, may be obligatory intermediates in retroviral DNA integration. It was, in fact, relatively easy to move from temperate phages to retroviruses. Conceptually, they are quite similar: Both recombine genetically with their host's DNA via specific mechanisms, both can establish productive infections or more benign relationships with the cells that they enter, and (not surprisingly) we now know that both are able to transduce host genes. In the case of the retroviruses, this property opened up a whole new area of cancer research, based on the study of transduced oncogenes. In the fall of 1975, thanks to a sabbatical with Saburo Hanafusa (at Rockefeller University), I became a retrovirologist—and thus transposed from one movable system to another.

As with λ, we began with two complementary approaches to study retroviral DNA synthesis and its integration. We initiated a structural analysis of unintegrated and integrated proviral DNA, anticipating that a knowledge of DNA organization would provide clues to the mechanism of integration. We also studied the biochemistry of retroviral DNA synthesis. Thanks to a familiarity with the prokaryotic vector systems, we were among the first to clone and sequence the terminal and junction regions of unintegrated and integrated DNAs as well as unoccupied integration sites in host DNA (Ju and Skalka 1980; Hishinuma et al. 1981). Our results, and similar work by others, revealed a remarkable resemblance between retroviruses and other prokaryotic and eukaryotic transposable elements. The most striking similarities were not, as it turns out, with λ but with a hierarchy of bacterial transposable elements, including insertion sequences, composite bacterial transposons, and the bacteriophage Mu (Fig. 2). As we noted at the time, the long terminal repeats (LTRs) at the ends of the DNA of retroviruses and retrotransposons play an important role in the control of viral gene expression. Many have strong promotors and some of these respond to tissue-specific transcription factors. They also contain polyadenylation signals. Furthermore, like the bacterial IS elements, they can exert biologically significant effects on host genes in the vicinity of their integration sites. The shared properties of these elements also include short inverted terminal repeats that contain strictly conserved TG. . .CA dinucleotides near or at their ends, and the creation of

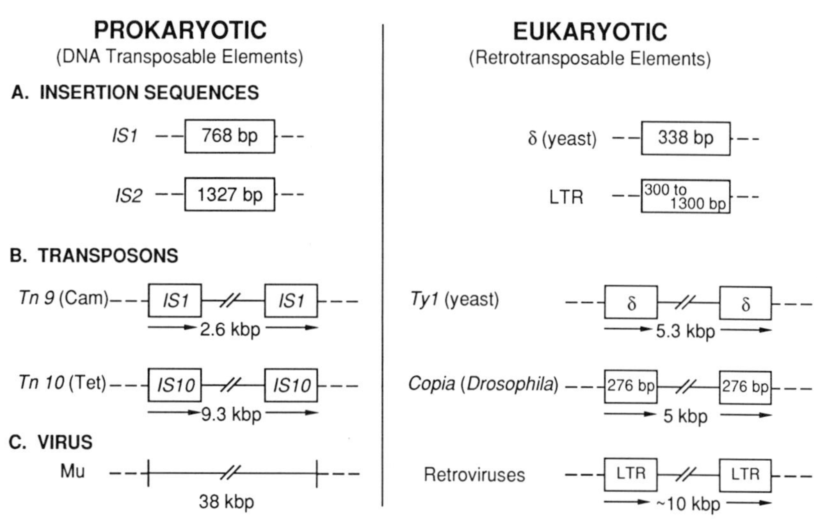

Figure 2 Hierarchy of transposable elements in prokaryotes and an analogous organizational arrangement of the components of eukaryotic retrotransposable elements. Solid lines represent the element, and dashed lines indicate the flanking cellular sequences. Approximate lengths are given in base pairs (bp) or kilobase pairs (kbp). Arrows show the orientation of terminally repeated sequences. (Adapted from Ju et al. 1982.)

short direct repeats (dr) of host target DNA sequences that flank the transposed (or integrated) elements (Fig. 3). Presumably, all of these elements share common molecular mechanisms for transposition, whose details must account for these unique features. Elucidation of these mechanisms now commands the bulk of our attention.

In 1980, while studying endogenous reverse transcription in avian retroviral particles, we devised a new way to permeabilize virions (with the honey bee toxin peptide, melittin), one that allowed synthesis in vitro of viral DNA that was virtually identical to that formed when virions infected cells (Boone and Skalka 1980, 1981a,b). Here, again, we discovered features that hinted at an intimate interrelationship between mechanisms of DNA synthesis and its recombination. Endogenously synthesized avian retroviral DNA has a structure that is most provocative to those familiar with recombination models. It consists of one continuous (–) strand that is complementary to the RNA genome and a discontinuous, or segmented, (+) strand from which single-stranded branches extend. Discovery of these branches provided the first evidence that reverse transcriptase could catalyze strand-displacement synthesis: We hypothesized that the nascent end of one (+) strand segment displaced its adjacent neighbor as DNA polymerization proceeded (Fig. 4). When viewed in the electron microscope, some duplexes appeared to exchange single (+) strand segments in reciprocal fashion (Junghans et al. 1982a). We proposed that since retrovirions each contain two RNA genomes, which can be reverse-transcribed simultaneously, such exchanges might partially explain the very high frequency of homologous recombination that is typical of these viruses (Fig. 4) (Junghans et al. 1982b).

Thanks to an enormous increase in the DNA sequence database, and the current widespread interest in transposition and retrotransposition, insights

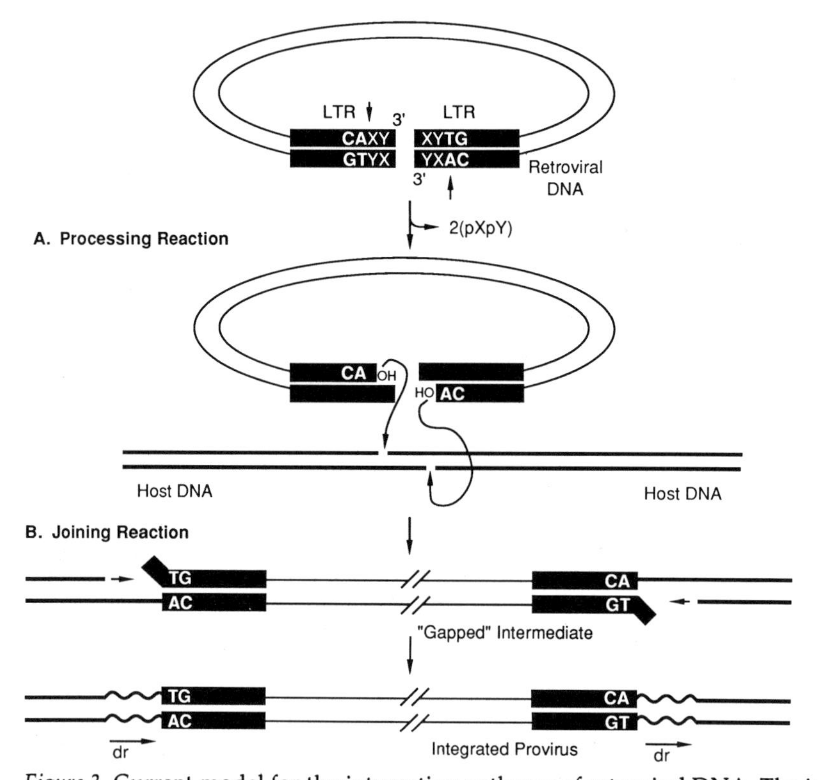

Figure 3 Current model for the integration pathway of retroviral DNA. The integration pathway can be separated into two temporally and biochemically distinct steps. (*A*) Linear viral DNA is shown at the top with ends juxtaposed. LTRs are represented by black rectangles. The first step, *Processing*, takes place in the cytoplasm of infected cells and involves single-strand endonucleolytic cleavages near the 3′ ends of both strands of the linear viral DNA. This nicking occurs after an invariant CA dinucleotide that is usually part of a longer terminal inverted repeat. (*B*) The second step, *Joining*, occurs in the cell's nucleus and involves ligation of the new CA3′ hydroxyl ends of viral DNA to 5′ phosphate ends of host DNA that is cleaved in a staggered fashion in what are likely to be energetically linked reactions (Engleman et al. 1991). Choice of the host target site is sequence-independent, but the length of the staggered cut is specific for each retrovirus. The retroviral integrase is both necessary and sufficient for both of these steps, and the reactions are independent of an external source of energy (Craigie et al. 1990; Katz et al. 1990). The immediate product is a "gapped intermediate," which may be repaired by host cell enzymes. The final product, a fully integrated provirus, is shorter than the unintegrated viral DNA (usually by 2 base pairs on either end) and is flanked on either end by a short (usually 4–6 base pair) direct repeat (dr) of host DNA. Temin (1980) has suggested that the loss of sequences from the ends of the viral DNA may be analogous to the excision from a larger DNA molecule that occurs during transposition of bacterial transposons and bacteriophage Mu. The conserved TG...CA dinucleotides at the ends of the provirus and the generation of short direct repeats of target DNA are clearly analogous, and likely reflect similar features of these transposition events.

into the inter-relatedness of reverse transcription and transposition have expanded greatly (for review, see Xiong and Eickbush 1990). We now know that reverse transcriptase has ancient origins in some predecessor microorganism that existed before the divergence of prokaryotes and eukaryotes, that many diverse elements in eukaryotic DNA either carry their own reverse transcriptase or bear signs of having passed through an RNA intermediate,

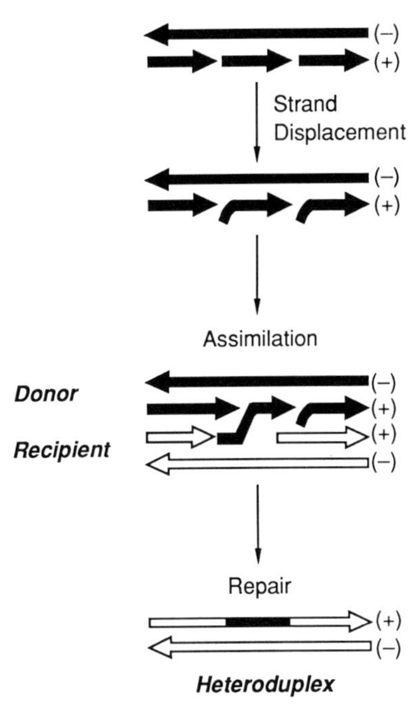

Figure 4 Strand displacement/assimilation model for homologous recombination in retroviruses. (*Top*) Retroviral DNA containing one continuous (–) strand that is complementary to the viral RNA genome (cDNA) and a segmented (+) strand. The segmentation reflects the fact that (+) strand DNA synthesis can be initiated at multiple sites on the (–) strand template using, as primers, short oligomers of RNA produced upon digestion of the genomic RNA template by the RNase H activity associated with reverse transcriptase. RNase H digestion and (+) strand DNA synthesis start before (–) strand synthesis is completed (Boone and Skalka 1981b), and all three activities are probably coordinated (Junghans et al. 1982a). Strand displacement synthesis produces (+) strand single-stranded branches, which may then anneal to (–) strand DNA (cDNA) synthesized from a second RNA genome encapsidated in the same (pseudodiploid) retrovirion, here labeled the Recipient. If the two genomes are genetically distinct, such exchanges could produce recombinant DNA duplexes. Since only (+) strand information would be exchanged, a strong prediction of this model is that the immediate products should be exclusively heteroduplex. Although conceptually simple, this prediction has not yet been tested because of the technical limitations inherent in the system. (Adapted from Boone and Skalka 1992.)

and that retroviruses, like temperate phages, are indeed genetic mosaics that include "modules" of information which existed long before the viral genomes were assembled (for a brief review, see Temin 1989). Indeed, emerging evidence suggests that replicating RNAs and ribozymes were progenitors of DNA genomes and some protein enzyme systems. If so, then some form of reverse transcriptase must have played a key role in the origin of DNA genomes.

The two other enzymes encoded in all nondefective retroviruses (and retrotransposons) also have recognizably ancient origins. The protease, which these elements use to process the precursor for the proteins that comprise and are contained within its core, is both structurally and functionally related to the large class of cell-derived aspartyl proteases like pepsin and renin (for a brief review, see Skalka 1989). The integrase, which is the only protein required for

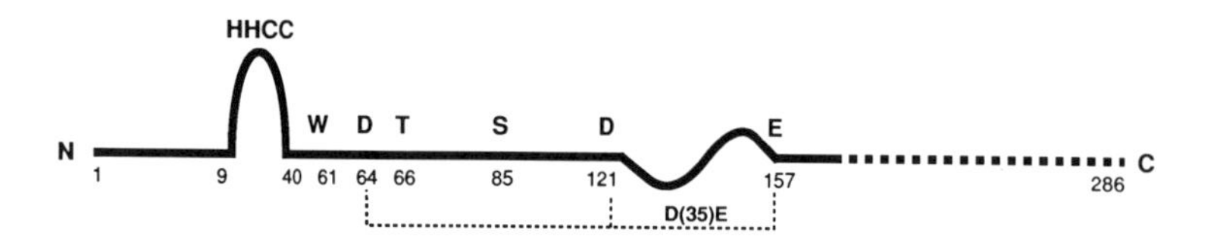

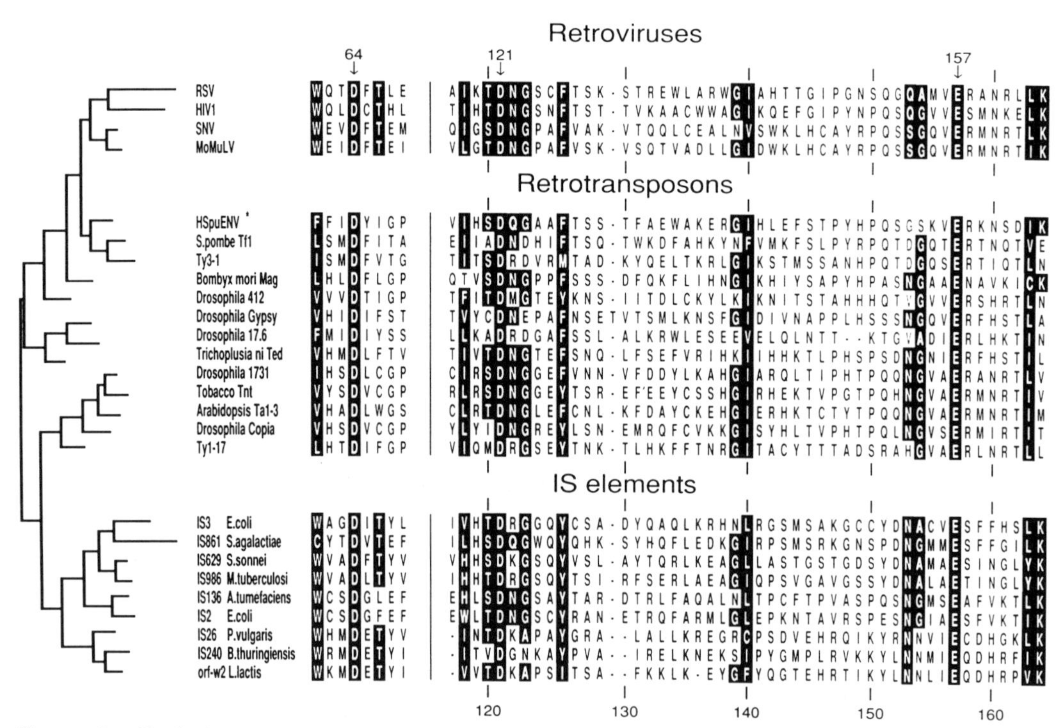

Figure 5 Evolutionary conservation in the integrase (transposase) proteins of eukaryotic retroviruses and retrotransposons and bacterial IS element. (*Top*) A linear representation of the integrase protein of the avian Rous sarcoma virus (RSV). The heavy line represents the evolutionarily conserved region. Computer-assisted sequence alignment of over 80 such sequences reveals at least 11 invariant or highly conserved residues indicated, by a single letter code, in their relative locations above the line (Khan et al. 1990; Katzman et al. 1991; Kulkosky et al. 1992). Conserved structural features include a potential Zn^{+} finger (HHCC) domain, shown as a loop at the amino terminus, and second region of structural conservation, D(35)E, which together with a short upstream segment including another D is also present in bacterial IS elements. (*Bottom*) Comparison of amino acid sequences in the D,D(35)E region. Representative sequences are grouped by phylogenetic relatedness as shown by the tree on the left. Numbers at the top of the alignment correspond to residues in RSV integrase. Invariant or highly conserved residues are highlighted by black or shaded backgrounds. Computer-assisted modeling predicts conserved structure in the region corresponding to RSV amino acids 121 to 157. The evolutionary conservation of this extended domain suggests that the component residues may be involved in activities that are shared by these proteins: DNA recognition, cutting, and joining. Mutational analysis has shown that the invariant D and E residues are essential for RSV and human immunodeficiency virus (HIV) integrase function (Kulkosky et al. 1992). On the basis of similarities with enzymes that catalyze analogous reactions, we have proposed that these residues may participate in coordination of the metal cofactor (Mn^{++} or Mg^{++}) required for the catalytic activities of integrase. (Adapted from Kulkosky et al. 1992.)

retroviral integration (Craigie et al. 1990; Katz et al. 1990), leaves hallmarks reminiscent of bacterial transposons and IS elements (cf. Fig. 3) and includes regions of extensive amino acid sequence homology with the transposases of a number of these elements (Fig. 5) (Fayet et al. 1990; Khan et al. 1990; Rowland and Dyke 1990; Katzman et al. 1991; Kulkosky et al. 1992).

Many threads now connect the various genetic components of transposons and retrotransposons, across species and even Phyla, so that what emerges today is a picture that contrasts sharply with the static view of genetics that prevailed when Barbara first "saw" jumping genes. This current view confirms her remarkable vision of the fluidity of genomes and of its profound significance to function and evolution. Our ability to trace such threads not only between transposable elements, clusters of genes, and individual genes, but even within structural domains of genes promises ultimately to provide new insights into how functional domains are used, conserved, and reused in different rearrangements to regulate and create new activities. We can look forward to the not-too-distant day when it will be possible to "read" a gene's function and its relationship with other genes directly from its sequence.

Final Thoughts

In this brief essay, I have described some of the weaving in only a few small sections of the rich tapestry that is beginning to emerge from our understanding of the molecular details and biological significance of genetic transposition. As scientists we believe, of course, that all of the secrets and principles of the natural world are there—simply waiting for us to uncover them. In making our discoveries, we must stand upon each other's shoulders to get the next new view. Even so, it is often difficult to see much beyond the neighboring trees. Yet, our scientific world is also populated by some of the most extraordinary individuals, like Barbara, whose imagination or special insight allows them to see far beyond and to open truly pioneering paths. Such visions are often breathtaking, sometimes difficult to accept, but ultimately an inspiration to us all.

One of the fondest memories from my early days at Cold Spring Harbor Laboratory is another view, that from the living room window of the Hooper House apartment where my husband Rudy and I lived. Every once in a while, I would catch a glimpse of Barbara on her periodic solitary walks along the sea wall. Occasionally, she would stop at the back porch of Hooper to chat with our first child, Jeanne. It was a touching scene, the diminutive scientist and the tiny little girl, in congenial but serious conversation. Jeanne has since grown up and is now a pioneer of another sort. I cannot help but think that she owes much to a spirit passed on, if only indirectly, by this very special scientist.

References

Boone, L.R. and A.M. Skalka. 1980. Two species of full-length cDNA are synthesized in high yield by melittin-treated avian retrovirus particles. *Proc. Natl. Acad. Sci. 77:* 847.

———. 1981a. Viral DNA synthesized *in vitro* by avian retrovirus particles permeabilized with melittin. I. Kinetics of synthesis and size of (–) and (+) strand transcripts. *J. Virol. 37:* 109.

———. 1981b. Viral DNA synthesized *in vitro* by avian retrovirus particles permeabilized with melittin. II. Evidence for a strand displacement mechanism in (+) strand synthesis. *J. Virol. 37:* 117.

———. 1992. Strand-displacement synthesis by reverse-transcriptase. In *Reverse transcriptase* (ed. S. Goff and A.M. Skalka). Cold Spring Harbor Laboratory Press, Cold Spring Harbor, New York. (In press.)

Cairns, J., G.S. Stent, and J.D. Watson, eds. 1966. *Phage and the origins of molecular biology*. Cold Spring Harbor Laboratory, Cold Spring Harbor, New York.

Craigie, R., T. Fujiwara, and F. Bushman, 1990. The IN protein of Moloney murine leukemia virus processes the viral DNA ends and accomplishes their integration *in vitro*. *Cell 62:* 829.

Daniels, D.L., J.L. Schroeder, W. Szybalski, F. Sanger, and F.R. Blattner. 1983. Appendix I. A molecular map of coliphage lambda. In *Lambda II* (ed. R.W. Hendrix et al.), p. 473. Cold Spring Harbor Laboratory, Cold Spring Harbor, New York.

Engelman, A., K. Mizuuchi, and R. Craigie. 1991. HIV-1 DNA integration: Mechanism of viral DNA cleavage and DNA strand transfer. *Cell 67:* 1211.

Enquist, L.W. and A. Skalka. 1973. Replication of bacteriophage λ DNA dependent on the function of host and viral genes. I. Interaction of *red, gam,* and *rec*. *J. Mol. Biol. 75:* 185.

Fayet, O., P. Ramond, P. Polard, and M. Chandler. 1990. Functional similarities between retroviruses and the IS3 family of bacterial insertion sequences? *Mol. Microbiol. 4:* 1771.

Gianni, A.M., D. Smotkin, and R.A. Weinberg. 1975. Murine leukemia virus: Detection of unintegrated double-stranded DNA forms of the provirus. *Proc. Natl. Acad. Sci. 72:* 447.

Greenstein, M. and A. Skalka. 1975. Replication of bacteriophage λ DNA: *In vivo* studies of the interaction between the viral gamma protein and the host *rec*BC DNAse. *J. Mol Biol. 97:* 543.

Guntaka, R.V., B.W.J. Mahy, J.M. Bishop, and H.E. Varmus. 1975. Ethidium bromide inhibits the appearances of closed circular viral DNA and integration of virus specific DNA in duck cells infected by avian sarcoma virus. *Nature 253:* 507.

Hishinuma, F., P.J. DeBona, S. Astrin, and A.M. Skalka. 1981. Nucleotide sequence of acceptor site and termini of integrated avian endogenous provirus *ev*-1: Integration creates a 6 bp repeat of host DNA. *Cell 23:* 155.

Ju, G. and A.M. Skalka. 1980. Nucleotide sequence analysis of the long terminal repeat (LTR) of avian retroviruses: Structural similarities with transposable elements. *Cell 22:* 379.

Ju, G., F. Hishinuma, and A.M. Skalka. 1982. Nucleotide sequence analysis of avian retroviruses: Structural similarities with transposable elements. *Fed. Proc. 41:* 2659.

Junghans, R.P., L.R. Boone, and A.M. Skalka. 1982a. Products of reverse transcription in avian retrovirus analyzed by electron microscopy. *J. Virol. 43:* 544.

———. 1982b. Retroviral DNA H structures displacement/assimilation model of recombination. *Cell 30:* 53.

Katz, R.A., G. Merkel, J. Kulkosky, J. Leis, and A.M. Skalka. 1990. The avian retroviral IN protein is both necessary and sufficient for integrative recombination *in vitro*. *Cell 63:* 87.

Katzman, M., J. Leis, and A.M. Skalka. 1991. Evidence for a covalent intermediate between the avian retroviral integration protein (IN) and nicked substrate DNA. *Proc. Natl. Acad. Sci. 88:* 4695.

Khan, E., J.P.G. Mack, R.A. Katz, J. Kulkosky, and A.M. Skalka. 1990. Retroviral integrase domains: DNA binding and the recognition of LTR sequences. *Nucleic Acids Res. 19(4):* 851.

Kulkosky, J., K.S. Jones, R.A. Katz, J.P.G. Mark, and A.M. Skalka. 1992. Residues critical for retroviral integrative recombination in a region that is highly conserved among retroviral/retrotransposon integrases and bacterial IS transposases. *Mol. Cell. Biol.* (in press).

Rowland, S.-J. and K.G.H. Dyke. 1990. Tn552, a novel transposable element from *Staphylococcus aureus*. *Mol. Microbiol. 4:* 961.

Skalka, A. 1966. Regional and temporal control of genetic transcription in phage lambda. *Proc. Natl. Acad. Sci. 55:* 1190.

———. 1969. Nucleotide distribution and functional orientation in the DNA of phage φ80. *J. Virol. 3:* 150.
———. 1971. Origin of DNA concatemers during growth. In *The bacteriophage lambda* (ed. A.D. Hershey), p. 535. Cold Spring Harbor Laboratory, Cold Spring Harbor, New York.
———. 1974. A replicator's view of recombination (and repair). In *Mechanism and recombination* (ed. R.F. Grell), p. 421. Plenum Press, New York.
———. 1989. Retroviral proteases: First glimpses at the anatomy of a processing machine. *Cell 56:* 911.
Skalka, A. and P. Hanson. 1972. Comparisons of the distribution of nucleotides and common sequences in DNA from selected bacteriophages. *J. Virol. 9:* 583.
Skalka, A., E. Burgi, and A.D. Hershey. 1968. Segmental distribution of nucleotides in the DNA of bacteriophage lambda. *J. Mol. Biol. 34:* 1.
Skalka, A., M. Poonian, and P. Bartl. 1972. Concatemers in DNA replication: Electron microscopic studies of partially denatured intracellular lambda DNA. *J. Mol. Biol. 64:* 541.
Smith, M.G. and A.M. Skalka. 1966. Some properties of DNA from phage-infected bacteria. *J. Gen. Physiol.* (Part 2) *49:* 127. (Also published in book form by Little, Brown and Co., Boston, 1966.)
Sogo, J.M., M. Greenstein, and A. Skalka. 1976. The circle mode of replication of bacteriophage lambda: The role of covalently closed templates and the formation of mixed catenated dimers. *J. Mol. Biol. 103:* 537.
Temin, H.M. 1980. Origin of retroviruses from cellular moveable genetic elements. *Cell 21:* 599.
———. 1989. Retrons in bacteria. *Nature 339:* 254.
Xiong, Y. and T.H. Eickbush. 1990. Origin and evolution of retroelements based upon their reverse transcriptase sequences. *EMBO J. 9:* 3353.

INTRANUCLEAR SYSTEMS CONTROLLING GENE ACTION AND MUTATION

Barbara McClintock

Department of Genetics, Carnegie Institution of Washington,
Cold Spring Harbor, New York

The organization of the chromosomes in the working nuclei of an organism is the standard of reference to which all observed genetic alterations of nuclear origin must be referred. It represents an evolved and necessarily integrated system functioning to control metabolic pathways and developmental patterns. Changes in this organization give rise to mutant expressions. For many studies, it is either favorable or necessary to select those mutants that give clear-cut phenotypic expressions and regular inheritance patterns, even though it is realized that many are not of this type. The latter are not so sharply defined with respect to phenotypic expressions, stability, inheritance patterns, or combinations of these. All gradations exist in this respect between these two categories of mutants, and thus it is not possible to separate them into distinct classes with regard to possible differences in their mode of origin.

In recent years, maize geneticists have investigated a number of cases of instability of expression of genetic components at previously known or at otherwise unknown loci in the maize chromosome complement. In the past, relatively little concentrated attention has been given by geneticists to such expressions. There are few adequate analyses of the behavior of so-called mutable genes or unstable loci, although literature citations could be given that would indicate the general prevalence of this phenomenon in many different organisms (1, 2). However, renewed studies of this phenomenon in maize have made evident the presence of previously unsuspected chromosome elements whose modes of operation in controlling gene action and in inducing mutation will be the subject of this report.

Controlling Elements and the Gene

It is generally believed that the chromosomes carry distinct elements or units, called genes, each of which serves a specific function in the control of cellular metabolism. If this view is accepted, then the elements to be described do not appear to operate in the same manner as the genes. Instead, they serve to modify and control the action of the genes themselves, and they do so in very precise ways. The term controlling element is therefore applied to them. There are different kinds of controlling elements, distinguishable from one another by their distinctive modes of control of gene action and mutation. The mutations they induce resemble those appearing spontaneously or produced by mutagenic agents: alleles with different qualitative or quantitative expressions, alleles showing pleiotropic effects, dominants giving unanticipated phenotypes, dominants that are semilethal, recessives that are viable or semilethal or lethal, and pseudoalleles. Thus, the kinds of change in gene action that these elements can induce are extensive. Because a controlling element may move from one location to another within the chromosome complement without losing its identity in the process, the same element can modify the expression of many different genes or the same gene may be modified in expression by different controlling elements. A large amount of information on the nature and mode of action of controlling elements has been acquired in recent years, much of which is yet unpublished. It will be possible here to give only an outline of this evidence and of the conclusions drawn from it.

In general, controlling elements behave in inheritance as Mendelizing units. The location of one such element within a particular chromosome may be determined by ordinary genetic and cytogenetic techniques. For some of them, it is possible to learn when changes in location are likely to occur in the life cycle. Also, methods of identifying individual plants with altered locations of such elements have been developed.

Knowledge of the presence in the chromosome complement of genetically identifiable elements capable of moving from one location to another poses a distinct problem. It is concerned with the possible kinds of genetic elements that may be present in the nucleus and with our ability to distinguish one kind from another. The cytologist has made us aware of different structural and functional components of the chromosomes, such as centromeres, nucleolus organizers, heterochromatic knobs, and other large heterochromatic components occupying specific locations in the chromosomes, as well as those heterochromatic components that may be distributed along the chromosomes. Also, we are aware of the distinctive chromomeric patterns that characterize each chromosome of an organism at different stages of the mitotic cycle in certain tissues.

We have not yet been able to associate those units of inheritance, detectable by mutant expressions, with identifiable unit structures in the chromosomes. Neither have we been able to determine the nature of the modification of the presumed unit that is responsible for the mutant expression. We are aware that the controlling elements behave as genetically identifiable units of inheritance and also that their primary mode of action appears to differ from that of other identifiable genetic units. This suggests the presence in the nucleus of two different classes of elements, each of which is concerned in its own way with the control of phenotypic expression. Therefore, not only to serve as a working hypothesis but also to facilitate descriptions of observations and experimentation in terms that will allow ready comprehension of their nature, we will assume that there are at least two different types of basic genetic elements carried in the chromosomes: those that give rise to the primary products of gene action, the genes; and those that control the mode of operation of these primary gene elements, the controlling elements. These latter control rates of action of the primary gene elements, their times of action in development of the organism or in the life of the cell, and also, possibly, the association of a particular gene element with another such element or elements to form a unit of function.

In the past, the controlling elements have been referred to as units. However, the use of the term "unit" with regard to members of either of the above two classes requires qualification. It refers only to the fact that both the controlling elements and those gene loci that have been modified in some genetically identifiable way may show regular inheritance patterns and thus be followed as "units" in progeny tests. No inferences regarding linear dimensions, subcomponents, duplication of elements, or organizational independence of the two different elements are implied by its use.

Types of Controlling Elements and Their Modes of Operation

The "Ds - Ac" Two-Element System

The most detailed knowledge of the behavior and mode of operation of controlling elements has come from studies of an integrated system composed of two elements which, in the past, have been referred to as the *Ds* - *Ac* two-unit system (3, 4). The first of these two units was given the symbol *Ds* for Dissociation because, when first discovered, it was noted that breaks occurred in the chromosome at the locus of *Ds* which divided the chromosome into two components: an acentric and a centric component. These breaks occurred only when another independently located element was also present in the chromosome complement. This latter element was given the symbol *Ac* for Activator since its presence was necessary for breaks to occur at the site of *Ds*. This type of behavior was unique, and thus an extensive study was commenced to determine the nature of action of these two genetic elements.

It was soon realized that the responses of *Ds* to *Ac* were conditioned by the amount of *Ac* present in the nucleus: the higher the concentration, the later the time in the development of a tissue when breaks would occur at *Ds*. Again it was found that breaks did not occur in all cells of a tissue but only in some of them. The distribution of such breaks produced a distinctive pattern. Thus, not only the time during the development of a tissue, but also the cells within a tissue in which breaks at *Ds* would occur were controlled by *Ac*.

The original location of *Ds* was readily determined by ordinary genetic and cytogenetic techniques. During these studies, it was noted that an occasional gamete produced by a plant having *Ds* at a known location would carry it at a new location. Studies were then commenced to determine the mode of appearance of *Ds* at these new locations. From these studies it was concluded that *Ds* could transpose from one location to another in the chromosome complement but would do so only if *Ac* was present and only at those times during the development of a tissue when breaks could also occur at the site of *Ds*. In other words, transposition of *Ds* was one of the types of response of *Ds* to the presence of *Ac*.

It was then discovered that the behavior of *Ac* was, in many respects, much like that of *Ds*. It could transpose from one location to another within the chromosome complement, and also it sometimes could cause breaks to occur at the locus at which it resided. It was also noted that coincidental transposition or alteration of both *Ds* and *Ac* often occurred in the same cell, which suggests integration in the time of modification of both elements. However, no alterations at the site of *Ds* would occur in the absence of *Ac.*

It was realized early in these studies that a chromosome break at the site of *Ds* was only one of several identifiable types of modification that *Ds* induces or that occur to it. Other results were the loss or inactivation of *Ds* or its transposition to new locations without an accompanying visible modification in chromosome organization, or readily identifiable modification of the *Ds* element itself unaccompanied by change in location or by visible modification of the chromosome. These changes effect alterations in the type of response of *Ds* to *Ac,* and also their relative frequencies, in subsequent cell and plant generations. These alterations have been called changes in state of *Ds.* A number of different states of *Ds* have been recognized, each showing a characteristic type of behavior in the presence of *Ac.* When first discovered, the state of *Ds* gave rise to many breaks at the site of *Ds.* Altered states soon appeared, some of which gave rise to few such breaks or none at all. Instead, *Ds* now responded to *Ac* by frequent removals from the locus or by other types of change that were unaccompanied by any detectable chromosomal aberration.

Early in the study of this two-element system it was discovered that transposition of *Ds* to the locus of a known gene affected the action of that gene by either immediately or subsequently altering its mode of expression (4-6). An immediate alteration in genic action, such as total or partial suppression, could be detected readily. In the presence of *Ac* subsequent modifications occurred at the site of *Ds* in its new location and these, in turn, were made evident by altered expression of the associated genic materials. New mutation-like changes in action of the genes appeared. In some cases, the mutation-producing event was accompanied by removal of *Ds* from the locus, and the mutant expression was thereafter stable in the presence of *Ac.* In other cases, the mutation-producing event did not remove *Ds,* and stability of the mutant expression was maintained only in the absence of *Ac.* Return of *Ac* to the nucleus provided the means for further changes at the site of *Ds,* and many of these were accompanied by other recognizable changes in action of the associated genic materials, that is, in subsequent gene mutation.

In the course of this study, the *Ds* - *Ac* system of control of genic expression appeared at five different known gene loci, and at four of them more than one inception occurred (6). The phenotypes associated with the action of these genes differ. Three reflect anthocyanin pigment formation, one is concerned with starch composition, and the fifth is related to morphological organization of the kernel. Since this two-element system could operate to control the action of genes that appear to be quite dissimilar, it was expected that it could also operate to control the expression of other genes and at any known locus. The insertion of the *Ds* element at the locus of the gene was all that was required to initiate a sequence of mutation-type changes in action of the genic materials there located. Tests of this hypothesis were then conducted with two selected gene loci, both known to be concerned with anthocyanin pigment formation. These two loci were chosen because of ease of detection of changes in action of their genes. The methods employed for these tests have been given elsewhere (6). It is necessary to mention here only that expectations were fulfilled. Under the given conditions of the experiments, cases of control of gene action by this integrated two-element system appeared at each of the selected loci.

Recently, Nuffer (7, 8) discovered in one plant of a culture a system of control of gene action that is basically the same as that outlined above. It was first identified on an ear produced by self-pollination. A newly arisen recessive mutation at a previously unknown locus was carried in one chromosome 1 of the plant that produced this ear. It effected a modification in anthocyanin pigmentation in both kernel and plant and was designated bronze$_2$ (bz_2). On this ear, three-quarters of the kernels that were homozygous for bz_2 showed instability of expression of the bronze phenotype: mutations to the dominant, Bz_2, occurred in some cells during development of these kernels. Subsequent study revealed the presence of an independently located dominant factor, designated Modifier (*M*), that was responsible for control of mutation at bz_2, and that had a detailed resemblance to *Ac* in its mode of control and general behavior. Tests were then conducted to determine whether or not *M* and *Ac* might be the same element. These tests were positive, and they established the similarity or identity of *M* and *Ac*.

The general similarity in behavior of *Ds* and *Ac* carries the inference that the *Ac* element alone could effect change in genic action in much the same manner that *Ds* does. This inference proved to be correct and evidence supporting it will be summarized in the next section.

Direct Control of Gene Action by the "Ac" Element

The P^{vv} Case

The first discovered case of what now appears to be *Ac* control of gene action and mutation involves the *P* gene in chromosome 1. Both somatic and germinal mutations occur at the locus of *P* when the unstable P^{vv} allele is present. These mutations affect pigment formation in the pericarp and cob tissues. Those occurring in the germ line may be isolated. Some mutants give the full dominant red - pericarp expression in subsequent generations. Others give allelic types of this expression, and still others behave as stable recessives.

The behavior of P^{vv} was examined by Emerson and his students in the early years of maize genetics (8-12). Only recently, however, was it discovered by Brink and his students (13-15) that the instability it shows is a reflection of the presence at this *P* locus of a separable gene-controlling element. This element was called Modulator (*Mp*) because it controlled not only the mutation process itself but also the pattern of these mutations. This pattern was found to be an expression of the dose of *Mp* present in the nucleus. Because of transposition of *Mp* from the *P* locus to other locations within the chromosome complement, it was possible to obtain P^{vv}-carrying individuals having *Mp* units in addition to the unit present at P^{vv}. It then appeared that the larger the dose of *Mp*, the later the time during development of the pericarp tissue when mutations occurred. This dosage expression was so similar to that given by *Ac* that tests were conducted to determine whether or not the *Mp* unit would activate *Ds*. Such tests were positive: *Mp* activates *Ds* in the same manner that *Ac* does (15). This suggests that *Mp* and *Ac* are the same or similar elements.

P^{vv} is present in a number of different strains of maize derived from widely divergent geographic areas. All the strains carrying it that have been examined gave positive tests of activation of *Ds* -- the standard test for the presence of *Ac* (15). Strains with other types of modification of the *P* locus did not show the presence of *Ac*. It may be concluded that *Ac* is not an element which has recently appeared or been introduced but rather one that probably resides in all maize germ plasm and could be made evident in many if not all strains. (For modes of uncovering controlling elements, see author's previous publication (3).)

Ac Control of Mutation at the Bronze Locus in Chromosome 9

Direct *Ac* control of gene action at the bronze locus in chromosome 9 appeared in a Cold Spring Harbor culture when *Ac* was transposed to this locus (16). (*Bz*, deep red or purple pigmentation in kernel and plant; *bz*, recessive allele, gives bronze-colored pigmentation.) The initial phenotypic expression following this insertion was similar to that of the known recessive, *bz*. Mutations occur, which are found to be associated with modifications induced by the *Ac* element. The majority of them give either the dominant, *Bz*, expression or a stable recessive, *bz*. A few of them, however, give intermediate levels of pigmentation. Some of the *Bz* mutants are stable; others, however, are not. The latter undergo subsequent mutations to an unstable recessive, a stable recessive, or a stable dominant. Tests of the various mutants have indicated that the stable ones, either those giving *Bz* or those giving *bz*, appear when *Ac* is removed from the immediate vicinity of the bronze locus. The unstable mutants have retained *Ac* at or close to the locus of bronze, and their instability is an expression of its presence and subsequent action. In some of these latter cases it was possible to determine that the mutation-producing event was associated with a modification of the *Ac* element which, however, did not remove it from the immediate vicinity of the bronze locus.

In all essential respects, control of gene action and of mutation at the P^{vv} locus and at the mutable bronze locus, described above, is the same. In both cases, the type of genic action and mutation is an expression of the presence and behavior of the *Ac* element at the locus.

Spread of Mutational Change Along the Chromosome Induced by "Ds" and the Relation Between Transposable and Nontransposable Controlling Elements

In a study aimed at determining the positions in the short arm of chromosome 9 that *Ds* may enter, two cases of its insertion just to the left of the Sh_1 locus were isolated (sh_1, shrunken endosperm). Subsequent study of these two cases has yielded information of singular significance. It was possible to learn the extent to which mutational change induced by *Ds* may spread along the chromosome, and to relate readily transposable controlling elements to those that do not normally transpose or that do so rarely.

The genetic markers *I*, Sh_1, and Bz_1 are located in the short arm of chromosome 9 and are aligned in this order with reference to its free end (*I*, dominant inhibitor of aleurone color). The standard cross-over value for the *I* to *Sh* interval is approximately 4, and for the *Sh* to *Bz* interval it is approximately 2. When *Ds* is inserted just to the left of *Sh*, changes it induces in gene substances located within the *I* to *Bz* interval may be detected. Altogether, 57 different alterations induced by *Ds* in this region were examined, of which 38 produced *sh* mutants. Twelve of the alterations affected both *Sh* and *Bz* and gave rise to the double mutant, *sh bz*, and 7 of them altered the gene substances located between *Ds* and *I* and including the latter. The origin and behavior of these mutants have been described elsewhere (4), and only a brief description of some aspects of their nature will be given here.

In none of the above-mentioned cases was *Ds* removed or its location obviously altered by the event occurring to it that induced the mutation-type change. *Ds*-induced reversions to *Sh* characterized some of the *sh* mutants, 3 of them being highly mutable. These reversions, however, were not accompanied by removal of *Ds*, which remained as before -- just to the left of *Sh*; and in the presence of *Ac* new mutations of the above described types again appeared. The more extended alterations -- those affecting both the *Sh* and *Bz* loci or the genetic materials located between *Ds* and *I* and including the latter -- behaved essentially as units in inheritance, and the *Ds* element was a component of this unit. Crossing over within the affected interval was either very much reduced or possibly eliminated. However, in several cases, reversions to normal gene action occurred, and in these cases the genes involved were those within the affected segment that were most distally located with respect to *Ds*. The reversions appeared only when *Ac* was also present, which indicates that the *Ds* element was responsible for them. In none of the examined cases was *Ds* removed or noticeably altered in location by the event that gave rise to the reversion, although some of them were associated with a change in state of *Ds*.

The mutation-type changes described above behave as alleles of one another. It is obvious that they fall into different categories: "isoalleles," alleles with "pleiotropic effects," and "pseudoalleles."

The behavior of *Ds* when it is inserted just to the left of *Sh* differs in one striking way from its behavior when inserted at some other loci: reversions of the mutant initially produced are not accompanied by the disappearance of *Ds*. This is in contrast to what occurs when *Ds* is inserted at several other loci. In these latter cases it is known that the majority of reversions are accompanied by removal of *Ds*. At present, there is no ready explanation for these differences in the behavior of *Ds* when inserted at different locations. The knowledge obtained from these studies does provide a basis for considering the similarities that may exist between readily transposable controlling elements and those that appear not to transpose or that do so rarely. It is now known that they need not represent separate classes of elements. And this knowledge raises the questions: are controlling elements present at all gene loci and are many of the so-called gene mutations the consequence of changes occurring to them?

The "Dt - a_1" System of Mutational Control

The classic example of the behavior of controlling elements is that of the *Dt* (Dotted) - a_1 system originally investigated by Rhoades (17-20) and subsequently utilized by a number of other maize geneticists. This system's mode of operation is so well known that a summary is unnecessary. Information recently acquired gives further evidence that relates the behavior of this system to those described above. Nuffer (21) discovered the presence of *Dt* in two South American strains of maize. *Dt* occupies a different position in the chromosome complement in each of these two strains, and neither position is the same as that occupied by the *Dt* element originally discovered by Rhoades. Again, it is clear that controlling elements need not occupy set positions in the chromosome complement but may be variously located. In this, as well as in some other respects, the *Dt* - a_1 system of control of gene action and mutation resembles the *Ds* - *Ac* system in its general mode of operation and behavior.

The "Spm - a_1^{m-1}" System of Mutational Control

Recently, another controlling system, unrelated to the *Dt* - a_1 or *Ds* - *Ac* systems and differing from them in its manner of control of gene action, has been investigated (16). It was isolated from an individual kernel that showed instability of genic expression at the previously normal behaving A_1 locus in chromosome 3. A two-element system is operating to control gene action and mutation at this modified A_1 locus, which is designated a_1^{m-1}. One element is at the locus of A_1. The second element, designated Suppressor-mutator (*Spm*), is independently located. When *Spm* is present, anthocyanin pigment formation in both kernel and plant is suppressed altogether until a mutation-inducing event occurs at a_1^{m-1} in some cells of the plant or kernel that allows pigment to be formed. The mutants so produced are stable thereafter, and they form a graded series of alleles with respect to the intensity of pigmentation each can express. In the absence of *Spm*, gene action at a_1^{m-1} is not totally suppressed. Uniformly distributed pigment appears in both kernel and plant, but no mutations occur. This expression of a_1^{m-1} is stable as long as *Spm* is kept from the nucleus. When *Spm* is returned, its Suppressor-mutator effect on a_1^{m-1} is again evident.

Like *Ac, Dt,* and *Ds*, *Spm* need not reside at one locus in the chromosome complement. It appears at new locations and disappears from previously known locations. Also like *Ac* and *Ds*, such changes in location occur in both somatic and germinal cells, and presumably they result from the transposition mechanism. Unlike *Ac* and *Dt*, however, *Spm* does not show pronounced dosage effects. The number of *Spm* elements present in a plant is not revealed by the pattern of mutation occurring at a_1^{m-1}. With *Dt*, added elements increase the frequency of occurrence of mutation at a_1, while added doses of *Ac* effect step-wise delays in time of occurrence of mutation at the various gene loci with which it may be concerned. In this a_1^{m-1} system, controls of time, type, and frequency of occurrence of mutation reside with the controlling element present at the A_1 locus, as indicated below.

In addition to the events at a_1^{m-1} that give rise to stable mutants, other modifications arise when *Spm* is present in the nucleus. These are expressed subsequently by changes in the time when mutations occur, the types of mutation, and the frequency of their occurrence, or by combinations of these. These modifications also alter the type of phenotypic expression that will appear in the absence of *Spm.* These striking changes of a_1^{m-1} are called changes in state. Such modifications occur far less frequently than those that give rise to stable mutants. Thus, it is possible to isolate and maintain a number of different states, each having its own characteristic mode of expression in the presence and absence of *Spm.*

A few examples will illustrate this. One state gives deep pigmentation in both kernel and plant in the absence of *Spm.* In the presence of *Spm,* no pigment appears until a mutation-producing event occurs at a_1^{m-1} in a cell of the plant whose progeny then exhibit pigmentation. Many of these mutations occur early in development, and the majority of the mutants so produced express the higher alleles of A_1. Another state behaves much like that just described in the absence of *Spm.* In its presence, however, mutations occur very late in development. Only dots of deep pigmentation appear in the kernel and only fine streaks in the plant, and few germinal mutations are recovered. Another state produces the same pattern of mutation in the presence of *Spm,* but in its absence the intensity of pigmentation is much reduced. Another state gives only very faint pigmentation in the absence of *Spm.* In its presence, mutations occur early in development but these give rise to mutants that express low levels of pigment intensity. In combinations and in inheritance behavior, each state acts as an allele of the other. By appropriate combinations of two different states in an individual, the independent action of each state in the presence of *Spm* is made evident by the distinctive types and patterns of mutation each produces.

Additional Evidence of Controlling Elements in Maize

Some indication of the scope and influence of controlling elements on gene action and mutation may be gained from the descriptions of their behavior given in the previous sections. Other cases of their influence in these respects are known, but time will allow only a brief mention of some of them. In a study to determine the factors responsible for the behavior of an unstable pale-green (chlorophyll) mutant, Peterson (22) discovered the presence of a separable controlling element which he called Enhancer (*En*) because its presence markedly increased the rate of mutation to green at the pale-green locus. In conformity with some other controlling elements here described, *En* does not maintain itself at any one locus in the chromosome complement. It appeared at new locations in the course of Peterson's study, and he interprets these changes in location as arising from transposition of the *En* element.

Another case of instability of expression at the A_1 locus in chromosome 3, designated a_1^{m-2}, appeared in the Cold Spring Harbor cultures (16). Although analysis of this case has not been as complete as those previously described, it is clear, nevertheless, that the mode of control of gene action and mutation differs from those for the other systems in a striking manner. The behavior of this control system is consistent with the following interpretation. A two-element system is involved, one element located at A_1 and one independently located. In the presence of the independently located element, a distinctive pattern of anthocyanin pigmentation appears in both plant and kernel which, in the kernel, is expressed by dots of the full A_1 phenotype in a lightly pigmented background. This background pigmentation is not uniform but presents a diffusely mottled appearance. Removal from the nucleus of the independently located element, either somatically or as a consequence of meiotic segregation, leads to a mutation-producing event at an a_1^{m-2}, and the mutants so produced are thereafter stable in expression. A few of them give the full A_1 expression, but the majority give lower levels of pigmentation with an intensity that can range from very faint to relatively dark. In all cases, however, this pigment is not uniformly distributed but presents a diffusely mottled appearance. Like *Spm,* the independently located controlling element of this system shows no distinctive dosage effects. Also, like *Spm* and some other controlling elements here described, it undergoes transposition.

Other cases of control of gene action and mutation by systems similar to those here described are now under investigation at several different laboratories. Dollinger is investigating several such cases (23). Mangelsdorf (24) has found that a large number of spontaneous mutants appear in inbred strains of maize into which one or more teosinte chromosomes had been introduced. Of these mutants, a number are unstable. Although analysis of individual cases is not extensive, Mangelsdorf suspects the presence of controlling elements responsible for the mutants observed.

Discussion

From the evidence summarized in this report, there can be no doubt about the presence in the maize chromosome complement of elements that control the action of genes and induce in them mutation-type responses. These elements are considered to be accessory to the gene and have been called controlling elements. The different elements may be distinguished from one another by their distinctive modes of control of gene action and mutation, and the latter is effected by alterations occurring to the elements themselves. The elements are capable of undergoing changes in location, and the process by which this is accomplished has been called transposition. When such transposition results in the insertion of a particular element at a known position, it may modify the action of the genic materials there located, both initially and subsequently, and it does so in a precise manner. Insertion of a different element at this same locus may also result in initial and subsequent modifications in gene action, but the manner in which these are expressed is not the same. In all cases, however, both the initial alteration in gene action resulting from insertion of a controlling element at a particular locus and the subsequent mutation-producing events that occur there are directly referable to the element that has become inserted at the locus.

The evidence indicates that individual controlling elements do not act in a strictly autonomous manner. Instead, the alterations they undergo appear to reflect the operation of an integrated system composed of several elements, and these, in turn, may be independently located within the chromosome complement. In some cases, the individual components comprising a system have been identified, and in all such cases one of them was at the locus of the gene whose changes in action were being observed. The types and patterns of mutation reflect the action of the system as a whole, although the mutation-inducing event itself arises directly from a modification induced by the particular element of the system that is at the locus of the gene. No interrelationships between the presently identified systems of controlling elements have yet been observed, although it is suspected that some relationships may be found in the future as tests of this are expanded. Such interrelationships are to be expected if the types of elements under consideration are essential components of all nuclei and if they are responsible for controlling when certain genes will become active and also their types of action. An orderly sequence of triggering responses of different systems could be anticipated to occur in much the same manner as they appear to be triggered within a system.

It has been shown that a particular controlling system may operate to produce mutations at a number of different gene loci and, conversely, that different systems may operate at one particular gene locus. In the latter case, some of the mutants produced by different systems may be alike in their phenotypic expressions. Knowledge of the mode of origin of such mutants allows discriminations to be made between them. In some cases, their differences are further expressed by their capacities for subsequent mutation under given conditions. For example, mutations at the A_1 locus that give rise to the recessive, a_1, have arisen from modifications at this locus induced by controlling elements belonging to quite different systems -- the *Dt,* the *Ac,* and the *Spm* systems and the system associated with a_1^{m-2}. On the basis of phenotype, these mutants are not distinguishable from one another. Differences among them are sometimes made evident by the types of response each may give when any one of the four above-mentioned elements is introduced into the nucleus by appropriate crosses. An a_1 mutant belonging to the *Dt* system will remain stable in the presence of *Ac, Spm,* or the independently located element associated with a_1^{m-2}. In the presence of *Dt,* however, numerous mutations may occur to the higher alleles of A_1, and their kinds and frequencies of occurrence are predictable. Similarly, an a_1 mutant belonging to the *Ac* system may undergo subsequent mutation in the presence of *Ac* but will not respond to *Spm,* etc. Thus, with regard both to the initial mutation and to potentials for subsequent mutation, each mutant reflects the operation of a particular controlling system irrespective of similarities in phenotypic expression of some of the mutations that these various systems may produce.

The mutants resulting from operation of the systems here described are like many of those that have arisen spontaneously or that have been produced by mutagenic agents. This suggests the following questions: Do many of these latter mutants arise from alteration or substitution of controlling elements, accessory to the genes? And are such elements normally associated with the genes? Obviously, no decisive answers may yet be given to these questions. Nevertheless, the evidence here outlined makes it necessary to consider such possibilities and to reconsider evidence obtained from other mutation studies that might be illuminating in this respect. One case in maize may be used as an illustration.

Previously in this report, evidence was given of the effective fixation of the *Ds* element following its insertion at one particular locus in chromosome 9. At this position, it can alter genetic materials located on either side of it. As a consequence of this, a number of different alleles have appeared. At the *R* locus, one of the most extensively studied of all known loci in maize (25-33), a similar situation may exist. A fixed controlling element (or elements) may be present at this locus, and changes occurring to it may be responsible for the production of the large number of known alleles. These alleles are distinguishable, one from another, by the type of effect each exerts on the expression of anthocyanin pigmentation in the plant tissues and in the aleurone layer of the kernel. They control whether or not anthocyanin pigment will appear in any one tissue and the location and intensity of this pigment within a tissue when it is formed. The identification, following specific types of crosses, of modifiers located elsewhere in the chromosome complement that effect marked changes in rates of "spontaneous" mutation at this locus also suggests the presence there of a fixed controlling element. For many years, however, this locus was considered to be most favorable for an examination of the nature of the gene itself and of its mutation potentials. In the light of our present knowledge, one would hesitate to draw such a conclusion.

The presence of controlling elements operating in the manner here described has not been reported in organisms other than maize, although their presence is to be suspected. The basic type of genetic phenomenon associated with their operation is known to occur in many widely divergent types of organisms. There is some evidence that suggests their presence in phage (34), bacteria (35), protozoa (36-38), and a number of higher plants and animals (1, 2, 39). In this regard, it may be recalled that the general prevalence of controlling elements in maize remained undetected for many years, although this organism received extensive genetic study by numerous investigators. Recognition of their presence in chromosome complements in general would make it necessary to incorporate them into some conceptual framework of genetic theory.

Discussion of controlling elements has been restricted in this report to those aspects of their behavior that are directly concerned with the mutation process, the topic of this symposium. Other aspects of considerable importance, such as the mode of transposition of controlling elements, the nature of the alterations that are associated with changes in constitution of these elements (changes in state or in quantity of a controlling element without change in location), inheritance patterns, modes of revealing controlling elements, interrelationships between elements of a given system, and so forth, cannot be discussed. Also, the possible developmental and evolutionary significance of these elements cannot be considered here. It is hoped, however, that the descriptions of their types and modes of behavior are extensive enough to suggest various aspects of their significance.

Summary

In recent years, maize geneticists have become aware of systems of control of gene action and mutation that reflect the activity of chromosome elements considered to be associated with but accessory to the gene. In this report, these are referred to as controlling elements.

Different controlling elements, and systems of interrelated elements, have been identified. These are distinguishable from one another by their distinctive modes of control of gene action and mutation.

A summary is given of the known types of controlling elements, and of systems of interrelated elements, and how each operates to modify gene action. The types of change that they induce in gene action resemble those arising spontaneously or those produced by mutagenic agents.

Because a controlling element may move from one location to another within the chromosome complement without losing its identity in the process, the same element may become associated with different genes. Thus, it can modify the action and induce subsequent mutation of a number of different genes and, conversely, the action and subsequent mutation of a particular gene may be controlled by different elements.

Although phenotypic expressions of mutants produced by action of different controlling elements at one particular gene locus may be quite similar, the mutants themselves are distinguishable from one another on the basis of their known origins and, in some cases, the predictable types of subsequent mutation each may undergo when a particular controlling element is introduced into the nucleus.

Evidence is given to show that controlling elements may effect simultaneous change in action of more than one gene when such genes are sequentially located within a chromosome.

The possible presence of a controlling element at the locus of a well investigated gene in maize, which may be responsible for the appearance of the many known alleles of this gene, is discussed. The suggestion is made that controlling elements are normal components of the chromosome complement and that they are responsible for controlling, differentially, the time and type of activity of individual genes.

References

1. Stubbe, H., Biblio. Genetica, 10, 299 (1933).
2. Demerec, M., Botan. Rev., 1, 233 (1935).
3. McClintock, B., Cold Spring Harbor Symposia Quant. Biol., 16, 13 (1951).
4. McClintock, B., Carnegie Inst. Wash. Year Book, Nos. 45-53 (1946-1954).
5. McClintock, B., Proc. Natl. Acad. Sci., 36, 344 (1950).
6. McClintock, B., Genetics, 38, 579 (1953).
7. Nuffer, M.G., Maize Gen. Coop. News Letter, 27, 68 (1953).
8. Nuffer, M.G., Maize Gen. Coop. News Letter, 29, 59 (1955).
9. Emerson, R.A., Genetics, 2, 1 (1917).
10. Emerson, R.A., Genetics, 14, 488 (1929).
11. Eyster, W.H., Genetics, 9, 372 (1924).
12. Anderson, E.G., Eyster, W.H., Genetics, 13, 111 (1928).
13. Brink, R.A., Nilan, R.A., Genetics, 37, 519 (1952).
14. Brink, R.A., Genetics, 39, 724 (1954).
15. Barclay, P.C., Brink, R.A., Proc. Natl. Acad. Sci., 40, 1118 (1954).
16. McClintock, B., unpublished.
17. Rhoades, M.M., J. Genet., 33, 347 (1936).
18. Rhoades, M.M., Genetics, 23, 377 (1938).
19. Rhoades, M.M., Cold Spring Harbor Symposia Quant. Biol., 9, 138 (1941).
20. Rhoades, M.M., Proc. Natl. Acad. Sci., 31, 91 (1945).
21. Nuffer, M.G., Science, 121, 399 (1955).
22. Peterson, P.A., Genetics, 38, 682 (1953).

23. Dollinger, E.J., Genetics, 40, 570 (1955) and personal communication.
24. Mangelsdorf, P.C., Maize Gen. Coop. News Letter, 29, 23 (1955).
25. Stadler, L.J., Proc. Natl. Acad. Sci., 30, 123 (1944).
26. Stadler, L.J., Genetics, 31, 377 (1946).
27. Stadler, L.J., Am. Naturalist, 82, 289 (1948).
28. Stadler, L.J., Am. Naturalist, 83, 5 (1949).
29. Stadler, L.J., Portugliae Acta. Biol. Ser. A, Goldschmidt Vol., 785 (1950).
30. Stadler, L.J., Cold Spring Harbor Symposia Quant. Biol., 16, 49 (1951).
31. Stadler, L.J., Nuffer, M.G., Science, 117, 471 (1953).
32. Stadler, L.J., Emmerling, M., Science, 119, 585 (1954).
33. Stadler, L.J., Science, 120, 811 (1954).
34. Visconti, N., Symonds, N., Carnegie Inst. Wash. Year Book, No. 52, 222 (1953).
35. Treffers, H.P., Spinelli, V., Belser, N.O., Proc. Natl. Acad. Sci., 40, 1064 (1954).
36. Sonneborn, T.M., Advances in Genet., 1, 263 (1947).
37. Nanney, D.L., Proc. Natl. Acad. Sci., 39, 113 (1953).
38. Nanney, D.L., Caughey, P.A., Proc. Natl. Acad. Sci., 39, 1057 (1953).
39. Hollander, W.F., Quart. Rev. Biol., 19, 285 (1944).

Discussion

SAGER: I would like to ask about the spontaneous origin of some of these rate-controlling alleles. Have you ever crossed either *Ac* or *Ds* separately into a strain that was previously stable to determine whether the presence of *Ac* would increase the probability of *Ds* arising; and reciprocally whether the presence of *Ds* would lead to the origin of a new *Ac*?

McCLINTOCK: Strains carrying *Ds* but no *Ac* have been propagated for the past 7 years. No evidence of the appearance in them of an *Ac* element has been obtained. It is not practicable to determine whether or not *Ds* will appear anew in strains carrying *Ac*. Determination of the initial absence of *Ds* would require that each arm of every chromosome in the complement be scanned for its presence. Furthermore, appropriate genetic markers in some of these arms that would aid in the detection of *Ds* in the arm are not now known.

TAYLOR: I believe you mentioned that certain of the unstable *bz* loci change to stable recessives. Is that right?

McCLINTOCK: The unstable genic expression at the bronze locus associated with the presence there of either *Ds* or *Ac* will disappear whenever the *Ds* or the *Ac* element is removed from the locus. The phenotypic expression following this removal may be either that of the recessive, *bz*, or that of the dominant, *Bz*.

TAYLOR: Do you know anything about the nature of the *bz* locus in those cases?

McCLINTOCK: Our information on the nature of the change at the bronze locus is not precise enough to allow a definite answer to be given to your question. There is only suggestive evidence. Some years ago, it was found that tissues homozygously deficient for the chromatin composing this locus would express the recessive, *bz*, phenotype. However, the expression given by the known recessive mutant appears to represent some form of inhibited gene action rather than deficiency, as indicated in the discussion of the origin and behavior of instability at this locus when *Ac* was inserted there. Some of the events occurring to *Ac* remove this inhibition. A dominant expression results and if *Ac* is removed from the locus, the dominant expression is thereafter stable. The majority of events occurring to *Ac*, however, do not appear to alter this inhibitory condition. Those that remove *Ac* from the locus result in return to the stable recessive expression. Nevertheless, it is conceivable that some of the events occurring to *Ac* could remove the chromatin materials composing the locus. A stable recessive expression would be anticipated to appear if such occurs. However, we do not yet have any direct evidence of the presence of such deficiencies in the cases of the stable recessives that have been examined. On the basis of various types of evidence accumulated over the years, it is believed that many of the stable recessive expressions arise from some form of alteration or reorganization of the materials composing the locus and that these account for the inhibited or altered gene action.

BONNER: My question hinges around the action of *Ds* and *Ac*. Do you feel that *Ds* and *Ac* are integral parts of a gene in terms of the action of a unit of genetic material, or are these factors involved only in changes of state as opposed to changes in action of the material itself. A given unit of genetic material might have a given action, and the state of that action could be altered by *Ds* and *Ac*. Conversely the action of a given unit of genetic material might be altered by these two factors. I was wondering whether you had any thoughts on this matter?

McCLINTOCK: In terms of position, a controlling element is an integral part of the locus in the chromosome where the genic materials reside. Because controlling elements are transposable and can affect action of genic materials at various different loci, it is assumed that they are not integral components of the genic substances themselves.

FOX: It has been suggested that the phenomena which you describe are somehow associated with heterochromatin. Since heterochromatin is widely regarded as duplicated material, it might be suggested that the apparent shifts in position that you observe are actually the assumption by one heterochromatic block of a function which had previously been performed by a different block. I wonder, therefore, if there is any way of actually demonstrating that the change of position of these elements is associated with an actual transposition of material, or if it is possible that there is a transposition of activity rather than material?

McCLINTOCK: There is strong presumptive evidence that heterochromatin is associated with the phenomena here discussed. However, we do not have any direct visual evidence of transfer of a material substance. We do know that some alterations in the chromatin materials at the affected loci can occur, for chromosomal abnormalities sometimes accompany transposition of these elements. Duplications or translocations may arise. In the origin of these chromosomal abnormalities, two positions in the chromosome complement are involved. One of them marks the previously known location of the element, such as *Ds*, and the second marks its new location. It is clear from these cases that a process other than transfer of activity is associated with the transposition mechanism. A mechanism similar to that postulated to account for incorporation of the transforming principle or that which may accompany the transduction phenomenon may operate also in these cases.

SAGER: As I understand it, you have described two classes of controlling genes. There is a class which must be interposed into the locus to have its effect, and a class which can act at a distance. For instability to occur, must there always be one allele of each class operating, or can either of these classes act autonomously? Would you please cite examples of the latter if they occur? Also, could you say approximately how many instances have been studied so far of each of the various types of control systems?

McCLINTOCK: The behavior of *Ds* and *Ac* may be cited to answer your first questions. *Ds*, when inserted at a known gene locus, affects the action of the genic materials that are present there. Changes in this action occur but only, however, when *Ac* is also present in the nucleus and regardless of the location of *Ac*. In this respect, *Ds* is "nonautonomous" in its mode of action. *Ac*, when inserted at a known locus, also can affect the action of the associated genic materials and in a manner similar to that of *Ds*. Control of change in gene action resides with this *Ac*. In this respect, then, *Ac* is "autonomous" in its control of gene action. The answer to your last question will be found in the written account. There was no time today to list the known types of control systems and the number of known cases of their operation.

HOTCHKISS: I wonder whether it would be reasonable to synthesize some of these things by suggesting that with the ordinary genes we are accustomed to see their end effects in the soma, at various points in the adult plant and animal. We might consider these activators as genes that have their end effect in the nuclear mechanism, playing a part there somewhat analogous to "usual" genes but doing it at early or special times and therefore seeming to be a different class.

McCLINTOCK: The interpretation you suggest resembles in some respects that which I hold. We might say that within the nucleus there are two distinctly different classes of chromosome components, both of which exhibit Mendelian inheritance patterns. The primary gene elements, as defined earlier, function in the production of substances that reach the cytoplasm and affect metabolic processes that occur there and thus the phenotype. The controlling elements, on the other hand, do not give rise to products that ultimately reach the cytoplasm. The direct consequences of their activities are confined within the nucleus itself, for they serve to control the time and type of activity of the primary gene elements. They represent a strictly intranuclear system serving to modulate the activities of the primary genic components and in very precise ways. As a consequence of this, they effect modifications in phenotypic expression. They could be viewed as "genes" controlling nuclear processes in contrast to the other class of "genes" whose activities result in products that reach the cytoplasm.

HOTCHKISS: Might I mention the abnormal transformations we obtained when the Type III capsule-inducing DNA is put upon one special strain of pneumococcus? One out of very many strains which we have tested behaved differently from all the others. Where all the others gave a Type III capsule (and this means a specific kind of polysaccharide) this one gave not just a different polysaccharide but a series of encapsulated organisms, some of them close to the Type III and others a great distance away. It looks like a case where transforming agent in a new particular biological context behaves differently from the way it would in most others.

GOODGAL: Do you have a measure of the rate of movement of *Ac* or *Ds*? Have you any quantitative information?

McCLINTOCK: It is difficult to obtain meaningful figures regarding rates of transposition. For example, in plants having both *Ds* and *Ac*, the time during development of a tissue when transposition of these elements will occur depends on the dose of *Ac* present in the nuclei of the plant. If the dose is low, transpositions will occur relatively early in development. A number will occur in the sporogenous cells. Thus, a number of gametes will be formed carrying transposed *Ds* and/or *Ac* elements. In plants with such low doses of *Ac*, we have found that from 10 to 20 percent of the gametes they produce may carry a transposed element. If the dose of *Ac* is increased in the nuclei of the plant, the time of occurrence of transposition is delayed. Relatively few transpositions may occur in the sporogenous cells. Therefore, the percent of gametes carrying a transposed element is much lower. The occurrence of transposition also depends upon the respective states both of *Ds* and *Ac*. In any one study aimed at determining the rate of transposition either of *Ds* or of *Ac*, each of the mentioned conditions must be known and its contribution equated. Obviously, then, no one figure for rate of transpostion can be given.

FRANKEL: I think the point that Dr. Bonner made is a very relevant one. Is this a state which is normal in an organism suggesting a pattern of regular individual development, or is it something that one might interpret as a "disease" of the gene? If it were normal, then it should not be as readily reversible as it appears to be.

McCLINTOCK: It is my opinion that the phenomenon here discussed is a general one and that it depicts one of the mechanisms responsible for control of gene action during development. This opinion is based on the large amount of evidence given by other organisms that suggests the presence in them of such gene-controlling systems. Extensive analyses of the modes of operation of such systems have been made, however, only with maize.

MEHLQUIST: Dr. McClintock, you suggested that these phenomena are often seen in horticultural plants. Is it logical, then, to assume that these phenomena are associated with the high degree of heterozygosity which is usually characteristic of horticultural plants?

McCLINTOCK: A number of horticultural plants were selected for propagation because they exhibited the types of phenomena here discussed. The systems responsible for the phenotypes exhibited would be expected to induce a considerable amount of heterozygosity in these plants.

MULLER: What is your opinion about whether or not this phenomenon may be related to the production of regular patterns? If I remember correctly, in maize, you not only have the irregular patterns that we used to call "ever sporting," but also alleles that give regular patterns. I mean, for example, when you have all the cob colored and all the kernel white. That would be an example.

McCLINTOCK: With these systems, the production of regular patterns is the rule. When the mode of operation of a system is understood, what may appear at first sight to be an irregular pattern assumes, on closer analysis, a regular and expected pattern. The behavior of the *Ds* - *Ac* system may be used to illustrate this. With high doses of *Ac*, the time of occurrence of mutation-producing events at *Ds* is late in development of any one tissue. Consequently, a uniform pattern of small mutant spots appears. If the dose of *Ac* is low, changes in the *Ds* and *Ac* elements can occur early. Some of them result in a mutation-inducing event at *Ds*. A large sector of mutant tissue may thus be produced by a cell in which this occurred. Other early occurring changes of these elements may not result in a mutation-producing event at *Ds* nor in an otherwise detectible alteration of it, but they may give rise to a modification of *Ac*. Increase in dose of *Ac* in one cell and reduction in this dose or loss of *Ac* from the sister cell may be the consequence of this. The pattern of mutant spots in the tissue produced by each of these two sister cells will differ. This is because the *Ac* constitutions of their nuclei differ and it is these constitutions that control when mutation-inducing events will occur at *Ds*. If no *Ac* is present in the nucleus of one of the two sister cells, no mutant spots will appear in the tissue this cell produces. Viewed as a whole, the pattern of mutant spots in a tissue may seem to be irregular but careful analysis reveals its underlying regularity. Thus, the patterns produced by *Ac* may be strikingly regular or seemingly irregular depending upon the *Ac* constitution initially present in the zygote. Differential expressions in whole tissues of a plant, similar to the example you cite, likewise appear. This is made evident from studies of the systems associated with both a_1^{m-1} and a_1^{m-2}. With a_1^{m-1}, states have been isolated that give rise to pigment in the plant tissues but to no pigment in the kernel. The pattern of pigmentation produced by a_1^{m-2} strikingly illustrates the differential control of pigment formation in different tissues of the plant. Also, mutants may be obtained from it that exhibit pigmentation only in restricted regions of the plant tissues; none appears in the kernel.

RETIREMENT

I believe there is little reason to question the presence of innate systems that are able to restructure a genome. It is now necessary to learn of these systems and to determine why many of them are quiescent and remain so over very long periods of time only to be triggered into action by forms of stress, the consequences of which vary according to the nature of the challenge to be met.

B. McClintock 1978

Clockwise from lower left: Barbara McClintock, student, Norman Davidson, Charles Laird, Mary-Lou Pardue, student (Molecular Cytogenetics Course, CSHL ca. 1970)

Barbara McClintock and Ira Herskowitz (CSHL 1980)

Steve Dellaporta and Barbara McClintock (Uplands Farm, CSHL 1983)

INTRODUCTION

Transposable elements were discovered in maize long before their presence was recognized in other organisms. This discovery was the logical outcome of a series of observations of maize chromosomes, each of which revealed an unanticipated and significant aspect of their behavior. These studies began with my observations of the events occurring to ring-shaped chromosomes during chromosome replication. As initially reported in 1932 (*12*), the replication of ring-shaped chromosomes usually produces two sister chromatids that separate freely in the following mitotic anaphase. However, as the amount of chromatin in the ring increases, replication is increasingly accompanied by a sister-chromatid exchange that produces a double-size ring with two centromeres. In this case, when the centromeres of the two sister chromatids are pulled to opposite spindle poles in the following anaphase, two chromatid bridges are formed between them. These bridges come under increasing tension as the centromeres approach the poles. The tension eventually becomes sufficient to rupture the two bridges, with the break occurring at an unspecified place that is generally different in each chromatin strand. Each telophase nucleus therefore receives a linear chromosome whose two ends have just been created by rupture events. What happens next is extraordinary and proved to be highly significant for later studies: the two ruptured ends find each other and "fuse" (are permanently ligated together). This event reestablishes a ring-shaped chromosome with a single centromere, but one whose genetic composition usually is modified because of the non-equivalent locations of the breaks in the two anaphase bridges. In the period from 1932 to 1941, some of the consequences of these genetic modifications were examined (*17*, *18*, *19*, *23*). Two general aspects of chromosome behavior discovered in the 1932 study would turn out to be crucial for later transposition studies: the approach of broken ends of chromosomes towards one another within the nucleus and the precise ligation of the ruptured ends.

An additional aspect of chromosome behavior that was important for the discovery of transposable elements relates to the behavior of a single ruptured end of a chromosome after it enters a telophase nucleus. This end has no partner to fuse with. In the spring of 1937 the question was asked: What would occur to such a single broken end during chromosome replication, as deduced by viewing this chromosome in the following mitosis? A means of generating such chromosomes was available. It utilized plants that were heterozygous for an inversion in the long arm of chromosome 4 of maize. A crossover within the limits of the inverted segment during meiotic prophase produces a dicentric chromosome plus an acentric fragment, each missing a large segment of chromosome 4. The acentric fragment is lost from the nucleus, either during a meiotic mitosis or in a subsequent mitosis, because it lacks a centromere to direct it to a spindle pole. However, the dicentric chromosome attaches to the spindle, and the movement of its two centromeres towards opposite spindle poles at the first meiotic anaphase produces a single chromatin bridge that is ruptured as the tension on it increases. A chromosome with a single, newly ruptured end then enters each telophase nucleus.

The fate of such a chromosome could be observed by examination of the chromosomes of the generative cells that are formed after male meiosis. In the maize tassel, meiosis occurs in cells of anthers in the florets. The meiotic process initially gives rise to four haploid spores. Each spore then undergoes two mitoses. The first of these results in a tube nucleus and a generative cell;

the generative cell then divides to form the two sperms of the pollen grain that will function in fertilization. During the first of these two mitoses, the chromosomes are readily observed from prophase through telophase. Among the spores undergoing mitosis, those that had received a chromosome with a newly broken end could be detected. In them the event that had occurred to the ruptured end was revealed. Replication had produced two sister chromatids. However, instead of being free to separate from each other at anaphase, these chromatids were joined together at the end that had been ruptured in the previous meiotic mitosis. This gave rise to a newly formed dicentric chromosome made visible by a chromatin bridge at the spore anaphase. Rupture of this bridge could be observed. Again, a newly ruptured end entered each telophase nucleus. Descriptions of this sequence were reported in 1938 (*20*) and reviewed in 1984 (*73*)

It seemed imperative to determine whether such chromatid fusions at positions of previous rupture would continue in subsequent mitoses. For this purpose, it was necessary to produce functional pollen grains whose sperm nuclei had received a chromosome with a newly ruptured end. These were not provided by the pollen grains produced in the experiment just described, since they lacked a segment of the long arm of chromosome 4 that prevented their functioning. Fortunately, just when needed, a special type of chromosome modification was isolated that could provide such pollen grains. It was found during a study of the crossover products of a chromosome 9 that had undergone a complex rearrangement of segments. Although this study was not extensively reported until 1941 (*22*), experiments with the fortunate isolate obtained from it were started without delay and the results reported in 1939 (*21*). A breakage-fusion-bridge cycle was described that continued from initiation of the break at a meiotic anaphase through the development of the gametophytes and gametes, and then could be detected during development of the endosperm of kernels that had received one such gamete. However, in the zygote nucleus in these kernels, or more likely in the division that followed, this cycle ceased. The broken end had "healed" and was replaced by a new and stable telomere that behaved thereafter the same as any normal end.

The ability to obtain functional gametes that carry a chromosome with a single newly ruptured end also made it possible to learn whether fusions of the broken ends of two chromosomes would occur if each gamete contributed one such chromosome to the zygote. The fusion (ligation) of the ruptured ends of a ring-shaped chromosome in plant tissues, as described earlier, suggested that this might occur. However, this would contrast with the behavior of a single broken end introduced into the zygote, which was shown to "heal" shortly thereafter. Detection of kernels on an ear that had received one chromosome with a newly broken end from each parent requires some means of recognizing such kernels. A chromosome 9 with a similar type of organization as that used in the above described tests would be ideal providing that it carried detectable dominant alleles. While at the University of Missouri I had discussed this problem with Dr. L. J. Stadler. Shortly thereafter, he showed me some kernels on an ear in his stocks whose phenotypic expressions suggested to him the presence of a chromosome with just the right gene markers and chromosome organization. Cytological examination of plants derived from such kernels confirmed the presence of a chromosome with just the right structural organization. Crosses were then made between plants with similarly organized versions of chromosome 9 that carried contrasting genetic markers. On ears so produced there were kernels whose phenotypes revealed that a chromosome 9 with a newly ruptured end had been contributed by each parent. Fusions (ligations) occurred between these broken ends, forming a dicentric chromosome composed of the two chromosomes 9 joined together at their positions of previous rupture. Initial reports of this fusion and its consequences appeared in 1942 (*25*, *26*). It was an expansion of this test, conducted in the summer of 1944, that produced the plants whose progeny revealed the unexpected presence of transposable elements. Why this expanded test was conducted needs to be explained.

The behavior of ruptured ends of chromosomes during mitotic cycles gives rise, in a number of instances, to recessive "mutations." Each "mutation" reflects the presence of a viable homozygous deficiency, and each can be recognized by a specifically altered phenotype it produces. Such "mutations" were noted initially in studies of ring chromosomes where each modified phenotype reflected a change in the chromatin content of the ring. These phenotypes were described in publications dating from 1937 through 1944 (*17*, *18*, *19*, *23*). The newer type of breakage-fusion-bridge cycle likewise produced deficiencies that were viable when homozygous, and they likewise could be recognized by the modified phenotype associated with each. A large percentage of these were minute deficiencies located at the end of the short arm of chromosome 9. Their nature was described in the Carnegie Institution of Washington Year Books from 1942 through 1947 (*25*, *26*, *27*, *29*) and in *Genetics* in 1944 (*28*).

The many examples of mutant phenotypes, already determined to be reflections of minute deficiencies, suggested another possible means of obtaining such deficiencies distributed at many different sites along the short arm of chromosome 9. The start of the test was to grow plants from kernels that had received from each parent a newly broken end in the chromosome 9 short arm. Because of fusion of the two ends, a dicentric chromosome would form in the zygote, resulting in initiation of the chromosome type of breakage-fusion-bridge cycle. A viable plant would come from those cells in the seedling in which the broken ends had "healed," causing this cycle to cease. These plants, in turn, would be self-pollinated and their progeny examined for segregation of new mutants. A total of 677 kernels that received a newly broken end of chromosome 9 short arm from each parent were sown in the summer of 1944. The appearance of the seedlings that these kernels produced is described in volume 10 of the *Stadler Symposium* (*70*). The plants that survived were self-pollinated, and the ears produced were examined for kernels with mutant expressions. Then, forty such kernels from each ear were sown in a seedling bench in the greenhouse early in 1945, and the seedlings examined for segregation of new mutants. These appeared. However, the phenotypic expressions of some of them were totally unexpected. The segregants were variegated for distinctive grades and types of chlorophyll distribution in the young leaves. Between cultures the patterns of this varied widely, but within a culture the pattern among the segregants was the same.

Seedlings with variegated chlorophyll patterns from some of these cultures were transplanted to pots in order to observe the pattern of variegation in the new leaves as the plant developed. I soon recognized that the changes in patterns of variegation that appeared in sectors on these new leaves held the key to an understanding of the events that were responsible for initiating variegation in the first place. Most significant in this regard were twin sectors, obviously derived from sister cells, in which the pattern changes in the twins were reciprocals of each other. For example, a reduced frequency of mutations to give full chlorophyll expressions on the pale or white background in the surrounding leaf tissue was matched in the twin with a much increased frequency of such mutations. My conclusion from these twin sectors was that during a mitotic cycle one cell had gained some component that the sister cell had lost, and that this component was responsible for regulating (i.e., controlling) the mutation process: that is, its time and its frequency of occurrence in the plant tissues.

On examining kernels on ears of plants grown in the summer of 1944, one ear was found in which loss of genetic markers carried in the short arm of chromosome 9 was occurring in a patterned fashion. Here, also, twin sectors appeared in some of the kernels; these expressed reciprocal relationships with regard to the loss patterns, as if initiated by one cell gaining what the other cell had lost. This suggested that the chromatin loss patterns were controlled by the same general regulatory mechanism that was controlling the patterns of variegation among chlorophyll mutants in the leaves. Because of the clear expression of chromatin losses in these kernels, study of plants derived from them was initiated in 1945.

The discovery of an ear segregating kernels with distinctive patterns of coupled loss of gene markers proved to be most fortunate. These patterns reflected the presence in the parent plant of the Dissociation–Activator (*Ds–Ac*) system of transposable elements. The initial observations were reported in the *Carnegie Institution of Washington Year Book* No. 45, issued in December 1946 (*33*). The symbols *Ds* and *Ac* were applied to these elements, and a discussion of their modes of operation appeared in the *Carnegie Institution of Washington Year Book* No. 46, issued the following year (*34*). During 1947 transpositions of both *Ds* and *Ac* were recognized, including two instances of transposition of *Ds* to the *C* locus in chromosome 9 short arm. These transpositions were reported in the *Carnegie Institution of Washington Year Book* No. 47 (*35*), issued in December 1948, along with an insertion of a *Ds* into the *Wx* locus in chromosome 9.

The study of insertions of a *Ds* into known gene loci had progressed sufficiently by 1950 to warrant publication in a journal with a wide readership (*37*). The report was titled "The Origin and Behavior of Mutable Loci in Maize." It was clear from responses to this report that the presented thesis, and evidence for it, could not be accepted by the majority of geneticists or by other biologists. Genetics was still in an unformed state compared to the rapid changes in concepts that occurred subsequently in the 1950s and 1960s, and there was no clear notion of the nature of the gene. It remained a hypothetical unit until proven otherwise. A long report in the *Cold Spring Harbor Symposium* for 1951 represented a second attempt to present the "mutable loci" story as it was progressing at Cold Spring Harbor (*40*). The response to it was puzzlement and, in some instances, hostility. A third attempt to support the thesis of the origin of mutable loci in maize appeared in 1953 in the widely read journal *Genetics* (*42*). It was titled "The Induction of Instability at Selected Loci in Maize." This article appeared before copying machines practically eliminated requests for reprints. At the time, reprints were distributed to a selected few, and to others on request. In this instance I received a total of only three requests for this reprint! By then I had already concluded that no amount of published evidence would be effective. As a consequence, beginning in January 1949, those projects that had reached a state allowing conclusions to be drawn from them, or were in need of data assembly and comment, were treated to an unpublished written account, with tables of data, diagrams when needed, and a discussion of the significance of the findings. Only the highlights of these studies were reported in the annual Year Books of the Carnegie Institution of Washington.

In 1952 R. A. Brink at the University of Wisconsin and in 1953 P. A. Peterson, now at Iowa State University, each published their experiences with transposable elements in maize. Even so, I was not convinced that transposable elements were viewed, generally, with confidence of their reality. There were many vocal skeptics. Therefore, the method I had chosen to record data and conclusions from them was continued into the early 1960s. Some topics, however, were dealt with and illustrated in symposia articles (*46, 50, 63, 66, 69, 70, 71*). Also, beginning in 1958, special topics were discussed in the annual reports of the Carnegie Institution of Washington, as the titles indicate.

In retrospect, it appears that the difficulties in presenting the evidence and arguments for transposable elements in eukaryotic organisms were attributable to conflicts with accepted genetic concepts. That genetic elements could move to new locations in the genome had no precedent and no place in these concepts. The genome was considered to be stable, or at least not subject to this type of instability. A further difficulty in communication stemmed from my emphasis on the regulatory aspects of these elements. In the mid-1940s there was little if any awareness of the need for genes to be regulated during development. Yet it was just this aspect that caught my attention initially. Unquestionably, some genetic mechanism was controlling the patterns of gene expression, as clearly illustrated by the twin sectors in leaves and kernels. In these instances, the responsible mechanism appeared to be associated with some event occurring during a mitotic cycle. It was my intention in the summer of 1945 to attempt to find out what this regulatory event might be. It was not until fifteen years later that the regulation of gene action began to gain credibility due to the elegant experiments of Jacob and Monod that were carried out in bacteria. Their studies began a new era in genetics. The integrated systems of transposable elements in maize, which I called "controlling elements" because of their distinctive modes of regulating the expression of genes, turned out to represent a quite different mode of gene regulation from that described by Jacob and Monod. Only now, more than forty years after the discovery of transposable elements, are we beginning to understand enough about the ways that they can affect genes to decipher some intriguing new aspects of gene control from their study.

Barbara McClintock

Publisher's Note: The above introduction was written by Barbara McClintock to show how the concept of transposable elements evolved, and to comment on subsequent investigations of these elements. The papers in this volume were selected because of their relevance to this topic. For the discovery of "Mobile genetic elements" she received the Nobel Prize for Physiology or Medicine in 1983.

NUMBERED LIST OF PUBLICATIONS

(1) Randolph, L. F. and B. McClintock. 1926. Polyploidy in *Zea mays* L. *American Naturalist* 60: 99–102.

(2) Beadle, G. W. and B. McClintock, 1928. A genic disturbance of meiosis in *Zea mays*. *Science* 68: 433.

(3) McClintock, Barbara. 1929. A cytological and genetical study of triploid maize. *Genetics* 14: 180–222.

(4) ——— 1929. A method for making acetocarmin smears permanent. *Stain Technology* 4: 53–56.

(5) ——— 1929. A 2N-1 chromosomal chimera in maize. *Jour. Hered.* 20: 218.

(6) ——— 1929. Chromosome morphology in *Zea mays*. *Science* 69: 629.

(7) ——— 1930. A cytological demonstration of the location of an interchange between two non-homologous chromosomes of *Zea mays*. *Proc. Nat. Acad. Sci.* 16: 791–796.

(8) McClintock, Barbara and H. E. Hill. 1931. The cytological identification of the chromosome associated with the *R-G* linkage group in *Zea mays*. *Genetics* 16: 175–190.

(9) McClintock, Barbara. 1931. The order of the genes C, Sh, and Wx in *Zea mays* with reference to a cytologically known point in the chromosome. *Proc. Nat. Acad. Sci.* 17: 485–491.

(10) Creighton, Harriet B. and Barbara McClintock. 1931. A correlation of cytological and genetical crossing-over in *Zea mays*. *Proc. Nat. Acad. Sci.* 17: 492–497.

(11) McClintock, Barbara. 1931. Cytological observations of deficiencies involving known genes, translocations and an inversion in *Zea mays*. *Missouri Agricultural Experiment Station Research Bulletin* 163: 1–30.

(12) ——— 1932. A correlation of ring-shaped chromosomes with variegation in *Zea mays*. *Proc. Nat. Acad. Sci.* 18: 677–681

(13) ——— 1933. The association of non-homologous parts of chromosomes in the mid-prophase of meiosis in *Zea mays*. *Zeitschrift für Zellforschung und mikroskopische Anatomie* 19: 191–237.

(14) ——— 1934. The relation of a particular chromosomal element to the development of the nucleoli in *Zea mays*. *Zeitschrift für Zellforschung und mikroskopische Anatomie*, 21: 294–328.

(15) Creighton, Harriet B. and Barbara McClintock. 1935. The correlation of cytological and genetical crossing-over in *Zea mays*. A corroboration. *Proc. Nat. Acad. Sci.* 21: 148–150.

(16) Rhoades, M. M. and Barbara McClintock. 1935. The cytogenetics of maize. *Bot. Review* 1: 292–325.

(17) McClintock, Barbara. 1937. The production of maize plants mosaic for homozygous deficiencies: Simulation of the bm_1 phenotype through loss of the Bm_1 locus. *Genetics* 22: 200.

(18) ——— 1938. A method for detecting potential mutations of a specific chromosomal region. *Genetics* 23: 159.

(19) ——— 1938. The production of homozygous deficient tissues with mutant characteristics by means of the aberrant mitotic behavior of ring-shaped chromosomes. *Genetics* 23: 315–376.

(20) ——— 1938. The fusion of broken ends of sister half-chromatids following chromatid breakage at meiotic anaphases. *Missouri Agricultural Experiment Station Research Bulletin* 290: 1–48.

(21) ——— 1939. The behavior in successive nuclear divisions of a chromosome broken at meiosis. *Proc. Nat. Acad. Sci.* 25: 405–416.

(22) ——— 1941. The stability of broken ends of chromosomes in *Zea mays*. *Genetics* 26: 234–282.

(23) ——— 1941. The association of mutants with homozygous deficiencies in *Zea mays*. *Genetics* 26: 542–571.

(24) ——— 1941. Spontaneous alterations in chromosome size and form in *Zea mays*. *Cold Spring Harbor Symposia on Quantitative Biology* 9: 72–80.

(25) ——— 1942. The fusion of broken ends of chromosomes following nuclear fusion. *Proc. Nat. Acad. Sci.* 28: 458–463.

(26) ——— 1942. Maize genetics. *Carnegie Inst. Wash. Year Book* 41: 181–186.

(27) ——— 1943. Maize genetics. *Carnegie Inst. Wash. Year Book* 42: 148–152.

(28) ——— 1944. The relation of homozygous deficiencies to mutations and allelic series in maize. *Genetics* 29: 478–502.

(29) ——— 1944. Maize genetics. *Carnegie Inst. Wash. Year Book* 43: 127–135.

(30) ——— 1944. Breakage-fusion-bridge cycle induced deficiencies in the short arm of chromosome 9. *Maize Genetics Cooperation News Letter* 18: 24–26.

(31) ——— 1945. Neurospora. I. Preliminary observations of the chromosomes of *Neurospora crassa*. American Jour. *Botany* 32: 671–678.

(32) ——— 1945. Cytogenetic studies of maize and Neurospora. *Carnegie Inst. Wash. Year Book* 44: 108–112.

(33) ——— 1946. Maize genetics. *Carnegie Inst. Wash. Year Book* 45: 176–186.

(34) ——— 1947. Cytogenetic studies of maize and Neurospora. *Carnegie Inst. of Wash. Year Book* 46: 146–152.

(35) ——— 1948. Mutable loci in maize. *Carnegie Inst. of Wash. Year Book* 47: 155–169.

(36) ——— 1949. Mutable loci in maize. *Carnegie Inst. of Wash. Year Book* 48: 142–154.

(37) ——— 1950. The origin and behavior of mutable loci in maize. *Proc. Nat. Acad. Sci.* 36: 344–355.

(38) ——— 1950. Mutable loci in maize. *Carnegie Inst. of Wash. Year Book* 49: 157–167.

(39) ——— 1951. Mutable loci in maize. *Carnegie Inst. of Wash. Year Book* 50: 174–181.

(40) ——— 1951. Chromosome organization and genic expression. *Cold Spring Harbor Symposia on Quantitative Biology* 16: 13–47.

(41) ——— 1952. Mutable loci in maize. *Carnegie Inst. of Wash. Year Book* 51: 212–219.

(42) ——— 1953. Induction of instability at selected loci in maize. *Genetics* 38: 579–599.

(43) ——— 1954. Mutation in maize. *Carnegie Inst. of Wash. Year Book* 52: 227–237.

(44) ——— 1954. Mutations in maize and chromosomal aberrations in Neurospora. *Carnegie Inst. of Wash. Year Book* 53: 254–260.

(45) ——— 1955. 1. Spread of mutational change along the chromosome. 2. A case of *Ac*-induced instability at the Bronze locus in chromosome 9. 3. Transposition sequences of *Ac*. 4. A suppressor-mutator system of control of gene action and mutational change. 5. System responsible for mutations at a_1-m2. *Maize Genetics Cooperation News Letter* 29: 9–13.

(46) ——— Issued 1956. Intranuclear systems controlling gene action and mutation. *Brookhaven Symp. in Biol.* 8: 58–74 (Symposium held in June, 1955).

(47) ——— 1955. Controlled mutation in maize. *Carnegie Inst. of Wash. Year Book* 54: 245–255.

(48) ——— 1956. Mutation in maize. *Carnegie Inst. of Wash. Year Book* 55: 323–332.

(49) ——— 1956. 1. Further study of the a_1^{m-1} *Spm* system. 2. Further study of *Ac* control of mutation at the bronze locus in chromosome 9. 3. Degree of spread of mutation along the chromosome induced by *Ds*. 4. Studies of instability of chromosome behavior of components of a modified chromosome. *Maize Genetics Cooperation News Letter* 30: 12–20.

(50) ——— 1956. Controlling elements and the gene. *Cold Spring Harbor Symp. Quant. Biol.* 21: 197–216.

(51) ——— 1957. 1. Continued study of stability of location of *Spm*. 2. Continued study of a structurally modified chromosome 9. *Maize Genetics Cooperation News Letter* 31: 31–39.

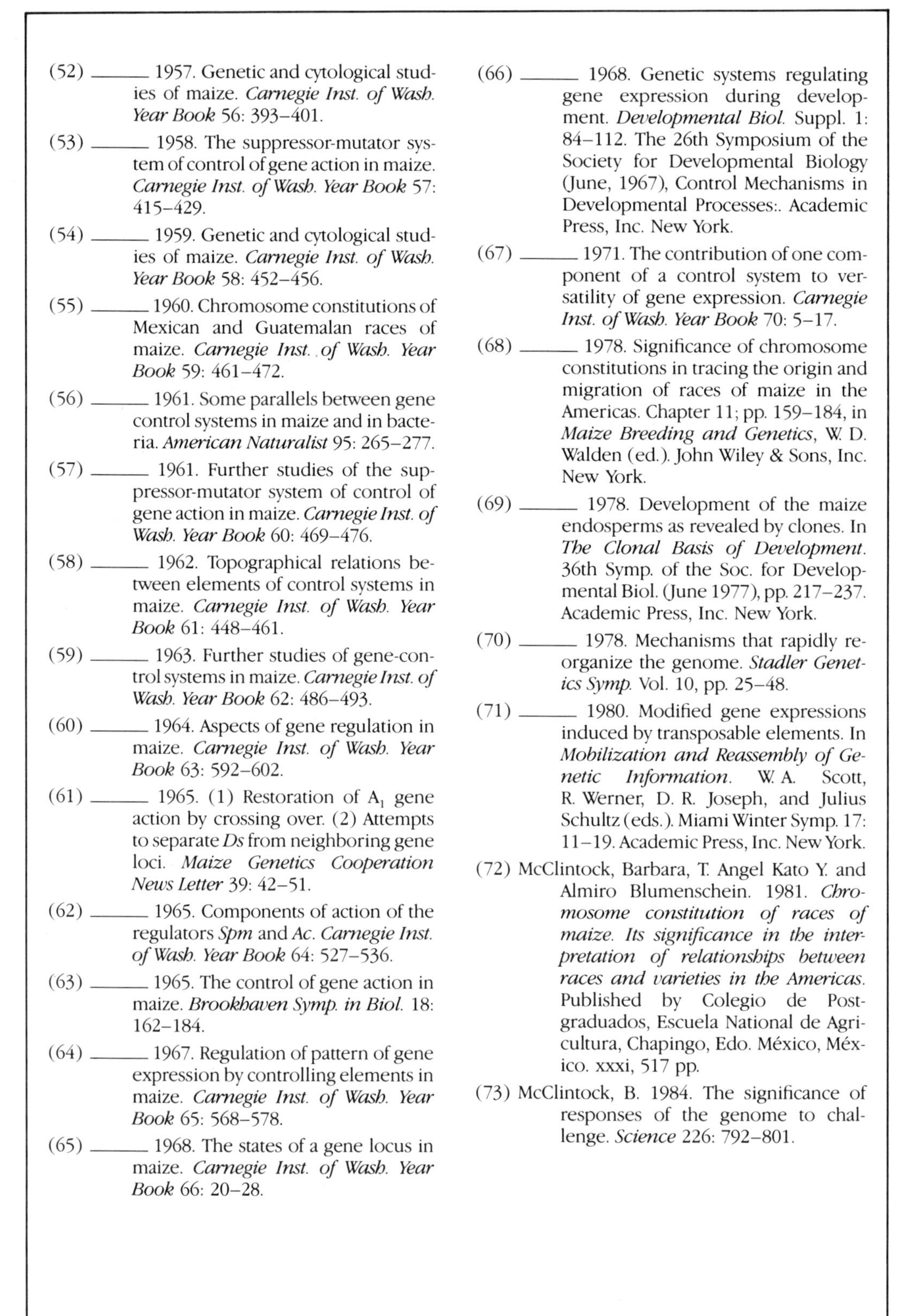

(52) ——— 1957. Genetic and cytological studies of maize. *Carnegie Inst. of Wash. Year Book* 56: 393–401.

(53) ——— 1958. The suppressor-mutator system of control of gene action in maize. *Carnegie Inst. of Wash. Year Book* 57: 415–429.

(54) ——— 1959. Genetic and cytological studies of maize. *Carnegie Inst. of Wash. Year Book* 58: 452–456.

(55) ——— 1960. Chromosome constitutions of Mexican and Guatemalan races of maize. *Carnegie Inst. of Wash. Year Book* 59: 461–472.

(56) ——— 1961. Some parallels between gene control systems in maize and in bacteria. *American Naturalist* 95: 265–277.

(57) ——— 1961. Further studies of the suppressor-mutator system of control of gene action in maize. *Carnegie Inst. of Wash. Year Book* 60: 469–476.

(58) ——— 1962. Topographical relations between elements of control systems in maize. *Carnegie Inst. of Wash. Year Book* 61: 448–461.

(59) ——— 1963. Further studies of gene-control systems in maize. *Carnegie Inst. of Wash. Year Book* 62: 486–493.

(60) ——— 1964. Aspects of gene regulation in maize. *Carnegie Inst. of Wash. Year Book* 63: 592–602.

(61) ——— 1965. (1) Restoration of A_1 gene action by crossing over. (2) Attempts to separate *Ds* from neighboring gene loci. *Maize Genetics Cooperation News Letter* 39: 42–51.

(62) ——— 1965. Components of action of the regulators *Spm* and *Ac*. *Carnegie Inst. of Wash. Year Book* 64: 527–536.

(63) ——— 1965. The control of gene action in maize. *Brookhaven Symp. in Biol.* 18: 162–184.

(64) ——— 1967. Regulation of pattern of gene expression by controlling elements in maize. *Carnegie Inst. of Wash. Year Book* 65: 568–578.

(65) ——— 1968. The states of a gene locus in maize. *Carnegie Inst. of Wash. Year Book* 66: 20–28.

(66) ——— 1968. Genetic systems regulating gene expression during development. *Developmental Biol.* Suppl. 1: 84–112. The 26th Symposium of the Society for Developmental Biology (June, 1967), Control Mechanisms in Developmental Processes:. Academic Press, Inc. New York.

(67) ——— 1971. The contribution of one component of a control system to versatility of gene expression. *Carnegie Inst. of Wash. Year Book* 70: 5–17.

(68) ——— 1978. Significance of chromosome constitutions in tracing the origin and migration of races of maize in the Americas. Chapter 11; pp. 159–184, in *Maize Breeding and Genetics*, W. D. Walden (ed.). John Wiley & Sons, Inc. New York.

(69) ——— 1978. Development of the maize endosperms as revealed by clones. In *The Clonal Basis of Development*. 36th Symp. of the Soc. for Developmental Biol. (June 1977), pp. 217–237. Academic Press, Inc. New York.

(70) ——— 1978. Mechanisms that rapidly reorganize the genome. *Stadler Genetics Symp.* Vol. 10, pp. 25–48.

(71) ——— 1980. Modified gene expressions induced by transposable elements. In *Mobilization and Reassembly of Genetic Information*. W. A. Scott, R. Werner, D. R. Joseph, and Julius Schultz (eds.). Miami Winter Symp. 17: 11–19. Academic Press, Inc. New York.

(72) McClintock, Barbara, T. Angel Kato Y. and Almiro Blumenschein. 1981. *Chromosome constitution of races of maize. Its significance in the interpretation of relationships between races and varieties in the Americas.* Published by Colegio de Postgraduados, Escuela National de Agricultura, Chapingo, Edo. México, México. xxxi, 517 pp.

(73) McClintock, B. 1984. The significance of responses of the genome to challenge. *Science* 226: 792–801.

*Kernels and Colonies: The Challenge of Pattern**

JAMES A. SHAPIRO
University of Chicago, Chicago, Illinois 60637

The bacteria I study and a series of lucky circumstances led me to make the acquaintance of Barbara McClintock in the late 1970s. The story began in the fall of 1964 when I arrived at Corpus Christi College, Cambridge, on a Marshall Scholarship to study Part II Biochemistry. I had just graduated college with a B.A. in English and a smattering of premed courses, and the biochemistry course seemed to be ideally suited to my medical school goals of the time. Fortunately (although it did not seem that way initially), I was soon informed that the course was full and that I would have to wait until the next fall to take it. What to do for a whole year? I had had a genetics course as an undergraduate; so I went to the Genetics Department (still housed in a temporary WWII structure) and asked if I could do research. A bemused Professor John Thoday said yes, and thanks to the flexibility of the Cambridge system, I soon found myself studying the role of transcription in the spontaneous mutation of the *Escherichia coli gal* operon under the sporadic, mercurial, and unofficial tutelage of Sydney Brenner. Some of my most pleasant memories of the 2 years I spent in the Genetics Department at Cambridge involved hours in the library, rummaging through old journals and books, tracing references from one fascinating paper to another. It was then that I first came across Barbara McClintock in the pages of a *Cold Spring Harbor Symposium on Quantitative Biology*. Needless to say, I lacked the background to understand the meaning of her work.

Nothing significant ever came of my transcription-mutation experiments in Cambridge. When I moved to the laboratory of Bill Hayes in London to complete my Ph.D. research on the structure of the *gal* operon in 1966, routine characterization of the spontaneous gal^- mutants isolated in those studies kept puzzling me. A number of the mutations were pleiotropic and affected two or three *gal* cistrons at once. Some were clearly deletions, but others were more troublesome. They reverted spontaneously and mapped as point mutations, but they did not respond to suppressors or mutagens and displayed much stronger polarity than the *amber* and *ochre* mtuations that I investigated in collaboration with Sankhar Adhya, who was simultaneously doing his Ph.D. on the *gal* operon at the University of Wisconsin (Adhya and Shapiro 1969; Shapiro and Adhya 1969). Sometime during the writing of my thesis, it occurred to me that the properties of these puzzling mutations could be

*This paper to dedicated to the memory of Nancy Cole, my colleague of over 15 years. She was a woman of great courage who mastered many skills needed to make the beauty of bacterial colonies visible to others.

explained as the results of extra DNA inserted into the *gal* operon, which would be expected to create strongly polar chain-termination events far from the next cistron; moreover, this idea could be tested by incorporating the mutations into λ*dgal* particles and measuring their density—an increased density would reveal the presence of additional DNA (Shapiro 1967).

The following year (1967–1968), I enjoyed a fascinating postdoctoral stint in the laboratory of François Jacob at the Institut Pasteur. Scientifically, culturally, and politically, it was a heady time of change for an impressionable midwesterner in his early twenties. Despite my conviction that isopycnic density gradient centrifugation was a technique far too arcane and sophisticated to be mastered by someone with my meager scientific background, I overcame my terror and learned how to perform CsCl gradients, prepared all the relevant λ*dgal* phages, and tested the insertion hypothesis. Early in 1968, positive results showed that the idea was correct, and I then spent a few months performing all the controls and showing that increased density was genetically linked to the extreme polar mutations (Shapiro 1969).

The political turmoil of the late sixties and early seventies diverted my attention from thinking about the meaning of the *gal* insertions for several years. I gave the λ*dgal* phages carrying those mutations to Mike Malamy, and he and Szybalski's group showed by heteroduplex mapping that the insertions belonged to specific classes of genetic elements that could insert into more than one locus, and the term IS elements was coined to label these *i*nsertion *s*equences (Fiandt et al. 1972). In 1975, while working on *Pseudomonas* hydrocarbon oxidation, I attended a plasmid meeting at Squaw Valley. The meeting turned out to be crucial in directing me back to the insertion problem. There were dramatic presentations on the roles of IS elements in F-chromosome interactions by Norman Davidson's group and on antibiotic resistance transposons by several workers, including Nancy Kleckner (Tn*10*), Doug Berg (Tn*5*), and the Datta, Cohen, and Falkow groups (Tn*3*). It became clear from these bacterial studies that a general phenomenon of major importance for DNA evolution was being uncovered. I was excited about this remarkable confluence of new results. Feeling a sense of involvement in the field because of my work on the *gal* insertions, I contacted Sankhar Adhya and the late Ahmad Bukhari, who had been thinking along similar lines from his work on bacteriophage Mu, and the three of us organized the May, 1976 Cold Spring Harbor meeting on DNA Insertion Elements, Plasmids, and Episomes. We only expected a small group of maybe three dozen people to show up, but the actual attendance was over 140. Connections between bacteria, yeast, *Drosophila*, and maize were made at the meeting (Bukhari et al. 1977). It was no longer possible to doubt the prophetic nature of McClintock's prediction that analogs of maize controlling elements would be found in other organisms. From that time on, I knew it was essential to learn as much as I could about her work and her understanding of the significance of the phenomena revealed by transposable elements.

Pattern as a Guide to the Unknown

One characteristic of Barbara McClintock's science is that her work continually directs us toward future challenges. In large measure, her ability to anticipate and exemplify new directions has come from an appreciation of the meaning of pattern. Pattern always indicates an underlying system of control over the material under observation. Based on keen visualization, her research has repeatedly aimed at elucidating those controls. Her studies on maize chromosome morphology and behavior were major contributions to our understanding of heredity in the early decades of modern genetics (see, e.g., Sturtevant and Beadle 1939). Puzzling over the meanings of kernel variegation in X-irradiated stocks led her to the discovery of ring chromosomes, the breakage–fusion–bridge cycle, repair of chromosome ends, and (ultimately) transposable elements (McClintock 1941, 1950). Systematic analysis of maize plants carrying mutable loci and of variants identified by novel kernel or plant patterns was the basis for her revolutionary discoveries on the mechanisms of genome reorganization, intranuclear networks controlling gene expression, and the evolution of genetic regulatory systems (McClintock 1987).

Clonal and Nonclonal Patterns in Maize Kernels

One of the major points that McClintock has repeatedly illustrated in her studies is that there are two major forms of organization in living organisms: clonal and nonclonal. In clones, groups of cells are linked by descent from a common ancestor, and the component cells in a clone can share properties by inheriting a particular genomic configuration. In the past several decades, our knowledge of genome structure, replication, and transmission has increased enormously. However, the patterns we observe in life are not all clonal. Many of the patterns involve coordinated groupings of cells descended from different ancestors, and these patterns must arise from the action of regulatory systems that do not operate by genomic transmission.

One particular kernel photograph illustrates this point very well (Fig. 1), and it is significant that McClintock has published this particular image twice (1965, 1978). This kernel began development with two mutable alleles: *wx-m7* at the *Waxy* locus encoding an enzyme needed for biosynthesis of starch inside the endosperm and *a-m3* at the *A* locus encoding an enzyme needed for anthocyanin biosynthesis in the outer aleurone layer covering the endosperm. *Waxy* expression can be visualized by iodine staining. The unstable alleles resulted from insertion of controlling elements belonging to the *Ac-Ds* family: an autonomous but initially inactive *Ac* at *wx-m7* and a dependent *Ds* at *a-m3*. The clonal structure of the kernel can be seen in the deep-staining wedge-shaped interior sector displaying recovery of high levels of *Waxy* expression. This sector was overlaid by a specked spot, indicating simultaneously recovered *Ac* activity in the progenitor of the corresponding aleurone layer cells. A second zone of *Waxy* expression is visible in this kernel as a lighter-staining rim extending all around the kernel. This zone was composed of cells from many different clones, and yet they all displayed a similar phenotype. What controlled this phenotype? How did the presence of *Ac* in this particular *Waxy* allele lead to this pattern of expression? These are questions we cannot yet answer, but McClintock's material gives us two major assists along the way to a solution: It defines the problem clearly and visually, and it indicates that we can use our knowledge of maize genetics to isolate and analyze specific components of the regulatory system.

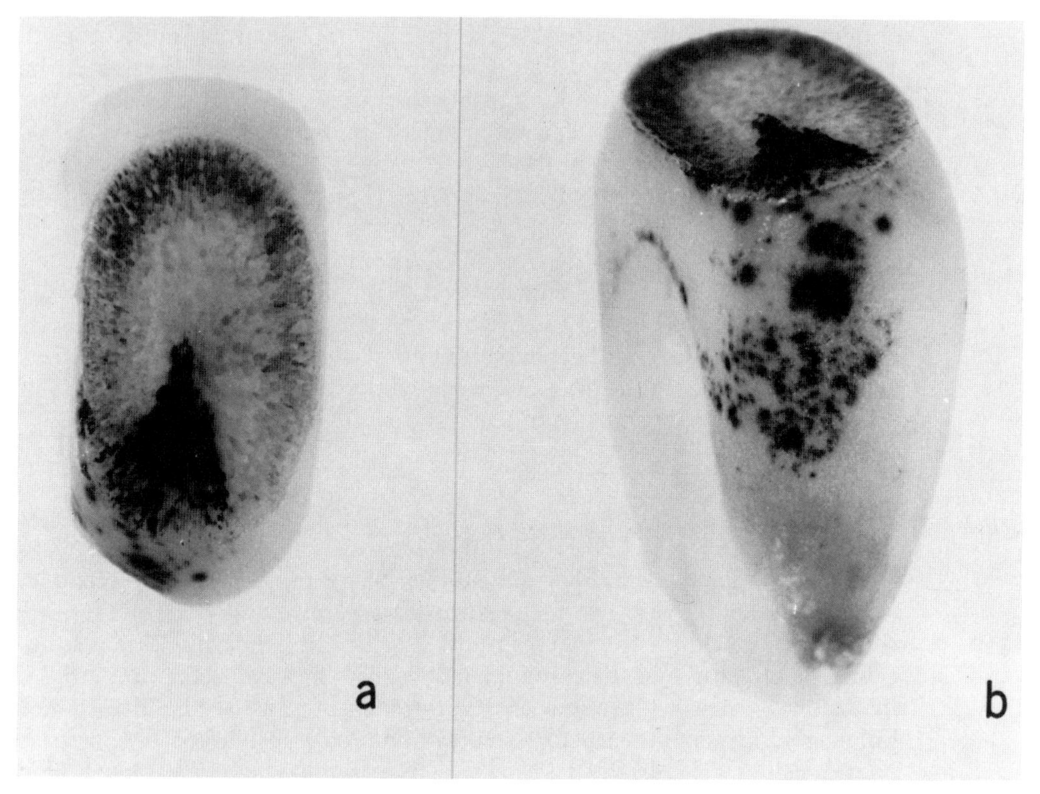

Figure 1 Maize kernel displaying clonal and nonclonal patterns of *Waxy* expression revealed by I-KI staining in the endosperm and a clonal patch of *Ac* activity in the aleurone layer.

"From the ear parent the kernel received *a-m3* in chromosome 3 and *wx-m7* in chromosome 9 whose associated *Ac* had an *m* component that initially was completely silent. In a cell, early in endosperm development, it became capable of acting. The V-shaped sector within the endosperm composed of very dark and some lightly stained cells represents the progeny of this cell as does the area with dark spots of anthocyanin pigment in the outermost layer of the endosperm (the aleurone layer). Note the absence of such spots in areas of endosperm in which the *m* component of *Ac* was silent. Note, also, the different intensities of stain in cells not included within the V-shaped sector, and the pattern these cells present on the exposed surface of the endosperm." (McClintock 1978)

Transposable Elements and lacZ Fusions

My exposure to Barbara McClintock's science has made me particularly sensitive to the broad applicability of her approach, characterized by its emphasis on the biology of the system under study and the importance of the most simple observation. When I met Barbara McClintock in the late 1970s, the transposable element field and my own research were focused on the molecular mechanisms of transposition (Shapiro 1979). My discussions with her made me aware of the further need to understand how these elements were integrated into overall organismal biology—in other words, how they operated and how they were regulated—and led me to search for bacterial systems that would permit investigations similar to hers. As described below, it turned out to be surprisingly easy to find such bacterial systems.

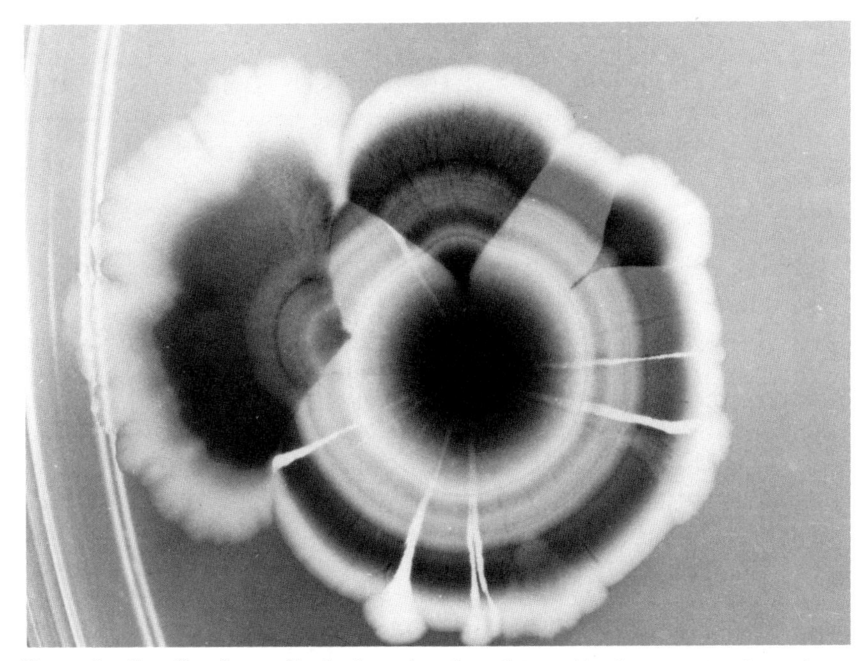

Figure 2 *E. coli* colony displaying clonal and nonclonal patterns of β-galactosidase activity on indicator agar due to differential replication of a Mu*dlac* element.

I was very fortunate in being able to follow the lead of my colleague Malcolm Casadaban, who had pioneered the use of bacteriophage Mu to make a wide variety of *lacZ* fusions in living cells. Casadaban's original system used a Mu prophage as portable genetic homology to place *lacZ* and other coding sequences in the proper orientation for selecting gene fusions (Casadaban 1976). When I studied the kinetics and genetic control of how *araB-lacZ*-coding sequence fusions formed by the original Casadaban procedure, it became apparent that the Mu prophage played an active role in the fusion process and that Mu DNA rearrangement functions were activated by stress (Shapiro 1984a; Shapiro and Leach 1990). These results illustrated the versatility of transposable elements in restructuring the genome and their responsiveness to what McClintock has called "genome shock" (McClintock 1984).

One of the major advantages of using *lacZ* fusions is that indicator media incorporating histochemical stains for β-galactosidase activity permit the direct visualization of gene expression patterns. When I tried to use some of Casadaban's more sophisticated Mu*dlac* fusion transposons to study hydrocarbon oxidation in *Pseudomonas* on one such indicator medium, I was amazed to find clonal and nonclonal patterns of gene expression in my colonies, and *E. coli* (Mu*dlac*) colonies displayed analogous patterns (Shapiro 1984b,c). These patterns expressed themselves as sectors and concentric rings (Fig. 2). They opened a whole new area of research to me—multicellular interactions in bacterial populations (Shapiro 1988)—and they demonstrated the universality of the questions McClintock raised about nonclonal control of gene expression (McClintock 1978). A further surprise came later when Pat Higgins and I discovered that the concentric rings in colonies harboring Mu*dlac* elements were not the result of periodic changes in transcription of fusions to particular operons but were, in fact, due to periodic episodes of Mu*dlac* activation and replication in some cells at particular stages of colony development (Shapiro and Higgins 1989). Thus, one of McClintock's most important demonstrations—the developmental control of transposable element activity—was just as true of *E. coli* as it was of *Zea mays*.

Nonclonal Patterns and Self-organization in Bacterial Colonies

Study of colony development has revealed an unexpected degree of cellular differentiation and multicellular organization (Shapiro 1985, 1987). At the microscopic level, this process is very complex, involving many interactions between individual cells and cell groups (Shapiro and Hsu 1989; Shapiro and Trubatch 1991). Nevertheless, despite this microscopic complexity, the overall patterns are quite regular and robust to perturbations of the growth substrate. Thus, bacterial colonies are excellent material for studying the general problems of pattern formation and self-organization in complex systems. The underlying microscopic complexity of colony development notwithstanding, it is remarkably easy to produce mathematical models that generate computer diagrams very similar to certain macroscopic aspects of colony patterns (M. Magnasco, pers. comm.). This modeling success raises a problem common to all complex self-organizing systems: Why do a large number of different microscopic events produce a regular, predictable macroscopic pattern?

Although the ease of modeling might be interpreted to suggest that some kind of automatic process is involved, experiments with *E. coli* colony patterns point in a different, more biological direction (Shapiro 1992). Using a stable *lacZ* fusion to the 18-minute region of the *E. coli* chromosome, colony pattern formation can be observed as a series of rather fine concentric rings on β-galactosidase indicator medium. These rings result from differential gene expression. At the suggestion of Marc Lavenant, one of our graduate students, I tested what happened when two colonies inoculated at different times grew into each other. The result was always the same: The rings in the two colonies came into register with each other, even though the colonies were not growing synchronously (Fig. 3). Together with the appropriate controls (colonies inoculated far apart at different times), this result demonstrated that nearby colonies influenced each other, presumably through diffusible signals in the agar substrate. Thus, the coherence of the ring patterns reflects the operation of long-range, multicellular regulation, rather than the perfectly synchronized execution of some invariant algorithm in each cell. Colonies are composed of numerous distinct clones (Shapiro 1984c). Now we need to find out how the different clonal populations are kept together to form coherent nonclonal patterns.

Problems for the Future

McClintock's emphasis on nonclonal patterns and the programmatic nature of genomic events during development points to the next major challenge in genetics: identifying what controls the structure and action of the genome. It has become traditional to think of the genome as controlling all aspects of cellular and organismal function, and this way of thinking was enshrined in the central dogma of molecular biology with its assertion of a one-way flow of

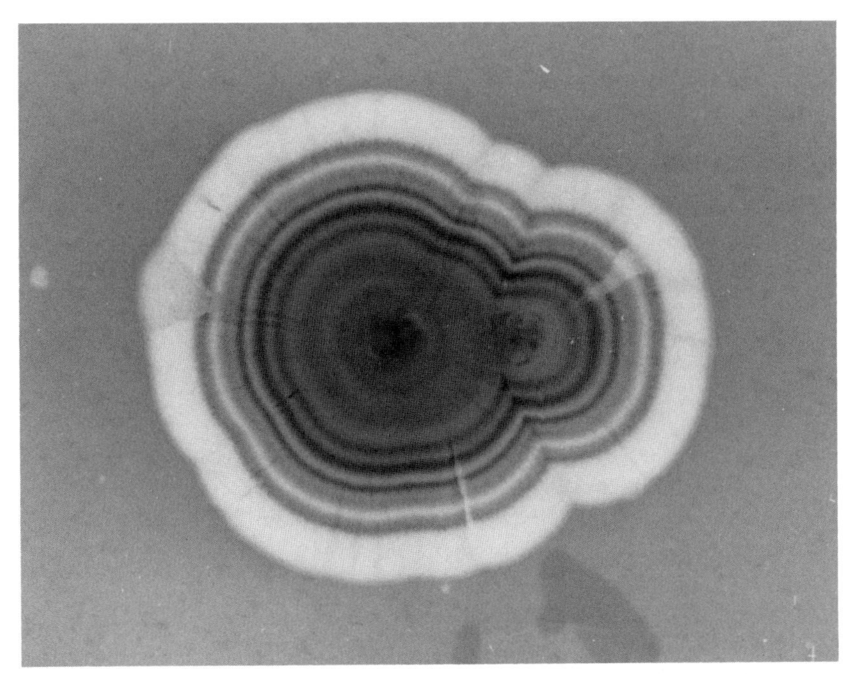

Figure 3 Communication between two fused colonies carrying a stable *lacZ* fusion. A toothpick sample from the center of the larger colony on the left was inoculated about 1–2 mm from the growing edge on the right after 29 hr of incubation. This photo was taken 5 days after the first colony was inoculated. Note how the concentric rings have come into alignment at the boundaries between the two colonies.

information from DNA to RNA and protein. Nowadays, however, the central dogma is no longer tenable because we know about proteins that reorganize DNA molecules, about retrotransposons, and about countless different regulatory systems that control expression of genetic information. As McClintock has pointed out, one genome can produce more than one organism, for example, during insect development. The maize kernel itself contains two distinct biological entities—the embryo and the endosperm—with essentially the same content of genetic information (differing only in ploidy). Thus, although the genome is essential for cell function and multicellular development, it is itself part of a larger whole and is subject to a kind of control that is not explained by our current thinking.

A great fuss has been made over Barbara McClintock's characterization of herself as a mystic. In my opinion, this characterization is central to her creative genius as a scientist. To her, mystic does not mean someone who mystifies. (Indeed, the real mystification occurs when we try to explain phenomena like nonclonal patterns by inadequate concepts simply because they are currently fashionable—in effect, by drawing hypercycles.) Instead, for Barbara McClintock, a mystic is someone with a deep awareness of the mysteries posed by natural phenomena. The courage to say, "I do not understand," and the courage to investigate the unexplainable are at the heart of her tremendous success as a scientist. It is possible in the future that she will be seen as the central figure in twentieth century biology. Both chronologically and intellectually, her career bridges the transition from the naturalists and embryologists of the last century through the geneticists of this century to the biologists of the next century.

Acknowledgments

My research on transposable elements has been supported by the National Science Foundation and the National Institutes of Health. In addition, I am deeply grateful to the National Science Foundation for generously supporting my research on pattern formation in bacterial colonies, even when the validity of this work was not widely accepted.

References

Adhya, S. and J.A. Shapiro. 1969. The galactose operon of *E. coli* K-12. I. Structural and pleiotropic mutants of the operon. *Genetics 62:* 231.

Bukhari, A.I., J.A. Shapiro, and S.L. Adhya, eds. 1977. *DNA insertion elements, plasmids and episomes.* Cold Spring Harbor Laboratory, Cold Spring Harbor, New York.

Casadaban, M.J. 1976 Transposition and fusion of the *lac* genes to selected promoters in *Escherichia coli* using bacteriophages lambda and Mu. *J. Mol. Biol. 104:* 541.

Fiandt, M., W. Szybalski, and M.H. Malamy. 1972. Polar mutations in *lac, gal,* and phage λ consist of a few IS DNA sequences inserted with either orientation. *Mol. Gen. Genet.* **119:** 223.

McClintock, B. 1941. Spontaneous alterations in chromosome size and form in *Zea mays. Cold Spring Harbor Symp. Quant. Biol. 9:* 72.

———. 1950. The origin and behavior of mutable loci in maize. *Proc. Natl. Acad. Sci. 36:* 344.

———. 1965. The control of gene action in maize. *Brookhaven Symp. Biol. 18:* 162.

———. 1978. Development of the maize endosperm as revealed by clones. *Symp. Soc. Dev. Biol. 36:* 217.

———. 1984. The significance of responses of the genome to challenge. *Science 226:* 792.

———. 1987. *The discovery and characterization of transposable elements.* Garland, New York.

Shapiro, J.A. 1967. "The structure of the galactose operon in *Escherichia coli* K-12." Ph.D. thesis, University of Cambridge, United Kingdom.

———. 1969. Mutations caused by the insertion of genetic material into the galactose operon of *Escherichia coli. J. Mol. Biol. 40:* 93.

———. 1979. A molecular model for the transposition and replication of bacteriophage Mu and other transposable elements. *Proc. Natl. Acad. Sci. 76:* 1933.

———. 1984a. Observations on the formation of clones containing *araB-lacZ* cistron fusions. *Mol. Gen. Genet. 194:* 79.

———. 1984b. Transposable elements, genome reorganization and cellular differentiation in Gram-negative bacteria. *Symp. Soc. Gen. Microbiol.* (Part 2) *36:* 169.

———. 1984c. The use of Mu*lac* transposons as tools for vital staining to visualize clonal and non-clonal patterns of organization in bacterial growth on agar surfaces. *J. Gen. Microbiol. 130:* 1169.

———. 1985. Scanning electron microscope study of *Pseudomonas putida* colonies. *J. Bacteriol. 164:* 1171.

———. 1987. Organization of developing *E. coli* colonies viewed by scanning electron microscopy. *J. Bacteriol. 197:* 142.

———. 1988. Bacteria as multicellular organisms. *Sci. Am. 256:* 82.

———. 1992. Concentric rings in *E. coli* colonies. In *Oscillations and Morphogenesis* (ed. L. Rensing). Marcel Dekker, New York. (In press.)

Shapiro, J.A. and S. Adhya. 1969. The galactose operon of *E. coli* K-12. II. A deletion analysis of operon structure and polarity. *Genetics 62:* 249.

Shapiro, J.A. and N.P. Higgins. 1989. Differential activity of a transposable element in *E. coli* colonies. *J. Bacteriol. 171:* 5975.

Shapiro, J.A. and C. Hsu. 1989. *E. coli* K-12 cell-cell interactions seen by time-lapse video. *J. Bacteriol. 171:* 5963.

Shapiro, J.A. and D. Leach. 1990. Action of a transposable element in coding sequence fusions. *Genetics 126:* 293.

Shapiro, J.A. and D. Trubatch. 1991. Sequential events in bacterial colony morphogenesis. *Physica D 49:* 214.

Sturtevant, A.H. and G. Beadle. 1939. *An introduction to genetics.* Saunders, Philadelphia.

Bacteriophage Mu: An Early Prokaryotic Controlling Element

MARTHA M. HOWE
University of Tennessee, Memphis, Tennessee 38163

The occasion of this volume has provided an unusual opportunity to think back to the formative years of Mu biology and the subtle but pervasive role that Barbara McClintock's work played in its development. From the beginning, when Larry Taylor's article in the 1963 *Proceedings of the National Academy of Sciences* reported the discovery of Mu and its unusual mutation-inducing properties, it was clear that Barbara's controlling elements provided an eminently relevant model for the study of this fascinating new virus. In that paper, Larry wrote "Phage Mu1's dual ability to occupy many chromosomal sites and to suppress the phenotypic expression of genes with which it becomes associated resembles the 'controlling elements' of maize more closely than any previously described bacterial episome. Although the system analyzed by McClintock in maize is much more complex than the present system, it is interesting to note the similarities. First, controlling elements move from one position to another in the plant chromosomes. Second, when they are located close to a recognizable gene, they may, among other things, modify its phenotypic expression sometimes to the point of completely suppressing its normal function."

Larry Taylor, Ahmad Bukhari, and Jim Shapiro kept that model before Mu researchers as we industriously pursued the genetic dissection of Mu transposition and an analysis of the genes and processes contributing to Mu's life as a virus. I am chagrined to realize that the impact of Barbara's work had not yet penetrated my thinking at the time of my 1972 Ph.D. dissertation on Mu; when I checked today, there was no citation to McClintock to be found. Nevertheless, as a postdoc at Cold Spring Harbor Laboratory, I remember my delight in finding the Carnegie Institution of Washington Year Book in the library there. I copied and read many of Barbara's papers. I have to admit that, having focused in graduate school on the molecular genetics of bacteria and viruses, it was not easy for me to follow all of the arguments, but the big picture was crystal clear. I recall a real sense of satisfaction in knowing what Larry, Jim, and Ahmad had been talking about.

Our knowledge of Mu grew both quickly and at a snail's pace—quickly in that most of us were busy at the laboratory bench sharing the excitement of each new day's results, and slowly in that there were too few of us. Researchers of the same day who were studying bacteriophage λ kept asking such sophisticated questions while we were still working out the basics. But it wasn't long before we could do respectable (read sophisticated) molecular biological and genetic experiments with Mu too.

I will not describe here the development of the Mu field since Ariane Toussaint has done a wonderful job of that in her "A History of Mu" in the 1987 book *Phage Mu* (published by Cold Spring Harbor Laboratory Press), which was written to honor Ahmad Bukhari after his sudden death in 1983. Suffice it to say that Barbara's controlling elements were an excellent paradigm for guiding our investigations. Rather, I will turn to a more personal perspective based on my interactions with Barbara while I was at Cold Spring Harbor (1973–1974) and during discussions with her upon return visits to "The Lab" for Phage meetings, a sabbatical (1981), and the Protein Purification course (1991).

The picture I have of Barbara is one of a scientist absolutely enthralled and fascinated by Nature and the multitude of tricks Nature has up her sleeve. As a postdoc I never visited her laboratory, but I do remember wonderful discussions out behind Demerec when she was cracking walnuts on the stoop. Cold Spring Harbor has a number of walnut trees, and Barbara considered it a real find when she could snatch them up before they were ferreted away by an enterprising squirrel or smashed to bits by the traffic on Bungtown Road. There she was, hands stained brown from removing the heavy green husks, with a huge wooden block that she used to break the shells and a small can or bowl to hold the precious meats. The path from my apartment (formerly her Greenhouse laboratory) to Demerec Laboratory led right by her, and I would stop and visit for a minute before going on. It always amazed me that this scientist whom I held in such awe could be interested in me and my experiments. I think it was those discussions that gave me the courage in subsequent years to knock on her laboratory door to visit when I came back to "The Lab."

My visits with Barbara are a wonderful mixture of science, hers and mine, and reflections on the changes in the way science is done. She is always full of excitement about her own latest project and about the molecular perspectives that are being gained in laboratories working with her material. She shares the wisdom of her years and the phenomenal breadth of her scientific knowledge, invariably introducing me to a new and unusual organism or a paper that I had not yet seen. To me, Barbara is an inspiration, not only because of her scientific accomplishments, but because of her insatiable fascination with Nature and her exquisitely honed powers of observation. It is that same fascination and approach to science that I try to pass on to my students as she has to me.

Discovery of the Bacterial Transposon Tn10

DAVID BOTSTEIN
Stanford University School of Medicine, Stanford, California 94305

I cannot remember exactly the date when I first met Barbara McClintock, but the occasion I remember clearly. It was at a meeting in Cold Spring Harbor in the late 1960s or very early 1970s. I came to a session late and counted myself fortunate to find a seat at the very back of the Bush Lecture Hall. The room's back wall is not vertical; it slopes a little inward, so that sitting in a chair leaning against it was surprisingly comfortable. I found that I could hear remarkably well despite the distance from the podium. I began to sit in this last row habitually. I noticed that a small elderly woman always sat in this last row, and we often sat next to each other. I knew enough about genetics and about Cold Spring Harbor to realize that this must be Barbara McClintock. I debated whether I should introduce myself to this legendary person; I chose to respect her privacy.

Our first conversation, initiated by Barbara, had the flavor of a Sherlock Holmes and Watson dialogue, a flavor that has characterized most of my conversations with Barbara since. One day, as we both sat in the rear of the lecture hall, she leaned over to me and asked, in a whisper, whether I would mind answering a personal question to satisfy her curiosity. I answered that of course I would. She asked, was I by any chance hard of hearing? I answered in astonishment that indeed I had a modest hearing loss. How did she know? She answered that she had noted my preference for the back row. She had observed also that I was often a participant in the scientific discussions of papers that, in those days, were a major feature of Cold Spring Harbor meetings; most of the other talkative types preferred a seat near the front of the room. She had formed the hypothesis that I must be sitting in the rear of the hall for some reason, and the possibility had suggested itself that I, like she before me, had noticed that the acoustics were best at the very rear of the auditorium.

After that conversation, Barbara and I habitually sat together in the back of the lecture hall. In those days, she attended virtually every session of every meeting I came to. She began to ask me about details she missed, as her hearing was less reliable than mine. She was utterly unlike the other scientists of her generation that I had had the opportunity to meet. She had obviously retained not only her quickness of mind, but also a youthful eagerness to think about things in new ways. One could readily see the qualities that had made her a legend in her own time. What impressed me the most was her intense interest in all matters genetic and her ability to find connections between the contemporary research she was hearing and her vast store of information about all of genetics, beginning, of course, at the very beginning. After all, this woman had lived through everything that had ever happened in our discipline except for Mendel; what she did not herself discover, she learned as it happened.

Curiously, however, we never discussed, in these early days, the research then going on in my own laboratory, research that was shortly to intersect her then neglected (but not forgotten) discoveries and insights. My group participated in the rediscovery and the genetic and molecular characterization of bacterial transposons. It was this rediscovery of transposable elements in a molecularly accessible system that, more than any other factor, gave impetus to the belated appreciation of Barbara's much earlier work on transposable elements in maize.

Transduction of Tetracycline Resistance by Phage P22

I never set out to study transposition; the phenomenon presented itself to me unbidden. One morning in the summer of 1970, I received a call from Salvador Luria's secretary, Nancy Ahlquist. It seemed that Salva was hosting a distinguished visitor from Japan. Through some misunderstanding, Salva had some other appointments that day that could not be deferred. Would I mind filling in for Salva for a few hours? Nancy said Salva had chosen me because his visitor, Professor Tsutomu Watanabe, was interested in *Salmonella* genetics and had an interesting story to tell about transduction. I said that of course I did not mind, just give me a few moments to defer some appointments of my own. I was then a new Assistant Professor working on the temperate *Salmonella* phage P22. When Zinder and Lederberg (1952) discovered generalized transduction, the vehicle turned out to be P22. Indeed, although I am slightly embarrassed even today in admitting that the name Watanabe meant nothing to me at the time, doing a favor for Salva was a welcome opportunity. Hearing anything at all about P22 transduction, on which I was actively working, was a bonus.

Tsutomu Watanabe, of course, turned out to be the discoverer of transmissible antibiotic resistance in bacteria. It was he who had discovered and learned the most about the properties of the R-factors, the plasmids that carry the drug resistance genes. He had established their similarity in genetic properties to the F-factor (cf. Watanabe 1963). The story Watanabe had to tell me was fascinating. By growing P22 on a strain of *Salmonella* that harbored the R-factor 222, he had produced a transducing lysate that allowed the specialized transduction of resistance to tetracycline. He was totally stumped, however, by the properties of the transducing lysates: Some of them resembled specialized transduction of the *gal* or *bio* operons by coliphage λ, but there were differences, one of which he recognized was very significant. When transduction was carried out at a high multiplicity of infection, one produced transductants that were fully immune lysogens which, when induced, produced another lysate. However, the lysates contained virtually no plaque-forming phage; the very tiny numbers of phage present in the lysate were normal-seeming P22 that could not transduce *tet*^{R} at all. When one transduced at low multiplicity of infection, one obtained many fewer transductants, but these rare transductants were often not immune to superinfection and, when attempts were made to induce lysates, no phage or transducing activity was obtained. If one used a mixture of normal P22 helper and the active lysate at low multiplicity transducing lysate in transduction experiments, one increased the frequency of *tet*^{R} transductants, but the lysates produced from these transductants, unexpectedly, never contained the helper but behaved just like the lysates of transductants made at high multiplicity.

Professor Watanabe, a distinguished-looking gentlemen in his sixties, had no reason to have heard of me. I was still a beginner. Unknowing, he had arrived at exactly the right place to get an answer for his riddle. I had published only three papers (all in the *Journal of Molecular Biology*; in those days, it was still true that large numbers of publications were not required to get a good job if the quality was high), and all were about the structure and replication of P22 DNA. The burden of my story was that the P22 genome is circularly permuted and terminally repetitious, as had been predicted for coliphage T4 by Streisinger (Streisinger et al. 1967). Hearing Watanabe's story, I was able, right then, to form a hypothesis: The gene(s) for tetracycline resistance was physically inserted into the P22 genome, making it too large to fit in a single phage head. If this had happened in λ, one would only get a transducing lysate if there were a compensating deletion, because the linear λ genome has fixed ends. But, because P22 filled its head by headful packaging from a concatemeric intermediate (the Streisinger model), all the DNA of an oversize genome carrying a large insertion would be packaged nevertheless, but not all in the same particle. Two particles from such a lysate could cooperate to bring in two incomplete but permuted copies of the oversize genome, which could then recombine to make the circular DNA intermediate that I had already shown was likely to be essential for both DNA replication and lysogeny. High-multiplicity infection with such particles with incomplete genomes would produce *tet*^{R} lysogens; addition of helper at low multiplicity would do just as well, but the lysogen produced would be a *tet*^{R} lysogen indistinguishable from the high-multiplicity transductants from which normal phage were rarely produced. The low rate of normal phage would be explained by deletions that remove the inserted element.

Even in 1992, when virtually all details of this phenomenon are well understood, this idea is not so simple to explain. The reader may well imagine that between this complexity, the lack of any concept in bacterial genetics that even vaguely resembles what we now call "transposition," and the language difficulty, it was not easy to communicate the substance of my hypothesis. However, I was able to make a series of experimental predictions, the outcomes to some of which Professor Watanabe already knew. Between the accuracy of my guessing and my not inconsiderable level of excitement, he must have recognized that something was up. We agreed to a collaboration that, because Watanabe's laboratory was not at all molecular, consisted of his sending strains along with what information he had about the origin of the original transducing lysates. Our contribution was to test my predictions about the DNA and the cooperativity of growth and lysogenization.

My laboratory contained, in addition to myself, a technical assistant and a graduate student, Russell Chan. Russell was working on deletion mapping the genome of phage P22. My thesis advisor, Myron Levine, had isolated many conditional-lethal mutations of P22; I had since made an even larger collection of nonsense (amber) mutants. Complementation tests by coinfection were difficult with P22 because of the very high rate of recombination. The genome is on the order of 1000 cM in length; this nets to about 20% recombination across a single kilobase of DNA! We had already found that opposite ends of apparently single complementation groups were nearly unlinked. Russell and I had therefore decided to make a deletion map of P22. We had begun to use a laborious genetic selection scheme to make deletions into the integrated prophage; this scheme worked and Russell had isolated some deletions by the time Watanabe came to my office. Watanabe's transducing lysate immediately offered us a way to make many more prophage deletions very quickly, because if the origin of transducing particles was indeed an oversize genome producing circularly permuted headfuls that could not circularize, then lysogenization at very low multiplicity of infection (i.e., virtually all infected cells contain only one input genome) should not allow the formation of tetracycline-resistant lysogens. Selection for tetracycline resistance should select for genomes that had shortened by deletion to compensate for the insertion of the extra material encoding the drug resistance, restoring terminal repetition and the ability to circularize. This in turn meant that some phage genes had to go, and in principle, any genes not required for establishment of lysogeny (i.e., most of the genes of the phage) could be deleted. If deletion occurred more or less at random, we would be able to make a classic deletion map this way.

Russell quickly succeeded with this idea. He used a Watanabe lysate (called P22Tc10) at low multiplicity and found tetracycline-resistant survivors that retained some immunity (meaning phage genes were present) but not complete immunity (meaning some were absent). These made a truly wonderful deletion map of the phage when used in marker-rescue experiments. Russell also did many other experiments to support the hypothesis that transduction was the consequence of an oversize genome created by insertion of extra DNA.

First Indications of a Transposition Mechanism

There were, however, unexplained phenomena. One that preoccupied us at the time was that the prophage deletions isolated by Russell all had one endpoint in common. All the deletions apparently started at one end of the prophage. The simplest (and ultimately wrong) interpretation that occurred to us was that somehow the phage site-specific integration and excision enzymes were making the illegitimate joints resulting in the deletion. Today we realize that the element was causing the deletions. More puzzling was that most of the tetracycline-resistant survivors at low multiplicity had no phage genes at all, they were just stably tetracycline-resistant and apparently contained no P22 DNA. Today, of course, we know that these were the result of a simple transposition of the element (now called Tn*10*) from the phage genome to the bacterial chromosome. Watanabe had noticed this phenomenon before, and not just with tetracycline resistance. In transduction experiments using an R-factor donor and a recipient containing no plasmid, he had obtained drug-resistant survivors that had the drug genes but no detectable R-factor genes (or, as we showed later, P22 genes). Furthermore, he had mapped these solo genes to a number of different positions on the genome of *Salmonella*, distant from the phage attachment site. We reproduced and amplified these results, finding several different sites at which the drug resistance genes had landed.

All of this began to convince Russell and me that what was happening was transposition as we now understand it: movement of the drug resistance genes from one genome to another by some kind of illegitimate recombination. Our first opportunity came when it was time to publish the results of our collaborations; as agreed, Watanabe sent a manuscript he had prepared on the R-factor transduction results and Russell prepared another one on the phage properties (including prophage deletions). I read Watanabe's draft, which contained no model or explanation for the results. I added a Discussion in which we proposed, essentially, transposition. Watanabe was sent the two drafts and no reply came for months. Finally, Luria was consulted. He read the manuscripts and suggested that perhaps our interpretation was too radical for Watanabe. He offered to mediate. He succeeded in fashioning a compromise, which meant that the transposition hypothesis was still presented but considerably toned down so that it would not seem too radical. Out of respect for Watanabe and Luria, Russell and I went along, and the two papers were published in 1972 in *Virology* (Chan et al. 1972; Watanabe et al. 1972). Looking back on them, one can discern the outline of the transposition hypothesis, but it is put forward very timidly. I was disappointed.

In retrospect, it is clear that even a vigorous exposition of the transposon hypothesis would have been to no avail, given the evidence we had then. Our scientific community was simply not yet fully prepared for the idea: At this time, only a very small cadre of enthusiasts were interested in bacteriophage Mu, for which more convincing genetic evidence of random insertion had been available since 1963 (Taylor 1963). And, of course, Barbara McClintock's movable elements, although not forgotten, were still not appreciated. Indeed, the most direct evidence for the prematurity of our idea was the fact that Watanabe's results with the element now called Tn*10* had themselves been anticipated; during the preparation of the *Virology* papers, we discovered that Bruce Stocker had published, in 1964, a short paper in *Nature* describing the acquisition by P22 of the ability to transduce tetracycline resistance at high frequency concomitant with the loss of the ability to grow normally (Dubnau and Stocker 1964). There is no doubt that Stocker had encountered Tn*10*. I realize now that we needed much stronger evidence to convince anybody of our ideas.

Proving the Existence of Tn10

In the next 2 years, we obtained irrefutable evidence, both genetic and physical, for the movable element now called Tn*10*. Bik-Kwoon Tye, another Massachusetts Institute of Technology graduate student, joined my laboratory and began to look at DNA heteroduplexes in the electron microscope. It is important to remind younger readers that restriction enzyme technology was still several years away. In heteroduplexes between P22 DNA and DNA from the particles in a P22Tc10 lysate, Bik saw the now familiar stem-and-loop structures showing that the insertion had a long (1400 bp) inverted repetition at its ends. We showed these at a meeting, and Phil Sharp came up very excitedly with the news that he had seen identical structures in heteroduplexes between an R-factor and the F-factor; he gave me a preprint of his paper (Sharp et al. 1973) then in press. Bik and Russell examined a second, independent P22 transducing derivative (P22Tc106) given to us by Watanabe, and Bik found that it had the same stem-and-loop structure inserted at a different site in the P22 genome.

We also accumulated genetic evidence. Just as Russell was defending his thesis, Nancy Kleckner joined us as a postdoc. She took up an experiment that Russell had just begun: to show directly that the transduction of tetracycline resistance by P22Tc10 lysates at low multiplicity was indeed caused by transposition of the element. The idea was to screen for auxotrophic mutations caused when the element inserts itself into bacterial genes and then to show genetically that the two phenotypes, auxotrophy and drug resistance, were the consequence of the same insertion mutation. This line of research proved very fruitful indeed, and by the fall of 1974, we had submitted a very complete paper (Kleckner et al. 1975) to the *Journal of Molecular Biology* that clearly described all the essential properties of the element now called Tn*10*.

Barbara McClintock's Extraordinary Insights

Just at this time, I took my sabbatical year at Cold Spring Harbor Laboratory. My intention was to work on yeast with Gerry Fink and John Roth. As it turned out, we did a lot with yeast that year, but transposons were a persistent subtext. John and I spent a great deal of time thinking about how insertion mutations might be used in bacterial genetics, ideas that occupied both our laboratories for many years thereafter. It was also at this time, just after I came to Cold Spring Harbor, that I gave Barbara a copy of the Kleckner et al. manuscript. It was the first time that we had ever discussed the subject of transposable elements. In retrospect, it is amazing that we had never discussed our work before. I guess I never brought it up because I was taken in, despite myself, by the notion that eukaryotic and prokaryotic illegitimate recombination mechanisms would be radically different from each other. My expectation was that viral genomes might be similar to phage, but chromosomal events would involve chromatin and thus be necessarily different. Barbara was, not surprisingly, entirely unaware of our work.

The manuscript, entitled "Mutagenesis by Insertion of a Drug-Resistance Element Carrying an Inverted Repetition," had four authors (Nancy, Russell, Bik, and myself). Nancy had written the first draft, and we had revised it several times after soliciting suggestions from several colleagues, including John and Gerry. It had already begun an expectedly complicated course through the review process. Barbara's comments were unique. She read it with an insight, not unexpected, that none of my colleagues could match.

I mentioned above that my conversations with Barbara had had the flavor of dialogue between Sherlock Holmes and Watson. She would say cryptic things about genetics or the real world; behind these apparently *ex cathedra* remarks, there would usually be a long chain of reasoning. To her, the conclusions she arrived at were "elementary," and I often found myself wishing I could figure them out without asking. I am sure that a nonfictional Watson would have had the same desire to avoid asking Holmes how he arrived at his conclusions. After Barbara had read our paper, this dialogue became intensely technical.

- Barbara asked why I had given such prominence to the inverted repetition in the title of the paper. For this I had the answer that it was the inverted repetition that allowed the definition of a physical "element" that remains the same when it transposes; the stem and loop defined the physical entity in the electron microscope. She accepted this point, and indeed recognized that the physical definition of the element was a substantial advance over what had been possible for her to do in maize, but still she shared her intuition that the extensive inverted repetition was likely to be a nonessential detail. She was, of course, right.
- Barbara asked whether we had looked for inversions and other rearrangements. I gave Barbara the deletion-mapping paper, which, she agreed, was consistent with the idea that the element might produce deletions. After I explained some of the details of phage DNA packaging, she urged me to look for inversion figures in electron microscopy photographs of the DNA in lysates; these I found, although they were never published.
- Barbara suggested that the elements might destroy the donor genome when they transpose. This question had already engaged Nancy's interest. It was, however, many years before Nancy could experimentally address the fate of the donor chromosomes, and the answer turns out to be very complex (for an excellent comprehensive review, see Kleckner 1989).
- Barbara suggested what Nancy and I had already come to call the "symmetry experiment," aimed at testing whether the "inside" ends of the inverted repetitions (now called the IS*10* elements) were active in transposition. Again, it took some years to show that they are, although as Kleckner's group has shown, the left and right IS*10* elements are not functionally identical (cf. Kleckner 1989).
- Barbara asked whether the element (it still had no name then) had a tendency to transpose to positions near itself on the genome. I believe this question is still open for Tn*10*.

Barbara's insight was remarkable. Even though I knew that she had been down much the same road 20 years earlier and already believed in transposable elements, I was amazed. She had assimilated the novel technology and gotten straight to the elemental issues. These issues, of course, were the same ones that had occupied her. She had no trouble seeing the strengths of the arguments in our paper—or the weaknesses. I instantly regretted my ignorance of her work, but she assured me that I would not have saved any time or work had I known it. I would have had to prove everything for myself again anyway. I remain embarrassed, but I think now that she was probably right.

Beyond her scientific insights, Barbara understood that there would be irrational resistance to our ideas. This was comforting and encouraging at a time when Nancy Kleckner and I were getting our first taste of it. Indeed, one of the reasons we had held off publication for so long was because of the intense skepticism, bordering on hostility, that had greeted our experiments on some occasions. Barbara told me that she had stopped publishing, although she had continued writing manuscripts. She pointed out that resistance was preferable to incomprehension or—worse yet—to having one's work ignored. By the time our paper appeared in 1975, we had the company of several other groups that had obtained evidence for transposition (Hedges and Jacob 1974; Berg et al. 1975; Gottesman and Rosner 1975; Heffron et al. 1975). Acceptance of the notion that there were transposable genetic elements in bacteria was quite rapid after that. And perhaps it was this that led the genetics community to rediscover Barbara McClintock's work of a quarter century earlier with its elegant experimental proofs of transposition, its original concepts of gene regulation in development, and its evidence of radical, yet regulated, genomic change, whose implications for evolution have yet to be fully understood.

References

Berg, D.E., J. Davies, B. Allet, and J.-D. Rochaix. 1975. Transposition of R factor genes to bacteriophage λ. *Proc. Natl. Acad. Sci. 72:* 3628.

Chan, R.K., D. Botstein, T. Watanabe, and Y. Ogata. 1972. Specialized transduction of tetracycline resistance by phage P22 in *Salmonella typhimurium.* II. Properties of a high-frequency transducing lysate. *Virology 50:* 883.

Dubnau, E. and B.A.D. Stocker. 1964. Genetics of plasmids in *Salmonella typhimurium. Nature 204:* 1112.

Gottesman, M.M. and J.L. Rosner. 1975. Acquisition of a determinant for chloramphenicol resistance by coliphage lambda. *Proc. Natl. Acad. Sci. 72:* 5041.

Hedges, R.W. and A.E. Jacob. 1974. Transposition of ampicillin resistance from RP4 to other replicons. *Mol. Gen. Genet. 135:* 31.

Heffron, F., C. Rubens, and S. Falkow. 1975. The transposition of a plasmid DNA sequence which mediates ampicillin resistance: Molecular nature and specificity of insertion. *Proc. Natl. Acad. Sci. 72:* 3623.

Kleckner, N. 1989. Transposon Tn*10*. In *Mobile DNA* (ed. D.E. Berg and M.M. Howe), p. 227. American Society for Microbiology, Washington, D.C.

Kleckner, N., R.K. Chan, B.K. Tye, and D. Botstein. 1975. Mutagenesis by insertion of a drug-resistance element carrying an inverted repetition. *J. Mol. Biol. 97:* 561.

Sharp, P.A., S.N. Cohen, and N. Davidson. 1973. Electron microscope heteroduplex studies of sequence relations among plasmids of *Escherichia coli.* II. Structure of drug resistance (R) factors and F factors. *J. Mol. Biol. 75:* 235.

Streisinger, G., J. Emrich, and M.M. Stahl. 1967. Chromosome structure in phage T4. III. Terminal redundancy and length determination. *Proc. Natl. Acad. Sci. 57:* 292.

Taylor, A.L. 1963. Bacteriophage-induced mutation in *E. coli. Proc. Natl. Acad. Sci. 50:* 1043.

Watanabe, T. 1963. Episome-mediated transfer of drug resistance in *Enterobacteriaceae.* VI. High-frequency resistance transfer system in *Escherichia coli. J. Bacteriol. 85:* 788.

Watanabe, T., Y. Ogata, R.K. Chan, and D. Botstein. 1972. Specialized transduction of tetracycline resistance by phage P22 in *Salmonella typhimurium.* I. Transduction of R factor 222 by P22. *Virology 50:* 874.

Zinder, N.D. and J. Lederberg. 1952. Genetic exchange in *Salmonella. J. Bacteriol. 64:* 679.

McClintock (1933): Implications for Meiotic Chromosome Pairing

NANCY KLECKNER
Harvard University, Cambridge, Massachusetts 02138

My first direct interaction with Barbara McClintock came in the early years after the discovery of transposable elements in bacteria when I came to Cold Spring Harbor Laboratory to give a lecture in the summer course on Bacterial Genetics. While I was primarily interested in the mechanism of Tn*10* transposition, she was interested in the effects of transposons and of partial transposon excisions on gene expression in the hope that analysis of Tn*10* might provide molecular insights into the nature and consequences of controlling elements in maize.

My real intellectual contacts with Barbara, however, began only some 10 years later when I found myself working in a second area in which we shared a common interest: the pairing and recombination of homologous chromosomes during meiosis. Our first conversation on this subject occurred when I was at Cold Spring Harbor for a meeting. At that time, visiting women scientists above a certain age were housed on the second floor of a house on Bungtown Road whose first floor is Barbara's home. We ran into one another one afternoon in the doorway. I explained that I was now working on this new subject and that I had become intrigued by the phenomenon of "interlocks," the cytological manifestation of chromosomes that have become nonspecifically entangled during the course of meiotic chromosome pairing. Barbara invited me to visit her office the following day: Having long ago identified the occurrence of interlocks as a problem of fundamental importance, she had collected a series of papers on the subject from the literature and wanted to discuss the problem further.

In subsequent visits, we began to discuss other observations that bear on the problem of chromosome pairing. On one occasion, Barbara gave me a copy of a paper that she published in 1933 (McClintock 1933). This paper describes a number of situations in which 2-by-2 association occurred between nonhomologous parts of chromosomes rather than between homologous parts, as is normally the case. This 1933 paper is routinely cited as the primary reference for the occurrence of such nonhomologous associations. However, McClintock's ideas regarding the implications of her observations are still very provocative in their implications for current thinking about meiotic chromosome pairing.

A Force for 2-by-2 Association

McClintock's view

She examined maize chromosomes in acetocarmine smears of microsporocytes. This method permits both visualization (by light microscopy, of course) and individualization of all ten chromosomes; it is of course McClintock's development of this method that paved the way for her far-reaching contributions to maize cytogenetics. The fundamental finding presented in the 1933 paper is that "there is a tendency for chromosomes to associate 2-by-2 in the prophase of meiosis whether or not the parts associated are homologous." In molecular terms, it is now clear that the intimate 2-by-2 association observed by McClintock corresponds to formation of a specific structure, the synaptonemal complex (SC); the tendency for 2-by-2 association therefore represents a tendency for SC to form. An important aspect of McClintock's paper is that she discusses her observations in terms of the existence of a "force" for 2-by-2 association (SC formation), a formulation that places great weight on the importance of the process. McClintock summarizes her observations (in part) as follows:

> If the two chromosome complements of a diploid plant are structurally similar, association is most frequently confined to homologous parts of chromosomes.... In plants with unbalanced chromosome complements (monosomies, trisomics, deficient rod-shaped chromosomes, deficient ring-shaped chromosomes) it is not possible to have all parts homologously associated 2-by-2. In these cases, a wide range of synaptic configurations results. Each type of unbalance produces its own range of synaptic associations. In general, it can be stated that there is a tendency for associations to occur between homologous parts leaving the parts without homologous sections with which to synapse to form non-homologous associations or to remain unassociated.... When the two complements are structurally dissimilar (inversions and translocations), conflicting forces responsible for 2-by-2 synapsis may result in the association of non-homologous parts of chromosomes in those regions representing the structural dissimilarity.... [Moreover,] Association between non-homologous parts of chromosomes appears to be as intimate, in many cases, as that between homologous parts.

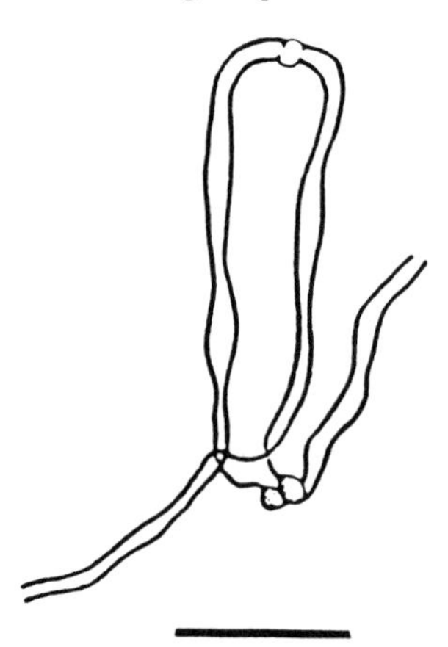

Figure 1 Mid-prophase of meiosis in a microsporocyte of *Zea mays* visualized by light microscopy of acetocarmine smears. Synaptic configuration resulting from homologous associations between a chromosome and its homolog possessing a pericentric inversion. Each line represents one homolog. (*Camera lucida* drawing reprinted from McClintock 1931.) Bar, 10 μm.

In the particular case of an inversion heterozygote, chromosomes that associate faithfully according to chromosomal homology form a characteristic inversion loop (Fig. 1). However, some chromosomes associate without respect to homology in the inverted region to form a simple rod in which the inversion segments are associated nonhomologously; other chromosomes associate with a combination of homologous association and asynapsis within the inversion loop.

It must be noted that the nature of the force(s) by which homologous parts of chromosomes attract one another during meiotic prophase is still not understood. However, there is currently much enthusiasm for the view that this force includes a recognition of homology at the level of DNA/DNA interactions, some or all of which may be identical with early steps in recombination (Maguire 1977; Moses et al. 1984; Smithies and Powers 1986; Carpenter 1987; Roeder 1990; Kleckner et al. 1991).

Potential general implications

McClintock's notion of a powerful force that demands 2-by-2 associations is very attractive to me personally. Coming to the problem from a different point of view, we have suggested that a positive driving force for SC formation not only exists, but plays a critical role in meiosis (Kleckner et al. 1991). We have suggested that the evolutionary "raison d'etre" of the synaptonemal complex is as a mechanism for sensing tangles and eliciting an appropriate cellular response that results in their elimination. According to this model, the driving force for SC formation is the central critical property of the SC; a block to the force for SC formation posed by the presence of a tangle might directly generate an appropriate molecular signal that triggers the necessary biological response. As part of this model, we point out that the existence of a force for SC formation might account not only for nonhomologous synapsis, but also for two other SC-related phenomena: synaptic adjustment and resolution of interlocks. As an example, I consider in detail here the specific case of synaptic adjustment.

Visualization of the synaptonemal complex in the mouse by Moses and colleagues (for review, see Moses et al. 1984) revealed that two homologous chromosomes differing by an inversion, deletion, or duplication may first undergo SC formation in the pattern corresponding to the underlying chromosomal homology and then subsequently undergo a readjustment of SC formation to yield a single continuous rod of synaptonemal complex. In the case of an inversion, the paired chromosomes first contain an appropriate inversion loop. Then, as the pachytene stage progresses, the loop becomes progressively smaller until it ultimately disappears altogether (Fig. 2), leaving a simple rod of continuous SC in which the inversion segments are paired nonhomologously within the rod. Moses has suggested that this phenomenon is related to the occurrence of successive homosynaptic and heterosynaptic pairing phases observed in triploid and tetraploid *Bombyx* oocytes.

There now seems to be good evidence that synaptic adjustment does not occur in maize (Maguire 1981; Anderson et al. 1988). However, at the biological level, the existence of synaptic adjustment (and similarly the occurrence of heterosynaptic pairing in *Bombyx*) would seem to be a very striking confirmation of the existence of a force that causes chromosomes to

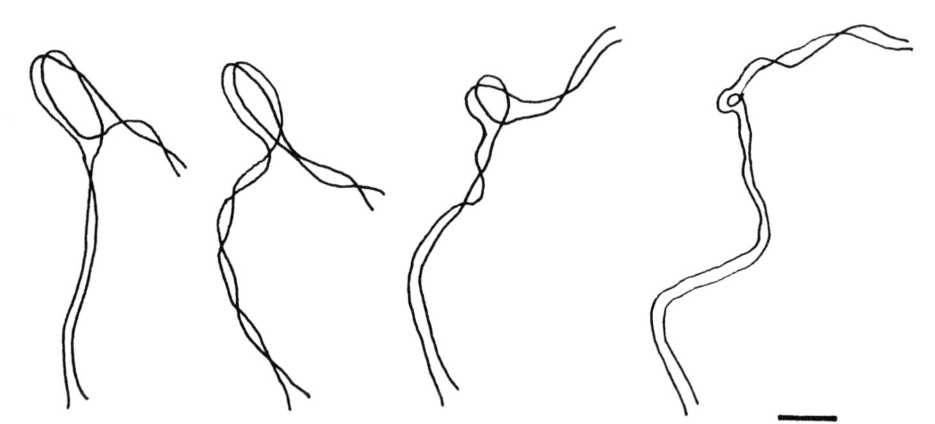

Figure 2 Synaptic adjustment in an inversion loop in the mouse visualized by electron microscopy of silver-stained surface-spread preparations. Each line corresponds to a lateral element that connects the two sister chromatids of one homolog. As adjustment progresses, the inversion loop gradually decreases in size. By late pachytene, the loop is completely eliminated and is replaced by a "straight" SC in which homosynapsis persists outside the inversion while the inversion itself is heterosynapsed (not shown). Bar, 1 μm. (N.K. drawings corresponding to electron micrographs in Fig. 5 of Moses et al. [1984], with apologies to the authors for possible overspecification of SC paths, made here for the sake of clarity.)

associate 2-by-2. According to this view, the same force exists in both maize and the mouse; the two organisms simply differ in the precise way in which it is manifested. In the case of maize, nonhomologous regions (or homologous regions in which the force of homologous attraction is not "strong enough," see below) associate as best they can; in the case of mouse, homologous associations in the inversion loop "dissolve" and are replaced by heterologous associations that eliminate discontinuities in the synaptonemal complex.

It might be argued that the tendency for 2-by-2 association is a simple consequence of the fact that SC formation is an assembly process that, like bacteriophage morphogenesis, is intrinsically driven by the relevant protein/protein interactions and need not have any broader functional significance. The existence of a process as complex as synaptic adjustment makes, if anything, an even more compelling argument that the need to maximize 2-by-2 association is a general and fundamental feature of meiotic chromosome metabolism than does the occurrence of nonhomologous association in the absence of such adjustment.

Possible Relationships between the Forces for 2-by-2 Association and for Homologous Association

McClintock's proposal that the force for 2-by-2 associations can predominate over the force for homologous association

McClintock's paper presents a specific set of ideas regarding the relationship between the forces of 2-by-2 association and the forces of homologous alignment. Most critically, she suggests that in certain cases, the failure of homologous association is not due to failure of homologous parts to attract

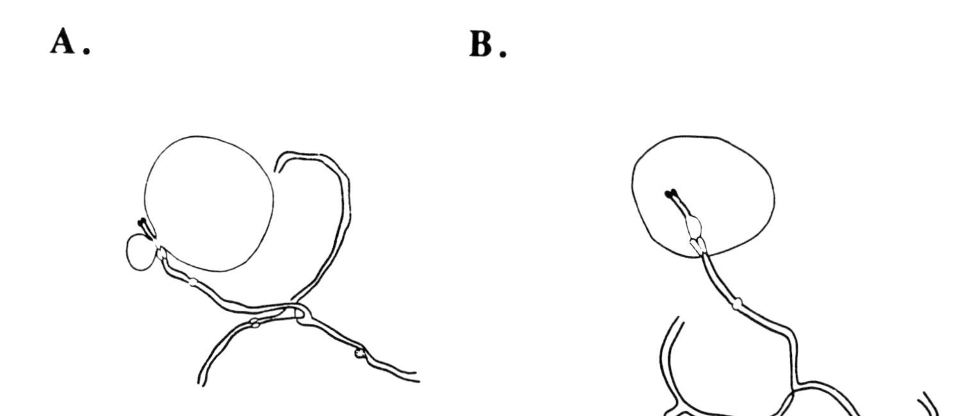

Figure 3 Synaptic forms observed for a reciprocal translation between chromosome 4 and chromosome 6 of *Zea mays*. (*A*) Typical X-shaped configuration resulting from homologous associations. (*B*) H-shaped configuration from a sporocyte of the same plant; inferred to represent continued 2-by-2 association beyond the region of homology (method is the same as that in Fig. 1). Magnification, 1450x. (Reprinted from McClintock 1933.)

one another but rather to...."the more powerful force demanding 2-by-2 associations which can supersede that of homologous attraction when these forces are not working in unison." More particularly

> In balanced complements, under normal conditions, the dominant requirement of the prophase period is a 2-by-2 association of chromosomes. Orderliness in producing this [condition] is maintained by the forces of homologous attraction which brings similar parts of homologous chromosomes into contact and in a negative way works against the association of non-homologous parts of chromosomes. [Thus] it is clear that in structurally balanced diploids the factors responsible for 2-by-2 association of homologous parts, although relatively complex, will result in the correct alignment of chromosomes.... In unbalanced or structurally dissimilar complements..., homologous attraction likewise plays a dominant role in governing the type of 2-by-2 association. Here, however, there [can be] a conflict between the forces of homologous attraction and the forces producing 2-by-2 associations between closely adjacent parts of chromosomes [and in such situations, the force for 2-by-2 association can predominate].

Conflicts between the two types of forces are manifested in the synaptic configurations produced by both translocation heterozygotes and inversion heterozygotes. In the case of a translocation heterozygote, the synaptic configuration that faithfully mirrors chromosomal homology is an X structure whose four arms are of specific length (Fig. 3A). McClintock observes two additional configurations, H structures (Fig. 3B) and X structures whose junctions are not at the appropriate position (not shown). She infers that "associations which commenced between homologous parts tend to continue beyond the region of homology into the region of non-homology."

In the case of an inversion heterozygote, McClintock inferred the existence of a competition between the two types of forces from variations in the proportions of different synaptic configurations according to the size of the inversion.

> In these cases, the larger the relative size of the inverted segment, the greater the combined force of homologous attraction in this region. Large inner segment inversions give a high frequency of inversion synaptic configurations. Small inner segment inversions show vastly more rod configurations than inversion configurations.... The combined effective force of homologous attraction in smaller inversions is less than in the longer inversions with the result that fewer inversion synaptic configurations are produced.

Reconsideration of McClintock's proposal

McClintock's notion that the force for 2-by-2 association is stronger than the force for homologous association is very provocative. The notion of such a competition presumably contributed to her general characterization of the tendency for 2-by-2 association as a "force" to begin with. As stated, McClintock's formulation implies that 2-by-2 association should actively overcome a homologous association, rather than occurring only when homologous interactions are absent: Thus, "failure of homologous association" (in cases where an appropriate homologous partner is present) "is not due to the failure of homologous parts to attract one another, but rather to the mechanism by which chromosomes start their intimate association and to the more powerful force demanding 2-by-2 associations which can supersede that of homologous attraction when these forces are not working in unison." Although I am attracted to the general notion that there is indeed a driving force for SC formation, the specific relationship of such a force to the forces of homologous attraction, more specifically the ability of SC formation to supersede such forces, is worth reconsidering in more detail in light of subsequent observations and ideas about meiotic chromosome pairing.

For this analysis, I will assume that the forces of homologous attraction exert themselves prior to the occurrence of 2-by-2 association, homologous or nonhomologous, and that these forces are in fact physical connections between homologous chromosomal segments. These assumptions were made by McClintock explicitly or implicitly in her discussion of her data, for example, "Attraction forces between homologous parts of chromosomes produces a movement of these parts toward one another. When association commences between any two homologous parts, 2-by-2 association of adjacent parts may follow and continue along the chromosomes." These assumptions are supported by a number of subsequent types of observations (Moens 1969; Zickler 1977; Albini and Jones 1987; Kleckner et al. 1991). If we make these assumptions, McClintock's proposal would imply that nonhomologous SC formation occurs despite the presence of, and overriding, preexisting physical connections.

In thinking about such a possibility at a more detailed level, two different scenarios can be envisioned that differ with respect to the distribution of physical connections along a pair of homologous chromosome segments. In one scenario, the density of homologous physical connections per unit length would be quite high, such that the strength of any homologous interaction (and the number of physical connections) would be precisely proportional to the extent of homology. In this case, when SC forms between nonhomologous regions instead of between available homologous segments (as in the extension of translocation joints and the formation of rod structures in inversion heterozygotes), SC formation would indeed be required literally to overcome or ignore preexisting homologous connections.

A second type of scenario, however, must also be considered. Specifically, the density of physical connections per unit length may be quite low, possibly one or a few per chromosome. This is the expected situation if the physical connections are provided by recombinational interactions, since each chromosome probably undergoes only one or a few such interactions. If this type of situation applies, it is possible to explain McClintock's observations in a way that does not imply a direct, local competition (at the molecular level) between SC formation and (preexisting) physical connections. Rather, at the local molecular level, 2-by-2 association could in fact be limited to regions in which homologous interactions had not occurred and extension of nonhomologous association might actually be blocked by the presence of an homologous physical connection. In this case, the molecular situation would be somewhat different from McClintock's characterization of the phenomenon as a whole.

This second scenario accounts for the structures formed by inversion heterozygotes as readily as the first scenario. The probability that inverted regions will pair homologously should reflect the probability that there has been a physical connection within the region. In fact, Maguire's subsequent cytogenetic analysis of inversion heterozygotes (and other types of aberrances) in maize meiosis has demonstrated that "the frequency of homologous synapsis at pachytene closely matched the frequency of occurrence of crossing over in the rearranged region" (Maguire 1981). Indeed, Maguire interpreted her results many years ago to mean exactly that initiation of homologous SC formation might require prior or coincident commitment to crossing over (see, e.g., Maguire 1965, 1977). It is of some historical interest that her interpretation, which now seems not only reasonable, but to some of us very probable, was met with active disapproval by many members of the meiosis community (see, e.g., Rhoades 1968; Stern 1977).

The second scenario is also compatible with the behavior of translocation heterozygotes. If homologous physical connections are sparsely distributed along chromosomes, and if nonhomologous SC formation cannot override such connections, then nonhomologous SC formation within a translocation heterozygote might extend beyond the homology-specified junction only as far as the first physical connection. In fact, McClintock anticipated the existence of such a block:

> When two homologous regions synapse 2-by-2, association continues to travel along the chromosomes bringing the successive regions of the chromosomes together unless some counter force is being applied to prevent this association.... In the case of reciprocal translocations, opposition to this continued association would be produced by forces of attraction in the regions beyond the breaks acting to pull these sections towards their homologous counterparts in the other two chromosomes.

Although the basis for this proposal is not given, it seems possible that McClintock inferred the existence of blocks to continued association from the simple fact that nonhomologous alignment in the translocation heterozygote usually extended for only a limited distance beyond the homologous junction point. It would be equally interesting to consider the relationship of preexisting homologous connections to SC formation in the case of synaptic adjustment. This problem has been considered previously (see, e.g., Moses et al. 1984), but may merit reconsideration in light of recent observations and speculations.

Importance of Chromosome Disposition for the Pattern of 2-by-2 Association

McClintock's 1933 paper contains the following additional deduction, which I find so remarkable that I simply quote it here in its entirety

> The type of synaptic configuration (observed in a trivalent) probably depends upon the position the chromosomes occupy in the nucleus and thus the distance a chromosome must travel in order to reach an homologous part. If there is competitive synapsis, homologous parts of the nearest two will associate, the homologous parts of the third will remain unassociated or participate in a non-homologous association. This judgement is based upon the fact that related cells very often show similar configurations. Since associated cells are often related in division, it is not improbable that the chromosomes in their nuclei occupy relatively similar positions with respect to one another. On the basis of competitive 2-by-2 association, such a condition predisposes the chromosomes to similar synaptic configurations.

A Personal Note

I am embarrassed to admit that I did not read Barbara's paper until very recently, in the course of preparing a contribution to this 90th birthday volume. I was especially chagrined to realize that I did not specifically cite her views on a force for 2-by-2 association in our own speculations on the subject (Kleckner et al. 1991). However, I have found at least a tiny excuse for my ignorance: In reviewing the literature on nonhomologous associations published since 1933, there are numerous citings of Barbara's paper as the fundamental documentation of nonhomologous association of chromosomes, but I have not encountered any reference to her general ideas regarding the implications of these observations. As a relative newcomer to the field, it is difficult for me to evaluate the reasons for this deficiency. However, I suspect that at least in some respects, Barbara may once again simply have been too far ahead of her time.

References

Albini, S.M and G.H. Jones. 1987. Synaptonemal complex spreading in *Allium cepa* and *A. fisulosum*. I. The initiation and sequence of pairing. *Chromosoma 95:* 324.

Anderson, L.K., S.M. Stack, and J.D. Sherman. 1988. Spreading synaptonemal complexes from *Zea mays*. I. No synaptic adjustment of inversion loops during pachytene. *Chromosoma 96:* 295.

Carpenter, A.T.C. 1987. Gene conversion, recombination nodules, and the initiation of meiotic synapsis. *BioEssays 6:* 232.

Kleckner, N., R. Padmore, and D.K. Bishop. 1991. Meiotic chromosome metabolism: One view. *Cold Spring Harbor Symp. Quant. Biol. 56:* 729.

Maguire, M.P. 1965. The relationship of crossover frequency to synaptic extent at pachytene in maize. *Genetics 51:* 23.

———. 1977. Homologous chromosome pairing. *Philos. Trans. R. Soc. Lond. B Biol. Sci. 277:* 245.
———. 1981. A search for the synaptic adjustment phenomenon in maize. *Chromosoma 81:* 717.
McClintock, B. 1931. Cytological observations of deficiencies involving known genes, translocations and an inversion in *Zea mays. Mo. Agric. Exp. Stn. Res. Bull. 163:* 1.
———. 1933. The association of non-homologous parts of chromosomes in the mid-prophase of meiosis in *Zea mays. Z. Zellforsch. Microsk. Anat. 19:* 191.
Moens, P.B. 1969. The fine structure of meiotic chromosome polarization and pairing in *Locusta migratoria* spermatocytes. *Chromosoma 28:* 1.
Moses, M.J., M.E. Dresser, and P.A. Poorman. 1984. Composition and role of the synaptonemal complex. *Symp. Soc. Exp. Biol. 38:* 245.
Rhoades, M.M. 1968. Studies on the cytological basis of crossing over. In *Replication and recombination of genetic material* (ed. W.J. Peacock and R.D. Brock), p. 157 Australian Academy of Science, Canberra
Roeder, G.S. 1990. Chromosome synapsis and genetic recombination: Their roles in meiotic chromosome segregation. *Trends Genet. 6:* 385.
Smithies, O. and P. Powers. 1986. Gene conversions and their relation to homologous chromosome pairing. *Philos. Trans. R. Soc. Lond. B Biol. Sci. 312:* 291.
Stern, H. 1977. Concluding remarks. *Philos. Trans. R. Soc. Lond. B Biol. Sci. 277:* 371.
Zickler, D. 1977. Development of the SC and the "recombination nodule" during meiotic prophase in the seven bivalents of the fungus *Sordaria macrospora* Auersw. *Chromosoma 61:* 289.

Twenty-five Years of Transposable Element Research in Köln

HEINZ SAEDLER
Max-Planck-Institut für Züchtungsforschung, Köln, Germany

PETER STARLINGER
Institut für Genetik der Universität zu Köln, Köln, Germany

Barbara McClintock's work during the 1950s was reportedly not properly recognized. This may be so in general, but it was not true for either one of the authors of this paper. One of us (P.S.) as a graduate student in the Max Planck Institute in Tübingen was assigned to read the 1951 Cold Spring Harbor Symposium and discovered her seminal paper as early as 1952. The other (H.S.) had to give a seminar on her work in a course held by Carsten Bresch in Cologne in 1964. In addition, Josef Straub, then director of the Botany Institute and one of the founders of the Cologne Institute for Genetics, had known Barbara McClintock during her research stay in Freiburg in 1933 and he lectured regularly on her work.

Still, when we encountered bacterial insertion elements in the 1960s, it was unintentional, and, in fact, it took us a while to understand the relationship of our insertions with *Ac* and *En/Spm*. Since then, much of our scientific work has been on the study of transposable elements, first in bacteria and later in plants. We found this work to be always interesting, sometimes frustrating, and occasionally rewarding. However, one additional experience during this time was our continued contacts with Barbara McClintock. She was a source of inspiration, both by her personality and through her scientific insights. We gratefully use this occasion of her 90th birthday to express our appreciation and our gratitude.

We cannot try to review here all studies of transposable elements with the methods of molecular genetics. Rather, we describe our own work, as we recollect it. We show how we were influenced by Barbara McClintock's work (her papers are still an indispensable source of information for everybody in our laboratories) and how much we learned from the now sizeable community of researchers on transposable elements.

The Discovery of IS Elements

This section is compiled by both H.S. and P.S. Insertion sequence (IS) elements in *Escherichia coli* were discovered simultaneously by J. Shapiro and by our laboratory in Köln, where we were joined in our work by Elke Jordan, who is now at the National Genome Research Center. The aim of both laboratories was the study of polar mutations in bacterial operons, which was an interesting research topic in the early 1960s. To isolate such mutations, we, as well as Shapiro, employed selection against active galactokinase in mutants unable to metabolize galactose-1-phosphate. Accumulation of this metabolite is lethal and can be avoided by a mutation in the regulatory or uptake genes, by the kinase gene itself, or by a polar mutation located upstream in the galactose operon. Such mutants were easily obtained, and we began to study them by the usual methods of bacterial genetics. This led to a surprise: The mutants had a high reversion rate to wild type, and thus they could not be deletions. On the other hand, their reversion rate could not be increased by a number of mutagens known to cause either base substitutions or frameshift mutations. This made it unlikely that our mutants were caused by these types of alterations of DNA structure, and thus we had to look for other types of mutations, e.g., chromosomal aberrations. These had not been well described in bacteria at this time, and it was also not clear which methods would be suitable to detect them in those early days before cloning and sequencing technology was available.

As deletions seemed to be excluded, duplications, inversions, insertions, or translocations—well known from eukaryotic cytogenetics—remained a possibility (Jordan et al. 1967; Shapiro 1967). A test for an inversion was difficult, because the genetic map was not dense enough to allow the detailed mapping experiments capable of showing an altered order of genetic markers. Therefore, we decided to first test for the presence of additional DNA in the mutated genes, which could be caused by either a duplication or an insertion, and then, if present, to study this additional DNA to determine whether it was a duplication of sequences belonging to the galactose operon or whether new DNA sequences were inserted there.

How could the presence of additional DNA be detected? We would have to investigate a DNA molecule of known length and ask whether this length increased in the known mutants. This posed two different problems. How could we obtain a segment of DNA of known length from the large *Escherichia coli* chromosome? Transducing phages offered an answer. Each of the phages is characterized by the loss of a defined segment of DNA and by the uptake of a defined segment of bacterial DNA. These defective transducing phages can be propagated in the presence of wild-type phage, and thus large amounts of DNA molecules of identical lengths can be isolated. Genetic recombination between the galactose operon carried by the transducing bacteriophage λ and the mutated galactose operon carried by the chromosome of the mutant bacteria would transfer the mutation to the known phage, which would thus be isogenic with its progenitor, but for the presence of the mutation. The two phages could then be compared.

The length of the DNA could have been measured in the electron microscope or by the sedimentation in a sucrose gradient, but given the length of the λ DNA molecule and our lack of knowledge about the size of any additional DNA segment, we did not consider these methods to be sensitive enough. Heteroduplex mapping had not yet been invented. We thus decided to use the very sensitive technique of equilibrium centrifugation of bacteriophage λ particles in CsCl. This was possible because of the well-known fact that the density of a bacteriophage was determined by the average of the relative contributions of its protein (low density) and its DNA (high density). When the protein was constant, an alteration in the amount of DNA

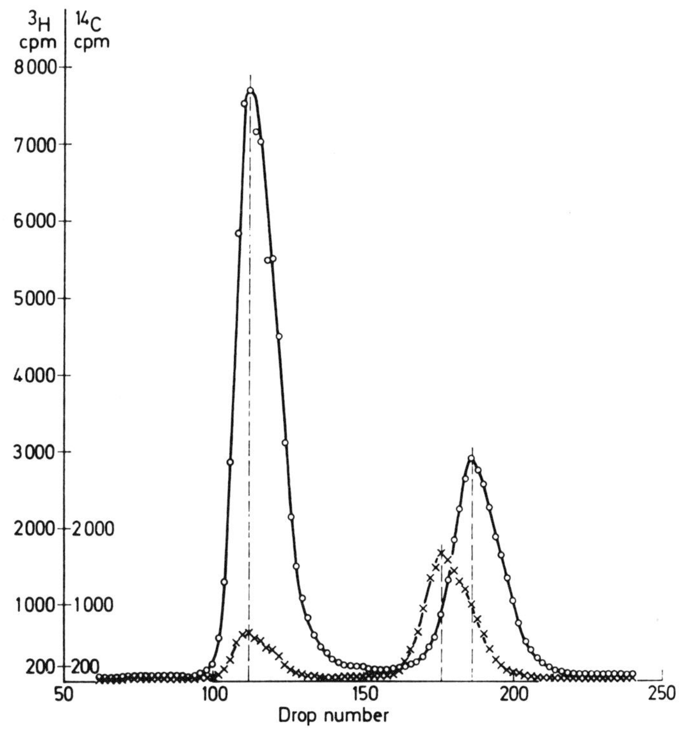

Figure 1 CsCl density gradient configuration of a mixture of lysates from strain *N10* λ,λ*dg*) labeled with ^{3}H (*open circles*) and strain *N10* (λ,λdg_{N102}) labeled with ^{14}C (x). (Reprinted, with permission, from Jordan et al. 1968.)

should translate itself into a potentially measurable alteration in the density of the two phages.

To avoid absolute measurements, we compared phages by labeling with different radioisotopes. The result was quite clear: One of the polar mutants carried an additional amount of DNA that we estimated to be on the order of between 1.5 kb and 1 kb, indicating an increase in the DNA amount of the phage by approximately 1–2%. To confirm this, we selected revertants to the wild-type Gal$^+$ phenotype on the phage and subjected them to CsCl centrifugation. These revertants banded indistinguishably from wild type (Fig. 1). Shapiro (1969) independently found the same result at the same time and, as the two studies confirmed each other, we accepted this result, even though the effect was small (Jordan et al. 1968).

The next step was to try to distinguish between duplications and insertions, and we chose RNA-DNA hybridization. We transcribed mutant DNA in vitro with *E. coli* RNA polymerase (in a very naive manner, hoping that every sequence would be transcribed regardless of the presence of promoters, etc.). This RNA was exhaustively hybridized to DNA of the isogenic wild type. We then asked whether a specific fraction hybridizing only to mutant DNA would remain and found that this was indeed the case. We took this as evidence for translocation of material that came from outside of the *gal* region (Michaelis et al. 1969).

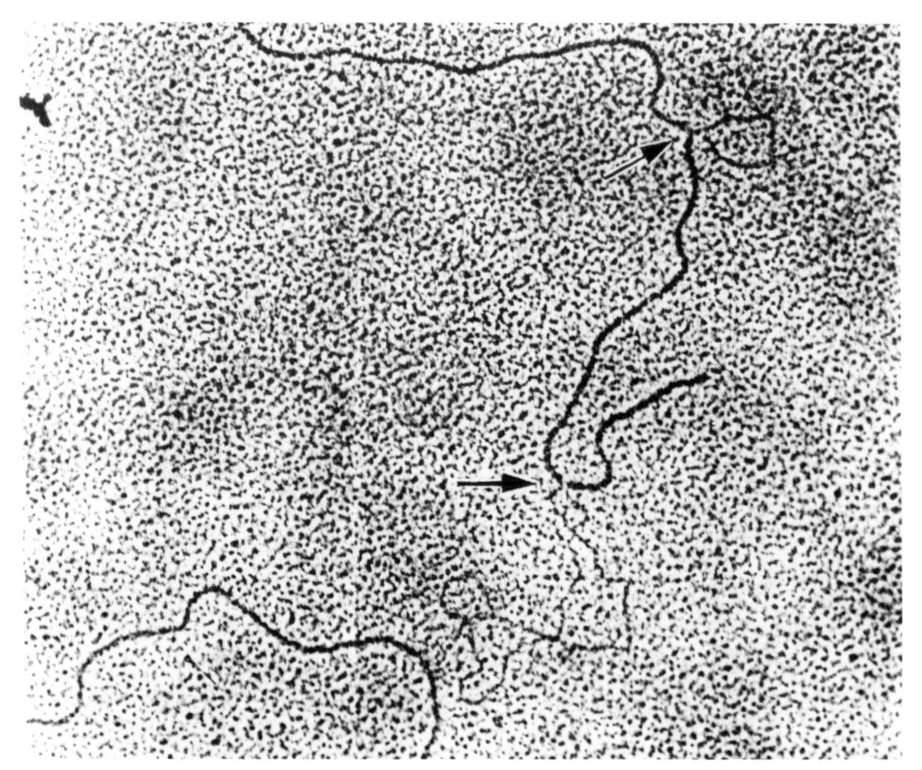

Figure 2 Shown is a DNA heteroduplex molecule between *λdgOPin141*/*λdg161*. The large single-stranded DNA loop indicates distal *gal* DNA sequences deleted in mutant 161, whereas the smaller single-stranded DNA loop represents the approximately 800-bp-long IS*1* insertion characteristic of mutation *galOPin141*. (Reprinted, with permission, from Hirsch et al. 1972a.)

These experiments demonstrated that the insertion-specific DNA was the same in two different mutants. This opened the possibility that the foreign DNA found in different independent mutants was not randomly translocated there, but also belonged to specific classes. We followed this up by using the then newly introduced techniques of heteroduplex mapping and were indeed able to show that the mutants isolated by us belonged to only a few different classes (Fig. 2) (Hirsch et al. 1972a). Collaborations with Philip Brachet from Paris and with Waclaw Szybalski and Michael Malamy from the United States showed that these insertion classes showed up in independent sets of mutants obtained both in bacteriophage and in the lactose operon, respectively. Malamy's lactose operon mutants had been isolated even before our galactose operon mutants. They had not been interpreted in our way because they responded to frameshift mutagens in a manner that was not yet understood (Hirsch et al. 1972b; Malamy et al. 1972).

By this time, it dawned on us that we were seeing something similar to the transposable genetic elements that had been described long ago by Barbara McClintock. This late realization was not due to the fact that we had not been familiar with McClintock's work. We had often discussed it in seminars because we found her experiments very interesting. We did not draw the parallel to bacteria, however, until it became clear that we were seeing preformed elements rather than random transpositions (Starlinger and Saedler 1972). At this time, the laboratory of Heinz Saedler was separated from that of Peter Starlinger.

Further Studies on IS Elements

This section is compiled by H.S. After the IS elements had been shown to be discrete DNA entities, we began to study various properties of IS*1* and IS2. The burning questions in those days were related to the origin of these elements. Specifically, were IS elements constituents of the *E. coli* chromosome or were they infectious agents?

IS elements are normal constituents of the E. coli chromosome and of some plasmids

We attempted to solve the above question by DNA hybridization. This was not such an easy task since the Southern blotting technique had not yet been developed. Bacteriophage λ DNA containing the IS element was thus loaded on small (5-mm) filters and hybridized to in-vivo-labeled *E. coli* DNA sheared by sonication to small (500-bp) fragments. As a control, λ DNA that did not carry insertion elements was used, such that the difference between these hybridizations indicated the amount of IS DNA sequences present in the labeled DNA material. The system was properly calibrated for quantitation and the results were striking. We observed about eight copies of IS*1* and about five copies of IS2 as chromosomal components of the *E. coli* strains used (Saedler and Heiss 1973). Many years later, these experiments were confirmed by others using Southern blotting techniques and restriction endonucleases.

During the experiments in 1973, we noticed that IS*1* and IS2 were natural components not only of the *E. coli* chromosome, but also of some of its episomes or plasmids. We found IS2 on the fertility factor F of *E. coli* and IS*1* on the multiple drug resistance factor R100. At that time, the inspection of DNA heteroduplex molecules in the electron microscope had been developed to fine art in Norman Davidson's laboratory at the California Institute of Technology. In 1973, during a 2-week stay in his laboratory, we localized the IS elements on F^+ and R factors, and two papers emerged from this collaboration (Saedler et al. 1974; Hu et al. 1975).

IS2 can serve as an element to control gene expression

We had previously isolated an IS2 mutation in the galactose operon that had been localized by genetic and electron microscopic studies in the promoter region of the galactose operon, thus conveying a Gal^- phenotype to the bacterium. Genetically, we had also isolated Gal^+ revertants from this mutation located on an F*gal* factor that now expressed the galactose genes constitutively. Heteroduplex analysis revealed that the *gal* operon had been fused to the IS2 copy resident on the F factor, which was integrated in an orientation opposite to that of the original mutation. This indicated that IS2 had a controlling function, which depending on its orientation of integration, could either turn off or turn on the activity of the adjacent genes (Saedler et al. 1974).

IS1 is an important component of some multiple antibiotic resistance factors

Another finding from our collaborative effort with the Caltech group related to IS*1* and R factors and actually had a high impact since it interconnected two fields: IS research and medical (genetical) research on multiple drug resistance factors. It became quite clear from our heteroduplex studies that two IS*1* elements were localized on R factors bracketing the r-determinant carrying the resistance genes such that through homologous recombination between the two IS*1* elements, the resistance transfer factor (RTF) and the r-determinant could separate and/or fuse. This process accounted for the many diverse rearrangements, including amplification of antibiotic resistance genes and some others observed by clinical microbiologists (Hu et al. 1975; Ptashne and Cohen 1975).

IS1 promoter deletions

We showed in an extensive genetic analysis that IS*1* causes deletions of adjacent genetic material to either side in a very peculiar manner. In the deletions, the IS*1* element was always retained such that multiple rounds of this process were observed. The consequence was that in these rearrangements, new genetic material was fused to IS*1*. This enormous instability of genetic material adjacent to IS*1* very likely reflected the high frequencies of IS*1* transposition, but in most cases, these events apparently were abortive, thus generating deletions. This behavior is typical for IS*1* and was not observed with other IS elements to such an extent (Reif and Saedler 1975).

IS1 can cause inversions

In a collaborative study with Guy Cornelis from Leuven, Belgium, we analyzed Tn*951*, which is a transposon from *Yersinia enterocolytica* that carried a lactose operon. The 16.7-kb Tn*951* clearly belonged to the TnA family of transposons, but curiously enough, it also contained an IS*1* element located in front of *lacI*. Selecting Lac$^-$ variants yielded not only IS*1*-induced deletions, but also transposition events of IS*1* into *lacZ*. The Tn*951* of these Lac mutants now contained two IS*1* copies in either direct or inverse orientation. Among the mutants analyzed, in one case, IS*1* had integrated into *lacZ*, but the genetic material between the two IS*1* elements was also inverted. All structures were compatible with the proposed replicative nature of IS*1* transposition and actually were a simple proof of it (Cornelis and Saedler 1980). By far, this was not the end of IS-mediated rearrangements. After having sequenced IS2 (Ghosal et al. 1979), a novel type of DNA rearrangement that led to altered IS2 elements became apparent.

IS2 mini-insertions

IS2 was integrated in the promoter region of the *gal* operon in polar orientation, which allowed selection of Gal$^+$ revertants. Some of these revertants expressing the Gal enzymes constitutively had small insertions of 108 bp within IS2 at the *gal* proximal end. DNA sequencing led to the proposal of their origin by slippage replication at small inverse sequence repetitions within IS2 (Ghosal and Saedler 1978). These mini-insertions had highly symmetrical sequences that were prone to recombinational breakdown. This occurred independent of *recA* in a sequential manner with very high frequencies (~10%) at 9-bp-long direct sequence duplications (Ghosal and Saedler 1979). The original IS2 and the IS2 that had undergone the process of replicational increase and corresponding recombinational decrease in size now differed by a small sequence inversion, which indicated that IS elements not only induced DNA rearrangements in their neighborhood, but also were subject themselves to DNA rearrangements.

When we moved from Freiburg, Germany, where most of the studies concerning the mini-insertions were carried out, to the Max-Planck-Institut für Züchtungsforschung in Cologne, we changed fields and started working on plant transposable elements in *Antirrhinum majus* (snapdragon) and *Zea mays* (corn).

Further Studies on IS Elements

This section was compiled by P.S. During the 1970s, a number of studies aimed at the more detailed chemical investigation of IS elements were carried out. Isolation of IS elements was attempted by a specific heteroduplex procedure in which only the IS element formed a double strand, whereas the remaining DNA was still single-stranded and thus digestible with a single-strand-specific nuclease-like S1. Terminal nucleotides were determined, but this was now the time when restriction analysis, cloning, and sequencing emerged and superseded these other methods. By this time, an increasing number of laboratories had begun to study IS elements as well as the larger transposons that contain IS elements plus additional genes, usually conferring resistance to an antibiotic. It was this latter class of transposable elements that attracted the most interest, both because of their importance in medical microbiology and because they could be used for mutagenesis at an increasing scale.

Our laboratory became interested in the question of insertion specificity. Without going into detail, it turned out that these can be quite different for different insertion elements and each case must be elucidated separately. We also directed our attention to a particular IS element, IS4, that we isolated, cloned, and sequenced (Klaer et al. 1981). One of our findings was surprising: IS4 gave rise to the formation of deletions that removed not only the element itself, but also DNA segments of varying lengths to both sites of the element (Habermann and Starlinger 1982). Hypotheses could be advanced to explain this phenomenon, but they need not be repeated here, as we did not continue this research far enough to either prove or disprove them. The reason was a shift in our interest, which was more and more directed toward the study of plant transposable elements.

Work on Maize Transposable Elements Ac and Ds

This section was also compiled by P.S. The techniques of molecular biology of eukaryotic cells became more and more refined during the 1970s, and this gave us the courage to attempt to work with McClintock's transposable elements in plants. The isolation of these elements was not straightforward. By the end of the 1970s, the usual method to isolate a gene was through the identification and isolation of either the mRNA transcribed from this gene or a cDNA clone obtained from the mRNA. Identification of the message could be achieved if the gene product, the protein, was known and antibodies could be raised against it. All of this was not possible for a transposable element. The work on bacterial elements had shown that these encode one or more proteins necessary for their own transposition. Even in bacteria, however, these proteins are present only in minute amounts, and their characterization was slow. Therefore, we chose another approach. We searched the literature and obtained much help from discussions with Barbara McClintock and Peter A. Peterson to identify a gene that would have a product present in large amounts in the cell. In addition, we sought a gene in which mutants were known that had been caused by the insertion of a known transposable element. Our choice was the sucrose synthase gene of *Zea mays*. The enzyme is present at levels of about 1% among the proteins in the developing endosperm. McClintock had isolated mutants caused by insertion of the transposable element *Ds*, a nonautonomous relative of transposable element *Ac*, and she kindly provided us with these materials.

We then had to try to isolate the sucrose synthase gene via the protein and its antibodies, the mRNA, and finally the cDNA. This proved possible both in the laboratory of B. Burr and F. Burr in Brookhaven and in our laboratory (Geiser et al. 1980; Burr and Burr 1981). It was then possible to isolate a genomic clone both from the wild type and from a mutant carrying a *Ds* insertion (Geiser et al. 1982). The clone from the mutant contained part of a very large *Ds* element that we later found to be 30 kb. This large element consisted of mutated *Ac* elements (*Ds* elements in McClintock's nomenclature) flanking a long sequence of non-*Ac* DNA. The whole 30-kb sequence behaved like a transposable element that could be excised from its site of insertion. Its properties were elucidated by Hans-Peter Döring from our laboratory in collaboration with Nina Fedoroff (Courage-Tebbe et al. 1983).

One of the *Ds* elements was sequenced and shown to be a complicated structure derived from the transposable element *Ac*. This "double *Ds*," as we called it, was made from a transposable element with 11-bp inverted repeats and caused an 8-bp target-site duplication; however, the structure consisted of two identical elements, of which one was inserted into the other in an inverted orientation (Döring et al. 1984). It is this double structure that gives rise to the chromosome breaks discovered by McClintock with some of her *Ds* elements, but not with the *Ac* elements (Döring and Starlinger 1984). The autonomous *Ac* element, first isolated and sequenced by Nina Fedoroff, is 4.5 kb long (Fig. 3) (Fedoroff et al. 1983; Pohlman et al. 1984). It contains a single gene with four introns and a long and possibly functionless leader sequence and encodes a protein of 807 amino acids (Kunze et al. 1987). The protein binds to DNA, where it recognizes an AAACGG hexamer that is present in multiple copies at both ends of *Ac*, but inside of the termini by about 80 bp (Kunze and Starlinger 1989). Deletion of these internal, AAACGG-containing regions decreases or abolishes transposition. This was shown by B. Baker and G. Coupland, who together with Fedoroff designed an in vivo test for *Ds* excision based on the observation by Baker, Fedoroff, and Schell (Baker et al. 1986) that

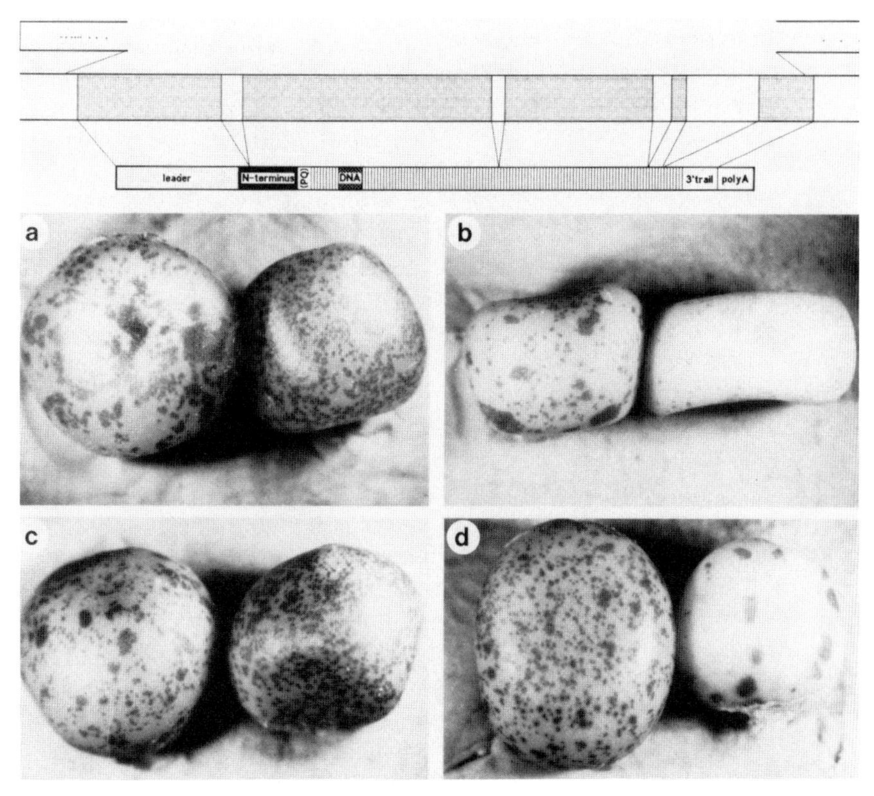

Figure 3 Sketch (*top*) shows *Ac* diagrammatically. The middle bar depicts the DNA structure including the four introns. The lower bar depicts the mRNA and the coding region of the protein in which the DNA-binding region, the dispensable amino terminus, and the functionally important Pro-Gln repeat are highlighted. The upper row shows the ends of *Ac* with the small arrows symbolizing the AAACGG hexamers. It should be noted that sequences similar to, but not identical with, the hexamers are abundant in the vicinity of the hexamers themselves and that their DNA-binding capacity has not been elucidated in detail. The kernels show the excision pattern of the *Ds* element *bz-m2(DI)* under the influence of either one (*left*) or two (*right*) copies of *Ac* inserted at different positions. (*a*) *bz-m2::Ac*; (*b*) *Ac up*; (*c*) *wx-m9::Ac*; (*d*) *wx-m7::Ac*.

Ac transposes in tobacco (Baker et al. 1987; Coupland et al. 1988, 1989). It will be interesting to learn how the transposition complex is composed and which protein cleaves the DNA at the termini. Until now, binding of the *Ac* protein to the inverted termini or to the region between these and the AAACGG region has not been detected, neither could the binding of non-*Ac* proteins of maize be demonstrated with certainty (R. Kunze; H.A. Becker; both unpubl.). Still, these more terminal regions must be important, as point mutations there can abolish *Ds* excision in *cis* (S. Chatterjee, unpubl.). We will also have to understand the role of the amino terminus of the *Ac* protein, as its removal is compatible with the element's function in tobacco, and the excision of *Ds* elements is even increased in a transient assay in *Petunia* protoplasts (Li and Starlinger 1990; Houba-Herin et al. 1990).

How is transposition rate controlled? Transposition occurs much less than once per cell generation and thus is much rarer than even DNA replication, not to mention other biochemical reactions. If several *Ac* and *Ds* elements can be observed in the same cell, usually no more than one of them is excised within a given cell generation (Heinlein and Starlinger 1991). This does not seem to be due to the frequency of *Ac* transcription and translation (Fusswinkel et al. 1991), and it cannot be due to an exclusive property of *Ds* (e.g., its inaccessibility to *Ac* protein) because a given *Ds* element reacts quite differently toward different *Ac* insertions. Even if these *Ac* insertions are located in the same gene (*waxy*) and have an identical DNA sequence (Pohlman et al. 1984; Müller-Neumann et al. 1984), the excision patterns they cause with a *Ds* element located in another gene can be quite different (Schwartz 1984; Heinlein and Starlinger 1991). The situation is even worse when *Ac* mRNA and *Ac* protein are compared in seedlings and in endosperm—no differences were observed between the two *Ac* insertions in the *waxy* gene. Is it conceivable that the expression of different *Ac* insertions is different in the very early endosperm, where the excisions occur and where we cannot measure RNA and protein as yet, due to the inaccessibility and small amounts of the tissue? If this is not so, understanding the differences between the transactions of different *Ac* insertions will be difficult.

Not only is *Ac* expression rare, but it is regulated. As known from Barbara McClintock's early work, two copies of *Ac* delay and sometimes decrease transposition, when compared with the effects of only one copy of *Ac*. This is not so under all circumstances, however. Manfred Heinlein has compared several *Ac* and *Ds* elements and has seen that the excision rates of *Ds* elements are characteristic for each *Ac* element used, as well as for the time during development. The rates also differ for the *Ac* dose employed, and two copies can cause either less or more excision than one copy, depending on the pair of elements employed. As mentioned above, the effects of a single *Ac* copy as well as the dosage effects can be strikingly different when two *Ac* elements located in the *waxy* gene are compared. If this is a position effect, positions can have different influences over short distances. In the case of the two *Ac* insertions in the *waxy* gene, the distance is only 2 kb. Thus, *Ac* action may become a very sensitive probe for changes in chromatin structure over short distances, e.g., for nucleosome phasing.

Work on Transposable Elements from Antirrhinum majus

This section was compiled by H.S. The phenomenon of paramutation is widespread in plants and has been amenable to molecular analysis in *Antirrhinum majus*.

Paramutation in A. majus

In the early 1980s, we isolated the *nivea* locus encoding chalcone synthase (*chs*), an enzyme involved in anthocyanin biosynthesis, from *A. majus*. An unstable mutation at that locus isolated in the 1930s by Kuckuck (1936) was found to be caused by the integration of a 15-kb transposable element that we named *Tam1* (Bonas et al. 1984). This element is inserted in the promoter region of the *chs* gene. The mutation caused by this insertion reverts at a high rate: 10–20% of the progeny are revertants. This instability is also observed somatically, resulting in a highly variegated flower phenotype: red spots and stripes on an almost colorless background. When this mutation (termed *niv-53*) is crossed to a stable white *nivea* mutant, 100% of the progeny are variegated, as expected. However, Harrison and Carpenter (1973) had already shown that when *niv-53* is crossed to one particular stable white *nivea* mutant (*niv-44*), a mixture of F_1 phenotypes is seen, which most importantly gave no variegated F_2 progeny. All F_2 progeny are white. It is this lack of segregation that characterized the above cross as involving paramutation. The *niv-53* allele formally undergoes a mutation under the influence of *niv-44* in the direction of the *niv-44* allele. Paramutation thus seems to be directed by in vivo mutagenesis. This phenomenon has been best described genetically at the *R* locus of maize, but it also exists in *A. majus* as shown above and most importantly could be analyzed because of the molecular availability of the alleles involved. *niv-44* has been a stable white mutant for more than 70 years now. Molecular analysis of *niv-44* showed that another transposable element, *Tam2*, is integrated within exon 2 of the *chs* gene (Upadhyaya et al. 1985). Apparently, this 5-kb element is defective in that it does not encode a functional transposase but has highly structured ends. The terminal 13-bp sequence is identical to that of *Tam1*, which otherwise does not share much homology. However, this observation was relevant to the analysis of the phenomenon of paramutation. First of all, it was relatively easy to show by Southern blotting that no gross alterations at the *nivea* locus occurred in paramutant progeny. These were molecularly perfect heterozygotes. Even though the F_2 progeny did not reveal phenotypic differences, molecularly, they were of the expected homozygous and heterozygous types. Because of the sequence identity at the ends of *Tam1* and *Tam2*, we thought that something other than mutation might be going on (Krebbers et al. 1987). To test for possible physiological interactions between these two *nivea* alleles characterized by the insertions of *Tam1* and *Tam2*, respectively, a mutant allele of *Tam1* was used that abolished the mobility of *Tam1*; 5 bp were deleted at one end of *Tam1*, thus preventing movement of the element. Because of this, the phenotype was stably white. This allele was called *niv-46*. If the stable white *niv-46* is crossed to the stable white *niv-44*, one expects exclusively white progeny. Surprisingly, however, variegated flowers appear. Variegated branches are seen even on plants with mostly white flowers. Clippings and regeneration of new flowers from variegated branches, as well as seeds derived from selfing of the variegated flowers always resulted in white flowers. The phenomenon was clearly transient, restricted to particular stages of development. Apparently, the system escaped from paramutagenicity at these stages, thus resulting in variegated flowers. We occasionally observed a very early event leading to a full red flower. Molecular analysis revealed that *Tam2* had excised, restoring an active *chs* gene but leaving a diagnostic footprint (see below) behind. Apparently, *Tam1* transposase, which could not act on *Tam1* ends because of the 5-bp deletion in one of them, could recognize the *Tam2* ends, excise, and transpose this element (Hehl et al. 1987; Hudson et al. 1987). So far so good, but should this *trans*-activation not increase variegation in F_1 or in subsequent progeny and why is *trans*-activation transient? An easy assumption is that *niv-44* contains a suppressor inhibiting or interfering with *Tam1* transposase. To test for such a negatively acting factor, a *niv-44* line, in which *Tam2* and part of *nivea* had been deleted, *niv-4432*, was used in a cross with *niv-53*. Clearly, *niv-4432* was still paramutagenic; hence, the factor could not have been encoded by *Tam2* at the *nivea* locus. Sequence analysis of the *Tam2* had already demonstrated that it was unlikely that this element encoded a protein. However, *niv-44* contains quite a number of *Tam2* elements in its genetic background, some of which are transcribed into RNA and hence could encode the postulated inhibitor. If the line *niv-44* contained a few genes encoding a suppressor of *Tam1* transposase activity, then indeed the lack of segregation of the variegated phenotypes in selfs of *niv-53* x *niv-44* crosses could be explained (Krebbers et al. 1987). Hence, a crossing program was initiated to segregate out the suppressor genes in order to find out whether or not the suppressor locus is identical with one of the many *Tam2* elements present in this line.

We favor the assumption that a *Tam2* element encodes a suppressor because in the *En/Spm* system of maize, we have shown molecularly that a truncated En-encoded protein can severely reduce *En*-transposase-mediated excision (Cuypers et al. 1988). If this would also apply to the *Tam1/Tam2* situation, it would indicate that paramutation in this instance is not a mutational phenomenon but a physiological interaction of different but related transposable elements. Currently, more than ten transposable elements have been isolated molecularly from *A. majus*. Except for *Tam3*, they belong to the CACTA family of elements, the prototype of which is *En/Spm* of *Zea mays* which we have been studying now for a decade.

Work on the En/Spm Tranposable Element System from Zea mays

This section was compiled by H.S. Molecular characterization of the *En/Spm* element system of maize in the 1980s was quite a challenge because this system had genetic features not shared by other elements. A second reason to work on this system was that in the 1970s, Patricia Nevers and I had not only tried to read all of McClintock's papers concerning the *Spm* system, but also tried to teach this system to students. This in itself was a challenge, and we soon realized that in order to make the students remember the facts, we had to develop a model for *Spm* action. However, since our experience was in bacteria and the structure of eukaryotic genes was not yet known, the outcome of this exercise as published in *Nature* (Nevers and Saedler 1977) naturally missed the molecular reality as we now know it. In any case, the value of the model was mnemotechnical in that all of the mutants McClintock had described could easily be deduced from the model.

The nature of the *Spm* system became apparent in the following decade when molecular cloning of plant genes became feasible. In the 1950s, Barbara McClintock and P.A. Peterson independently had discovered this element system. The name *Spm* (suppressor-mutator) coined by McClintock already indicated that the element encodes two functions. She, as well as Peterson, who termed the element *En* (enhancer of mutability) and later showed that the elements were genetically identical, had seen that the system consisted of two units: an autonomous element encoding all functions for transposition and a defective element, which was unable to transpose unless complemented by an autonomous *En/Spm*. We have cooperated with P.A. Peterson on the analysis of this system for more than 10 years.

In the early 1980s, we succeeded in cloning a defective element (*dSpm* or *I* in the nomenclature of Peterson) from the *waxy* locus (Schwarz-Sommer et al.

1984). Peterson's success in moving *En* into *wx* also made available the autonomous *En* element (Pereira et al. 1985). Sequence comparisons of the autonomous and defective elements substantiated the genetic notion of a two-component system.

Suppressor

The system to test for suppression consists of the *a1-m1* mutant (*5719A1*). This mutant features purple kernels in the absence of *Spm*, but variegated kernels are seen in the presence of *Spm* element somewhere else in the genome. This phenotype arises by suppression as seen by the colorless background and colored spots that are formed in response to mutator, restoring *a1* gene expression. Molecularly, the autonomous *En/Spm* element encodes two RNAs: a 2.5-kb transcript encoding *tnpA* (Pereira et al. 1986; Cuypers et al. 1988) and a 6-kb RNA encoding *tnpD* (Pereira et al. 1986; Masson et al. 1989). A problem arose of which of the two molecularly identified functions corresponded to McClintock's suppressor and mutator, respectively.

The clearest experiment to demonstrate that *tnpA* has suppressor function comes from studies using artificial reporter systems in transgenic tobacco protoplasts. Transient expression of the bacterial glucuronidase gene (GUS) expressed from a strong viral plant promoter (35S) was monitored. A *dSpm* element was inserted between the 35S promoter and the GUS gene such that in the absence of *tnpA*, glucuronidase was produced, whereas in the presence of *tnpA*, very little enzymatic activity could be measured. The study also indicated that for suppression-specific target sequences within, the ends of *dSpm* are required (Grant et al. 1990). Not only is *tnpA* the suppressor, but it also seems to be a multiple-purpose protein. The functions of *tnpA* are as follows: (1) a DNA-binding protein recognizing a 12-bp sequence motif scattered throughout the ends of *En/Spm*, however, only if not methylated (Gierl et al. 1988); (2) only the tail-to-tail binding motif close to the terminal inverse repeats (TIR) is required for suppression (Grant et al. 1990); (3) elevates *En* transcription (autoregulation); (4) required, but not sufficient for excision (Frey et al. 1990); and (5) serves as a glue to bring both ends of the element together (see Mechanism of transposition, below).

Mutator

This function requires *tnpA* and *tnpD*. McClintock had already isolated *Spm*^{w} alleles that still caused suppression but did not yield excision. Molecular analysis revealed that they had deleted part of *tnpD* (Masson et al. 1987; Gierl et al. 1988). The clearest evidence that *tnpA* and *tnpD* are needed for McClintock's mutator function comes from transgenic experiments.

Again, the responding allele was a *dSpm* element inserted within a GUS gene, thus causing a GUS^{-} phenotype in the transgenic tobacco plant. However, only when crossed to a plant expressing both *tnpA* and *tnpD* cDNAs was GUS variegation in the leaves observed (Fig. 4). Hence, for excision, both *TnpA* and *TnpD* are required (Frey et al. 1990). This experiment also shows that for excision of *dSpm*, only three components are needed: (1) intact ends of the element, (2) TnpA, and (3) TnpD. In a subsequent study, it was shown that these components are also sufficient for transposition (Cardon et al. 1991).

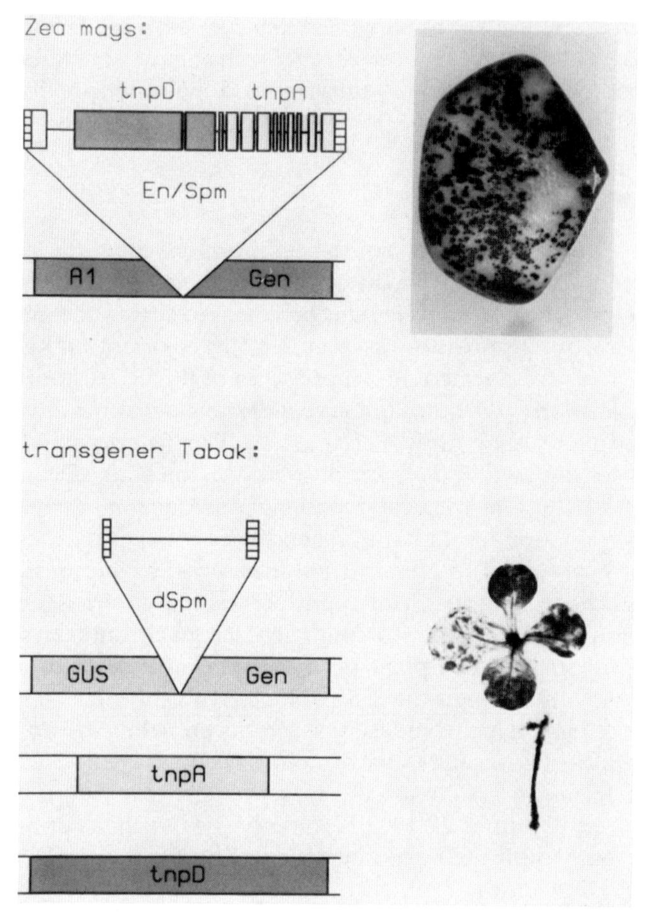

Figure 4 Structure of the *En/Spm* autonomous transposable element system integrated at the *A1* locus of *Zea mays* is shown. Hatched boxes indicate the highly structured and symmetrical ends of the element, and the different shades of the exons reveal the two different proteins encoded by the element. Excision of the element from the *A1* locus leads to spotted kernels. In the lower part of the figure, it is shown that only three components are required for transposition in transgenic tobacco.

Mechanism of transposition

Like other transposable element systems, plant elements generate small sequence duplications at their target site upon integration. However, in the excision process, the wild-type sequence is neither regenerated nor is the duplicated segment retained; rather, a whole array of sequence variations are produced as seen first by Sachs et al. (1983). To achieve this, certain rules must be followed (Saedler and Nevers 1985). According to these rules, staggered nicks are produced at each end of the integrated element, and nonelement-encoded functions, such as polymerase, exonuclease, and ligase, are then responsible for the repair of this structure. This eventually leads in the majority of cases to a novel DNA sequence and hence to a mutation. These mutations can be rather complex in structure and can lead to quite a number of additional amino acids in the protein if the transposon had been integrated in an exon (Schwarz-Sommer et al. 1985). Such a mechanism of excision seems to be specific for plants. According to the model, the two ends of the element must come into physical contact in order to facilitate excision and healing of the chromosome. How can this be achieved?

In the case of *En/Spm*, we had shown previously that *TnpA* is a DNA-binding protein (Gierl et al. 1988) that recognizes subterminal repeats. If the repeats of both ends are aligned properly, a zipper-like arrangement becomes apparent. Thus, *TnpA* could mediate the interaction between the two ends and could function as a glue (Frey et al. 1990). The importance of the *TnpA*-binding motifs to facilitate the interaction of the termini of the element comes from internal deletions removing motifs at one end and thereby affecting the frequency of excision. An extreme case is McClintock's a_2-*m1* state II. Whereas a 2241-bp long *dSpm* is integrated in the intron-less *A2* locus in a_2-*m1*, a further deletion of 900 bp occurred in the state II version, thus removing all *TnpA*-binding motifs at one end of the element without, however, affecting the TIR. This allele is colored in the absence of *Spm* but fully suppressed in the presence of *Spm* and does not show any spots. The 1341-bp *dSpm* apparently now is taken as an intron in the absence of *Spm*. This insert is removed via splicing, leading to an altered A2 protein with seven additional amino acids that apparently do not obstruct A2 function (Menssen et al. 1990).

In vitro studies are currently being carried out to substantiate the assumption that *TnpA* serves as a glue. Once the complex is formed, *TnpD* is hypothesized to bind at the TIR and to cleave endonucleolytically. Healing of the chromosome might occur subsequently at times, when the complex is still not totally dissociated. All of this is speculation and must be verified biochemically.

What Would We Like to Know?

Certainly, a number of questions concerning the biochemistry and physiology of plant transposable elements should be answered.

1. The mechanism of transposition is far from being clear. The nature and role of the proteins involved and of their domains should be clarified. In the end, a detailed knowledge of the transposition complex and of the interaction and catalytic steps would be desirable.

   Regarding control and regulation of transposition rates as well as the mechanism of transposition, both *Ac* and *En/Spm* seem to differ from other transposons. In bacteria, in yeast, and in *Drosophila*, at least in the case of the *P* element, the same basic principle seems to be implemented: From one gene, two similar proteins are produced, of which one is the transposase and the other is either the direct inhibitor of transposition or a repressor of the transposase gene. Neither of the two plant transposable elements described here seems to work this way. In the case of *Ac*, only one protein is known, and whether its inhibitory amino terminus just keeps transposition low at all times or whether it has a regulatory role is not yet known. In the case of *En/Spm*, two proteins are made, but instead of opposing each other, they are both required for transpositon. However, in this latter (*En/Spm*) system, a truncated element exists in certain maize lines that encodes a product consisting of amino-terminal *tnpD* and carboxy-terminal *tnpA* sequences that strongly reduces excision rates (Cuypers et al. 1988). Whether or not such a product could also result from an autonomous *En/Spm* element via differential splicing remains to be shown.

2. Why do organisms carry transposons? The biolgical role of these elements in plants is not obvious. Rather, the biology of plant transposable elements implies that at least too much transposition is of disadvantage for the plants. This would be an explanation for the large number of nonauton-

Isolation of new homeotic loci

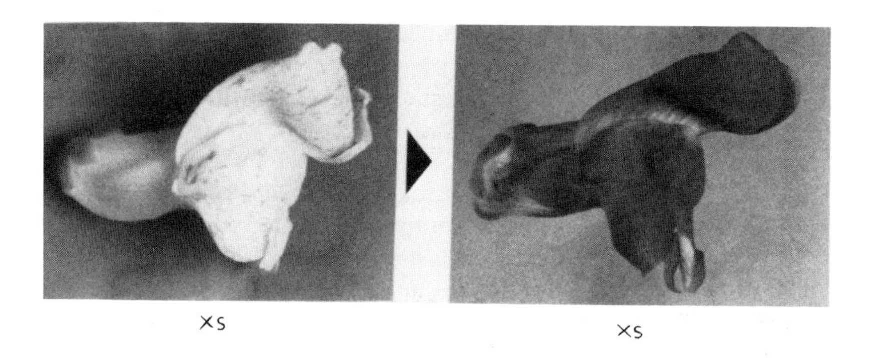

Frequency: 10^{-2}

(-at present: 8 candidates)

Figure 5 Spotted *Antirrhinum* flower pattern (*upper left*) is due to the integration of the *Tam1* transposable element into the promoter region of the chalcone synthase gene. Excision leads to colored flowers. The thus liberated element can now integrate into other genes. Upon selfing of the colored revertants, new abnormal flower phenotypes segregate with a frequency of about 1%. To date, 15 candidates have been isolated. Most of the new mutants are unstable and hence good candidates for newly tagged genes.

omous elements found for both *Ac* and *En/Spm* (and also for other transposable elements). It is conceivable that the presence of too many autonomous elements is bad for the plants and that inactive forms are therefore selected, whether they are caused by mutation or by epigenetic changes. If this were the only explanation, inactivated *Ac* or *En/Spm* elements could be explained. However, the situation is more complicated than that at least in the case of *Ac*. In addition to deletion mutants of *Ac*, there are elements like *Ds1* and *Ds2*, which carry sequences completely unrelated to *Ac* (Sutton et al. 1984; Merckelbach et al. 1986). These elements multiply within maize until they form families with many members, and these change and thus show a sequence evolution of their own. Is this simply the behavior of "selfish DNA"? Do they drift or do they evolve toward more pronounced selfishness? Or do they confer an advantage to the plants that we have not yet been able to determine? Unfortunately, we do not see an experiment that could answer these questions, but we hope such experiments will be designed eventually.

3. In addition to how transposable elements evolve within a given species, we would like to know how they were acquired early on. Were they always part of their host species or were they obtained by horizontal gene transfer? Two families of elements have been studied in more detail in this respect: *Ac* (*Zea mays*), *Tam3* (*Antirrhinum majus*), and *hobo* (*Drosophila melanogaster*) have similar ends and share homologies in their coding sequences as do the CACTA elements *Tam1* (*A. majus*), *En/Spm* (*Z. mays*), and *Tgm1* (*Glycine max*) (Nacken et al. 1991). Does this suggest that they derived from a donor long after these species had separated from a common progenitor?
4. Apart from an exploration of the biology of transposable elements, we anticipate that they will be useful in "tagging" experiments for the isolation of new genes. An example of this is shown in Figure 5 in a tagging program of genes involved in flower development in *A. majus*. This technique, first used in bacteria by Kleckner et al. (1975) and in plants by Fedoroff et al. (1984) and subsequently amply used by many laboratories including our own, promises to yield the starting materials for many investigations in cell biology and in development (Table 1).

Acknowledgments

In these last 25 years, we have been associated with many colleagues from Germany and from other countries and with many Ph.D. and diploma students. Not all of them are mentioned in the list of references. However, our work could not have been done without them and we acknowledge all of their scientific contributions as well as their pleasant company while they stayed with us. Our work was financed by the Max Planck Gesellschaft, the Land Nordrhein-Westfalen, the European Community, the Bundesministerium für Forschung und Technologie, and the Deutsche Forschungsgemeinschaft, mostly through Sonderforschungsbereiche 74 (Molekularbiologie der Zelle) and 274 (Der modulare Aufbau des genetischen Materials). Part of this article was published in the *Abschlussbericht des SFB74* and is reprinted here with permission of Verlag Chemie, Weinheim, Germany.

Table 1 Genes Cloned or Identified with Transposable Elements

Gene	Function	Element
Zea mays		
A1	NADPH-dep. reductase	En, Mu
A2	anthocyanin pathway	En, rcy
Bz1	UDP-glycosyltransferase	Ac
Bz2	anthocyanin pathway	Ds2, Mu
C1	regulatory gene	Spm, En
C2	chalcone synthase	Spm
hcf-106	chloroplast development	Mu
Kn1	regulatory gene	Ds2
O2	regulatory gene	Ac, Spm
P	regulatory gene	Ac
R	regulatory gene	Ac
Vp1	regulatory gene	Mu
A. majus		
deficiens	regulatory gene	Tam7
delila	regulatory gene	Tam2
floricaula	regulatory gene	Tam3
globosa	regulatory gene	Tam7, Tam9
incolorata	antocyanin pathway	Tam1
olive	chloroplast development	Tam3
pallida	NADPH-dep. reductase	Tam3

For review, see Gierl and Saedler (1992)

References

As we have not attempted a review of bacterial and plant transposable element research, we have quoted mainly papers from our own laboratories and papers of others are included in those cases, where our own work depended on their studies or was carried out simultaneously.

Baker, B., J. Schell, H. Lörz, and N. Fedoroff. 1986. Transposition of the maize controlling element "*Aktivator*" in tobacco. *Proc. Natl. Acad. Sci. 83:* 4844.

Baker, B., G. Coupland, N. Fedoroff, P. Starlinger, and J. Schell. 1987. Phenotype assay for excision of the maize controlling element *Ac* in tobacco. *EMBO J. 6:* 1547.

Bonas, U., H. Sommer, and H. Saedler. 1984. The 17-kb *Tam1* element of *Antirrhinum* induces a 3-bp duplication upon integration into the chalcone synthase gene. *EMBO J. 3:* 1015.

Burr, B. and F. Burr. 1981. Detection of changes in maize DNA at the *shrunken* locus due to the intervention of *Ds* elements. *Cold Spring Harbor Symp. Quant. Biol. 45:* 463.

Cardon, G.. M. Fey, H. Saedler, and A. Gierl. 1991. Transposition of En/Spm in transgenic tobacco. *Maydica 36:* 305.

Cornelis, G. and H. Saedler. 1980. Deletions and an inversion induced by a resident IS*1* of the lactose transposon Tn*951*. *Mol. Gen. Genet. 178:* 367.

Coupland, G., B. Baker, J. Schell, and P. Starlinger. 1988. Characterization of the maize transposable element *Ac* by internal deletions. *EMBO J. 7:* 3653.

Coupland, G., C. Plum, S. Chatterjee, A. Post, and P. Starlinger. 1989. Sequences near the termini are required for transposition of the maize transposon *Ac* in transgenic tobacco plants. *Proc. Natl. Acad. Sci. 86:* 9385.

Courage-Tebbe, U., H.P. Döring, N. Fedoroff, and P. Starlinger. 1983. The controlling element *Ds* at the shrunken locus in *Zea mays*: Structure of the unstable sh-m5933 and several revertants. *Cell 34:* 383.

Cuypers, H., S. Dash, P.A. Peterson, H. Saedler, and A. Gierl. 1988. The defective En-I102 element encodes a product reducing the mutability of the En/Spm transposable element system of *Zea mays*. *EMBO J. 7:* 2953.

Döring, H.P. and P. Starlinger. 1984. Barbara McClintock's controlling elements: Now at the DNA level. *Cell 39:* 352.

Döring, H.P., E. Tillmann, and P. Starlinger. 1984. DNA sequence of transposable element dissociation in maize. *Nature 307:* 127.

Fedoroff, N., D. Furtek, and O. Nelson. 1984. Cloning of the *bronze* locus in maize by a simple and generalizable procedure using the transposable controlling element *Activator* (*Ac*). *Proc. Natl. Acad. Sci. 81:* 3825.

Fedoroff, N., S. Wessler, and S. Shure. 1983. Isolation of the transposable maize controlling elements *Ac* and *Ds. Cell 35:* 235.

Frey, M., J. Reinecke, S. Grant, H. Saedler, and A. Gierl. 1990. Excision of the En/Spm transposable element of *Zea mays* requires two element-encodes proteins. *EMBO J. 9:* 4037.

Fusswinkel, H., S. Schein, U. Courage, P. Starlinger, and R. Kunze. 1991 Detection and abundance of mRNA and protein encoded by transposable element *activator* (*Ac*) in maize. *Mol. Gen. Genet.* **225:** 186.

Geiser, M., H.P. Döring, J. Wöstemeyer, U. Behrens, E. Tillmann, and P. Starlinger. 1980. A cDNA clone from *Zea mays* endosperm sucrose synthase mRNA. *Nucleic Acids Res. 8:* 6175.

Geiser, M., E. Weck, H.P. Döring, W. Werr, U. Courage-Tebbe, E. Tillmann, and P. Starlinger. 1982. Genomic clones of a wild type allele and a transposable element-induced mutant allele of the sucrose synthase gene of *Zea mays* L. *EMBO J. 1:* 1455.

Ghosal, D. and H. Saedler. 1978. DNA sequence of the mini-insertion IS2-6 and its relation to the sequence of IS2. *Nature 275:* 611.

———. 1979. IS-61 and IS2-611 arise by illegitimate recombination from IS2-6. *Mol. Gen. Genet. 176:* 233.

Ghosal, D., H. Sommer, and H. Saedler. 1979. Nucleotide sequence of the transposable DNA element IS2. *Nucleic Acids Res. 6:* 1111.

Gierl, A. and H. Saedler. 1992. Plant transposable elements and gene tagging. *Plant Mol. Biol.* (in press).

Gierl, A., S. Lütticke, and H. Saedler. 1988. TnpA product encoded by the transposable element En-1 of *Zea mays* is a DNA binding protein. *EMBO J. 7:* 4045.

Grant, S.R., A. Gierl, and H. Saedler. 1990. En/Spm encoded tnpA protein requires a specific target sequence for suppression. *EMBO J. 9:* 2029.

Habermann, P. and P. Starlinger. 1982. Bidirectional deletions associated with IS4. *Mol. Gen. Genet. 185:* 216.

Harrison, B.J. and R. Carpenter. 1973. A comparison of the instabilities at the *nivea* and *pallida* loci in *Antirrhinum majus. Heredity 31:* 309.

Hehl, R., H. Sommer, and H. Saedler. 1987. Interaction between the *Tam1* and *Tam2* transposable elements of *Antirrhinum majus. Mol. Gen. Genet. 207:* 47.

Heinlein, M, and P. Starlinger. 1991. Variegation pattern caused by transposable element *Ac. Maydica 36:* 309.

Hirsch, H.J., H. Saedler, and P. Starlinger. 1972a. Insertion mutations in the control region of the galactose operon of *E. coli. Mol. Gen. Genet. 115:* 266.

Hirsch, H.J., P. Starlinger, and P. Brachet. 1972b. Two kinds of insertions in bacterial genes. *Mol. Gen. Genet. 119:* 191.

Houba-Herin, N., D. Becker, A. Post, Y. Larondelle, and P. Starlinger. 1990. Excision of a *Ds*-like maize transposable element (*Ac*Δ) in a transient assay in *Petunia* is enhanced by a truncated coding region of the transposable element *Ac. Mol. Gen. Genet. 224:* 17.

Hu, S., E. Ohtsubo, N. Davidson, and H. Saedler. 1975. Electron microscope heteroduplex studies of sequence relations among bacterial plasmids: Identification and mapping of the insertion sequences IS*1* and IS2 in F and R plasmids. *J. Bacteriol. 122:* 764.

Hudson, A., R. Carpenter, and E.S. Coen. 1987. *De novo* activation of the transposable element *Tam2* of *Antirrhinum majus. Mol. Gen. Genet. 207:* 54.

Jordan, E., H. Saedler, and P. Starlinger. 1967. Strong-polar mutations in the transferase gene of the galactose operon in *E. coli. Mol. Gen. Genet. 100:* 296.

———. 1968. O^0 and strong-polar mutations in the *gal* operon are insertions. *Mol. Gen. Genet. 102:* 353.

Klaer, R., S. Kühn, E. Tillmann, H.J. Fritz, and P. Starlinger. 1981. The sequence of IS4. *Mol. Gen. Genet. 181:* 169.

Kleckner, N., R.K. Chan, B.K Tye, and D. Botstein. 1975. Mutagenesis by insertion of a drug-resistance element carrying an inverted repetition. *J. Mol. Biol. 97:* 561.

Krebbers, E., R. Hehl, R. Piotrowiak, W.-E. Lönnig, H. Sommer, and H. Saedler. 1987. Molecular analysis of paramutant plants of *Antirrhinum majus* and the involvement of transposable elements. *Mol. Gen. Genet. 209:* 499.

Kuckuck, M. 1936. Über vier neue Serien multipler Allele bei *Antirrhinum majus. Z. Indukt. Abstammungs. Vererbungsl. 71:* 429.

Kunze, R. and P. Starlinger. 1989. The putative transposase of transposable element *Ac* from *Zea mays* L. interacts with subterminal sequences of *Ac. EMBO J. 8:* 3177.

Kunze, R., U. Stochaj, J. Laufs, and P. Starlinger. 1987. Transcription of transposable element *Activator* (*Ac*) of *Zea mays* L. *EMB0 J. 6:* 1555.

Li, M. and P. Starlinger. 1990. Mutational analysis of the *N*-terminus of the protein of maize transposable element Ac. *Proc. Natl. Acad. Sci 87:* 6044.

Malamy, M.H., M. Fiandt, and W. Szybalski. 1972. Electron microscopy of polar insertions in the *lac* operon of *Escherichia coli. Mol. Gen. Genet. 119:* 207.

Masson, P., G. Rutherford, J.A. Banks, and N.V. Fedoroff. 1989. Essential large transcripts of the maize Spm transposable element are generated by alternative splicing. *Cell 58:* 755.

Masson, P., R. Surosky, J. Kingsbury, and N.V. Fedoroff. 1987. Genetic and molecular analysis of the Spm-dependent a-m2 alleles of the maize a locus. *Genetics 177:* 117.

Menssen, A., S. Höhmann, W. Martin, P.S. Schnable, P.A. Peterson, H. Saedler, and A. Gierl. 1990. The En/Spm transposble element of *Zea mays* contains splice sites at the termini generating a novel nitron from a dSpm element in the A2 gene. *EMBO J. 9:* 3051.

Merckelbach, A., H.P. Döring, and P. Starlinger. 1986. The aberrant *Ds* element in the adh1-2F11::Ds allele. *Maydica 31:* 109.

Michaelis, G., H. Saedler, P. Venkov, and P. Starlinger. 1969. Two insertions in the galactose operon having different sizes but homologous DNA sequences. *Mol. Gen. Genet. 104:* 371.

Müller-Neumann, M., J. Yoder, and P. Starlinger. 1984. The DNA sequence of the transposable element *Ac* of *Zea mays* L. *Mol. Gen. Genet. 198:* 19.

Nacken, W.K.F., R. Piotrowiak, H. Saedler, and H. Sommer. 1991. The transposable element *Tam1* from *Antirrhinum majus* shows structural homology to the maize transposon En/Spm and has no sequence specificity of insertion. *Mol. Gen. Genet. 228:* 201.

Nevers, P. and H. Saedler. 1977. Transposable genetic elements as agents of gene and chromosomal rearrangements. *Nature 268:* 109.

Pereira, A., H. Cuypers, A. Gierl, Z. Schwarz-Sommer, and H. Saedler. 1986. Molecular analysis of the En/Spm transposable element system of *Zea mays. EMBO J. 5:* 835.

Pereira, A., Z. Schwarz-Sommer, A. Gierl, I. Bertram, P.A. Peterson, and H. Saedler. 1985. Genetic and molecular analysis of the enhancer (En) transposable element system of *Zea mays. EMBO J. 4:* 17.

Pohlmann, R., N. Fedoroff, and J. Messing. 1984. The nucleotide sequence of the maize controlling element *Activator. Cell 37:* 635.

Ptashne, K. and S.N. Cohen. 1975. Occurrence of insertion sequences region on plasmid deoxyribonucleic acid as direct and inverted nucleotide sequence duplications. *J. Bacteriol. 122:* 776.

Reif, H.J. and H. Saedler. 1975. IS*1* is involved in deletion formation in the gal region of *E. coli* K12. *Mol. Gen. Genet. 137:* 17.

Sachs, M.M., W.J. Peacock, E.S. Dennis, and W.J. Gerlach. 1983. Maize *Ac/Ds* controlling elements—A molecular view point. *Maydica 28:* 289.

Saedler, H. and B. Heiss. 1973. Multiple copies of the insertion-DNA sequences IS*1* and IS2 in the chromosome of *E. coli* K12. *Mol. Gen. Genet. 122:* 267.

Saedler, H. and P. Nevers. 1985. Transposition in plants: A molecular model. *EMBO J. 4:* 585.

Saedler, H., H.J. Reif, S. Hu, and N. Davidson. 1974. IS*2*, a genetic element for turn-off and turn-on of gene activity. *Mol. Gen. Genet. 132:* 256.

Shapiro, J.A. 1967. "The structure of the galactose operon in *E. coli.*" Ph.D. thesis, University of Cambridge, United Kingdom.

———. 1969. Mutations caused by the insertion of genetic material into the galactose operon of *Escherichia coli. J. Mol. Biol. 40:* 93.

Schwartz, D. 1984. Analysis of the *Ac* transposable element dosage effect in maize. *Mol. Gen. Genet. 196:* 81.

Schwarz-Sommer, Z., A. Gierl, H. Cuypers, P.A. Peterson, and H. Saedler. 1985. Plant transposable elements generate the DNA sequence diversity needed in evolution *EMBO. J. 4:* 591.

Schwarz-Sommer, Z., A. Gierl, R.B. Klösgen, U. Wienand, P.A. Peterson, and H. Saedler. 1984. The Spm (En) transposable element controls the excision of a 2-kb DNA insert at the wx m-8 allele of *Zea mays. EMBO J. 2:* 1021.

Starlinger, P. and H. Saedler. 1972. Insertion mutations in microorganisms. *Biochimie 54:* 177.

Sutton, W.D., W.L. Gerlach, D. Schwartz, and W.J. Peacock. 1984. Molecular analysis of *Ds* controlling element mutations at the *Adh1* locus of maize. *Science 223:* 1265.

Upadhyaya, K., H. Sommer, E. Krebbers, and H. Saedler. 1985. The paramutagenic line niv-44 has a 5kb insert, *Tam2,* in the chalcone synthase gene of *Antirrhinum majus. Mol. Gen. Genet. 199:* 201.

Obsession with Sequences

NIGEL D.F. GRINDLEY
Yale University, New Haven, Connecticut 06510

What was it that first attracted me to the field of DNA transposition? In early 1975, immediately after the Asilomar Conference on Recombinant DNA and Its Implications, a conference on Bacterial Plasmids was held at Squaw Valley in California, sponsored by the ICN-UCLA Symposia. As a postdoctoral fellow working on plasmid replication in Bill Kelley's laboratory at Carnegie Mellon University, it seemed to be an obviously relevant meeting to attend and a good opportunity to get together for some premeeting skiing at Lake Tahoe with my old friend David Finnegan (then a postdoc with David Hogness and, along with Gerry Rubin, soon to become intimately involved in the transposition field himself with the first molecular characterization of the *copia* element in *Drosophila*). The dominant memory of the Squaw Valley meeting is of endless discussions of the issues raised at Asilomar and of the proposed restrictions on the construction and use of recombinant DNA. However, I also recollect hearing for the first time about the discovery of transposable drug resistance elements on bacterial plasmids (from talks by Naomi Datta, Fred Heffron, Nancy Kleckner, and Doug Berg) and being infected by the excitement that this had engendered. The discovery was made especially exciting for me since it made immediate sense of a series of intriguing but puzzling observations made in E.S. Anderson's laboratory at the Central Public Health Laboratories, London, where I had done the research for my Ph.D.

Once back in Pittsburgh, I was fully occupied with my project investigating aspects of plasmid replication, but I could not get the transposition phenomenon out of my mind. Shortly before the Squaw Valley meeting, Roger Hendrix at the University of Pittsburgh had introduced me to Joan Steitz, and we had talked about the possibility of my joining her laboratory at Yale, starting in October when I had completed 2 years of study with Bill Kelley but still had 1 year to run of my postdoctoral fellowship. One of the many attractions of Joan's laboratory was its strength in nucleic acid sequencing. Although primarily an RNA laboratory, the new procedures for sequencing DNA were just becoming available in color-coded photocopies from Alan Maxam and Walter Gilbert at Harvard, and Joan was very keen that somebody should get the technology going in her laboratory. Thus, the time seemed ripe for me to ask a biological question that could only be answered through DNA sequencing and, at the same time, to gain expertise in a brand new technology. My thoughts immediately returned to transposition—perhaps by looking at the junctions between transposable elements and their target sites, I might gain some insight regarding the elements themselves and the transposition process.

As materials, I was attracted by the bacterial insertion sequence (IS) elements chiefly for practical reasons. Ideally, what I needed for my proposed project was an abundant supply of three DNA molecules, one carrying a target gene and two containing distinguishable insertions of the same element into this target. The research literature contained several examples of what I wanted—distinct insertions of IS*1*, IS2, and IS*3* in the *lac* or *gal* operons. Moreover, the target genes and insertions were available on specialized bacteriophage λ transducing phages, which could be prepared pure and in high quantity. I picked IS*1* as my first choice not only because of the potential availability from several different sources, but also because IS*1* was known to be at both junctions between the cluster of drug resistance genes (the "r-determinant") and the vector (replication and transfer) portion of a variety of composite drug resistance plasmids, such as R1, R6, and R100, and thus seemed likely to play a role in transposition of resistance genes. In July of 1975, I wrote a letter to Joan describing my ideas in some detail but also stating that I was not irrevocably committed to working on the problem. In fact, by the time I moved to Yale in October, I had become totally intrigued by the behavior of some mutants of a small multicopy plasmid that replicated uncontrollably at elevated temperature, and wondered if I should not sequence some of these (and the wild-type). Only when I found out that Deepak Bastia, a postdoctoral associate of Sherman Weissman in the same department at Yale, was sequencing the replication origin of the ColE1 plasmid did I finally decide on the IS project. Immediately I started trying to get the necessary sets of strains and finally found a willing source in Michael Malamy at Tufts University—soon a set of three strains lysogenic for λd*gal*::IS*1* or λd*gal* arrived.

A Long Learning Curve

The next several months were spent learning and applying the procedures for growth and purification of bacteriophage λ, DNA preparation, etc. Being a complete novice to lambdology (and in a lab of equal novices), I "rediscovered" all sorts of interesting phenomena including the requirement for Mg^{++} to keep phage with oversized genomes from spontaneously lysing—the first time I dialyzed my phage preparations, the two IS*1* derivatives lysed, whereas the wild-type phage with its smaller genome survived intact.

With milligrams of DNA in hand, identifying the relevant portion of the λd*gal* genome became the next task. Those were the early days of restriction enzymes. Few enzymes were commercially available; in fact, by comparison with the riches of today, few enzymes were even known. Fortunately, just a short distance across Long Island Sound at Cold Spring Harbor Laboratory, Richard Roberts was busy collecting and discovering as many restriction enzymes as he could. Even more fortunately, Rich had an "open freezer" policy, making his stock available to those who knew of it and could spend a day or two in his laboratory. Two visits to Cold Spring Harbor resulted in the identification of a few enzymes that generated target sequences and IS-target junctions of a manageable size for attempts at sequencing. These visits also introduced me to Ahmad Bukhari and his colleagues who inhabited the space adjacent to Rich Roberts and were working to unlock the mysteries of the giant transposing bacteriophage Mu. Regrettably, however, I failed to come across Barbara McClintock at that time, nor indeed was I aware of her work, since to me, anything bigger than a bacterium was a closed book.

In the summer of 1976, Ahmad and two of the pioneers in the IS element field, Jim Shapiro and Sankar Adhya, organized a meeting at Cold Spring Harbor devoted to "DNA Insertions." The field of transposable elements had kicked into high gear—a great deal of new information was described and the atmosphere was very exciting. I was somewhat relieved to find no obvious indication that my own project was in danger of being surpassed by competition, although it was clear that Peter Starlinger's group in Cologne was pursuing similar plans. By then I had constructed a reasonably detailed restriction map of the two IS*1* insertions in *galT*, and although I did not give a talk at the meeting, I was asked to submit a manuscript for publication in the book that resulted.

It is hard now, in the era of kilobases-per-day, to understand why it took so long to proceed from a restriction map to a known DNA sequence. All of the DNA fragments were obtained only from the 50-kb λd*gal* transducing phages; pBR322 was as yet unheard of and the idea of subcloning into a small plasmid never arose. Of the seven restriction enzymes used in the sequencing strategy, only three were commercially available (from the infant New England BioLabs). Two of the remainder I made myself, one other was made by Maggie Rosa, a fellow postdoc in Joan's group, and the final one was traded from a neighboring laboratory group. All sorts of baroque procedures were used: reverse phase chromatography on RPC-5 for partial purification of restriction fragments; strand separation of restriction fragments; "wandering spot" analysis (two-dimensional fractionation of partial digestion products using snake-venom phosphodiesterase and DNase I) to confirm sequences at 5′ ends; and even pyrimidine tract analysis to improve confidence in the DNA sequence obtained. The procedures devised by Alan Maxam and Walter Gilbert were still undergoing development and fine tuning; yet to come were the days of thin polyacrylamide gels and ^{25}S-labeling, and I do not think that we ever derived sequences much beyond 100 bases from the labeled end.

Despite the novelty of most of the procedures, the work proceeded reasonably well, punctuated by the occasional setback. Sankar Adhya, with whom I had discussed my research at the DNA Insertions meeting, provided me with a third IS*1* insertion in *gal* which I added to the project. By late 1976, I had obtained sequences at both of the ends of one IS*1* insertion and found that they were nearly identical, with 18 of the 23 terminal base pairs the same. This immediately suggested that both ends were recognized by the same enzyme. Through a bit of bad luck, the first target segment sequenced terminated only seven base pairs from the site of insertion, providing insufficient information to discern much of interest regarding the target and the events accompanying the insertion event. Somehow, the fact that I had the terminal sequences came to be known to the organizers of the DNA Insertions meeting and I was invited to incorporate them into an appendix for the soon to be published book *DNA Insertion Elements, Plasmids, and Episomes*.

Success at Last

Competition was by now starting to heat up. That winter, at a meeting in Jackson Hole, I heard that Eiichi Ohtsubo, at SUNY, Stony Brook, was also sequencing IS*1*. Then, in the spring of 1977, soon after publication of the Cold Spring Harbor book, I got a phone call from Jeffrey Miller at Harvard. Two students working with Jeff, Michelle Calos and Lorraine Johnsrud, were sequencing mutants of the *lacI* gene and had come across two IS*1* insertions. Would I object if they proceeded to sequence these mutants too? We agreed to pursue our parallel courses and keep each other fully informed of progress. By the summer, substantial progress had been made and we both were invited to give short talks at the Gordon Conference on Biological Regulatory Mechanisms. The ultimate goal, however, still proved elusive—neither group yet knew a full set of sequences describing a complete insertion event. Then, just a few days after the Conference I got another call from Jeff—the first *lacI*::IS*1* insertion fully characterized was associated with duplication of a 9-bp target sequence, with one copy at each end of the inserted DNA. I confess that my initial reaction was disappointment; Jeff's group had beaten me to the punch. However, I had data completing a set for a *gal*::IS*1* insertion either under film or even awaiting analysis, and the day after Jeff's call, I was able to call him back confirming the target duplication. By mid fall, all of our data were complete: Michelle and Lorraine had sequenced two different IS*1* insertions in *lacI* and I had data for three insertions in *galT*—all were associated with 9-bp target site duplications. We agreed to publish our findings back to back.

The target duplication was just the sort of result I had hoped for when starting on the sequencing project, since it raised specific questions regarding the mechanism of IS*1* insertion. Two distinct origins of the duplication were immediately apparent: It could be copied from the target during the insertion process or one copy could be brought into the target as part of the incoming transposon. A precedent for the latter possibility was provided by the integration of bacteriophage λ during lysogen formation.

Genetic experiments had long suggested that the phage and bacterial attachment sites each contained a short identical sequence, and DNA sequencing by Wilma Ross and Arthur Landy had shown very recently that this "common core" was a 15-bp sequence. Both phage and bacterial sites carried one copy of this sequence, whereas the inserted phage genome was flanked by direct repeats of it, one at *attL* and the other at *attR*; the analogy to the IS*1* insertions with their flanking 9-bp direct repeats was obvious. However, each of the five insertions sequenced were at different targets, so it was clear that IS*1* did not carry a unique 9-bp "common core" sequence. The diversity could have resulted from the multiple copies of IS*1* in the bacterial chromosome, each of which, through accumulation of mutations, might be associated with a different 9-bp core sequence, or the process of insertion might not demand total identity between target and IS-associated core sequences. I was more attracted by a different possibility, however: Insertion of IS elements did not follow the bacteriophage λ model, but rather used a mechanism that actually copied the short target sequence presumably by making a pair of single-strand breaks at the target, separated by nine base pairs.

Meetings and Models

In the spring of 1978, a meeting devoted to bacterial plasmids was held in Berlin. The opening session was on transposable elements and much had been accomplished since the meeting 2 years earlier at Cold Spring Harbor Laboratory. I was due to give the first talk but nearly missed it (a restless night and jet lag caused me to oversleep and I was only awakened by shouts from the street below my hotel room where the bus to the meeting site was waiting to leave). The description of the consequences of IS*1* insertion at the DNA sequence level was the first such characterization of a transposition event (other than λ integration), although Nancy Kleckner, who had been characterizing Tn*10* insertions at Harvard, received news by phone from her laboratory that Tn*10* also generated 9-bp target site duplications. These molecular details prompted people to think more about possible molecular mechanisms of transposition, and an informal workshop was convened to discuss them. If the short duplications flanking IS*1* were indeed copied from a single target sequence by the "staggered nick" mechanism we had already suggested, just a single strand at each end of the element could be joined to the cleaved target DNA—the other target strand would first have to be replicated before any joining event. David Sherratt, from Glasgow, drawing an analogy to the transfer of single-stranded plasmid DNA during bacterial conjugation, suggested that one strand of a transposon might be transferred from a donor site into a target site. An attraction of this idea was that it would result in duplication of the transposon; replication of bacteriophage Mu was known to be coupled to its transposition, and evidence obtained recently by Fred Heffron suggested that insertion of the transposon Tn*3* was also a duplicative process. We played about briefly with this idea, filling scraps of paper with sketches trying out all the possible DNA strand polarities. The general idea appeared to be plausible. Immediately after the meeting, I visited Cambridge, England, with Doug Berg (my roommate in Berlin) and tried the model out on Sydney Brenner who seemed to be quite attracted by it.

Back in the United States, I made the move from Yale to set up my own laboratory at the University of Pittsburgh. Looming ahead was the Cold Spring Harbor Symposium on Quantitative Biology. Since I had no new data to talk about, I decided to discuss the transposition model and feverishly set to work, putting together convincing drawings and sending drafts of a manuscript back and forth across the Atlantic to Dave Sherratt. Ever since my undergraduate days in the Genetics Department at Cambridge University, I had held the view of Cold Spring Harbor Laboratory as the Mecca of Molecular Biology, with attendance at the annual Symposium as the ultimate pilgrimage. Now my wildest dreams were reality—not only had I been invited to talk, but to my great surprise the model turned out to be a big hit as well. I relished my brief moment of glory.

From IS1 to γδ

The move to Pittsburgh coincided with the end of my dalliance with IS*1*. Although I wanted to pursue the investigation of IS transposition, the multiple copies of IS*1* apparent in all *Escherichia coli* laboratory strains made the isolation of transposition mutants an unattractive approach and I wanted to use genetic methods. Little did I know that the IS*1* transposase would turn out to be *cis*-acting, making my fears groundless. I picked on IS*903* as a similar but more suitable alternative; it too was known to make a 9-bp target duplication, but it had the added advantages over IS*1* in that it was not present in the standard laboratory strains of *E. coli* and was part of a composite transposon associated with a drug resistance gene (for kanamycin phosphotransferase) facilitating selection for transposition events.

In my last few weeks as a postdoc at Yale, the seeds were sown for a project with yet another transposon that has come to dominate my research career. It all started with Mark Guyer who had been a postdoctoral fellow in Norman Davidson's laboratory at the California Institute of Technology, where a DNA element, called γδ was discovered. This element, a component of the F fertility plasmid, was found at the junction between F and host genomic sequences in certain F′ plasmids. Moving to Martin Gellert's laboratory at the National Institutes of Health, Mark had obtained convincing evidence that γδ was a transposon able to form transient cointegrates between F and the nonmobilizable plasmid pBR322. The final consequence of this cointegrate formation was insertion of γδ into pBR322. Mark realized his collection of pBR322::γδ insertions was a valuable resource for sequence analysis of γδ and its insertion process, but he was not interested in the actual sequence determination himself. (Looking back, there is some irony in this, since Mark is currently a senior official in, and cheerleader for, the project to sequence the entire human genome; perhaps the success of our collaboration convinced Mark that he was better at recruiting sequencers than being one of them.) Knowing of my involvement with IS sequencing, Mark contacted me to ask if I was interested in getting involved. I said I was—somewhat—so he sent me the collection of plasmids. It was not long before a fresh young graduate student, newly arrived in Joan's laboratory for a first year rotation, was convinced to sequence the junctions of γδ and pBR322. The student was Randy Reed, and his rotation, in which he managed to show that γδ insertions were associated with duplications of a 5-bp target sequence, was the start of his remarkably successful association with γδ. Subsequently, making *Eco*RI linker insertions into γδ, Randy showed that in addition to the transposase, there was a second γδ-encoded recombination function; in its absence, γδ transposition gave rise exclusively to stable cointegrates and no single insertions were seen. This extra recombinase, later called resolvase, was the product of the *trpR* gene, which in the closely related Tn*3* transposon was known to encode a repressor of transposase but not (at that time) a recombinase. Although Fred Heffron and his colleagues, in a thorough and elegant analysis of Tn*3*, had found that certain deletions within the transposon gave rise to stable cointegrates and had postulated the existence of an internal resolution site, the involvement of the *tnpR* gene product in the resolution process had been missed through a bit of bad luck. For the transposition assay, Fred had made use of the F factor which, through its resident copy of γδ, provided in *trans* a cointegrate resolution function that was able to act on Tn*3* cointegrates as well as γδ cointegrates. Randy went on to locate the resolution site and to develop an in vitro system for the resolvase-mediated site-specific recombination. By then, I was back at Yale as an Assistant Professor and was located in the space adjacent to Joan Steitz's laboratory where Randy was working. We immediately renewed our 2-year interrupted collaboration and I was fortunate to witness the excitement of the unfolding story of resolvase and to play a role in the discovery that when Mg^{++} ions were omitted from the recombination reaction, resolvase cleaved its substrate at the crossover site, becoming covalently linked to the 5′ ends of the DNA.

Epilogue

Looking back now I see my research and accomplishments in the field of specialized recombination as resulting from a series of fortuitous associations with some remarkable scientists. E.S. Anderson, my thesis advisor in London, had sown the seeds for my later interest with transposable DNA by introducing me to the genetics of drug resistance plasmids. Coming to the United States from England resulted from a short visit to Bill Hayes' Microbial Genetics Unit at Edinburgh University and a lucky and enjoyable encounter with Bill Kelley, an American postdoc in the Unit. Once I completed my Ph.D. thesis, Bill invited me to join his laboratory in Pittsburgh, where he taught me biochemistry and enzymology and a real enthusiasm for research. A chance introduction to Joan Steitz by Roger Hendrix took me to Yale. Joan provided the most stimulating research environment I could have imagined; although her research was focused on nucleic acids, notably RNA, she encouraged everyone to pursue his or her own interests. As a result, the members of her group all seemed to work on something quite different—from tRNA in the silk gland of *Bombyx mori* to ribosomal RNA in *E. coli*. Joan supported my foray into transposable DNA with extraordinary enthusiasm, giving me the courage to branch out into something unfamiliar and showing far more confidence in my abilities and judgment than I had myself. Of almost equal importance, Joan was a fully fledged member of the nucleic acid information network. She seemed to know immediately of everything relevant that was going on elsewhere and anything we achieved was quickly made known to the broader community. Joan's friendship with Wally Gilbert gave us instant access to the evolving protocols for DNA sequencing, and Nigel Godson, another Yale friend, told me of Rich Roberts' collection of restriction enzymes and arranged for me to visit Rich's laboratory. Without these two advantages, my initial project would have ended up as one of the dozens of others that later characterized transposon insertions at the DNA sequence level. Later on, my entire involvement with the γδ transposon and its resolvase rested on a chance phone call from Mark Guyer, Randy Reed's initial interest in it for a rotation project, and my subsequent return to Yale from the University of Pittsburgh. Being in the right place at the right time made all the difference.

My research in the mid 1970s proceeded in shameful ignorance of Barbara McClintock, founder of the transposition field. My introduction to the field came entirely through the discoveries made in bacteria, particularly with drug resistance plasmids, and I was too busy, or myopic, to seek out precedents from further afield. The growing popularity of the subject, accompanied by an increasingly high profile, soon changed that, starting with the 1976 DNA Insertions meeting. Even though the meeting was held at Cold Spring Harbor, her own home turf, Barbara remained somewhat reclusive; however, her work received the acclaim that had long eluded her. It was as if the bacterial discoveries, easily demonstrated and assimilated, lent credence to her conclusions borne of the abstruse and complex field of maize genetics and cytology. Barbara was quickly raised to heroine status, although still largely a heroine *in absentia* to most of us. Nevertheless, her founder status was celebrated in my own laboratory by Ted Weinert, my first graduate student at Yale, who decorated his bench with variegated Indian corn (still there today), and we were all delighted on October 10, 1983 to hear the news that Barbara had been awarded the Nobel prize. The award gave our field a new respectability and helped convince others that we were not pursuing an esoteric and irrelevant offshoot of biology. It also gave us bacterial types the "relevance" that seemed more and more to be required for successful grant applications. It is a regrettable sign of the times that today *Ac-Ds* has been supplanted by HIV-IN in our grants (but not in our hearts).

*The Revenge of the Mayans**

GUENTER ALBRECHT-BUEHLER
Northwestern University Medical School, Chicago, Illinois 60611

Couldn't the Swedes tell a traitor when they saw one? By presenting the Nobel prize to Barbara McClintock, the Swedish Academy invited a Trojan horse into the solid edifice of modern biological thought. I am not speaking of the fact that McClintock's instruments of observation were entirely inappropriate. Obviously, she merely used her eyes, a knife, a simple microscope, and a camera, but not a single radioisotope or gel. To decorate such scientists may send the wrong message to the youngsters in science. They may think that a scientist's mind matters or that there is science without sequence homologies. Yet, it does not constitute treason.

I am likewise not overly concerned with the fact that McClintock published practically the entire Nobel work in the yearbook of the Carnegie Institution of Washington, although rewarding such a person sends a dangerous message, too, because the yearbook of the Carnegie Institution, no matter how meritorious, is not peer-reviewed! I submit that one should not encourage behavior like this and instead should show more respect for the feelings of the editors of the "power" journals. Besides, it was not necessary for McClintock to avoid peer review. Even if 30 years ago there were no peers around willing to accept her idea of a maize genome that programmed itself, she should not have taken the easy way out. Like the rest of us, she should have submitted her papers over and over again and waited patiently until someone found the same thing in *E. coli*. Afterward, many converted reviewers would have supported one or the other publication of her data as confirmations of an important *E. coli* finding. Still, publishing in a nonreviewed journal does not constitute treason, either.

But there is a much more serious point. If my sources are correct, Alfred Nobel's testament specified people who had made the most important advancement in science. Science! Remember? As in Aristotle, Galileo, Darwin, Einstein, Curie. They were all Europeans. McClintock is not. She is an American. Sure, there are many great American scientists, but they all seem to be firmly rooted in the European tradition of seeking laws and explanations. Explanations, those wonderful inroads to application, manipulation, even mastering, and finally: Conquest! Ha!

McClintock does not subscribe to this philosophy. She does not want to conquer Nature; rather, she wants to observe her with deep humility, to marvel, to integrate, to grow. She does not believe that something as infinitely creative as Nature obeys any laws or that scientists have ever observed more

*With my apologies to the experts of pre-Columbian history, I have lumped the prehistoric Mexicans, the Mayans, the Toltecs, and the Aztecs into one people and called them all "Mayans."

than limited relationships. She loves to uncover these relationships, though. The more subtle and enigmatic, the better. Irresponsible! She seriously describes our undeniable progress toward the complete mastering of the universe as "fun and games." That is high treason. I know where her attitude comes from. She has been brainwashed by the Mayans, those people who stuck feathers in their hair and awaited the arrival of some plumed serpents. We trusted that our Castilian friends had taken care of the feather people through the superiority of our culture, religion, and muskets. Now they may return riding on the spirit of this woman.

It is not entirely clear how the feather people got hold of her mind, but I suppose that Barbara McClintock would have been an easy prey for any culture that worshipped a plant. With profound religious feelings, the Mayans observed and integrated every phase of the life of maize from its planting, to the breaking of the pericarp, to the harvest, and they celebrated them in special ceremonies. No wonder she was intrigued. More importantly, she may have practically encountered the feather people. After she had conceived the completely novel idea of "cyto-archeology," she began to identify the origin of maize and to track its migrations across Central and South America by studying the geographical distributions of cytogenetic markers. Here, under the microscope and in the form of knobs on chromosomes, the migrations of the Mayans came to life. Somehow at this time, the creators of maize recruited her as the instrument of their revenge.

Today, she carries on their legacy of un-European, unorthodox, and reverent plant breeding that clearly lacked the required agricultural sobriety. Every reasonable civilization bred one of the local grasses into a genetically stable polyploid. The Europeans turned *Triticum m.* into wheat, the Asians turned *Oryza p.* into rice, and the Africans came up with sorghum. What did the American feather people do? They turned *Teosinte* into maize, but hear this: They kept it deliberately unstable. They liked the colorful variegations of the kernels that came up in every crop! I cannot believe this botanical childishness. They kept the creative forces in the genome of their food plant alive. Appalling! That is another of McClintock's problems. She likes creativity. She fosters it and feeds it with her own. In a word, she is dangerous. Worse, she is fearless, independent, and immensely insightful, and she lacks every bit of sentimentality. In all the years I have known her and compared her assessment of scientific, political, social, or cultural situations with mine, we agreed in about 5% of the cases. In all the others, she was right. I resent that in a woman. Of course, I would resent it in a man too, but I never met one who radiated more truth than she, who struck me as equally capable of deception as a forest, a mountain range, or a glacier. On the other hand, it is easy for her to see the truth so clearly. I spotted her cheap trick right away. All she does is observe the world as intensely, oblivious of herself and ready to commit her full power, as a cat stalking a prey. At the same time, she displays about the same degree of personal vanity and desire to impress as a Bengal tiger stirred up by a photo safari.

We must also criticize her for placing altogether too much emphasis on plants. After all, what are we, Dahlias? We are animals, right? Why should we care what shrubs do? All right, there may be one or the other slightly significant contribution of plants to our lives. It may even be a bit disturbing that plants can learn to live just fine without us animals, whereas we would choke and starve without them. This kind of situation could be misinterpreted as a master-servant relationship, where we play the role of the dependents. Fortunately, everybody knows that plants are just dumb, green plastic backdrops of our splendid existence, meant by God to serve us. After all, who has ever heard of a master who has no brain and is forced to stay in one place, while the would-be servants are splendid thinkers and free to move about as they please?

Sure, there are some other mildly redeeming facts about plants. Without the angiosperm, the dry lands on Earth would be just that. There would be no grasses for humans to invent agriculture, settle down, and develop music, cathedrals, and science. In fact, large mammals like us might never have evolved in the first place without the opportunity to eat and store the rich food supply with which the plants equipped their angiospermical embryos. But obviously, plants are too primitive to take proper care for their young, and hence, we had every right to take them for granted, eat them along with their food supply, and become brilliant.

Other than that, plants are outright dangerous. Try to think of any food that makes eating irresistible which is not derived from plants! And as if plant inventions like chocolate (Ah, chocolate!), coffee, wine, and spices would not do enough damage to the neat and trim appearance of human bodies, the plants provide us with addictive substances, too. Keep in mind who seduces our brains with tricky substances like nicotine, cocaine, or morphine! So, let us not be too sentimental about some other slightly beneficial plant products like digitalis, aspirin, and five million other pharmaceuticals. Look at the inner city problems all over the United States and tell me that these problems were not caused by plants!

Plants are no good for research either. McClintock was clearly way off the mark with her continued pleas to use plants for studies of development. Isn't it much more challenging to establish the lineages of cells in animal embryos, where countless cells evade tagging by roaming freely all over the place? It would be far too easy to read cell lineages off the bodies of plants that keep every progeny cell in physical connection with its ancestor and thus turn their bodies into historical records of genetic and developmental events. Only hardship can make us grow, and therefore heroically inclined developmentalists have to reject Barbara McClintock's suggestions.

Most troublesome, however, is the fact that McClintock takes advantage of her recent increase in notoriety and begins to speak of irresponsible concepts like "genome shock" as if it were a fundamental mechanism of evolution. We don't need that. We know exactly how evolution works: It is nothing but random base-pair mutations followed by natural selection. That's all it takes to generate dinosaurs, mimicry, the hammer orchid, and podocytes. Genome shock, really! As if a genome could feel distress, recognize complex problems, seek intelligent solutions, and have a life of its own! Even if she were right 30 years ago, this time she is wrong. Cells, let alone genomes, are not intelligent. I know that for a fact.

Ultimately, I must admit, it is not any particular bit that disturbs us. It is the whole picture that threatens European tradition of science. Minds like McClintock's may have created a whole new attitude. In her footsteps, future scientists may replace dreams of conquest with reverence, prefer the logic of creativity to the logic of analysis, and value integration more than dissection.

We should also be alarmed that McClintock is a woman. Her followers may learn from her to care little about the kind of immortality bestowed by names inscribed in stone which we men covet so much, but feel deeply responsible for everything alive. The very thought makes me cringe.

"Please Come to My Laboratory for Better Coffee, Fresh Orange Juice, . . . Conversation"

BRUCE M. ALBERTS
American Cancer Society Research Professor of Biochemistry
University of California, San Francisco 94143-0448

I cannot remember exactly when I first met Barbara McClintock. I have spent so many profitable hours in her large laboratory room at Cold Spring Harbor Laboratory that, in my memory, many of our meetings have been joined together, as if they were part of the same visit. Barbara once reported to one of my students about me that she had "never met anyone with so little sense of space or time." This statement, made by a person with remarkable gifts for observation, is now engraved on the back of a gift watch from my laboratory; it could certainly explain my complete failure as an historian.

Bruce, Sunday morning.

Please come to my laboratory for

1. Better coffee
2. Fresh orange juice
3. Oatmeal or eggs.
4. Toast or English muffins (or both)
5. Marmalade or honey (or both)
6. Conversation

Barbara

Note from Barbara, sent May 1978, while at Cold Spring Harbor Laboratory.

My own excuse for the memory lapse is that Barbara has not changed at all in the 20 years that she and I have been friends. As I write this, I have just returned from yet another 5 hours of intense conversation with Barbara. This visit was different only because, knowing that I would be writing this brief tribute to her, I asked Barbara questions designed to refresh my memory about past discussions. Whenever we retouched old topics, she rekindled the same intense excitement about science that drew me to her 20 years ago. As before, I left exhilarated. Why? An attempt to answer this question for myself inspires this essay.

Although apparently not attracted to the professional literature of neurobiology, Barbara is fascinated by the brain and what she views as the three-dimensional nature of all thought processes: the dispersed manner in which information is stored, how we sift through this information to come up with interesting new associations and ideas, and how this processing continues "automatically" without our conscious involvement. Many of her opinions about the above topics have been derived from observations of her own mind, which, like most of the rest of biology, Barbara has examined closely and with consummate skill.

In the case of the brain, it is obvious that a large network of complex interactions between many different individual elements (nerve cells) creates something that is very much more profound than the sum of its parts—for example, human consciousness, including the joy that comes from certain types of understanding. Barbara's life has been intimately connected to this fact in several ways. First and foremost, for 70 years she has devoted herself to using her mind to analyze complex patterns, armed with the realization, already clear to her as an undergraduate at Cornell, that she was uniquely gifted at being able to derive new understandings from such efforts.

Barbara is by far the most "inner-directed" person whom I have met. To me, she seems to want nothing more from life than the tremendous pleasure she derives from making new connections, and exercising old ones, in her mind. For about 45 years, this pleasure came from intense, full-time scientific investigation, most of it on the maize (corn) plant. Year after year, 10 months of meticulous planning would culminate in six 7-day weeks of 24-hour days, caring for 3000–5000 carefully crossed maize plants. These efforts led to her pioneering discovery and analysis of transposable elements, for which she was awarded the 1983 Nobel prize for Medicine or Physiology. This physically demanding life-style ended shortly before I met Barbara. But for as long as I have known her, Barbara has derived the same type of pleasure from reading and analyzing the work of others, which is done with nearly the same intensity that she applied to her maize work. Her reading provides her tremendous opportunities for discovery because Barbara carefully selects information for storage, and then lets the new data "reverberate" throughout a carefully prepared mind. The result is often a new connection or insight concerning the nature of living systems—a new node in the very large network of biological information that Barbara has been developing and nurturing for the past 70 years.

What has resulted from all this reading and thinking is a unique, very rich view of the world—a view that is much too rich for a simple linear readout. During one of our conversations many years ago, Barbara made an extremely interesting connection between art, thought, and language. We can all accept that the highly intricate patterns of three-dimensional connections between the nerve cells in our brains form the physical basis for the huge collection of potential thoughts that constitute human consciousness. Barbara suggests that, even without our conscious involvement, these connections are continually being adjusted to create new, more refined arrangements—the driving force being a search for more "interesting" relationships. She views great art as an attempt to mimic the interesting relationships in our brain as a pattern of shapes and colors. The interconversion between complex three-dimensional thought processes and a much simpler two-dimensional painting is necessarily incomplete. But it can be a better representation than language, which can only provide a one-dimensional, linear readout of the actual thoughts in our heads.

This analysis made a deep impression on me for several reasons. First of all, like many scientists, I consciously search for new, interesting relationships in my own work. Where does the strong urge that motivates this search come from? As an undergraduate at Harvard, I was fascinated by a lecture on Chinese poetry by Archibald MacLeish, stressing the pleasure derived from the juxtaposition of two disparate images: poetry as "the space between two distantly related words" (MacLeish 1961). I was therefore ready to accept the notion that humans could be motivated by an attempt to form particularly pleasurable patterns of connections in our brains. Second, this view of consciousness helped to explain the pleasure that I derive from my long talks with Barbara. Although language is inadequate to express what is really in her mind, our long rambling discussions of biology produce numerous one-dimensional hints of the fascinating way in which Barbara has organized her image of the world. Reconstructing the image from these hints, like filling the space between the words of a Chinese poem, helps to explain the intellectual stimulation that often survives afterward for days.

What have I learned? When I met Barbara, I was a chemist who viewed cells as little more than a test tube full of a complex mixture of proteins and nucleic acids. Barbara therefore had a lot of biology to teach me—about the development of the maize plant, Queen Anne's lace, ladybird beetles, plant galls, stick insects, and mimicry in plants and animals, to name a few of the major topics that we have touched on. But most of all, I think that I have begun to see what Barbara means by "a feeling for the organism." Barbara looks at living cells and organisms differently than anyone else I know. From her, I have come to realize that sophisticated groups of interacting molecules can have "system properties" that are hard to decipher from the sum of the individual parts. Many of us would agree that consciousness is a qualitatively new property that emerges from a complex network of brain cells. By analogy, Barbara would argue that the result of the complex network of interactions between the many cytoskeletal proteins in the cell cytoplasm, to pick but one example, may not be understandable without new paradigms for dealing with the system as a whole. She has reached this conclusion in part from her careful observations of the global response of a living cell to a small local perturbation (McClintock 1984). More revolutionary is her analogous view of organisms. Here, the interacting units are cells instead of molecules, and included among the "system" properties that Barbara is intrigued with is the ability of tissues and organisms to rearrange from one ordered functional state to another, with only a small perturbation in the cells that form them. Such an ability would obviously have major evolutionary implications, indicating that organisms have been formed in a way that allows mutation to cause sudden changes in their shape and other properties. Barbara is fascinated by this idea, which explains many of her particular interests in biology.

Consider plant galls, for example. Galls are abnormal outgrowths of plant tissues caused by certain parasitic insects, fungi, and bacteria (Meyer 1987). In response to a chemical signal from the parasite, the plant forms a capsule that encloses it. The structure formed by the plant is different for different parasites, can be complex, and is not part of the normal plant repertoire. Barbara takes this as evidence that the plant is preprogrammed in a way that allows its various tissues to form a variety of new and potentially useful arrangements after a change in gene expression caused by simple signals. A plasticity of this type is presumably present in organisms because it was selected for during the course of evolution: Systems of interacting cells with this property have been the favored survivors of natural selection.

In Barbara's view, a hint that animals, as well as plants, have this property comes from studies of mimicry, wherein (for example) a change in one or a few genes allows a species of butterfly to adopt the different coloration and wing patterns of an unrelated poisonous species, when it invades a geographical region containing the second species (Wickler 1968).

It would be dishonest to imply that Barbara's many friends only derive intellectual stimulation from her during their visits to Cold Spring Harbor. Barbara is also inspiring in a second way. She reminds us of the pure curiosity and excitement about science that we all experienced as young students, and she shows by her example that it is possible to maintain this curiosity throughout one's lifetime as a scientist. I return to the ever-present pile of mail to be answered and forms to be filled out much more impatient with these distractions, and ever more determined to reverse the seemingly inevitable march from scientist, to scientist-administrator, to dean. For this, no less than for all the rest, Barbara, I thank you.

References

MacLeish, A. 1961. *Poetry and experience.* Houghton Mifflin, Boston.

McClintock, B. 1984. The significance of responses of the genome to challenges. *Science 226:* 792.

Meyer, J. 1987. *Plant galls.* Gebrüder Borntraeger, Berlin.

Wickler, W. 1968. *Mimicry in plants and animals.* McGraw-Hill, New York.

Transposable Elements (Ty) in Yeast

GERALD R. FINK
Whitehead Institute for Biomedical Research and Massachusetts Institute of Technology, Cambridge, Massachusetts 02142

Each spring in the early 1970s heralded a visit to Cornell University by Barbara McClintock. Barbara had been appointed an Andrew D. White Professor, which meant that she spent a week or two in residence on the Ithaca campus every year for 5 years. As I recall, the only major condition of her professorship was that she present an annual lecture, but Barbara plunged immediately into the intellectual and social life of the campus each time she appeared.

She seemed to relish talking to students and faculty—slots in her busy schedule were full from the day she arrived. Her small office was down the hall from my laboratory in Bradfield Hall. I often watched her in animated discussions with graduate students on topics ranging from current research, to politics, to the latest books on her expansive reading list (in those days, she was particularly keen on publications in the field of acupuncture). Although she was seventy, her energy never flagged. Her gusto and vivacity were contagious and seemed to provide momentary rejuvenation to even the most lifeless graduate student. She emerged from her office at the end of a busy day ready for the evening's festivities. She spent most of her annual stipend on a party for the students, and rumor had it that she gave whatever was left over to the Cornell alumni fund.

When Barbara was not in her office, she often could be found wandering along the rim of one of the spectacular gorges that slice through the Cornell campus on their way down to Lake Cayuga. Barbara had spent both her undergraduate and graduate student years at Cornell, so she knew her way around. She seemed to find pleasure in walking through the woods and ambling through libraries that she had frequented 50 years earlier. She also took an intense interest in campus politics, especially the stormy political events surrounding the Vietnam debacle. It was the early 1970s, the days of the armed takeover of Willard Straight Hall (the student union), which was pictured on the cover of *TIME* magazine. Barbara spent hours on the steps of "the Straight" debating with the various political factions that had set up booths to politicize both students and faculty. She returned from these forays full of stories about her encounters. She often expressed surprise about how little the various insurgent groups had read about the subjects they were promoting—revolution, civil disobedience, pacifism, and nihilism.

I will never forget the anxiety and trepidation that preceded my own first encounter with Barbara; little did I know that she would become a constant source of inspiration and encouragement for more than 20 years. I was scheduled to meet with her at the end of her first week in 1970. I vacillated for days about whether I should talk about my work or hers. The possibility for disaster was great. I had just come to Cornell as an assistant professor, what did I have to say to the great McClintock?

Intrigued by gene regulation, I was trying to determine whether the elegant regulatory mechanisms worked out by Jacob and Monod for the catabolism of lactose in *Escherichia coli* applied to yeast. Did the bacterial operons, operators, and repressors exist in yeast? Since *Saccharomyces* does not utilize lactose as a carbon source, I had turned my attention to histidine biosynthesis to test whether the insights from *E. coli* could be applied to eukaryotic systems. Much of my work entailed either isolating *his4* mutations or purifying the HIS4 protein, because this system seemed to be the most likely to represent the eukaryotic equivalent of a bacterial operon. Since I had just set up my laboratory, I did not have much in the way of results. Mostly I had technical problems, mapping discrepancies in the fine structure of *HIS4* and difficulties in purifying the HIS4 enzyme. How could I waste our distinguished visitor's time with these mundane details?

The other alternative was to eschew my own work and focus on McClintock's work. This alternative was also not very appealing. The truth is that I did not feel sufficiently confident to go into any detail about transposable elements. Educating oneself about McClintock's work in 1970 was not an easy task, even though she had published the critical experiments more than 20 years earlier. Her seminar on "Maize Controlling Elements" delivered at the beginning of that crucial week did not raise my confidence. Although I got the take-home message, I had difficulty understanding the critical experiments that led her to her conclusions. My presumption was that I had lost the thread of the argument because of my ignorance of maize biology.

The narrative literature of that era was not much help either. Neither textbooks nor reviews provided the foundation necessary to understand Barbara's work on transposable elements. The best genetics text of that time, *General Genetics* by Srb, Owen, and Edgar (copyright 1965), included her work in a chapter entitled "Extrachromosomal and Epigenetic Systems." This chapter presented a grab bag of nonconventional systems of inheritance that had been reported for many different organisms (the petite mutation in yeast, episomes in bacteria, chloroplast inheritance in *Chlamydomonas*, and cytoplasmic states in *Paramecium*). Including Barbara's work among the genetic curios of the time only heightened the potential for misunderstanding—some of the systems were clearly nonchromosomal and others were so poorly investigated that they could not be fit into any consistent body of information. *General Genetics* did emphasize two critical points: Barbara's system contained two controlling elements (*Ac* and *Ds*) and "certain controlling elements characteristically undergo transposition from one location in the genome to another."

It was clear that I would have to go back to Barbara's original papers to make any further progress. I decided to try to understand one aspect of the work that had been puzzling me for years. I had always been confused by the instability of the elements; they seemed to be like Mexican jumping beans. How did they stay still long enough for Barbara to map them? A careful reading of her papers revealed that, like all good geneticists, she had reliable tester stocks, some lines containing only *Ds* and others containing *Ac*. Since *Ds* could not be activated without *Ac*, a line without *Ac* contained *Ds* in a specific chromosomal location. This *Ds* could be activated by intercrosses with an *Ac* line. In the presence of *Ac*, *Ds* had a high tendency to cause chromosomal breakage at the *Ds* site. This site, the *Ds* chromosomal location, could be identified either cytologically or genetically. Although today this may not seem to be a very profound understanding, at the time, it provided me with sufficient ballast to make a decision. With my confidence bolstered by my newly acquired understanding of *Ac/Ds*, I decided that I would discuss Barbara's work instead of my own.

On the day of the meeting, the discussion started out quite well. To my delight, Barbara acknowledged that I had acquired a basic understanding of the relevant corn genetics. As an apparent reward for my achievement, she pulled out exquisite photographs of maize chromosomes. (To my knowledge, these remarkable photomicrographs have never been published.) She then began to regale me with some of the intricacies concerning chromosome movement. She started to go into high gear and for a short while I actually kept up. Eventually, however, I was completely lost. In a moment of exhaustion, I blurted out a polite, "Yes, I think I understand." She winked at me and said, "You think you do, but you don't." Then she rescued me with the offer, "Let's discuss your work."

Despite my trepidations about discussing my work on histidine regulation, Barbara seemed to be really interested. I do not remember the details of our conversation, but something I said must have led her to believe that histidine regulation bore some relationship to her work on the controlling elements in maize. Barbara was convinced that the maize elements were intimately involved in regulation of corn development and so she was intrigued by any insights into gene regulation. I was flattered and relieved to get out of corn and back into yeast. I went on for a while, but suddenly, whatever magic I had used to capture her interest vanished. Barbara had listened patiently, but once she realized that I was mucking around in biochemistry and physiology and had no information on the control regions of the genes I was studying, she lost interest in the details. Yeast genetics and molecular biology simply did not have the resolution in those days to provide any meaningful information about promoters and operators.

Barbara then asked me a question that would have profound consequences for my work for the next 10 years: "Do you think that there are transposable elements in yeast?" At first, I stuck to the prejudice of the time—I thought that transposable elements, or "McClintock elements," were something special to corn, so I never even thought about such things in yeast. Unperturbed by my provincialism, she asked me whether I had come upon any histidine mutants that were unusually unstable. Although we had over 200 *his4* mutations, none was particularly unstable. She asked how we had made the existing set of mutations. When I told her that we had made most of them with ethylmethane sulfonate she laughed. "Oh no, no," she said. "You don't want to look for transposable elements with chemical mutagens, look with X-rays or find spontaneous mutations. If you want to find transposable elements, find unstable mutations."

Barbara's prescription for finding yeast transposable elements presented several technical problems. Both X-ray-induced and spontaneous mutations were extraordinarily rare in yeast. Indeed, we had never found a *his4* mutant of either type. Fortunately, just at this time, Dick Snow (Snow 1966) at Davis worked out a procedure that allowed us to kill off the wild-type yeast cells in a population, thereby permitting a screen for spontaneous histidine auxotrophs. Using this procedure, my technician Cora Styles and I devoted ourselves to accumulating spontaneous *his4* mutants. Although the Snow procedure worked, it required enormous effort to collect a large number of independent spontaneous histidine auxotrophs. To increase our output, we enlisted several unsuspecting undergraduate students and ultimately had four people involved in the search. After about 2 years, we had accumulated over 1500 spontaneous and independent histidine auxotrophs. We then began the laborious task of sifting through them to find any that were particularly stable or unstable.

Mutant hunts are fun, but I must admit that when none of the first 400 or 500 spontaneous mutants in our collection showed any of the desired characteristics, I became discouraged. This initial group of spontaneous mutants was no different from the set we had obtained so easily by chemical mutagenesis. I was ready to give up, but Cora, blessed by an incomparable zeal for thoroughness, systematically sifted through the next 1000 strains. When we reviewed the new data, two of the 1421 mutants analyzed seemed to stand out from the rest. These two, *his4-912* and *his4-917* (the 912th and 917th among the strains we had obtained), reverted at frequencies 1000-fold higher than those of any other mutants in our collection. In 1975, after 3 years of analysis, we had uncovered only two unstable *his4* mutations. My elation at our good fortune must seem risible when viewed by today's standards, but for mutant hunters of the time, it was not so bad—we had two curiosities in the bag.

Given the limitations of yeast molecular biology in the early 1970s, we had only a few ways to characterize our new alleles. Fortunately, we had a relatively good genetic map of *HIS4* and a collection of nonsense alleles from which we had deduced the direction of translation of *HIS4*. By 1977, our presumed body of knowledge about *his4-912* and *his4-917* consisted of the following: (1) They were extremely unstable; (2) they mapped at the beginning of the *HIS4* gene, "proximal" in genetic terms to any of the known structural gene mutations; (3) they reverted to growth on minimal medium, but neither *his4-912* or *his4-917* ever gave rise to true wild-type His^+ revertants (indeed, many of the revertants grew slowly on minimal medium at 36°C and not at all at 23°C); and (4) the His^+ revertants themselves were unstable and gave rise to revertants that grew yet better in the absence of histidine.

In the next few years, we were able to show that *his4-912* and *his4-917* were indeed mutations caused by transposable elements. What is amazing in retrospect is that I was able to convince graduate students and postdoctoral fellows to commit so much time and effort to these two mutations when we knew so little at the beginning. In those days, however, genetic instability meant a lot. Moreover, as each new piece of information came in, the evidence that *his4-912* and *his4-917* were something special became more compelling—I had new bait for the hook to catch the next unsuspecting student or postdoc. For example, Debbie Chaleff, a graduate student in the laboratory, performed a heroic genetic analysis to show that some of the revertants of *his4-912* were chromosomal aberrations, specifically translocations and inversions.

The appearance of chromosomal aberrations among the revertants buttressed my notion that these two mutations were like McClintock's unstable elements. By this time, Barbara had become like the Oracle at Delphi for my laboratory. I would call her up at Cold Spring Harbor and relate our most recent intimations of transposition in yeast. When Barbara heard about the chromosomal aberrations, she became certain that *his4-912* and *his4-917* were caused by insertion elements. She encouraged me to send students down to Cold Spring Harbor to discuss these genetic anomalies in more detail. During the 3 weeks that I taught the Cold Spring Harbor summer Yeast Course, we had many discussions about my unstable mutants. Although she made numerous suggestions, the one we pursued was to look for newly arising mutations among the His^+ revertants of *his4-912* and *his4-917*. Her thought was that the extreme instability of these strains represented hops out of *HIS4* and into new locations in the genome. Perhaps we would find new mutations associated with these hops and thereby provide direct evidence for the transposition idea.

Cora and I spent a considerable amount of time looking for new mutations among the revertants, but without any success. I was rather disheartened by this result and not very optimistic about the future of these unstable mutants. The limitations of classical gene manipulation in yeast gave us only shadow evidence that all of these "traits" were a consequence of transposable elements. In the absence of cytology and recombinant DNA methods, I was left with the disappointing possibility that I would never "see" the entities behind these Platonic images. In fact, I had begun to write the whole story up as the weird behavior of "putative promoter mutants in yeast." Just in time, the tools necessary to solve the puzzle appeared on the horizon.

First, Albert Hinnen, a postdoctoral fellow in my laboratory, discovered yeast transformation. Shortly before he left to return to Switzerland, he presented me with a clone of the yeast *HIS4* gene that he had obtained by brute force screening of an integrating library in *Saccharomyces* (plasmids with sequences that permitted autonomous replication in yeast did not yet exist). Of course, once we had the *HIS4* gene, it was possible to use the new techniques of recombinant DNA to examine the molecular structures of *his4-912* and *his4-917*. Although the path seems pretty straightforward today, the route did not seem clear at the time. It is difficult to convey how important the success of each experiment was in goading us on to try the next, more speculative step. Shirleen Roeder, a new postdoctoral fellow in the laboratory, quickly showed that the *his4-912* mutation resulted from transposition of Ty1 (Roeder and Fink 1980), a family of repetitive yeast DNA sequences first identified in Ron Davis' laboratory (Cameron et al. 1979). This finding spurred Tom Donahue and Phil Farabaugh to sequence the *HIS4* gene. Once he had the wild-type sequences, Phil went after the *his4-912* sequences. On the basis of the comparison between the wild-type sequence and that of the *his4-912* mutant, it became clear that *his4-912* resulted from insertion of Ty1 into the 5′-noncoding region of *HIS4* (Farabaugh and Fink 1980).

The Ty elements, like the newly discovered bacterial transposons, made duplications of the target DNA that flanked the insertion element. Because we had isolated *his4-912* as a spontaneous mutation from an isogenic wild-type strain with an unaltered *HIS4* sequence, it was clear that these mutations represented authentic transpositions of Ty into the *HIS4* region. Our next goal was to make the Ty elements move. Barbara's intuition that the instability of *his4-912* and *his4-917* was due to subsequent transposition was not correct. Shirleen had shown that most of the instability was a consequence of recombination between the direct repeats at the ends of the element (called δs), which caused the internal Ty sequences to pop out, leaving behind a single δ. Several of the postdoctoral fellows in the laboratory took a crack at finding subsequent transpositions of *his4-912* and *his4-917*, but even the heroic efforts of the first-rate geneticists Forrest Chumley and Fred Winston were to no avail.

In 1983, two new postdoctoral fellows, Jef Boeke and Dave Garfinkel, arrived in the laboratory and dedicated themselves to figuring out how to mobilize Ty elements. For 2 years, nothing they tried worked. Temperature shifts, ionizing radiation, exotic genetic backgrounds—none of these treatments roused the slumbering Ty elements. I was beginning to worry about their careers and tried to persuade them to shift to different projects so that they would have something to talk about when they hit the job market. Fortunately, they did not listen to me; they persisted and quite suddenly achieved success.

The technique they devised to mobilize a Ty element was to replace the Ty promoter with the controllable *GAL1* promoter in such a way that the Ty RNA could be induced to a high level by galactose (Boeke et al. 1985). They were very lucky and picked a "live" Ty for this construction (most of the Ty elements contain mutations that prevent transposition). This *GAL1*-promoted Ty transposed at high frequency on galactose. Once they had the transposition process under control, they rapidly completed a wonderful set of experiments showing that the Ty elements are retrotransposons, functionally similar to animal retroviruses, and transpose via an RNA intermediate (DNA→ RNA→DNA). Within a few months, they found that cells induced for Ty transposition accumulate a Ty-encoded reverse transcriptase as well as virus-like particles (Ty-VLPs; Garfinkel et al. 1985; Fink et al. 1986).

The last year that I taught the Cold Spring Harbor Yeast Course (1987), I presented the transposition work in the course lecture series. I wanted this to be an especially clear lecture because I knew that Barbara would be in the audience. Over the years, Barbara and I had developed such a close friendship that despite her other commitments, she was always present at my annual talk in the Yeast Course. A sure sign of her fondness for me was that she attended even when the focus was the biochemistry and physiology of histidine biosynthesis. But that year it was different; I could repay her fidelity with a talk on transposable elements. At the end of the lecture, as Barbara and I were walking back from the lecture hall along Bungtown Road, she began pressing me. "How do the Ty elements move?" she asked. At first, I thought that my lecture had been unclear, so I tried to restate what I had said earlier. I said, "We know the mechanism of movement, it is retrotransposition." But Barbara's only response was the now familiar challenge: "You think you do, but you don't."

Barbara had assimilated the mechanistic details from the talk, but her question went deeper. She recognized that we had mobilized the Ty elements artificially and she was asking about the "natural" inducers of Ty transposition. Only by identifying these natural inducers could we understand the importance of Ty elements in the biology and evolution of yeast. Although Ty elements have been studied intensively since then (Boeke and Sandmeyer

1991), we still do not have the answer to her question. Thus, Barbara has flung down the gauntlet for a whole new generation of yeast biologists.

References

Boeke, J.D. and S.B. Sandmeyer. 1991. Yeast transposable elements. In *The molecular and cellular biology of the yeast* Saccharomyces: *Genome dynamics, protein synthesis, and energetics* (ed. J.R. Broach et al.), vol. 1, p. 193. Cold Spring Harbor Laboratory Press, Cold Spring Harbor, New York.

Boeke, J.D., D.J. Garfinkel, C.A. Styles, and G.R. Fink. 1985. Ty elements transpose through an RNA intermediate. *Cell 40:* 491.

Cameron, R.R., E.Y. Loh, and R.W. Davis. 1979. Evidence for transposition of dispersed repetitive DNA families in yeast. *Cell 16:* 739.

Farabaugh, P.J. and G.R. Fink. 1980. Insertion of the eukaryotic transposable element Ty*1* creates a 5-base pair duplication. *Nature 286:* 352.

Fink, G.R., J.D. Boeke, and D.J. Garfinkel. 1986. The mechanism and consequences of retrotransposition. *Trends Genet. 2:* 118.

Garfinkel, D.J., J.D. Boeke, and G.R. Fink. 1985. Ty element transposition: Reverse transcriptase and virus-like particles. *Cell 42:* 507.

Roeder, G.S. and G.R. Fink. 1980. DNA rearrangements associated with a transposable element in yeast. *Cell 21:* 239.

Snow, R. 1966. An enrichment method for auxotrophic yeast mutants using the antibiotic 'Nystatin.' *Nature 211:* 206.

Controlling Elements, Mutable Alleles, and Mating-type Interconversion

IRA HERSKOWITZ
University of California, San Francisco, California 94143-0448

My first awareness of Barbara McClintock and controlling elements came from studies at the Oshima laboratory (Oshima and Takano 1971; Harashima et al. 1974) which concerned the enigmatic phenomenon of homothallism and mating-type interconversion in yeast. Just as "unstable genes" and "mutable alleles" had provided a challenge to geneticists of earlier eras (Demerec 1935; McClintock 1946, 1948), so did the bizarre genetic properties of the mating-type locus provide a challenge to us (Herskowitz 1988).

Genetic Instability at the Yeast Mating-type Locus: HML and HMR Are Proposed to be Controlling Elements

Strains carrying the *HO* allele, one of two natural alleles (the other being the inactive, *ho* allele), exhibit a high frequency of mating-type switching: they change from one mating type to another (**a** to α or α to **a**) (for review, see Herskowitz and Oshima 1981). This switch in phenotype results from a change in the mating-type locus itself: from *MAT***a** to *MAT*α or from *MAT*α to *MAT***a** (Hawthorne 1963b; Takano and Oshima 1970). The change is remarkably frequent—as often as every cell division in competent cells (Strathern and Herskowitz 1979). The switching process also occurs in *ho* strains, but at much lower frequency. In their 1974 paper, Harashima et al. summarized the genetic data for three loci distinct from the mating-type locus that govern mating-type interconversion. All of these loci were defined by naturally occurring alleles. *HO* was necessary to switch in both directions (from to **a** to α and from α to **a**); the gene now known as *HML*α was required to switch from **a** to α, and the gene now known as *HMR***a** was required to switch for the reverse situation, from α to **a** (Table 1). What are the functions of *HO* and the *HM* loci?

Harashima et al. (1974) proposed that "the elementary structure of the mating-type locus for both **a** and α alleles is essentially the same" and that "the association of some kind of controlling element with this locus would cause the differentiation of two mating-type alleles. The mating-type locus on chromosome III would act as an affinity site for a controlling element." It was furthermore proposed that "the *HM*α and *HM***a** genes produce the specific controlling element; the association of an *HM*α element with the mating-type locus would form the **a** mating-type allele and the association of the *HM***a**

Table 1 Genes Required for Mating-type Interconversion

Gene name[1]	Required for switching from **a**→α	α→**a**	Gene function
HO	yes	yes	site-specific double-strand endonuclease; cleaves *MAT*
*HML*α [*HM***a**]	yes	no	silent version of *MAT*α information; an "α cassette"
*HMR***a** [*HM*α]	no	yes	silent version of *MAT***a** information; an "**a** cassette"

[1]Original nomenclature of Harashima et al. (1974) is in brackets. The *HML* locus is on the left arm of chromosome III and harbors an α cassette in standard strains; the *HMR* locus is on the right arm of chromosome III and harbors an **a** cassette in standard strains (see Herskowitz and Oshima 1981).

element with the mating-type locus would give rise to the α mating-type allele" (see Table 1).

Controlling elements had been mentioned in an earlier paper by Oshima and Takano (1971): "Our observations, however, do show some similarities to the specific transposable genic controllers of mutation in maize both in their locus-specific action and in their mode of modulation. McClintock (1957) has called these genic units 'controlling elements'." What did Oshima and Takano mean by "controlling elements"? They noted that "Similar genic elements have been suggested in both eukaryotes such as *Drosophila* (Green 1967, 1969a,b) and in prokaryotes such as *Salmonella* (Dawson and Smith-Keary 1963)." Although the molecular nature of controlling elements was not evident to us, it was clear that they were transposable genetic elements, capable of being moved from one position to another.

At the time, of course, what we knew about controlling elements of maize and mating-type switching in yeast was based exclusively on genetic analysis; none of the components had been cloned and characterized in molecular terms. There were, however, analogies that stood out at the time and that remain. The strongest analogy is that the mating-type locus is an "unstable gene"—*MAT***a** and *MAT*α are "mutable alleles"—in the same sense that the standard a_1 allele of maize (Rhoades 1938) is a mutable allele (Table 2). A further important analogy is that *MAT* and a_1 alleles are mutable only in the presence of specific unlinked loci, *HO* in yeast and *Dt* in maize (Rhoades 1938). We now know from Barbara McClintock's work that the mutable alleles of maize arise because of genetic instability at that locus. Furthermore, she showed that in two-component controlling element systems, one element could act in *trans* to bring about the instability of the mutable allele (McClintock 1948). Thus, *Ac* controls *Ds*, *Dt* controls the target element in the a_1 allele, and *HO* controls *MAT* (Table 2).

We now know that *HO* codes for an endonuclease that catalyzes the genetic change at MAT (for review, see Strathern 1988): It cleaves the mating-type locus, which first results in removal of the existing information (for example, *MAT***a**). The mating-type locus is then repaired using homologous DNA sequences elsewhere in the genome (near the ends of the chromosome), which results in creation of a *MAT*α allele. This process is thus a complex genetic

Table 2 Mutable Alleles and Regulators of Their Mutability

Mutable allele	Regulator
*MAT***a** or *MAT*α	*HO*
a_1 (standard allele)	*Dt*
$c^{m\text{-}1}$ (*Ds* insert)	*Ac*

rearrangement in which a block of information (a segment of 747 bp from *HML*α or a segment of 642 bp from *HMR***a**; Astell et al. 1981) is transposed to the mating-type locus, where it replaces the information that was previously resident. The information removed from the mating-type locus is lost forever (see, e.g., Rine et al. 1981).

A Variety of Possible Models for Mating-type Interconversion

Oshima and Takano's proposal that *HML*α and *HMR***a** are controlling elements and that their association with the mating-type locus determines cell type could be interpreted in a number of different molecular models. Several of these were based on the idea (Hawthorne 1963a) that the mating-type locus contains regulators for both **a** and α cell types and that mating type was determined by which of these regulators was used. One could imagine that these regulators were adjacent to each other and that the DNA region between them determined which regulator was used. One possible role for *HML*α and *HMR***a** is that they are transposable regulatory sites (containing a promoter or enhancer) that can integrate at the mating-type locus and thereby direct transcription leftward or rightward (Rine et al. 1979). McClintock (1957) has described maize controlling elements in similar terms.

Several other models are also possible. In one imaginative model proposed by Don Hawthorne (D. Hawthorne [unpubl.] cited in Holliday and Pugh 1975), it was hypothesized that the regulatory region of the mating-type locus contains a block of C residues that is subject to deamination and thereby conversion to T residues, which would lead to a change in transcription of the regulators at *MAT*. (A variation on this scheme involving DNA methylation is described by Holliday and Pugh [1975].) In this scenario, *HO* coded for the deaminating enzyme. The *HM* loci could be accessories or catalyze the reverse reaction. In another model (Hicks et al. 1977), the region between the **a** and α regulators was subject to inversion, and *HO* and its accessories *HML* and *HMR* were responsible for this site-specific inversion. In all of these models, the elementary structure of the mating-type locus is identical in both **a** and α cells.

The Controlling Elements of Mating-type Interconversion Are Genetic Cassettes

The answer to the structure of the mating-type locus and the nature of mating-type interconversion turned out to be "none of the above." *HML*α and *HMR***a** were uncloaked to be *coding information* for regulators of mating type. They were *structural genes*, not transposable regulatory sites, DNA-modifying enzymes, etc. In their natural position, at *HML* or *HMR*, these blocks of information were silent; we called them "genetic cassettes" (Hicks et al. 1977) because they were expressed only when they were plugged into the playback locus, the mating-type locus. Work from my laboratory (Kushner et al. 1979; Rine et al. 1979) and others (Hicks et al. 1979; Klar and Fogel 1979; Klar et al. 1979; Nasmyth and Tatchell 1980) confirmed this hypothesis. Thus, the "controlling elements" of Oshima proved to be coding information, "genetic cassettes" (Herskowitz and Oshima 1981).

There are numerous differences, of course, between the transposition of maize controlling elements and the transposition that occurs in yeast mating-type switching. The event catalyzed by *HO* is stereotyped: The same event occurs over and over again (a directed transposition of information from *HML* or *HMR* to one spot in the genome, *MAT*). In contrast, movement of *Ds* or other elements can occur at random or at least to many different sites (McClintock 1957). Second, the genetic rearrangement process in mating-type interconversion leads to activation of ordinarily cryptic information. I do not know of any examples in which an analogous situation occurs in *Ac* or *Ac-Ds* systems, although Barbara probably does. One could imagine that an *Ac* element has inserted into a chromosomal position where its transposase is no longer expressed because of silencing due to its neighborhood (the type of silencing that is exerted on the *HML* and *HMR* loci by the Sir proteins; Brand et al. 1985; Schnell and Rine 1986). This inactive *Ac* might have the phenotype of a *Ds* element: unable to move on its own but subject to movement by presence of a functional *Ac* element elsewhere in the genome.

The Yeast Ty Element Has Strong Analogies to Maize Controlling Elements

Although the focus of this paper is on mating-type interconversion, some discussion of another transposable element of yeast, the Ty element, is warranted. These are retrotransposable elements (Boeke et al. 1985) with the remarkable property that their insertion adjacent to another gene can cause the gene to come under special types of regulation. For example, association of a Ty1 element adjacent to *CYC7* can place this gene under control of a cell-type-specific regulator: Under such circumstances, transcription of *CYC7* is high in **a** and α cells but low in **a**/α cells (Errede et al. 1980). We now know that the Ty1 element provides some type of transcriptional enhancer to boost transcription of the *CYC7* gene. More importantly, the element contains a recognition site for the **a**/α-cell-specific regulatory protein, **a**1-α2 (Errede et al. 1987). We see from this example that association of a Ty1 element could, in principle, place many genes under control by **a**1-α2. Although the dozen or so genes (Herskowitz 1989) regulated by **a**1-α2 do not contain any homologies with the Ty1 element outside of the **a**1-α2-binding site, one can speculate (Errede et al. 1980) that these binding sites were "dropped off" by a Ty element. The effect of the Ty1 element on the *CYC7* gene provides a precise analogy to things that Barbara was describing, for example, in her 1961 paper, "Some parallels between gene control systems in maize and bacteria." Here, she discusses the fact that controlling elements were originally identified by their transposition but that "an element previously exhibiting transposition may become fixed in location" and continue to provide regulation of that gene.

The yeast analog to this statement is that **a**1-α2-binding sites might have been distributed throughout the genome by a Ty1 element which is no longer present or recognizable at that spot. Although Barbara was interested in transposition mediated by the *Spm* element (the "mutator" component of the element), I can see that another reason for her fascination with this element was the "suppressor" component (impressively displayed in Figure 2 of McClintock 1965): This was the part of the element that provided the *trans*-acting regulator of expression of the target gene (the analog to **a**1-α2).

Learning More About Controlling Elements and Barbara McClintock: Both Capable of Great Leaps in Genetics

Having seen that our studies of yeast mating-type interconversion had brought us into the "business" of genetic rearrangement in a process in which controlling elements were invoked, I set out over a few years to learn more about them. I first consulted my father's genetics text (Herskowitz 1973), which was one of the only genetics texts that dared to present this material. My father was familiar with Barbara and her work, having taken her course at the University of Missouri in 1941 as part of his M.S. degree studies. He describes her as an extremely conscientious teacher who underlined the key passages in the reading assignments to aid the students in the class (Irwin Herskowitz, paternal communication). It has often been said that people did not understand Barbara's work. My father has always maintained, however, that the genetics community was well aware of Barbara's work demonstrating the existence and providing characterization of transposable genetic elements. Larry Sandler (pers. comm.) said that Barbara "was invited to major meetings, for example, at Cold Spring Harbor. Anyone else saying what she said would have been readily dismissed, except that she had the reputation of always being correct." She was already renowned for her cytological demonstration of crossing over (Creighton and McClintock 1931) and studies of the behavior of ring chromosomes (McClintock 1938). Her work had been readily incorporated into *the* genetics text of its time (Sturtevant and Beadle 1939), well before controlling elements arrived on the scene. A further sign that Barbara was well appreciated by the genetics community was her election to the National Academy of Sciences in 1944. My father recalls that although geneticists understood the existence of controlling elements, they did not necessarily know what to make of them. Included in this group was H.J. Muller, who did not appreciate controlling elements. Little did Muller know that many of the mutants that fueled the early years of *Drosophila* genetics resulted from transposable genetic elements.

Manipulative Genetics the McClintock Way: Transposon Tagging with Controlling Elements

My efforts to learn more about Barbara's work were greatly facilitated by George Streisinger's nearly complete collection of Barbara's reprints, which were not easy to come by. Although Barbara had published a series of papers in *Genetics* and the *Proceedings of the National Academy of Sciences*, much of her work was published in what we would now regard as a somewhat informal place to publish, the annual report of the Carnegie Institution of Washington.

In reading through Barbara's papers, I was greatly aided by coaching from Flora Banuett, whose background in plant genetics and fascination with maize and its controlling elements led her to scrutinize the McClintock reprint collection carefully. My route to learning about controlling elements also included a couple of lucid presentations by Larry Sandler, who was a great fan of Barbara's. He was a masterful teacher and relished helping people understand material that was considered difficult. And, of course, since it was genetics, it was clearly in Larry's purview.

Although I found myself mystified when I began reading her papers, I was struck by how clear they were when I started at the beginning. It was then apparent what drove Barbara to ponder controlling elements. My suspicions were corroborated by subsequent conversations with her, in which she described how she was struck by the variegation in maize plants carrying a circular chromosome. After developing an understanding of this type of variegation (McClintock 1932, 1938), she was then in a position to appreciate the mystery of other types of variegation and try to understand them.

Reading through the articles from the Carnegie Institution of Washington Year Book was quite exciting. I could see from one installment to the next what she was thinking, what hypotheses were being formulated, and how they were subsequently tested. The logic of the analysis was crystal clear, and I did not have to wait one year between installments. A remarkable culmination to this set of experiments was when she said (and here I am paraphrasing) "OK, we see that this element can move progressively along one chromosome, from its original position to another, and then to another under the influence of *Ac*. If it can transpose *along* a chromosome, then it ought to be able to transpose to another chromosome." Her strategy for recognizing such an interchromosomal transposition event was to look for insertion of the *Ac* or *Ds* element into the A_1 or A_2 genes, which control kernel pigmentation. She designed a simple experiment to try to "trap" such elements, and it succeeded exactly as she predicted (McClintock 1953). This of course was the first use of "transposon tagging" to mark a gene with a defined genetic element and was manipulative genetics of the highest order. We made some attempts to see if we could get the yeast mating-type cassettes to hop into genes and inactivate them (the "errant cassette" experiment; J. Rine, unpubl.) but to no avail: Their movement is restricted—they move only to sites that contain homologous sequences.

More McClintock Memories

Barbara's visit to the University of Oregon in 1978 provided several memorable moments, perhaps the most notable coming after her seminar, when I attempted to ask a question. As part of the preface to my question, I said, "Let me see if I understand what you are saying about controlling elements" and then proceeded to draw a simple diagram on the board showing a structural gene as a rectangle on a line representing a chromosome. Within the structural gene, I drew a triangular insert to indicate the position of the controlling element. I then turned to Barbara to see if I was on track. Much to everybody's amusement (except mine) she said, "That's completely wrong." That was the end of my question but not the end of my quest to learn about controlling elements. (Barbara, I wasn't *completely* wrong, was I?)

Through her visits to Eugene and to Steamboat Springs, Colorado, where she gave the keynote address at the meeting on Genome Rearrangement that Mel Simon organized in 1984, and visits with her at Cold Spring Harbor, I came to learn a great deal more about controlling elements, plant biology, and many other things. Her huge fund of knowledge and her opinions on scientific matters (she refers to the plan to sequence the entire human genome as "knitting") and on scientists were helpful to me in planning conference sessions on several occasions.

Another Facet of Barbara

I have seen that Barbara is in tune not only with the natural world, but with the world of music as well. In Steamboat Springs, she described at dinner on the night of the banquet that she had been a member of a banjo band while she was a student at Cornell University. She had a Model T and used to drive the group and their instruments to the parties where they played. I asked what she enjoyed most about playing in this group, wondering whether she would comment on the music itself (such as the Charleston and other music from the Roaring Twenties) or whether she liked being the center of attention during a solo. Her response was that she enjoyed being part of a musical group, harmonizing with the others and playing along with them. What then happened made it very clear that she had not lost touch with music. Since this was the banquet evening, recorded music was playing and people were dancing. One of the conference participants, a young person whom I did not recognize, walked over to our table and asked Barbara if she would like to dance. She graciously accepted and gracefully danced to "Every Breath You Take" by the Police, commenting on her return that she had heard the song before but had never danced to it. One other musical memory connected with Barbara comes to mind. I gave her a copy of my brother Joel's record, "The DOUBLE Talking HELIX Blues" (a talking blues about DNA—"I wouldn't want to write a novel with four letters. . .I think I'll write a human being instead"), and she loved it. In characteristic fashion, she taped copies of it for others at Cold Spring Harbor to enjoy.

Stories and the Story Teller

The fact that my research crossed paths with Barbara's and encouraged me to read about her work and interact with her has greatly enriched my own scientific outlook. It brought me in touch with formal genetics of the highest order in an organism that would otherwise have remained foreign to me. Barbara has always been eager to share her knowledge—I experienced the same feeling as my father and many others—and can remember one conversation that began when she walked up to me while I was sitting on the grass near her laboratory. She asked me a question about the research going on in my laboratory, and I began by saying that we were studying the mating-type locus of yeast, and how there were two co-dominant alleles at this locus, etc. She took "two co-dominant alleles" as a cue to tell me two truly fascinating stories about loci (in white clover and in ladybird beetles) that have multiple co-dominant alleles. A particularly intriguing aspect of these loci is that they determine nonclonal patterns on the clover leaf (Davies 1963) or on the elytra (wingcovers) of the ladybird beetle (Tan 1946). I have also been fortunate to walk with Barbara through the rhododendron garden in Hendricks Park in Eugene and hear her comment on the floral patterns and muse on other wonders of nature. For me, Barbara represents a person who deeply appreciates the natural world and is committed to understanding it. She is a consummate geneticist who has masterfully exploited genetics and contributed immeasurably to it.

References

Astell, C.R., L. Ahlstrom-Jonasson, M. Smith, K. Tatchell, K.A. Nasmyth, and B.D. Hall. 1981. The sequence of the DNAs coding for the mating type loci of *Saccharomyces cerevisiae. Cell 27:* 15.

Boeke, J.D., D.J. Garfinkel, C.A. Styles, and G.R. Fink. 1985. Ty elements transpose through an RNA intermediate. *Cell 40:* 491.

Brand, A.H., L. Breeden, J. Abraham, R. Sternglanz, and K. Nasymth. 1985. Characterization of a "silencer" in yeast: A DNA sequence with properties opposite to those of a transcriptional enhancer. *Cell 41:* 41.

Creighton, H.B. and B. McClintock. 1931. A correlation of cytological and genetical crossing-over in *Zea mays. Proc. Natl. Acad. Sci. 17:* 485.

Davies, W.E. 1963. Leaf markings in *Trifolium repens*. In *Teaching genetics in school and university* (ed. C.D. Darlington and A.D. Bradshaw), p. 94. Oliver & Boyd, Scotland.

Dawson, G.W.P. and P.F. Smith-Keary. 1963. Episomic control of mutation in *Salmonella typhimurium. Heredity 18:* 1.

Demerec, M. 1935. Unstable genes. *Bot. Rev. 1:* 233.

Errede, B., M. Company, and C.A. Hutchison III. 1987. Ty1 sequence with enhancer and mating-type-dependent regulatory activities. *Mol. Cell. Biol. 7:* 258.

Errede, B., T.S. Cardillo, F. Sherman, E. Dubois, J. Deschamps, and J.M. Wiame. 1980. Mating signals control expression of mutations resulting from insertion of a transposable repetitive element adjacent to diverse yeast genes. *Cell 22:* 427.

Green, M.M. 1967. The genetics of a mutable gene at the white locus of *Drosophila melanogaster. Genetics 56:* 467.

———. 1969a. Mapping a *Drosophila melanogaster* "controlling element" by interallelic crossing over. *Genetics 61:* 423.

———. 1969b. Controlling element mediated transpositions of the white gene in *Drosophila melanogaster. Genetics 61:* 429.

Harashima, S., Y. Nogi, and Y. Oshima. 1974. The genetic system controlling homothallism in *Saccharomyces* yeasts. *Genetics 77:* 639.

Hawthorne, D.C. 1963a. Directed mutation of the mating type alleles as an explanation of homothallism in yeast. *Proc. Int. Congr. Genet. 11:* 34.

———. 1963b. A deletion in yeast and its bearing on the structure of the mating type locus. *Genetics 48:* 1727.

Herskowitz, I.H. 1973. *Principles of genetics*. Macmillan, New York.

Herskowitz, I. 1988. The Hawthorne deletion twenty-five years later. *Genetics 120:* 857.

———. 1989. A regulatory hierarchy for cell specialization in yeast. *Nature 342:* 749.

Herskowitz, I. and Y. Oshima. 1981. Control of cell type in *Saccharomyces cerevisiae*: Mating type and mating type interconversion. In *The molecular biology of the yeast* Saccharomyces: *Life cycle and inheritance* (ed. J.N. Strathern et al.), p. 181. Cold Spring Harbor Laboratory, Cold Spring Harbor, New York.

Hicks, J.B., J.N. Strathern, and I. Herskowitz. 1977. The cassette model of mating type interconversion. In *DNA insertion elements, plasmids and episomes* (ed. A. Bukhari et al.), p. 457. Cold Spring Harbor Laboratory, Cold Spring Harbor, New York.

Hicks, J., J.N. Strathern, and A.J.S. Klar. 1979. Tranposable mating type genes in *Saccharomyces cerevisiae. Nature 282:* 478.

Holliday, R. and J.E. Pugh. 1975. DNA modification mechanisms and gene activity during development. *Science 187:* 226.

Klar, A.J.S. and S. Fogel. 1979. Activation of mating type genes by transposition in *Saccharomyces cerevisiae. Proc. Natl. Acad. Sci. 76:* 4539.

Klar, A.J.S., S. Fogel, and K. MacLeod. 1979. *MAR1*—A regulator of the *HM***a** and *HM*α loci in *Saccharomyces cerevisiae. Genetics 93:* 37.
Kushner, P.J., L.C. Blair, and I. Herskowitz. 1979. Control of yeast cell types by mobile genes—A test. *Proc. Natl. Acad. Sci. 76:* 5264.
McClintock, B. 1932. A correlation of ring-shaped chromosomes with variegation in *Zea mays. Proc. Natl. Acad. Sci. 18:* 677.
———. 1938. The production of homozygous deficient tissues with mutant characteristics by means of the aberrant mitotic behavior of ring-shaped chromosomes. *Genetics 23:* 315.
———. 1946. Maize genetics. *Carnegie Inst. Washington Year Book 45:* 176.
———. 1948. Mutable loci in maize. *Carnegie Inst. Washington Year Book 47:* 155.
———. 1953. Induction of instability at selected loci in maize. *Genetics 38:* 579.
———. 1957. Controlling elements and the gene. *Cold Spring Harbor Symp. Quant. Biol. 21:* 197.
———. 1961. Some parallels between gene control systems in maize and in bacteria. *Am. Nat. 95:* 265.
———. 1965. The control of gene action in maize. *Brookhaven Symp. Biol. 18:* 162.
Nasmyth, K.A. and K. Tatchell. 1980. The structure of transposable yeast mating type loci. *Cell 19:* 753.
Oshima, Y. and I. Takano. 1971. Mating types in *Saccharomyces*: Their convertibility and homothallism. *Genetics 67:* 327.
Rhoades, M.M. 1938. Effect of the *Dt* gene on the mutability of the a_1 allele in maize. *Genetics 23:* 377.
Rine, J., J.N. Strathern, J.B. Hicks, and I. Herskowitz. 1979. A suppressor of mating type locus mutations in *Saccharomyces cerevisiae*: Evidence for and identification of cryptic mating type loci. *Genetics 93:* 877.
Rine, J., R. Jensen, D. Hagen, L. Blair, and I. Herskowitz. 1981. The pattern of switching and the fate of the replaced cassette in yeast mating type interconversion. *Cold Spring Harbor Symp. Quant. Biol. 45:* 951.
Schnell, R. and J. Rine. 1986. A position effect on the expression of a tRNA gene mediated by the *SIR* genes in *Saccharomyces cerevisiae. Mol. Cell. Biol. 6:* 492.
Strathern, J.N. 1988. Control and execution of homothallic switching in *Saccharomyces cerevisiae.* In *Genetic recombination* (ed. R. Kucherlapati and G.R. Smith), p. 445. American Society for Microbiology, Washington, D.C.
Strathern, J.N. and I. Herskowitz. 1979. Asymmetry and directionality in production of new cell types during clonal growth: The switching pattern of homothallic yeast. *Cell 17:* 371.
Sturtevant, A.H. and G.W. Beadle. 1939. *An introduction to genetics.* Reprinted by Dover, New York.
Takano, I. and Y. Oshima. 1970. Mutational nature of an allele specific conversion of the mating type by the homothallism gene HO_α in *Saccharomyces. Genetics 65:* 421.
Tan, C.C. 1946. Mosaic dominance in the inheritance of color patterns in the lady-bird beetle, *Harmonia axyridis. Genetics 31:* 195.

Thinking about Programmed Genome Rearrangements in a Genome Static State of Mind

JEFFREY N. STRATHERN
Frederick Cancer Research and Development Center, Frederick, Maryland 21702-1201

Shortly before I left Cold Spring Harbor Laboratory, Dr. McClintock began to complain about her eyesight. She was having difficulty seeing her journals and the computer screen. I was starting my first (and only) experiment in maize genetics and was just beginning to learn the phenotypes that are the manifestations of the intricate genetics mechanisms I hoped to unravel. Like so many others, I was drawn to the mutations that affect pigment synthesis and deposition during kernel development. The first step beyond the literature, and several years of discussions with Barbara, was to actually hold maize in my own hands and examine kernels myself. Barbara not only provided the necessary seed stock for the proposed experiment, but, most importantly, spent time showing me how to score the kernels. Which brings me back to her eyes. She could see features that escaped my detection. She could distinguish at a glance mutations in the anthocyanin pigment biosynthesis pathway by the ability of the accumulated intermediates to diffuse from one cell to another. But of course, her eyesight is not what makes her such a remarkable scientist, it is a lifetime of vision about the subtleties of genetics and development that places her in such high esteem.

I am asked to recall what I was thinking at the time that Ira Herskowitz, Jim Hicks, and I proposed the cassette model for mating-type switching in yeast. In particular, what influence did Barbara McClintock have on that thinking? I remember two impressions from that period that relate to Barbara. First is my fascination with the experiments defining presetting and erasure. I thought that those observations were the most interesting genetic phenomena in any of my readings. They focused my attention on aspects about controlling elements beyond their mobility. More about them later as they relate to my interactions with Barbara McClintock after moving to Cold Spring Harbor. The second impression has to do with wrestling with the concept of homothallic switching in a world that was just beginning to accept the view that the genome was subject to programmed rearrangements. She once assured me that any regulatory mechanism that I could imagine was no doubt in use somewhere in biology. Feel free to consider all possibilities.

We were forced to consider recombination as a means of regulation by several genetic observations about mating-type switching in homothallic yeast. Cell type in *Saccharomyces cerevisiae* is controlled by the *MAT***a** and

*MAT*α alleles of the mating-type locus. These alleles are codominant, reflecting the production of different regulatory proteins. Homothallic yeast (cells expressing the *HO* allele) are able to change mating type by changing *MAT***a** to *MAT*α and *MAT*α to *MAT***a**. The *MAT* alleles produced by homothallic switching are stable when crossed into a heterothallic (*ho*) strain. This then was the puzzle: By what mechanism could efficient interconversion of these functional alleles occur which would result in stably heritable alleles in the absence of the switching system? What was the nature of the *MAT* locus that it could stably express two different regulatory states? In the cassette model (Hicks et al. 1977), which can now be called the cassette mechanism, we proposed that homothallic switching involved replacing the allele at *MAT* by a replica of the alternate allele copied from an unexpressed site. The specific proposal for the sites of the unexpressed copies of *MAT***a** and *MAT*α information (now called the *HML* and *HMR* loci) was based on elegant genetics by the Oshima laboratory and a key observation by Don Hawthorne described below.

A series of beautiful genetic studies (notably, Takano and Oshima 1970; Naumov and Tolstorukov 1973; Harashima et al. 1974) had defined two genes in addition to *HO* that have roles in homothallic switching. One locus (*HM*α; now called *HMR***a**) was defined by an allele required to switch from *MAT*α to *MAT***a**, and the other locus (*HM***a**; now called *HML*α) was defined by an allele required to switch from *MAT***a** to *MAT*α. What were the roles of *HO, HML,* and *HMR* in the switching mechanism?

Don Hawthorne (1963) made an observation that was a focal point for consideration of any model for homothallic switching. In heterothallic (*ho*) yeast, the *MAT* alleles are very stable but can undergo rare interconversions. Don selected a rare switch from *MAT*α to *MAT***a** in a heterothallic strain and obtained a *MAT***a** allele associated with an extensive deletion to the right of *MAT* ("Hawthorne's deletion"). The implication was that the *MAT* locus contained both **a** and α information and that deletion of the α sequences could allow activation of the **a** allele. What was the normal mechanism by which these two sets of information at *MAT* were differentially activated?

Our experiments were being done at a time when most people viewed the genome as a static repository of genetic information that was accessed differentially in various cell types due to the action of *trans*-acting positive or negative regulators controlling transcription. In retrospect, that was a very simple view of gene regulation. Few alternative paradigms were available.

Proposals that covalent modifications of the DNA (e.g., methylation) could be heritable and act in a regulatory fashion had been advanced (Holliday and Pugh 1975). Applied to mating-type switching, this model suggests that expression of the **a** or α information from *MAT* depends on the state of modification of the *MAT* locus. This view leads to proposed roles for *HML* and *HMR* in the modification mechanism (e.g., a methylase and a demethylase). Hawthorne (pers. comm. cited in Holliday and Pugh 1975) proposed a conceptually related mechanism for directed base pair changes as a means of giving heritable changes of gene expression.

Phase variation in bacteria provided a simple reversible mechanism for how a DNA rearrangement could be used to alternate between the expression of two different sets of genes. In this mechanism, a segment of DNA is inverted by site-specific recombination. The two different orientations of the

DNA allow two different expression states for genes within or adjacent to the invertible segment. Models for yeast mating-type switching related to the flip-flop mechanism were presented by Brown (1976) and by Hicks and Herskowitz (1977). In this context, roles for *HML* and *HMR* as direction-specific recombinases (conceptually related to the roles of *int* and *xis* in λ) were entertained.

The second emerging paradigm for recombination as a means of regulation was the immunoglobin gene rearrangements associated with generating the immune response repertoire in mammals (Tonegawa et al. 1977). The mature gene is assembled by a series of deletions. Each deletion has several options, resulting in the tremendous repertoire of mature antibodies that can be produced by this mechanism. This pathway was enormously powerful but, as described then, had the feature that it involved irreversible changes. A programmed deletion mechanism would not adequately explain the ready reversibility of mating-type switching in yeast. It is of interest to note that avian immunoglobin diversity has recently been shown to derive from a cassette mechanism involving efficient gene conversion between unexpressed pseudogenes and the expressed heavy-chain locus (Thompson and Neiman 1987).

The other key observation that constrained attempts to make a model for homothallic switching was the ability of homothallic switching to heal mutations at *MAT*. Hicks and Herskowitz (1977) showed that homothallic switching readily converted a *matα1* mutation to *MAT***a** and then to *MAT*α. Hawthorne had communicated similar results to us in which he used a nonsense allele of *matα1*. It is possible to accommodate these results within the context of flip-flop and DNA modification mechanisms (Hicks and Herskowitz 1977). However, we began to entertain models in which cryptic copies of *MAT* were present which could serve as donors of the sequences required to heal the mutations. My wife Ann, working as a technician with Ira, showed that λ mutations could be healed by recombination with cryptic λ prophages (Strathern and Herskowitz 1975). The problem became how to accommodate Hawthorne's deletion, homothallic healing of mutations, and the roles for *HML*, *HMR*, and *HO* in a single model.

Of course, the other system suggesting genome rearrangements altering expression was the movable genetic elements in maize. After reading the descriptions of the maize elements in the "stable genome" intellectual environment, it was easy to come away with the impression that they reflected transposable elements foreign to the maize genome that were visible only because of their effects on maize genes. My training at the time included λ genetics (a movable genetic element with a very specific insertion "attachment" site), and, again, the world was just beginning to come to grips with bacterial insertion sequences and transposable elements.

We had the good fortune to have Barbara McClintock visit the Herskowitz laboratory in Eugene, Oregon (the date escapes me). Two transitions in thinking came about from discussions with Barbara. First, she emphasized the point that these elements were monitors of differentiation. The developmental timing of controlling element excision is allele-dependent. This feature was very clear from the phenotypes of the mutants. She even gave us glossy prints of figures showing relevant kernels to emphasize this characteristic. Some alleles give early excisions (large sectors), whereas others give only small sectors reflecting late events. The other point that she made was the suggestion that controlling elements might themselves reflect a differentiation mechanism based on heritable genetic rearrangements. In effect, her proposal was that controlling elements show sensitivity to the developmental state because they are related to programmed genetic rearrangements that are involved in normal differentiation. I distinctly remember the moment when I came to appreciate that proposal. Movable genetic elements might be a normal means of developmental regulation. Oshima and colleagues had recognized the applicability of McClintock's proposal to homothallic switching in suggesting that homothallic switching reflected the attachment of a controlling element to *MAT* as a means of differentiating the *MAT***a** and *MAT*α alleles (Oshima and Takano 1971).

I gather these comments were designed to relate perception. In some senses, it does not matter whether I understood Barbara completely or correctly, it only matters that in trying to understand her, my thoughts traversed new ground for me and helped us to come to the view that resulted in the cassette model. The key to tying together Hawthorne's deletion, homothallic healing, and roles for *HML* and *HMR* was suggested by the map position of *HMR* determined by Harashima and Oshima (1976). They showed that *HMR* (then called *HM*α) was on the right arm of chromosome III distal of *MAT* (Fig. 1). The cassette model derives from the realization that the right end of Hawthorne's deletion might be *HMR***a** and that rather than that deletion activating **a** sequences already at *MAT*, it might activate a cryptic copy of **a** sequences at *HMR***a** by fusing them to *MAT*. From the proposal that *HMR***a** was a cryptic copy of *MAT***a**, the proposals for *HML*α, *HMR*α, and *HML***a** as cryptic α, α, and **a** sequences, respectively, were straightforward extensions. The gene conversion mechanism of switching with its capacity to heal also follows from this proposal. Accumulating genetic and physical data to demonstrate that cell-type switching in homothallic yeast involved this specific genome rearrangement mechanism have provided exciting research experiences for Ira, Jim, and me, as well as several other investigators for more than 15 years. Research continues into the details of the switching mechanism and the mechanism of silencing *HML* and *HMR*. Mating-type switching now serves as a clear example of the principle proposed by McClintock that movable genetic elements could be used to activate regulatory genes in a preprogrammed fashion and thus control development.

While at Cold Spring Harbor, I ventured into maize genetics only once. I hoped to study one of the most interesting genetic observations from any organism: presetting and erasure (if you don't know this story, find it and entertain yourself). The presetting and erasure phenomena are described in the 1961–1964 yearbooks of the Carnegie Institution of Washington. I admit to having had a hard time wrapping my brain around those observations, but Barbara's visit to Eugene encouraged me to ponder them further. Briefly, there are mutant alleles of one of the anthocyanin pigment synthesis genes, *a1*, that respond to the *Spm* element. These alleles express pigment in the presence of an active *Spm* element and are colorless in the absence of *Spm*. McClintock described cases from plants heterozygous for *Spm* in which pigment was expressed in kernels that did not inherit the *Spm* element. The alleles had some mechanism to remember that they had passed through a cell expressing *Spm*. In McClintock's words "*Spm*, present in a plant, is involved in controlling

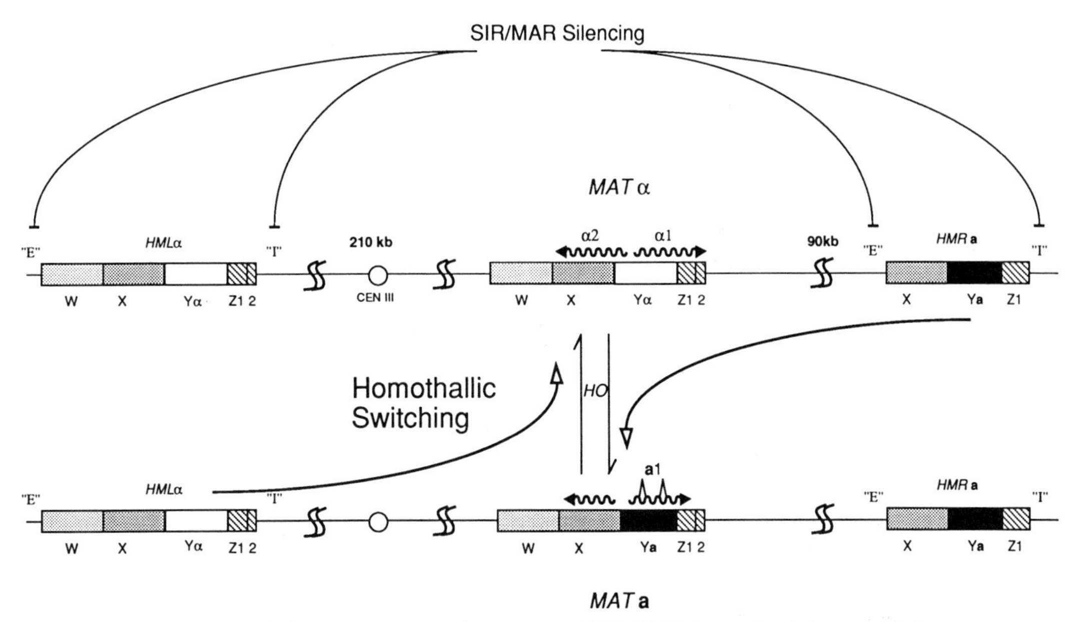

Figure 1 Organization of the cassettes on chromosome III. *HML* is on the left arm of chromosome III. *MAT* and *HMR* are on the right arm. W, X, Y, Z1, and 22 represent regions defined by homologies between *HML*, *HMR*, and *MAT*. The W region is 723 base pairs in length and is homologous between *HML* and *MAT*. The X region (704 base pairs) is found at *HML*, *HMR*, and *MAT*. *HML*, *HMR*, and *MAT* can have either the **a**-specific sequence Y**a** (642 base pairs) or the α-specific sequence Yα (747 base pairs). Z1 (239 base pairs) is found at all three cassettes, whereas Z2 (89 base pairs) is found only at *HML* and *MAT*. The three cassettes are orientated in the same direction. Homothallic switching is a gene-conversion-like process that replaces the expressed allele of *MAT* with a copy of a mating-type cassette from *HML* or *HMR*. The cassettes at *HML* and *HMR* are repressed by *trans*-acting regulators (SIR/MAR Silencing).

phenotypic expression in progeny kernels that do not receive *Spm*." In most cases, the expression lasted only through the development of the kernel after the segregation of the *Spm* element away from the responsive allele. The *a1* allele was said to be preset to a level of expression and erased before the next generation. Barbara could see in that residual expression reflections of whole regulatory mechanisms that could be involved in development. This vision was the other concept that she imparted to me on that visit to Eugene.

I could see no way to study presetting if erasure was absolute. However, in the 1964 yearbook, there was a particularly exciting observation. McClintock was describing the loss of the *a1* expression in subsequent generations. "On a few ears, however, produced by plants carrying state 7995 of $a1^{m\text{-}2}$, several kernels exhibited a pattern of anthocyanin distribution resembling that in the kernel from which the plant was derived." Barbara did not seem to need these "exceptional kernels" or failures to erase (my term) to convince her that an underlying genetically stable mechanism existed, but I clung to them and my attempts to understand them were the basis of conversations over several years with her. They were also the focus of my single maize experiment.

In response to a general question, Barbara McClintock most often responds with a description of specific results from her experiments. Fully understanding her responses can require more familiarity with her work than most people retain. She prefers that the questions be related to specific experiments. As a result, our conversations often evolved to descriptions of specific crosses. Much of our conversations were spent educating me as to the right way to ask the question. The most memorable for me was a series of conversations about how it was demonstrated that kernels exhibiting presetting did not have some residual *Spm* activity (perhaps an unintegrated copy). These led her to point out crosses in which an *Spm* responsive allele of *wx* was used as an internal control in these preset kernels to demonstrate that *Spm* activity was absent.

The feature that I chose to address was an observation by McClintock that presetting can give a wide variety of levels of expression of the responsive *a1* allele. Whereas the pigment intensity varies from kernel to kernel, within an individual kernel, all of the pigmented areas showed the same intensity. In the next generation, kernels that failed to erase the presetting resembled the phenotype of the specific preset kernel from which the plant was derived. "They did not show the wide range of intensities" apparent among the original preset kernels. This suggested that the mechanism of presetting was not identical to the mechanism of erasure. I hoped to gather more evidence for this hypothesis by gathering more failures to erase from single preset kernels. The reported results with $a1^{m\text{-}2}$ (state 7995) involved failures to erase through the maternal germ line. As a consequence, the number of events was limited to the number of kernels that could be produced by a single plant. Because failure to erase was rare, this limited the number of "exceptional kernels" to one to five per plant derived from a preset kernel. I hoped to monitor failure to erase through the pollen parent and hence be able to generate many more exceptional kernels.

Barbara provided the seed stock and guidance to do the required crosses. There were several memorable moments in the maize plots that I cherish. Most importantly, she made me recognize the price I paid by passing the preset allele through the pollen. Again this was an example of her seeing a feature of the biology that might have escaped others: She had observed that the distribution of the kernels showing failure to erase was not random on the cob. They were more often found at the base or the tip of the cob. Again, this was telling her something about the mechanism that would be lost in the experiment as I was doing it. She assured me that failure to erase occurred through the pollen parent, but commented that she wouldn't sacrifice the information inherent in the positional information on the cob.

In the end, she was right on all counts. Failure to erase appeared less common through the pollen parent so that despite having generated about ten times as many total kernels as one would get through the cob, we still obtained only about half a dozen kernels that retained the preset pattern. Second, there was no positional bias for the exceptional kernels, confirming the conclusion that the erasure did not occur after fertilization. Finally, there was the point of the experiment. Did all of the failures to erase derived from a single preset parent have the same phenotype? Once again, I had to rely on her eyes. The pigmentation in preset (or failure to erase) kernels is unevenly expressed. They have a mottled appearance. On any one kernel, the expression level is similar within the pigmented areas, but the intensity differs from one preset event to another. One must be able to separate the intensity of pigment production in the pigmented areas from the proportion of the kernel that is pigmented. My eyes tended to average the color over the surface of the kernel.

Thus, different degrees of mottling looked like different color levels. Barbara pointed out that the intensity of the pigmented spots (ignoring the colorless areas) was very similar on independent kernels that had failed to erase the same presetting event and that it resembled the intensity on the preset kernel that was the pollen donor. Again, to her this reflected a mechanism of genetic change that had inherent stability and a well-regulated mechanism for setting and changing levels of expression of genes. In these small mottled patches of color, she could see the tools of developmental regulation.

It has been 8 years since I left Cold Spring Harbor and any attempt to do maize genetics. I returned to the analysis of the mechanics of homothallic switching and genome rearrangement in general. However, I still have the exceptional kernels that came out of that summer with Barbara, and I still have the insights she shared of her exceptional vision of biology.

References

Brown, W.S. 1976. A cross-over shunt model for alternate potentiation of yeast mating type alleles. *J. Genet. 62:* 81.

Harashima, S. and Y. Oshima. 1976. Mapping of the homothallism gene HMα and HMa in *Saccharomyces* yeasts. *Genetics 84:* 437.

Harashima, S., Y. Nogi, and Y. Oshima. 1974. The genetic system controlling homothallism in *Saccharomyces* yeasts. *Genetics 77:* 639.

Hawthorne, D.C. 1963. Directed mutation of the mating type alleles as an explanation of homothallism in yeast. *Proc. Int. Congr. Genet. 7:* 34. (Abstr.)

Hicks, J.B. and I. Herskowitz. 1977. Inter-conversion of yeast mating types. II. Restoration of mating ability to sterile mutants in homothallic and heterothallic strains. *Genetics 85:* 373.

Hicks, J.B., J.N. Strathern, and I. Herskowitz. 1977. The cassette model of mating type interconversion. In *DNA insertion elements, plasmids and episomes* (ed. A. Bukhari et al.), p. 457. Cold Spring Harbor Laboratory, Cold Spring Harbor, New York.

Holliday, R. and J.E. Pugh. 1975. DNA modification mechanisms and gene activity during development. *Science 187:* 226.

McClintock, B. 1961. Further studies of the suppressor-mutator system of control of gene action in maize. *Carnegie Inst. Washington Year Book 60:* 469.

———. 1962. Topographical relations between elements of control systems in maize. *Carnegie Inst. Washington Year Book 61:* 448.

———. 1963 Further studies of gene-control systems in maize. *Carnegie Inst. Washington Year Book 62:* 486.

———. 1964. Aspects of gene regulation in maize. *Carnegie Inst. Washington Year Book 63:* 592.

Naumov, G.I. and I.I. Tolstorukov. 1973. Comparative genetics of yeast. X. Reidentification of mutators of mating-types in *Saccharomyces. Genetika 9:* 82.

Oshima, Y. and I. Takano. 1971. Mating types in *Saccharomyces*: Their convertibility and homothallism. *Genetics 67:* 327.

Strathern, A. and I. Herskowitz. 1975. Defective prophage in *Escherichia coli* K12 strains. *Virology 67:* 136.

Takano, I. and Y. Oshima. 1970. Mutational nature of an allele-specific conversion of the mating type by the homothallic gene HOa(α) in *Saccharomyces. Genetics 65:* 421.

Thompson, C.B. and P.E. Neiman. 1987. Somatic diversification of the chicken immunoglobulin light chain gene is limited to the rearranged variable gene segment. *Cell 48:* 369.

Tonegawa, S., N. Hozumi, G. Matthyssens, and R. Schuller. 1977. Somatic changes in the content and context of immunoglobulin genes. *Cold Spring Harbor Symp. Quant. Biol. 41:* 877.

The Role of McClintock's Controlling Element Concept in the Story of Yeast Mating-type Switching

AMAR J. S. KLAR
NCI-Frederick Cancer Research and Development Center, ABL-Basic Research Program, Frederick, Maryland 21702-2102

The paper in the field of yeast mating type that most influenced me was that by Isamu Takano, Takaaki Kusumi, and Yasuji Oshima (1973). This paper was handed to me in 1974 by one of the authors, Dr. Isamu Takano, who was a visiting scientist at Brandeis University (Waltham, Massachusetts) where I was a graduate student in the laboratory of Professor Harlyn O. Halverson. These authors showed in this paper that a particular strain of the budding yeast *Saccharomyces distaticus* was inefficient in switching for mating-type (*MAT*)α to *MAT***a**. Because it was an inefficient switcher, they named the allele *MAT*α-*inc* (*inc* for *inc*onvertible). The interesting result was that when it did switch, it produced the standard *MAT***a** derivatives. Subsequent switching of these *MAT***a** derivatives remarkably produced wild-type *MAT*α and not the *MAT*α-*inc* allele! To explain this odd result, they quoted their Controlling Element model published 2 years earlier (Oshima and Takano 1971). The key reference in the 1971 paper was that of Barbara McClintock (1957), a paper entitled "Controlling elements and the gene." Below is a part of the Discussion in their 1973 paper:

> For the explanation of the mating-type differentiation, we previously proposed the hypothesis (Oshima and Takano, 1971) that the elementary structure of the mating-type locus for both **a** and α is essentially the same. The association of some kind of controlling element with this locus would cause differentiation of two mating-type alleles. The mating-type locus on chromosome III would act as an affinity site for a controlling element. The association of an *HM*α-element to the mating-type locus would form the **a** mating-type allele and the association of an *HM***a**-element to the mating-type locus would give rise to the α mating-type allele. Another gene, *HO*, could be concerned with the effective association and removal of the controlling elements with their affinity site. According to this model, the α-*inc* allele could be explained as an α mating-type allele in which a controlling element from the *HM***a** gene was attached to its affinity site abnormally. The abnormal association of the *HM***a**-element at the mating-type locus could cause the insensitiveness of α-*inc* to the action of the *HO* gene and to the *HM*α-element.

As it turned out, they were right on target. What follows is my participation in this system that led to the establishment of the model. Since almost everything about Barbara McClintock's work is historical and this work involves transposable elements, it is most appropriate to reminisce about yeast in Barbara's birthday celebration proceedings. In addition, anything not written gets forgotten and what follows should help one to understand how the story developed.

Having read the above-quoted yeast papers, I talked to Isamu Takano and suggested that it would be worthwhile to define the hypothesized controlling element further. I was aware that Don Hawthorne (University of Washington, Seattle) had a *mat*α*-ochre* mutant that was sterile for mating, and thus I proposed testing whether the *mat*α*-ochre* allele could also be removed for good by switching. Isamu was so encouraging about the experiment that I promptly wrote for the *mat*α*-ochre* mutant strain. I did not receive the strain, perhaps because Don Hawthorne already knew (which I later found out) that the mutation disappears by switching.

The second incident happened when another visiting scientist, Dr. Johans van der Plaat (Gist-Brocades, Delft, Holland), asked me how to control the ploidy of industrial yeast strains. I suggested that one way may be to manipulate the *D* (i.e., *HO*) gene phenomenon. The *D* stands for diploidization (Winge and Roberts 1949). It was known that gene *D* containing *MAT***a** or *MAT*α haploid strains cannot be maintained as haploid cultures as they quickly become *MAT***a**/*MAT*α diploid cells due to *MAT* switching and subsequent mating of haploid cells. I questioned how a gene could be sensitive to the ploidy of the cell and stop its action of switching *MAT* once the cell has become diploid. The idea I wanted to test was if one makes a diploid that was a *MAT***a**/*MAT***a** or *MAT*α/*MAT*α, then the switching process may change one or both *MAT* loci to the opposite allele if the expression of mating type, as opposed to ploidy, was the reason for *D* gene action. The resulting diploids of opposite type could mate to generate tetraploid cells.

*The hm***a** *(HML***a***) Locus Performs the HMa (HMR***a***) Function*

After joining Seymour Fogel's laboratory at the University of California, Berkeley, for my postdoctoral training, I proceeded to test the effect of the *D* gene on diploid cells as a side project. The experiment worked beautifully. I have been working on homothallism since that time. While I was writing the *D* gene result as a small note for publication, I noticed a cleanly stated paradox and a possible solution to it. I thought my approach of investigating switching of diploid cells would cleanly solve the paradox.

In strains of several different interbreeding *Saccharomyces* species, such as *oviformis, cerevisiae,* and *chevalieri,* it was found that most could switch in both directions: some could switch *MAT***a** to *MAT*α but not the reverse, and others could only switch *MAT*α to *MAT***a**. Such genetic analyses identified two other loci, *HM***a** and *HM*α, which control the direction of switching (Takahashi 1958; Takano and Oshima 1967; Santa Maria and Vidal 1970; Naumov and Tolstorukov 1973). In particular, *HO, HM***a**, and *HM*α strains switched in both directions. The locus defined for the *MAT*α switching defect was designated naturally occurring variant (mutant?) *hm*α. Similarly, *HO, hm***a**, and *HM*α cells switched *MAT*α but not *MAT***a**. It seemed as if *hm***a** and *hm*α were defective forms of *HM* loci. It was found that *HO, hm***a**, and *hm*α strains obtained as segregants from a *HM***a** *hm*α x *hm***a** *HM*α cross, rather than being defective in switching in both directions, paradoxically switched in both directions, similar to the *HO HM***a** *HM*α strains. Naumov and Tolstorukov (1973) cleanly stated that perhaps *hm***a** is functionally equivalent to *HM*α and that *hm*α is equivalent to *HM***a**. To test this suggestion, we made a *MAT*α/*MAT*α, *HO*/*HO*, *hm***a**/*HM***a**, *hm*α/*hm*α diploid. Note that the strain does not contain *HM*α, which is normally required to switch *MAT*α. If *hm***a** were functionally equivalent to *HM*α, then the cells should change the *MAT*α allele on one or both chromosomes to *MAT***a**. This was indeed found. As a side issue, the results also showed that *HO* (*D*) is dominant to a naturally occurring mutant (variant) *ho* (Klar and Fogel 1977). This paper apparently went to Ira Herskowitz (then at the University of Oregon, Eugene) for review. Because they also had results showing that *HO* acts as a dominant marker, Ira called my advisor Seymour Fogel and suggested publishing both their results and our results in the same issue of *Genetics* (Klar and Fogel 1977; Hicks et al. 1977a). Since he knew that we were also working on mating-type switching, Ira visited our laboratory in April, 1977, where I first met him.

Genetic Proof of the Transposition Model: Discovery of the MAR1 Locus and the Wounding Experiment

In a chalkboard talk to Seymour Fogel's group, Ira described their *mat*α*-ste* mutations switching to *MAT***a** and *MAT***a** switching to wild-type *MAT*α$^+$ (Hicks and Herskowitz 1977), a result analogous to that of *MAT*α*-inc* (Takano et al. 1973) and *mat*α*-ochre* (D. Hawthorne, pers. comm. to Ira Herskowitz) mutations healing due to switching. These results led them to propose a specific version of the controlling element model, the Cassette Model, whereby the *HM***a** and *HM*α loci were suggested to contain the unexpressed *MAT*α- and *MAT***a**-coding information (Hicks et al. 1977b). That information was suggested to lack some essential element, e.g., a promoter, leading to their unexpression. Switching was proposed to involve transposition of a copy of this information to *MAT* where it would be expressed. We related to Ira our results showing that a defective *mat***a**^{0} could likewise be healed by switching (Klar et al. 1979b). Trying to explain another rearrangement, called the Hawthorne deletion, led Jeffrey N. Strathern (a graduate student with Ira Herskowitz) to suggest originally the cassette model is covered from a historical viewpoint elsewhere (Herskowitz 1988).

We were already working on a stock of a nonswitching strain 2180-1A (*MAT***a**, *ho*) that when struck for single colonies mainly produced sterile subclones. Since I was working on mating type, we decided to pursue the sterility defect. I proposed that this strain might allow the expression of *HM***a** and *HM*α loci causing the haploid to behave as a *MAT***a**/*MAT*α diploid; although in the cassette model, it was suggested that the *HM* loci could not be expressed because it lacked a positive regulatory site. A telling phenotype was that this strain exhibited incipient meiosis and sporulation even in a haploid state. Kathy MacLeod, a technician in Seymour Fogel's laboratory, crossed the sterile strain by "rare mating" with standard strains and picked up mutations elsewhere in the genome that allowed cells to mate. The mutations suppressed the sterile phenotype in a *MAT*-allele-specific manner. In that summer of 1977, I tested whether the mutations map at *HM***a** and *HM*α by crossing with strains containing naturally found *hm***a** and *hm*α alleles. The mutations mapped at *HM* loci! These were an exciting series of experiments which also showed that sterility could be suppressed by standard *hm*a and *hm*α alleles. We named the gene that silenced the *HM*a (and *hm*a) and *HM*α (and *hm*α) loci, *MAR1* (for mating-type regulator or master regulator). The *mar1* mutation was mapped to chromosome IV by Kathy MacLeod and was shown to be linked to the centromere-linked marker, *TRP1*.

The second most critical genetical prediction of the transposition model was that if the *HM* locus carried the mutation, then it should generate a mutant *MAT* locus after inserting the mutant controlling/cassette element into *MAT*. The first mutation we tested was *HM*α*'-1*, isolated from *mar1* mutant strains discussed above. The *HO* locus was crossed into the strain to produce a *MAT*α, *HO*, *HM*a, *HM*α*'-1* genotype. On August 5, 1977, we obtained a most exciting result: The strain was promiscuous in mating. It kept on switching between *MAT*α and defective *mat*a°. The switched products mated to produce *MAT*α/*mat*a° diploids. That genotype expressed an α phenotype since defective *mat*a° is recessive. Consequently, the diploid would switch again to produce a tetraploid that should continuously switch, mate repeatedly. These results genetically established that (1) *HM*a and *HM*α (and *hm*a and *hm*α) carry the *MAT*α and *MAT*a information, respectively, (2) information is kept unexpressed by the *trans*-acting *MAR1* function (Klar et al. 1979a), and (3) switching involves transfer of a copy of the *HM* locus to *MAT*, establishing the controlling element and the more specific cassette model (Klar and Fogel 1979).

In August, 1977, I attended the International Yeast Meeting at Cold Spring Harbor Laboratory. I vividly remember describing these results to Ira Herskowitz, Jeff Strathern, and Jasper Rine before my formal presentation at the meeting. After I described the *mar1* mutant result and that the *HMa'-1* mutant generates defective *mat*a°, Jeff Strathern got highly excited, got up and stomped the ground by the steps of Blackford Hall, and suggested that a shrine should be made for mating type right there. Jasper Rine (a graduate student with Herskowitz) said that another mutation he had been working with also allowed partial expression of silent genes. He thought that it was loosely linked to *TRP1* as well and suggested that it may be an allele of *MAR1* (called N at that time). When I tested it, it turned out to be another gene now called *SIR1* (silent information regulator), mutations of which only allow partial expression of *HM* loci; presumably, it plays a peripheral role in silencing the *HM* loci. After learning about the properties of the *mar1* mutant from my presentation, Anita Hopper (Hershey, Pennsylvania) suggested to James Haber (Brandeis University, Massachusetts) that her *cmt* (change of mating type) mutant also behaved like the *mar1* mutant. That suggestion turned out to be correct as Haber showed that *cmt* is another gene involved in regulation of silent *HM* genes. These three mutations were published simultaneously (Haber and George 1979; Klar et al. 1979a; Rine et al. 1979).

The Mating-type Work in the "Yeast Group" at Cold Spring Harbor Laboratory

After teaming up with James Hicks and Jeff Strathern at Cold Spring Harbor Laboratory in the summer of 1978, I isolated *amber* and *ochre* nonsense mutations in *HM*α. The mutants only produced mutant *mat*a° alleles possessing the specific nonsense defect originally found in the *HM* loci, erasing any doubt about the validity of the transposition model. A party was quickly organized to celebrate the result. Furthermore, a defective *HM***a** mutation also generated defective *mat*α, with both loci exhibiting an identical defect (Klar 1980). That summer, at a yeast meeting in Rochester, New York, we learned that the Herskowitz group had subsequently isolated mutations at *HM***a** that produced defective *mat*α by switching (Kushner et al. 1979). *HM***a** was named *HML* and *HM*α was named *HMR* at that meeting to indicate the location of these loci on the *left* or the *right* arm of chromosome III. We also made strains where mating type was genetically mapped directly to *HML* or *HMR* in *mar1* strains. For example, the mating type mapped to the *HML* (*HM* located in the left arm of the chromosome) locus in an "α" strain (*HM***a**[α], *mat***a**°, *HMa'-1*[**a**°] *mar1*) crossed with an "**a**" strain (*hm***a**[**a**], *mat***a**°, *HMa'-1*[**a**°] *mar1*) instead of at *MAT*. This was the first clear result showing that complete mating-type information resides at *HM***a** and *HM*α loci (Klar et al. 1980). These unpublished genetic results presented at the 1978 Rochester meeting convinced Dr. Kim Nasmyth (a postdoctoral fellow at the University of Washington, Seattle) to get into the act.

When my colleagues Jim Hicks and Jeff Strathern cloned the *HML*α locus, it was a pretty sight to find the same restriction-length polymorphism at *MAT* as well as at *HML* and *HMR*. At each locus, a restriction fragment with α information was about 100–200 bp longer than the fragment containing an **a** information (Hicks et al. 1979). Soon after, Nasmyth and Tatchell (1980) also cloned the *MAT* locus. The molecular work was very gratifying as it confirmed the conclusions derived earlier by genetical means and also made the system understandable to geneticists and nongeneticists alike.

As a part of the Yeast Group at Cold Spring Harbor Laboratory, we worked on several aspects of the system, primarily showing that (1) switching involved a gene conversion event initiated by a double-strand cut at *MAT*, (2) the *HO* gene encoded the endonuclease, and (3) the *MAR/SIR* control acts by transcriptional repression of the *HM* loci. Why the *HM* loci do not themselves switch was also shown to be under the *MAR/SIR* control. Once the break is made at *MAT*, the cells primarily switch to the opposite allele. It was shown that cells choose a particular donor on the basis of cell type, thus dictating the direction of switching. Many other investigators also contributed to these issues. During all of these studies, we all benefited from Barbara's interest in transposable genes. In those days when nothing was considered proof unless it was cloned and sequenced and abstract thinking was not taught, discussions with Barbara were most refreshing.

In any scientific story where a considerable number of investigators have participated, it is always interesting to find out how key ideas were developed and tested. In this context, one of the first models for switching suggested directed mutations of *MAT* by *HO* (Hawthorne 1963). For example, demethylation of adenine could generate an A-T to G-C change. Oshima and Takano (1971) argued that most prokaryotic mutators known at that time were not locus-specific and thus discounted such a model. They suggested instead that their observations showed some similarities to the specific transposable genetic controllers of mutation in maize in their locus-specific action and in their mode of modulation (McClintock 1957). Since Oshima and Takano participated in the discovery of *HM* loci, they assigned the controlling element function to the *HM* loci. The *MAT*α-*inc* healing result of Takano et al. (1973)

kept the model alive so that it could be tested rigorously by genetic and molecular means. Don Hawthorne had the *mat*α*-ochre* healing result for many years, a result that unfortunately remains to date unpublished. Had that result been published in due course, the field of transposable elements would have developed earlier. The generalities of Barbara McClintock's work would have easily been realized a decade earlier, and that could have influenced the timing of Barbara's Nobel award. The proponents of the cassette model knew the result of *mat*α*-ochre* mutation healing (and of course of the earlier *MAT*α*-inc* healing), and thus it was appropriate to suggest that controlling elements are silent genes of *MAT*. All in all, it has been an exciting textbook story. The system has become the "λ" of eukaryotes. Many researchers working outside of this field were first exposed to yeast literature because they had to learn studies on mating type. McClintock's controlling element concept was the key factor that helped propel the yeast mating-type field.

Fission Yeast Switching Is Just as Interesting

My colleagues wanted to move on to higher systems—Jim Hicks to corn and Jeff Strathern to tomatoes. I decided to dabble in the switching system of another distantly related yeast, the fission yeast *Schizosaccharomyces pombe*. This system turned out to be similar to that of the budding yeast in overall strategy, but all of the details are different. I described the *S. pombe* results to Barbara in a 45-minute appointment I had with her, which turned out to last more than 4 hours. I suggested that perhaps the pattern of switching in cell pedigrees was segregating with the particular strand of DNA and wanted to discuss with her how I planned to test this idea using *mat1* locus duplications. She was strong in her intuition and said that this idea had to be right, I must follow it. When the same experiment was described to Dr. James D. Watson, in his usual way, he wanted to know how soon I could get to the answer. If it turns out that way, Jim said, I would be "in," whatever he meant by that. The experiment was a big success. It did turn out that strains constructed to contain inverted duplication at *mat1* produced two cousin cells among four granddaughters that switched, in comparison to the standard strains that switch only one in four granddaughters of a cell (Klar 1990). The result showed that switching potential segregates with a specific strand of DNA according to the classical model of semi-conservative replication and segregation. Thus, the primary reason for cellular differentiation in fission yeast is the difference in the sequence of DNA chains that are complementary and not identical (Watson and Crick 1953).

Epilogue: Barbara Agrees to an Autograph

About 6 years ago, Dr. Yasuji Oshima attended the Yeast meeting at Cold Spring Harbor. He had bought *A Feeling for the Organism* (Keller 1982), a book about Barbara McClintock, and wanted it autographed for his daughter. I knew that Barbara did not like the idea of autographs and so went along with Dr. Oshima to see her. They wished each other well, and Oshima thanked Barbara for the controlling element idea of transposable genes in corn, an idea that he said he used to explain mating-type switching in yeast. Oshima requested the autograph. Barbara was very quick in saying that the book was not written by her, she never read it, and she saw no reason why she should autograph it. Oh boy! After a few tense seconds of silence, I jumped in and suggested that maybe Barbara would autograph her Nobel prize lecture from the *Science* journal, to which she agreed. I got one copy for my daughter, probably as a fee for negotiating. Barbara insisted on putting these papers in separate envelopes, a move I could not understand. While we were stepping out of her office, she laughed and said that this way we would not show the autographs to others attending the conference during a coffee break.

I thank all of the people whose names are mentioned above and who were active participants in developing the yeast story. I shall particularly treasure the insight and encouragement Barbara McClintock gave me during the course of this work.

References

Haber, J.E. and J.P. George. 1979. A mutation that permits the expression of normally silent copies of mating type information in *Saccharomyces cerevisiae*. *Genetics 93:* 13.

Hawthorne, D.C. 1963. A deletion in yeast and its bearing on the structure of the mating type locus. *Genetics 48:* 1727.

Herskowitz, I. 1988. The Hawthorne deletion twenty-five years later. *Genetics 120:* 857.

Hicks, J.B. and I. Herskowitz. 1977. Interconversion of yeast mating types. II. Restoration of mating ability to sterile mutants in homothallic and heterothallic strains. *Genetics 85:* 373.

Hicks, J.B., J.N. Strathern and I. Herskowitz. 1977a. Interconversion of yeast mating types. III. Dominance of the homothallism gene (*HO*) an action in diploids homozygous for the mating type locus. *Genetics 85:* 395.

———. 1977b. The cassette model of mating-type interconversion. In *DNA insertion elements, plasmids, and episomes* (ed. A. Bukhari *et al.*), p. 457. Cold Spring Harbor Laboratory, Cold Spring Harbor, New York.

Hicks, J.B., J.N. Strathern, and A.J.S. Klar. 1979. Transposable mating type genes in *Saccharomyces cerevisiae*. *Nature 82:* 478.

Keller, E.F. 1982. *A feeling for the organism: The life and work of Barbara McClintock*. W.H. Freeman, New York.

Klar, A.J.S. 1980. Interconversion of yeast cell types by transposable genes. *Genetics 95:* 631.

———. 1990. The developmental fate of fission yeast cells is determined by the pattern of inheritance of parental and grandparental DNA strands. *EMBO J. 9:* 1407.

Klar, A.J.S. and S. Fogel. 1977. The action of homothallism genes in *Saccharomyces* diploids during vegetative growth and the equivalence of *hm***a** and *HM*α loci. *Genetics 85:* 37.

———. 1979. Activation of mating type genes by transposition in *Saccharomyces cerevisiae*. *Proc. Natl. Acad. Sci. 76:* 4539.

Klar, A.J.S., S. Fogel, and K. MacLeod. 1979a. *MAR1*—A regulator of *HM*α and *HM***a** loci in *Saccharomyces cerevisiae*. *Genetics 93:* 37.

Klar, A.J.S., S. Fogel, and D.N. Radin. 1979b. Switching of a mating-type **a** mutant allele in budding yeast *Saccharomyces cerevisiae*. *Genetics 92:* 759.

Klar, A.J.S., J. McIndoo, J.B. Hicks, and J.N. Strathern. 1980. Precise mapping of the homothallism genes, *HML* and *HMR*, in yeast *Saccharomyces cerevisiae*. *Genetics 96:* 315.

Kushner, P.J., L.C. Blair and I. Herskowitz. 1979. Control of cell types by mobile genes—A test. *Proc. Natl. Acad. Sci. 76:* 5264.

McClintock, B. 1957. Controlling elements and the gene. *Cold Spring Harbor Symp. Quant. Biol. 21:* 197.

Nasmyth, K.A. and K. Tatchell. 1980. The structure of transposable yeast mating-type loci. *Cell 19:* 753.

Naumov, G.L. and I.I. Tolstorukov. 1973. Comparative genetics of yeast. X. Reidentification of mutators of mating-types in *Saccharomyces*. *Genetika 9:* 82.

Oshima, Y. and I. Takano. 1971. Mating types in *Saccharomyces:* Their convertibility and homothallism. *Genetics 67:* 327.

Rine, J., J.N. Strathern, J.B. Hicks, and I. Herskowitz. 1979. A suppressor of mating-type locus mutations in *Saccharomyces cerevisiae:* Evidence for and identification of cryptic mating-type loci. *Genetics 93:* 877.

Santa Maria, J. and D. Vidal. 1970. Segregacion anormal del "mating type" en *Saccharomyces. Inst. Nac. Invest. Agron. Conf. 30:* 1.

Takahashi, T. 1958. Complementary genes controlling homothallism in *Saccharomyces. Genetics 43:* 705.

Takano, I. and Y. Oshima. 1967. An allele specific and a complementary determinant controlling homothallism in *Saccharomyces oviformis. Genetics 57:* 875.

Takano, I., T. Kusumi and Y. Oshima. 1973. An α mating-type allele insensitive to the mutagenic action of the homothallism genes system in *Saccharomyces cerevisiae. Mol. Gen. Genet. 126:* 19.

Watson, J.D. and F.H.C. Crick. 1953. Molecular structure of nucleic acids: A structure for deoxyribose nucleic acid. *Nature 171:* 737.

Winge, O. and C. Roberts. 1949. A gene for diploidization of yeast. *C. R. Trav. Lab. Carlsberg Ser. Physiol. 24:* 341.

From Bacterial Flagella to Homeodomains

MELVIN I. SIMON
California Institute of Technology, Pasadena, California 91125

In celebrating Barbara McClintock, I want to describe how her ideas and experiments have provided a starting point and remain a continuing source of inspiration in our efforts to try to understand the function and evolution of genomes and regulatory mechanisms. However, I must first say something about her style of doing science. In an age of mega-laboratories that contribute incrementally to our understanding of genetics, Barbara's work as an individual experimentalist and her enormous creativity and insight stand in sharp contrast. They point to a way of doing science that engages the spirit as well as the intellect and involves the ability to organize ideas, insights, and experiments, rather than postdoctoral fellows, laboratory space, and grant requests.

My initial contact with Barbara McClintock's work came indirectly through an interest in gene stability in bacteria. There are many examples of bacterial systems where programmed changes in gene expression affect the ability of the organism to swim, to generate light, or to mount a virulent infection (Simon and Silverman 1983). These changes are characterized by frequencies that are much higher than the frequency of mutation. One early observation was that *Salmonella* could switch from having predominantly one type of flagellar antigen to a different type at a fixed frequency that was apparently genetically determined (Stocker 1949). When Joshua Lederberg and his student Tetsuo Iino applied the abortive transduction technique to the analysis of the genes that controlled the synthesis of *Salmonella* flagellar antigens, they found that the flagellar genes could exist in alternate "states" and that these controlled the phases of gene expression. Furthermore, they found that the regulatory state of the gene could be transferred along with the gene itself (Lederberg and Iino 1956). The phase variation apparently represented a relatively stable alternation in the expression of two different genes that encoded flagellar antigens. Lederberg realized that the behavior of this simple system had the landmarks and properties of the "controlling elements" that Barbara McClintock had described in maize (McClintock 1952). The detailed analysis of the genetic elements in the system was done by Tetsuo Iino and his students in Japan (Iino 1969). However, although the nature of the individual genetic elements that contributed to this system became clear, the molecular nature of the switch was obscure and its relationship to McClintock's transposing systems remained at the level of analogy.

It was another 20 years before it was finally possible to clone the "controlling" element in both the "on" and "off" states. By comparing the two states, we found that they were the result of a localized inversion of the DNA adjacent to one of the flagellar genes (Zieg et al. 1977). The inversion was driven by a site-specific recombination event that required the product of the Hin recombinase gene and two specific sites on either side of the promoter that controlled flagellar gene expression. It thus became clear that a site-specific recombination event was responsible for inverting a DNA segment and that phase variation correlated with the different inverted forms connecting or disconnecting the promoter from its downstream gene (Silverman and Simon 1980). How did this mechanism relate to transposition? The relationship began to emerge when it was discovered that the Hin recombinase was not the only site-specific recombinase that controlled gene expression. Similar and evolutionarily related systems were found associated with bona fide transposons such as Tn3, the transposing bacteriophage Mu, bacteriophage P1, and the cryptic e14 element (for review, see Simon et al. 1980). Barbara McClintock's work (McClintock 1965) provided a basis for a scenario that allowed us to rationalize the way in which these mobile recombinase modules might have evolved into controlling elements. The insertion of a transposon adjacent to a structural gene and subsequent rare recombination events driven by the recombinase that was part of the transposon could lead to the excision of most of the transposon but would leave the phase-controlling element in place. Environmental conditions that favored clones capable of phase shift could select for the survival of this recombinant. Support for this scheme was found by analyzing the sequences adjacent to the flagellar genes in *Escherichia coli*. We found sites at which possible transpositional insertions could have been involved in recombinational recruitment and the formation of the phase-controlling system (Szekely and Simon 1983). It started becoming clear to us how in bacteria, transposition and site-specific recombination might lead to the evolution of a simple program of alternating gene expression that could provide a clonal population with the means to survive dramatic environmental changes.

When we reread Barbara's papers and spoke with her at a meeting in Steamboat Springs in the early 1980s, it became clear that she had carried the notion much further. She suggested that transposition and the properties of transposable elements and attendant chromosomal rearrangements played a basic role in the evolution of the mechanisms that controlled the timing and patterns of gene expression required for development and differentiation. We began to think that although bacteria might lack the complexity and diversity of the regulatory mechanisms that she had observed in maize, the phase variation system may have retained the early ancestors, the anlage, from which some of these developmental regulatory systems evolved.

Work on phase variation and on the Hin recombinases has become much more oriented toward understanding the molecular structure of the recombinase, its mode of action, and its interaction with a specific enhancer protein and with DNA. We found that although the recombinase consisted of 189 amino acids, a small peptide of only 52 amino acids was sufficient to recognize the appropriate DNA-binding site specifically (Szekely and Simon 1983). Hin recombinase binds as a dimer to a 26-base-pair symmetrical site. The 52-amino-acid peptide binds to a half site, and we found that it had unique binding characteristics. It binds to both the major and minor grooves of the DNA, and the minor groove interaction is essential for binding to DNA (Sluka et al. 1990). Hin binding requires three specific base pairs in the major groove of the DNA-binding site and two base pairs in the adjacent minor groove. The arginine and proline residues identified as the portion of the peptide required for minor groove binding were found to be conserved among all the members of the Hin recombinase family (Mack et al. 1990). However, the most interesting correlation with the Hin DNA-binding site emerged from recent studies of regulatory systems in eukaryotes. The homeodomain proteins that are thought to be involved in the complex regulation of morphogenesis and development show the same pattern of DNA binding as the Hin recombinase. Thus, two members of the homeodomain family, *Antennapedia* and *engrailed*, have been found to preserve the same minor groove binding element as is found in the Hin recombinase family (Affolter et al. 1991). The three-dimensional structures of the homeodomain proteins that have been determined are consistent with the structure and mechanism of binding deduced and predicted by studies on Hin recombinase. The homeodomain proteins show 27% conserved amino acid sequence identity with Hin recombinase in the DNA-binding region. Thus, it is possible that the site-specific recombinases that evolved as part of the prokaryotic transposon system were incorporated through evolution into the structure of the regulators of cell lineage and morphogenesis.

As we acquire the tools to do the molecular archaeology and dig out these elements and compare them and their arrangement on genomes in a variety of organisms, we may begin to see the origins of the regulatory events that drive the evolution of complex organisms. It may take thousands of DNA sequences and rafts of computers before we can see this process as clearly as Barbara McClintock can see it in a yellowed sector on the green leaf of a clover plant.

References

Affolter, M., A. Percival-Smith, M. Müller, M. Billeter, Y.Q. Qian, G. Otting, K. Wüthrich, and W.J. Gehring. 1991. Similarities between the homeodomain and the Hin recombinase DNA-binding domain. *Cell 64:* 879.

Iino, T. 1969. Genetics and chemistry of bacterial flagella. *Bacteriol. Rev. 33:* 454.

Lederberg, J. and T. Iino. 1956. Phase variation in *Salmonella*. *Genetics 41:* 743.

Mack, D.P., J.P. Sluka, J.A. Shin, J.H. Griffin, M.I. Simon, and P.B. Dervan. 1990. Orientation of the putative recognition helix in the DNA-binding domain of hin recombinase complexed with the hix site. *Biochemistry 29:* 6561.

McClintock, B. 1952. Chromosome organization and genic expression. *Cold Spring Harbor Symp. Quant. Biol. 16:* 223.

———. 1965. The control of gene action in maize. *Brookhaven Symp. Biol. 18:* 162.

Silverman, M. and M. Simon. 1980. Phase variation: Genetic analysis of switching mutants. *Cell 19:* 845.

Simon, M.I. and M. Silverman. 1983. Recombinational regulation of gene expression. In *Gene function in prokaryotes* (ed. J. Beckwith et al.), p. 211. Cold Spring Harbor Laboratory, Cold Spring Harbor, New York.

Simon, M., J. Zieg, M. Silverman, G. Mandel, and R. Doolittle. 1980. Phase variation evolution of a controlling element. *Science 209:* 1370.

Sluka, J.P., S.J. Horvath, A.C. Glasgow, M.I. Simon, and P.B. Dervan. 1990. Importance of minor-groove contacts for recognition of DNA by the binding domain of hin recombinase. *Biochemistry 29:* 6551.

Stocker, B.A.D. 1949. Measurement of the rate of mutation of flagellar antigenic phase. *J. Hyg. 47:* 398.
Szekely, E. and M.I. Simon. 1983. DNA sequence adjacent to flagellar genes and the evolution of flagellar phase variation. *J. Bacteriol. 155:* 74.
Zieg, J., M. Silverman, M. Hilmen, and M. Simon. 1977. Recombinational switch for gene expresssion. *Science 196:* 170.

Discovery of Tc1 in the Nematode Caenorhabditis elegans

PHILIP ANDERSON
University of Wisconsin, Madison, Wisconsin 53706

SCOTT W. EMMONS
Albert Einstein College of Medicine, Bronx, New York 10461

DONALD G. MOERMAN
University of British Columbia, Vancouver, British Columbia, Canada, V6T 2A9

From today's perspective, the *Caenorhabditis elegans* transposable element Tc1 looks perfectly ordinary. Its size, its inverted repeat structure, its encoded proteins, and its effects on the *C. elegans* genome are typical of transposable elements in many organisms. All known *C. elegans* strains contain multiple copies of Tc1, and in many of these strains, the elements excise and transpose frequently (for review, see Moerman and Waterston 1989). Aside from its inherent interest as a mobile DNA element, Tc1 is an important tool for *C. elegans* investigators. Starting in the mid 1980s, Tc1 "tagging" and chromosome "walking" from Tc1-induced restriction-fragment-length polymorphisms (RFLPs) provided the first realistic strategies to clone interesting genes (see, e.g., Greenwald 1985; Moerman et al. 1986; Ruvkun et al. 1989). Continuing progress with the integrated *C. elegans* genetic and physical map (Coulson et al. 1986, 1988) and with DNA transformation (Stinchcomb et al. 1985; Fire 1986) offers additional strategies today. But without Tc1, the *C. elegans* field might just now be emerging from its premolecular age.

C. elegans investigators were motivated, at least in part, to find a transposon by their desire to clone genes and to study them molecularly. Many of the early *C. elegans* workers had backgrounds in biochemistry and molecular biology, particularly the molecular biology of bacteriophage. They were attracted to *C. elegans* by the hope that it would be possible to study molecular mechanisms underlying development or behavior in this simple multicellular animal. As they had done with phage, these workers wanted to use a combined genetic and molecular approach. This required building a bridge between the genetic world of mutations and the physical world of DNA. By the mid 1970s, transposons had been identified in and cloned from bacteria, yeast, and *Drosophila*. Their potential utility in joining genetic and physical maps was obvious. As cloning studies began in *C. elegans*, the exciting findings regarding *P* elements in *Drosophila* demonstrated just how valuable a well-characterized transposon might be. With this strong motivation and with a substantial body of background information, the discovery of transposons in *C. elegans* seemed inevitable. The only questions were how, when, and by whom.

As it turned out, four separate lines of investigation led to Tc1 almost simultaneously. The four paths moved independently and in parallel, but frequent communication between the investigators quickened the pace of progress. By early 1984, the four paths yielded an understanding—at least in outline form—of Tc1 and its genetic properties: (1) Scott Emmons and David Hirsh, while investigating the possibility of DNA rearrangement during somatic development, discovered that random genomic clones were frequently polymorphic in different strains of *C. elegans* (Emmons et al. 1979). One of these polymorphisms proved to be caused by Tc1 (Emmons et al. 1983). (2) Jim Files and David Hirsh, while cloning and analyzing the *C. elegans* actin genes (Files et al. 1983), discovered a nearby insertion/deletion RFLP. The inserted DNA proved to be Tc1 (Liao et al. 1983; Rosenzweig et al. 1983a). (3) Phil Anderson, attempting to identify active families of transposons directly, isolated and analyzed spontaneous mutations affecting a muscle myosin heavy-chain gene (Eide and Anderson 1985a). Anderson and colleagues successfully "trapped" a number of Tc1 elements within the myosin heavy-chain gene (Eide and Anderson 1985b). (4) Don Moerman and Bob Waterston, while screening for electrophoretic variation among muscle proteins, discovered that the "Bergerac" strain of *C. elegans* exhibits high spontaneous mutation rates (Moerman and Waterston 1984). The high mutability of Bergerac proved to be due to frequent transposition and excision of Tc1 (Moerman et al. 1986). Thus, the trail to Tc1 combined deliberate experimentation and serendipitous observations in varying proportions. But whatever the mix, all of these investigations benefited from the growing understanding of transposable elements in other organisms.

Molecular Approaches to Tc1

Two of the four paths to discovery of Tc1 began in David Hirsh's laboratory at the University of Colorado. The Hirsh laboratory had been isolating and studying temperature-sensitive developmental mutants of *C. elegans* for several years. When Scott Emmons arrived in 1977, after 2 years as a postdoctoral fellow in Don Brown's laboratory studying cloned *Xenopus* 5S genes, the Hirsh laboratory needed ways to study the mutants molecularly. Cloning the genes seemed of obvious value, but there was no known way to do it at the time. Emmons decided to plunge into a cloning project anyway, more or less "to just get started." Besides, he already knew how to do the still-new recombinant DNA methods.

Emmons decided to look for DNA rearrangements during development. Little was known at the time about molecular mechanisms of determination and differentiation, but stable rearrangement of DNA had long been an attractive idea (McClintock 1952). Immunoglobulin genes had recently been shown to undergo rearrangement during lymphocyte development (Brack et al. 1978). DNA inversion was demonstrated as the molecular switch of *Salmonella* phase variation (Zieg et al. 1977). In addition, it had been known for decades that somatic and germ-line chromosomes of certain organisms (including some nematodes) are strikingly different. During chromatin diminution of the nematode *Parascaris equorum*, for example, chromosomes are highly fragmented, and large segments of chromatin are discarded during early development (for review, see Wilson 1925). The process occurs only in somatic cells; germ-line nuclei are unaffected. Could such processes be widespread during development? A great deal of DNA rearrangement could be taking place on a scale too small to be seen cytologically. Sulston and Brenner (1974) had not detected differences between *C. elegans* germ-line and somatic DNAs using C_0t analysis, but many types of rearrangements would have been missed.

Emmons designed experiments that might reveal somatic rearrangement if it occurred. His method was simple. Random genomic clones were isolated and used as hybridization probes in genomic Southern blots. Restriction digests of germ-line and somatic DNAs were probed with these random clones. If any of the clones was at or near a site of somatic rearrangement, then the Southern blot pattern of the two DNAs would be different. Germ-line DNA was prepared from *C. elegans* sperm. Somatic DNA, contaminated with varying amounts of germ-line DNA, was isolated from hermaphrodites synchronized at various stages of postembryonic development. In newly hatched larvae, greater than 99% of such DNA comes from somatic nuclei. Thus, relatively pure germ-line and somatic DNAs could be compared. This experimental protocol was relevant not only to the question of developmental DNA rearrangements, but also to the general problem of cloning mutant genes. Emmons and Hirsh realized that certain types of mutations might be recognized by differences in the arrangement of restriction sites.

Emmons knew that this experiment was, at best, a long shot. DNA rearrangements would have to be widespread for detection. Hybridization probes were to be tested one by one, and each would detect only rearrangements in its immediate vicinity. Even a monumental effort would sample but a small portion of the *C. elegans* genome. As insurance against his expectation of negative results, Emmons decided to include two additional genomic DNAs in the blots. One DNA was obtained from *Caenorhabditis briggsae*, a nematode species thought to be closely related to *C. elegans* (Nigon and Dougherty 1949). The second DNA was obtained from *C. elegans* variety "Bergerac," a wild-type strain that is distinct from the standard "Bristol" strain. The Bristol strain, collected in Bristol, England sometime before 1959 (Nicholas et al. 1959), was adopted by Sydney Brenner as a laboratory standard. It is considered the canonical *C. elegans* wild type. The Bergerac strain, collected in Bergerac, France during the 1940s (Nigon 1949), proved to be a key player in the discovery of Tc1.

Emmons included Bergerac and *C. briggsae* DNAs in the Southern blots as positive controls. Although hoping otherwise, he fully expected that somatic rearrangements would not be detected. But, how far diverged from the canonical wild-type *C. elegans* would one have to look before detecting differences? Would *C. elegans* genomic clones hybridize to *C. briggsae* DNA? Would restriction maps of Bristol DNA be the same as those of Bergerac DNA? These may not sound like stimulating questions today, and they were not exactly earth-shaking at the time, but at least some useful information would come from the experiments. In addition, if RFLP differences between cross-fertile strains could be found, these could serve as genetic markers to provide the first links between the genetic map and DNA.

It is fortunate that Emmons included Bergerac DNA, for it led to his cloning of Tc1. Fifteen percent of the detected restriction fragments had differing sizes in Bristol and Bergerac DNAs (Emmons et al. 1979). The very first clone tested (clone pCe1; it was probably one of the first *C. elegans* recombinant clones anywhere) hybridized to several *Bam*HI fragments, including one estimated to be 7.0 kb in Bristol DNA and 8.7 kb in Bergerac DNA. It took some time for Emmons to sort out the origins of these hybridizing fragments. Clone pCe1 contains repetitive DNA, and the 7.0 kb/8.7 kb polymorphism results from hybridization of the repetitive DNA portion of pCe1. By late 1978, Emmons had established that the RFLP associated with a second random Bristol clone was an insertion/deletion polymorphism of approximately 1.7 kb (as opposed to a point mutation that destroyed or created a restriction site). However, cloning the Bergerac alleles of these polymorphisms (and, hence, Tc1) would await Emmons's move as an Assistant Professor to the Albert Einstein College of Medicine in the fall of 1979.

Shortly after Emmons discovered his RFLPs, and while he was still sorting out what they were, Jim Files discovered yet another insertion/deletion RFLP that was 1.7 kb in size. Files was also a postdoctoral fellow in David Hirsh's laboratory. His project was to clone and analyze the *C. elegans* actin genes. It was clear from early work by Henry Epstein and Bob Waterston (Epstein et al. 1974; Waterston et al. 1977, 1980; Zengel and Epstein 1980) that genetic analysis of *C. elegans* muscle was going to be a successful and productive line of investigation. As a prelude to genetic analysis of actin function, Hirsh decided to clone *C. elegans* actin genes (Files et al. 1983). After doing so, Files discovered a 1.7-kb insertion/deletion RFLP adjacent to the chromosome V actin gene cluster when he compared Bristol and Bergerac DNAs. It seemed striking that the number "1.7 kb" had come up three different times in the Hirsh laboratory: Two of Emmons' clones and the Files actin clone all detected insertion/deletion RFLPs of 1.7 kb. In all three cases, the larger of a polymorphic pair of restriction fragments was in Bergerac. By the end of 1979, it was clear to Emmons, Files, and Hirsh that there might be a unifying explanation for these frequent RFLPs. The Bergerac genome might contain many copies of a 1.7-kb repeated sequence (a transposon?) that is absent (or in differing positions) in the Bristol genome. Because the potential transposon sequences were in Bergerac, for which no libraries were available, three additional years would be needed to prove this point. Files was immersed in the molecular analysis of actin genes. Emmons was finishing his postdoctoral work with Hirsh and moving to the Albert Einstein College of Medicine. With his technician, Lewis Yesner, Emmons's first priority at Einstein was to construct Bergerac clone banks and to screen them with polymorphic Bristol clones. The first of several champagne corks saved by Emmons is labeled "Lewis's λ clone bank, 12/11/80."

By the middle of 1981, both Files and Emmons had cloned the Bergerac alleles of their polymorphic clones. At the Cold Spring Harbor Laboratory Worm Meeting of May, 1981, Files reported that the actin-linked 1.7-kb insertion in Bergerac was repetitive DNA. It hybridized to "approximately 20" restriction fragments of Bristol DNA and "a lot" of fragments of Bergerac DNA. Although not proved, a transposable element was certainly a prime suspect, and Files suggested this during his presentation. Emmons cloned the Bergerac allele of pCe1 (more precisely, the Bergerac 8.7-kb cross-hybridizing sequence) and the Bristol 7.0-kb fragment shortly after the Worm Meeting. He confirmed that the Bergerac insert was a dispersed repeated sequence, and he estimated the Bergerac copy number to be about 200. Tc1 was now in hand, although it was not yet called such. There was still no proof that it was mobile. That proof would wait 3 more years, until two different investigators succeeded in isolating Tc1-induced mutations.

Both the Hirsh and the Emmons laboratories were now hot on the trail of Tc1. During 1981 and 1982, both of these laboratories characterized the structure, sequence, and some of the dynamics of Tc1. We shall return to this storyline later, but before doing so, the activities of Phil Anderson and Don Moerman should be described. Although not directly working on Tc1 during this period, their interests converged with those of Emmons and the Hirsh laboratory during 1983 and 1984. Both Anderson and Moerman were approaching Tc1 from a genetic perspective.

Genetic Approaches to Tc1

Anderson arrived as a new postdoctoral fellow in Sydney Brenner's laboratory at the MRC Laboratory in Cambridge, England in January 1978. His initial project was to find extragenic suppressors of some of the existing muscle mutations. The hope was that interacting muscle proteins might be discovered or that additional informational suppressors might be identified. *sup-5*, later shown to be an amber tRNA suppressor (Wills et al. 1983), had already been discovered (Waterston and Brenner 1978). Anderson came to realize that there just were not very many muscle mutants of known etiology to work with. There were a handful of mutations affecting the *unc-54* myosin heavy-chain gene (MacLeod et al. 1977a,b) and a few affecting the *unc-15* paramyosin gene (Waterston et al. 1977), but that was it. Coming from a bacterial genetics background, Anderson felt this was a sad state of affairs. He decided that a better way was needed to isolate muscle mutants. He wanted to use a positive *selection* for mutants, as he was accustomed to doing in *Salmonella*.

After discussions with Brenner, Anderson adopted a "loss-of-dominance" scheme. Loss-of-function mutations affecting the *unc-54* myosin heavy-chain gene are fully recessive. Brenner described to Anderson an exceptional allele of *unc-54* that is strongly dominant. Heterozygotes are paralyzed, although not as severely as homozygotes. (Thus, this allele, *unc-54(e1152)*, is more precisely described as being incompletely dominant.) Loss-of-function mutations affecting *unc-54(e1152)* would convert a dominant heterozygote from a paralyzed animal to one with wild-type motility. The problem is that *C. elegans* is a self-fertilizing hermaphrodite. Just like Mendel's peas, one quarter of the selfed offspring of *unc-54(e1152)/+* heterozygotes are wild-type. A balancer chromosome was needed such that large populations of heterozygotes could be propagated. The needed balancer chromosome would be *unc-54(+)*, but it would be inviable when homozygous and would hopefully suppress crossing-over in the vicinity of *unc-54*.

It took Anderson about 1 year to generate the needed balancer and to demonstrate that the selection worked. Using the balancer chromosome, large populations of *unc-54* dominant heterozygotes could be grown. Following mutagenesis, most of the wild-type offspring contained new loss-of-function mutations introduced into the *unc-54(e1152)* gene copy (Anderson and Brenner 1984). These mutants were easy to obtain, and Anderson isolated several dozen of them.

At some point along the way, Anderson began to think about isolating spontaneous, rather than mutagen-induced, mutations. This was prompted by work emerging from maize, *Drosophila*, and yeast. Transposons were turning up in genes and proving to be the causes of spontaneous mutations (Green 1980; Fedoroff 1983; Roeder and Fink 1983). If Anderson could isolate spontaneous (rather than mutagen-induced) *unc-54* mutants, Southern blots would identify which of them were insertions. The *unc-54* gene was very close to being cloned in Brenner's laboratory (MacLeod et al. 1981), and Anderson knew that the necessary hybridization probe would be forthcoming. In an attempt to "trap" transposons in *unc-54*, Anderson turned to his selection for mutations. Rather than mutagenize the balanced heterozygotes, he picked spontaneous wild-type offspring. Alas, all of them were recombinants. The balancer chromosome just was not good enough. Crossing-over between the *e1152*-containing chromosome and the balancer chromosome generated wild-type chromosomes at frequencies higher than those of spontaneous mutations. Disappointed, Anderson spent the remainder of his time in Cambridge characterizing genetically and physically various mutagen-induced *unc-54* mutations (Anderson and Brenner 1984).

The 1981 Cold Spring Harbor Worm Meeting was an important one for Anderson's hunt for transposons. During a brief and somewhat frenzied conversation with Bob Horvitz on the stairs of Blackford Hall, Anderson was lamenting about his unsuccessful attempts to isolate spontaneous *unc-54* mutations. Horvitz replied, in essence, "See Joan Park. She is a graduate student in my lab and has a new way of isolating *unc-54* mutants. They occur as partial suppressors of *unc-105*(*n490*). And, she has even seen some spontaneous mutants."

Anderson quickly found Joan Park and learned that she had indeed isolated spontaneous alleles of both *unc-54* and the muscle-affecting gene *unc-22* as partial revertants of *unc-105*(*n490*). Park had isolated *unc-105*(*n490*) in a general screen for dominant mutations (Park and Horvitz 1986a). *n490* animals are strongly paralyzed due to sustained contractile rigor of muscle cells. Mutations in genes like *unc-54* and *unc-22*, which lead to flaccid paralysis, are partial suppressors of *n490*, yielding animals that have improved motility (Park and Horvitz 1986b). Park sent *unc-105(n490)* to Anderson shortly after the Worm Meeting. He picked a few revertants during his last few months as a postdoc in Cambridge. But Anderson did not even have time to find out what genes were involved, let alone whether any of them were transposon insertions into *unc-54*. A major assault on *n490* would be his first order of business when arriving as an Assistant Professor at the University of Wisconsin in the fall of 1981.

Between October and December of 1981, Anderson picked 185 spontaneous revertants of *n490*, 65 of which proved to be *unc-54* alleles. During late 1981 and throughout 1982, Anderson and Dave Eide, a graduate student who had joined Anderson's laboratory, analyzed the structures of these spontaneous mutations. Sadly, none of them proved to be transposon insertions (Eide and Anderson 1985a). This was a major disappointment for Anderson. He was sure that transposon insertions should be isolated among the spontaneous mutations. Month after month the blots continued; not a single insertion was found. Two mutations proved to be unstable, reverting to *unc-54*(+) spontaneously. For a while, there was cautious excitement that transposon insertions had been found. Southern blots showed that these two mutants had DNA "inserted" within *unc-54*. However, the inserts were about 10 kb in size and seemed too large to be transposons of the standard sort. Further analysis showed the mutations to be tandem duplications (Eide and Anderson 1985a).

Anderson's failure to isolate Tc1 insertions among the spontaneous mutations was due to the fact that *unc-105*(*n490*), like most *C. elegans* mutants, is descended from the Bristol wild-type strain. Although Bristol contains about 30 copies of Tc1, they do not transpose or excise at detectable frequencies in the germ line. Tc1 is highly active in the Bergerac germ line, however, as Anderson and Moerman would both discover a year or two later.

By the middle of 1982, Emmons and Files had established that the copy number of the repeated DNA now called Tc1 was high in Bergerac and low in Bristol. And, as described below, Emmons was reporting an extraordinarily high frequency of Tc1 excision in Bergerac somatic cells. Why didn't Anderson cross *n490* into Bergerac and isolate spontaneous *unc-54* mutations in 1982? There are two answers. The first is a geneticist's conceit. Anderson was suspicious of Tc1. Since Tc1 was identified only as a piece of cloned DNA, Anderson thought that it might well be a relic of a transposon. After all, he thought, the genomes of most organisms are probably full of dead and dying transposons. Anderson did not want a *dead* transposon, he wanted a *live* one. If Tc1 showed up causing any of his spontaneous mutations, all the better. But, if something other than Tc1 appeared, he would have something unique, and certainly more interesting genetically. Second, Anderson wanted to stay away from things that others were doing. Emmons and the Hirsh laboratory had "discovered" Bergerac, and it was their project. Anderson did not want to move in on someone else's potentially active strain; he wanted his own active strain.

Anderson thus took to the flower beds of Wisconsin. During late 1982, after finally deciding that there just were not any insertion mutations in Bristol, he collected several wild-type *C. elegans* strains. But before Anderson used any of the Wisconsin strains for spontaneous mutant hunts, he learned of strain DH424, a wild-type *C. elegans* collected by Louise Liao in California (Liao et al. 1983). Like Bergerac, DH424 was described as having a high copy number of Tc1. Anderson received DH424 from Liao, crossed *n490* into the background, and collected 31 spontaneous *unc-54* mutations. None of them proved to be insertions (Eide and Anderson 1985b). Anderson was beginning to doubt whether transposons would ever turn up. He was even more convinced now that Tc1 was a genomic remnant, one that may have had a vigorous prior life, but which was now senescent at best.

Don Moerman, meanwhile, had joined Bob Waterston's laboratory at Washington University as a postdoctoral fellow in early 1980. There he began experiments that would constitute the fourth line of investigation leading to Tc1. Moerman was not new to *C. elegans*. He had worked as a graduate student with David Baillie studying the muscle-affecting gene *unc-22*. Moerman characterized *unc-22* genetically and, most importantly, developed techniques that select for *unc-22* mutants in large populations (Moerman and Baillie 1979). Muscle contraction in *unc-22* mutants is abnormally regulated; homozygous mutants display a unique twitching phenotype. Nicotine, an agonist of acetylcholine, paralyzes wild-type animals by inducing contractile rigor but exacerbates twitching of *unc-22* heterozygotes or homozygotes. The conditional dominance imparted by nicotine and the ease of seeing twitching permit detection of very rare *unc-22* mutants.

Although continuing in the general area of muscle genetics, Moerman did not intend to work on *unc-22* in the Waterston laboratory. His project was to characterize the ATPase of a set of missense alleles affecting the *unc-54* myosin heavy-chain gene (see Moerman et al. 1982). This was going to be complicated, however, by the fact that *C. elegans* contains four different myosin heavy-chain isoforms (Epstein et al. 1974; Schachat et al. 1977; Waterston et al. 1982). Purifying only the *unc-54*-encoded isoform would be difficult. With a goal of possibly being able to eliminate the other isoforms mutationally, Moerman began a second project in the fall of 1981 to map genetically the genes encoding the non-*unc-54* myosin heavy-chain isoforms. He hoped that their genetic map positions would coincide with known muscle-affecting loci. If not, at least a genetic region would be defined in which to concentrate future mutant hunts. Moerman's strategy was to first identify electrophoretic variants of the myosin heavy chains and to then map them using standard linkage analysis. Bergerac seemed to be a likely source for the needed polymorphisms, because Emmons had already reported the high incidence of RFLPs in Bergerac (Emmons et al. 1979).

Moerman received Bergerac and began both genetic and biochemical experiments. He mutagenized Bergerac and isolated a collection of visible mutants, concentrating on muscle-defective mutants. If he was going to study muscle proteins in Bergerac, he wanted to be sure that the same genes and mutants were available. Moerman isolated a number of ethylmethane sulfonate (EMS)-induced *unc-22* mutants; they behaved just like typical Bristol *unc-22* mutants. Then he isolated a number of formaldehyde-induced *unc-22* mutants. Some of these mutants were genetically unstable; they reverted to *unc-22*(+) at high frequencies (10^{-3} to 10^{-4}). Moerman knew that formaldehyde could induce tandem duplications (Slizynska 1957), and he believed that the reversions resulted from unequal crossing-over between duplicated segments. It would not occur to him for several months that transposons might be the cause of the instability or that the mutational events were spontaneous, rather than mutagen-induced.

During this period, Moerman searched for the electrophoretic variants he needed. He isolated myosin proteins from Bristol and Bergerac, cleaved them with cyanogen bromide, and analyzed the resulting peptide fragments on two-dimensional gels. More than 100 peptides were resolved, so this was a sensitive screen for electrophoretic variants. Remarkably, the peptide patterns of Bristol and Bergerac were identical. This disappointing result put an end to the myosin heavy-chain mapping project. But there was still the puzzle of the unstable *unc-22* alleles. If Bristol and Bergerac were as divergent as the Emmons paper indicated, why weren't there any electrophoretic variants? The answer must be that the Bristol/Bergerac divergence is not caused by "point" mutations (the kind that yield electrophoretic variants), but rather by differences in repetitive DNA.

Could these repetitive DNA differences be related to the unstable *unc-22* mutants? By early 1982, Moerman was suspicious of the high proportion of unstable *unc-22* mutants following formaldehyde mutagenesis. When some of the "EMS-induced" mutants also proved to be unstable, he knew that something funny was going on. Moerman now questioned whether the mutants were mutagen-induced at all. In March 1982, Moerman looked for spontaneous *unc-22* mutants in Bergerac. He found them at the remarkably high frequency of 10^{-4}. Moreover, they were all unstable, reverting to *unc-22(+)* at frequencies between 10^{-3} and 10^{-4}. He looked for spontaneous mutants in Bristol. None were isolated in over 10^6 animals. Bergerac was a mutator strain!

Moerman genetically characterized the mutator activity of Bergerac during the spring and summer of 1982. He had named the mutator activity "omega," but this moniker would not stick. Moerman established that he could go through repeated cycles of mutant isolation, reversion, mutant isolation, etc. He was thinking of "flip/flop" mechanisms for mutation and reversion, along the lines of *Salmonella* phase variation, a popular paradigm for gene instability at the time. By June of 1982, Moerman established a number of genetic properties of omega: (1) in the presence of omega, spontaneous mutations occur at high frequencies; (2) in the presence of omega, the mutations are unstable; (3) omega is unlinked to *unc-22*; (4) replacing the Bergerac allele of *unc-22(+)* with the Bristol allele (in the presence of omega) still results in a high frequency of spontaneous mutations (this argued against a flip/flop mechanism); and (5) following an outcross to Bristol, the unstable *unc-22* mutations were still unstable. Moerman incorrectly concluded that reversion did not require omega. (Bergerac contains several loci that activate Tc1 transposition and excision [Mori et al. 1988]; a single backcross did not remove all of them.) With hindsight, these "facts" concerning omega sound straightforward, but several months would pass before Moerman got the omega story approximately correct. This was due in part to the fact that the "phenotype" being scored in these genetic experiments was elevated frequencies of mutation or reversion, which took a few generations to observe.

During this period, Moerman read the literature concerning "unstable" mutations, especially the work of Barbara McClintock on *Ac/Ds* in maize and of Mel Green on the *white* and *singed* loci in *Drosophila* (McClintock 1948, 1957; Green 1980). The similarity of "omega" to these systems was striking and led Moerman to think more explicitly about transposons being the cause of Bergerac mutability. But what transposon? Evidence was accumulating from the Hirsh and Emmons laboratories showing that Tc1 looked very much like a transposon. Moerman wrote to Emmons in August, 1982: "An entertaining hypothesis is that what I call omega is equivalent to what you call Tc1. I realize that this is a long shot, but it is easy enough to test." Unfortunately, it was not so easy to test. Moerman did not have an *unc-22* clone to use as a hybridization probe. He would have to use Tc1 as a probe, but there were no "zero copy" strains to eliminate Tc1 elements throughout the genome. Ultimately, Moerman and his collaborators used Emmons's Tc1 probe to clone *unc-22*, but it would take over a year to do it.

Moerman was thinking specifically of Tc1 because he learned of Emmons' and Hirsh's unpublished results in the *C. elegans* newsletter in January and July of 1982. The importance of the newsletter for advancing the pace of *C. elegans* research cannot be overstated. Known affectionately as The Worm Breeders Gazette (or just the Gazette), the newsletter is a continuing source of valuable information pertaining to all aspects of *C. elegans* research. The Gazette began in 1975, when the *C. elegans* field was small and fragile. The nematode research community embraced a philosophy of openness that dictated, "whatever you know, you put into the Gazette." This philosophy continues today. Abstracts report new techniques, mutants, or simply work in progress. Isolated observations, detailed descriptions, or technical minutia inappropriate for formal publication are contributed for all to share. It remains an essential part of the *C. elegans* research enterprise.

Molecular and Genetic Approaches Converge on Tc1

The stage was set in 1982 for rapid progress with Tc1. Four groups were closing in on it. Louise Liao and Brad Rosenzweig had begun to work on Tc1 in the Hirsh laboratory, having received the clone from Files. Emmons had cloned Tc1 at Einstein and was continuing his molecular analysis. Moerman was sorting out the "rules" of Bergerac mutability, and Anderson was isolating spontaneous *unc-54* mutants like crazy, but not finding any transposons. Before Anderson and Moerman would have any success with a genetic approach, the molecular work progressed rapidly. Rosenzweig and Liao reported in the January, 1982 Gazette that Tc1 had inverted repeat termini and that its copy number varied from 30 to 50 in different strains. Their abstract was cautiously titled "Is the Bergerac/Bristol DNA polymorphism a transposon?"

Emmons then made an important discovery, one that made Tc1 look very much alive. Using a unique sequence probe covering a Tc1 insertion site, he detected a remarkably high frequency of Tc1 excision. The probe hybridized to bands of 7.0 kb in Bristol DNA and 8.7 kb in Bergerac DNA. (The difference is Tc1.) Bergerac DNA, however, contained a faint band of 7.0 kb. Two other cloned sites of Tc1 insertion showed a similar phenomenon. In each case, a few percent of Bergerac DNA had the exact same size as Bristol DNA. One possible explanation was that, for each site, a few percent of Bergerac DNA no longer contained Tc1. This was remarkable. Excision of transposons was supposed to be rare and only detected by sensitive genetic reversion assays. Excision detected on a Southern blot would make Tc1 possibly the most active transposon known! Unfortunately, there were some other, less interesting, explanations as well. Could it be that the laboratory's cultures of Bergerac worms were contaminated with Bristol worms? Could sloppy procedures have cross-contaminated the DNAs, or the lanes on the gels? Emmons is now ashamed to recall the number of times he questioned Yesner's technical expertise.

To put these doubts to rest, and to gain information about the frequency of excision, Emmons and Yesner performed an experiment to test for the presence of the "excision band" in independent, small cultures. Eight single Bergerac hermaphrodites were allowed to produce children, grandchildren, and great grandchildren. This yielded just enough worms to do a Southern blot. DNA from each population was isolated and tested for excision. It was a Luria/Delbrück-style fluctuation test (Luria and Delbrück 1943). If cross-contamination of worm cultures was the explanation, none of the populations should have shown excision. If excision was rare but occurred in the germ line, some cultures might show excision and some not, depending on when the first event happened. If excision was frequent in most or all cells, all of the cultures should show the same faint band. The third result was obtained. Excision must be occurring at high frequency in most or all cells! Emmons and

Yesner would later demonstrate that the high frequency of excision is limited to somatic cells (Emmons and Yesner 1984). Emmons, along with his graduate student KeSan Ruan, and, independently, Ann Rose and Terry Snutch would also detect extrachromosomal linear and circular copies of Tc1 in Bergerac DNA (Rose and Snutch 1984; Ruan and Emmons 1984). Such extrachromosomal elements are presumably the products of high frequency somatic excision.

During early 1982, Emmons demonstrated that most of the approximately 200 copies of Tc1 in Bergerac have similar, if not identical, restriction maps. Tc1 repeats are inserted at unique sites with respect to Tc1 DNA but at variable sites with respect to the genomic DNA. This was the finding that finally convinced Emmons that Tc1 could be called a transposon. He reported this in the July, 1982 Gazette and proposed the name "Tc1" for the element. The second champagne cork in Emmons's collection is labeled "Paper sent to Cell-Tc1 9/14/82."

From his reading of the literature, Moerman realized that the hallmarks of a transposable element system are high frequencies of spontaneous mutations, genetic instability of those alleles, and sensitivity of these two properties to genetic background. So far, he had only demonstrated that the *unc-22* mutations exhibited the first two properties. In August 1982, Moerman repeatedly crossed several unstable *unc-22* mutations into a Bristol background; they no longer reverted spontaneously. A single cross back into Bergerac activated them anew (Moerman and Waterston 1984). This corrected his earlier conclusion that reversion did not require "omega." He and Waterston were now convinced that the mutator system in Bergerac was due to transposons. He wrote to Emmons in November, 1982: "I believe these observations further strengthen the argument that these events are transposon mediated. They are, I think, very similar to Engels' results with *P* elements and the *singed* locus ..., all of which is reminiscent of the *Ds* element in maize."

By early 1983, Moerman was looking for copies of Tc1 on Southern blots that cosegregated with *unc-22* and that might be the cause of the mutations. This was not easy. It was difficult to eliminate all of the *unc-22*-linked Bergerac Tc1 elements and to resolve over 30 bands of hybridization. In May, 1983, he identified *stP1*, a Tc1 RFLP that mapped to the left of *unc-22*, but no Tc1 elements within *unc-22* itself.

The May 1983 Worm Meeting was an important one for Tc1. A persuasive case was building that Tc1 was mobile, even though no one had yet seen a transposition event. Liao and Rosenzweig reported the sequence of Tc1. If ever a piece of DNA looked like a transposon, that was it. Tc1 was 1610 bases long, had 54-bp terminal inverted repeats, contained element-encoded open reading frames, and created a 2-bp target site duplication upon insertion (Rosenzweig et al. 1983b). Emmons reported the high frequency of Tc1 somatic excision. Both Emmons and Liao reported that different wild-type *C. elegans* strains have differing patterns of Tc1 hybridization. Thus, the element was mobile at some point during divergence of the strains. Moerman reported his results concerning mutability of *unc-22* in Bergerac. It was apparent that Tc1 might be the culprit. Anderson reported the absence of transposon insertions among 83 spontaneous and "heat-shock-induced" *unc-54* mutants in Bristol. His sobering conclusion was that if transposon insertions occur, they do so at frequencies less than 5×10^{-9}.

Anderson left the Worm Meeting convinced that Bergerac must somehow be special. There was just too much going on to ignore: the high copy number of Tc1, the extraordinary frequency of excision, the mutability of *unc-22*. It all just "smelled" of transposons. He thought, "We just have to do this in Bergerac. If Tc1—or anything else—is actively transposing in Bergerac, we can trap it in *unc-54*. No more looking for our own active strain; let's just do it in Bergerac!" He and Dave Eide returned to Madison, crossed *n490* repeatedly into Bergerac, and collected 18 spontaneous *unc-54* mutants. Even before the Southern blots could be done, several of the mutants reverted spontaneously to *unc-54*(+). Anderson repeated some of Moerman's genetic analysis of "omega" and established that the unstable *unc-54* mutations behaved just like Moerman's unstable *unc-22* mutants. By late 1983, Eide's Southern blots proved that 10 of the 18 spontaneous mutations contained approximately 1.6-kb insertions within *unc-54*. It would require another 6–9 months before all of the insertions were known to be Tc1, but this was certainly the breakthrough they had worked so hard for. The champagne was popped in about November of 1983.

After the 1983 Worm Meeting, Moerman continued a long and frustrating search for Tc1 elements that might be causing his unstable *unc-22* mutations. None were found. It was during this search that Guy Benian, another postdoctoral fellow in Waterston's laboratory, joined the project. Their November, 1983 Gazette abstract was downbeat: "We feel that Tc1 is unlikely to be responsible for generating the *unc-22* mutations . . . [but] we cannot rule out the possibility that the critical Tc1 element is in a band comigrating with other Tc1-containing segments." This was exactly the case, as they would soon learn. Their inability to find a Tc1 associated with *unc-22* led them to consider "walking" to *unc-22* from *stP1* using the approach pioneered in *Drosophila* (Bender et al. 1983). In the process of preparing strains for this project, they found that *stP1* was not only absent from Bristol, but also absent in Bergerac. *stP1* represented a new Tc1 insertion!

The transposition that gave rise to *stP1* demonstrated to Moerman and Benian that Tc1 was mobile. It induced them to renew their search for causative Tc1 elements in the unstable *unc-22* alleles. They finally identified a convincing Tc1-hybridizing band that cosegregated with *st137*, one of their unstable mutants. The element was present in all mutant-bearing strains and absent in isogenic revertants and in intragenic *unc-22*(+) recombinants. The cosegregating Tc1 was cloned in February, 1984, and the flanking *unc-22* DNA was subcloned shortly thereafter. Using this unique sequence probe, *st137* and a number of other *unc-22* alleles were shown to have Southern blot abnormalities. Waterston was not in the laboratory the night that the crucial autoradiogram was developed. But, when phoned, he came in to see it, and under his arm he carried a bottle of fine red wine. Moerman remembers, "We toasted success, but even as we drank I periodically reexamined the autoradiogram. But there was no mistake this time—and no sleep that night either." The results proved both that Moerman and Benian had cloned *unc-22* and that Tc1 was the cause of the unstable mutations (Moerman et al. 1986). Both Anderson and Moerman reported their identification of Tc1-induced mutants in the May, 1984 Gazette.

What were the important social and scientific components of this process of discovery? Serendipity was prominent. In at least three instances, the experiments that ultimately led to Tc1 were designed for other purposes. Unexpected observations were pursued, opening new lines of research. Good communication was also important. Both the Worm Breeders Gazette and the biennial *C. elegans* meeting were critical for conveying unpublished information among the workers involved. All of the participants shared their information, strains, and clones without hesitation. The response of one investigator to a request for materials from another investigator was not "Let's collaborate on that project," but rather, "Here's what you need; good luck with your experiments." Finally, the interplay between the molecular biologists and the geneticists kept both groups on track.

The work described here resulted in discovery both of a useful research tool and of a significant system of transposable elements. Both the usefulness and significance of Tc1 are likely to increase in the years ahead. DNA has recently been transplaced into the *C. elegans* genome following Tc1 excision (Plasterk and Groenen 1992). This promises new and powerful techniques for genetic manipulation of *C. elegans*. Tc1 elements have also been discovered in *Drosophila* (Harris et al. 1988; Henikoff and Plasterk 1988; Brezinsky et al. 1990). The presence of Tc1 homologs in both nematodes and arthropods suggests that this family of highly uniform 1.6-kb elements is widespread. Are they very ancient, or are they horizontally mobile? The implications for mechanisms of evolution are possibly profound.

References

Anderson, P. and S. Brenner. 1984. A selection for myosin heavy-chain mutants in the nematode *C. elegans*. *Proc. Natl. Acad. Sci. 81:* 4470.

Bender, W., M. Akam, F. Karch, P.A. Beachy, M. Peifer, P. Spierer, E. B. Lewis, and D. S. Hogness. 1983. Molecular genetics of the *bithorax* complex in *Drosophila melanogaster*. *Science 221:* 23.

Brack, C., M. Hirama, R. Lenhard-Schuller, and S. Tonegawa. 1978. A complete immunoglobulin gene is created by somatic recombination. *Cell 15:* 1.

Brezinsky, L., G. Wang, T. Humphreys, and J. Hunt. 1990. The transposable element Uhu from Hawaiian *Drosophila*—Member of the widely dispersed class of Tc1-like transposons. *Nucleic Acids Res. 18:* 2053.

Coulson, A., J. Sulston, and S. Brenner, Jr. 1986. Toward a physical map of the genome of the nematode *C. elegans*. *Proc. Natl. Acad. Sci. 83:* 7821.

Coulson, A., R. Waterston, J. Kiff, J. Sulston, and Y. Kohara. 1988. Genome linking with yeast artificial chromosomes. *Nature 335:* 184.

Eide, D. and P. Anderson. 1985a. The gene structures of spontaneous mutations affecting a *Caenorhabditis elegans* myosin heavy chain gene. *Genetics 109:* 67.

———. 1985b. Transposition of Tc1 in the nematode *C. elegans*. *Proc. Natl. Acad. Sci. 82:* 1756.

Emmons, S.W. and L. Yesner. 1984. High-frequency excision of transposable element Tc1 in the nematode *Caenorhabditis elegans* is limited to somatic cells. *Cell 36:* 599.

Emmons, S.W., M.R. Klass, and D. Hirsh. 1979. Analysis of the constancy of DNA sequences during development and evolution of the nematode *Caenorhabditis elegans*. *Proc. Natl. Acad. Sci. 76:* 1333.

Emmons, S.W., L. Yesner, K. Ruan, and D. Katzenberg. 1983. Evidence for a transposon in *Caenorhabditis elegans*. *Cell 32:* 55.

Epstein, H.F., R.H. Waterston, and S. Brenner. 1974. A mutant affecting the heavy chain of myosin in *Caenorhabditis elegans*. *J. Mol. Biol. 90:* 291.

Fedoroff, N.V. 1983. Controlling elements in maize. In *Mobile genetic elements* (ed. J.A. Shapiro), p. 1. Academic Press, New York.

Files, J., S. Carr, and D. Hirsh. 1983. Actin gene family of *Caenorhabditis elegans*. *J. Mol. Biol. 164:* 355.

Fire, A. 1986. Integrative transformation of *Caenorhabditis elegans*. *EMBO J. 5:* 2673.
Green, M.M. 1980. Transposable elements in *Drosophila* and other Diptera. *Annu. Rev. Genet. 14:* 109.
Greenwald, I. 1985. *lin-12*, a nematode homeotic gene, is homologous to a set of mammalian proteins that includes epidermal growth factor. *Cell 43:* 583.
Harris, L., D. Baillie, and A. Rose. 1988. Sequence identity between an inverted repeat family of transposable elements in *Drosophila* and *Caenorhabditis*. *Nucleic Acids Res. 16:* 5991.
Henikoff, S. and R. Plasterk. 1988. Related transposons in *C. elegans* and *D. melanogaster*. *Nucleic Acids Res. 16:* 6234.
Liao, L.W., B. Rosenzweig, and D. Hirsh. 1983. Analysis of a transposable element in *Caenorhabditis elegans*. *Proc. Natl. Acad. Sci. 80:* 3585.
Luria, S.E. and M. Delbrück. 1943. Mutations of bacteria from virus sensitivity to virus resistance. *Genetics 28:* 491.
MacLeod, A.R., J. Karn, and S. Brenner. 1981. Molecular analysis of the *unc-54* myosin heavy-chain gene of *Caenorhabditis elegans*. *Nature 291:* 386.
MacLeod, A.R., R.H. Waterston, and S. Brenner. 1977a. An internal deletion mutant of a myosin heavy-chain in *C. elegans*. *Proc. Natl. Acad. Sci. 74:* 5336.
MacLeod, A.R., R.H. Waterston, R.M. Fishpool, and S. Brenner. 1977b. Identification of the structural gene for a myosin heavy-chain in *C. elegans*. *J. Mol. Biol. 114:* 133.
McClintock, B. 1948. Mutable loci in maize. *Carnegie Inst. Washington Year Book 47:* 155.
———. 1952. Chromosome organization and genic expression. *Cold Spring Harbor Symp. Quant. Biol. 16:* 13.
———. 1957. Controlling elements and the gene. *Cold Spring Harbor Symp. Quant. Biol. 21:* 197.
Moerman, D.G. and D.L. Baillie. 1979. Genetic organization in *C. elegans*: Fine-structure analysis of the *unc-22* gene. *Genetics 91:* 95.
Moerman, D.G. and R.H. Waterston. 1984. Spontaneous unstable *unc-22* IV mutations in *C. elegans* var. Bergerac. *Genetics 108:* 859.
———. 1989. Mobile elements in *Caenorhabditis elegans* and other nematodes. In *Mobile DNA* (ed. D.E. Berg and M.M. Howe), p. 538. American Society For Microbiology, Washington D.C.
Moerman, D.G., G.M. Benian, and R.H. Waterston. 1986. Molecular cloning of the muscle gene *unc-22* in *Caenorhabditis elegans* by Tc1 transposon tagging. *Proc. Natl. Acad. Sci. 83:* 2579.
Moerman, D.G., S. Plurad, R.H. Waterston, and D.L. Baillie. 1982. Mutations in the *unc-54* myosin heavy-chain gene of *C. elegans* that alter contractility but not muscle structure. *Cell 29:* 773.
Mori, I., D. Moerman, and R. Waterston. 1988. Analysis of a mutator activity necessary for germline transposition and excision of Tc1 transposable elements in *Caenorhabditis elegans*. *Genetics 120:* 397.
Nicholas, W., E. Dougherty, and E. Hansen. 1959. Axenic cultivation of *Caenorhabditis briggsae* (Nematoda: Rhabditidae) with chemically undefined supplements comparative studies with related nematodes. *Ann. N.Y. Acad. Sci. 77:* 218.
Nigon, V. 1949. Les modalites de la reproduction et le determinism dusex chez quiques Nematodes libres. *Ann. Sci. Nat. Zool. Biol. Anim. 11:* 1.
Nigon, V. and E.C. Dougherty. 1949. Reproductive patterns and attempts at reciprocal crossing of *Rhabditis elegans* Maupas 1900, and *Rhabditis briggsae* Dougherty and Nigon, 1949 (Nematoda: Rhabditidae). *J. Exp. Zool. 112:* 485.
Park, E.-C. and H.R. Horvitz. 1986a. Mutations with dominant effects on the behavior and morphology of the nematode *Caenorhabditis elegans*. *Genetics 113:* 821.
———. 1986b. *C. elegans unc-105* mutations affect muscle and are suppressed by other mutations that affect muscle. *Genetics 113:* 853.
Plasterk, R.H.A. and T.M. Groenen. 1992. Targeted alterations of the *Caenorhabditis elegans* genome by transgene instructed DNA double strand break repair following Tc1 excision. *EMBO J. 11:* 287.
Roeder, G.S. and G.R. Fink. 1983. Transposable elements in yeast. In *Mobile genetic elements* (ed. J.A. Shapiro), p. 300. Academic Press, New York.
Rose, A.M. and T.P. Snutch. 1984. Isolation of the closed circular form of the

transposable element Tc1 in *Caenorhabditis elegans. Nature 311:* 485.
Rosenzweig, B., L.W. Liao, and D. Hirsh. 1983a. Sequence of the *C. elegans* transposable element Tc1. *Nucleic Acids Res. 11:* 4201.
———. 1983b. Target sequences for the *C. elegans* transposable element Tc1. *Nucleic Acids Res. 11:* 7137.
Ruan, K. and S.W. Emmons. 1984. Extrachromosomal copies of transposon Tc1 in the nematode *Caenorhabditis elegans. Proc. Natl. Acad. Sci. 81:* 4018.
Ruvkun, G., V. Ambros, A. Coulson, R. Waterston, J. Sulston, and H. Horvitz. 1989. Molecular genetics of the *Caenorhabditis elegans* heterochronic gene *lin-14. Genetics 121:* 501.
Schachat, F.H., H.E. Harris, and H.F. Epstein. 1977. Two homogeneous myosins in body-wall muscle of *Caenorhabditis elegans. Cell 10:* 721.
Slizynska, H. 1957. Cytological analysis of formaldehyde-induced chromosomal changes in *Drosophila melanogaster. Proc. R. Soc. Edinb. 66:* 288.
Stinchcomb, D.T., J.E. Shaw, S.H. Carr, and D. Hirsh. 1985. Extrachromosomal DNA transformation of *C. elegans. Mol. Cell. Biol. 5:* 3484.
Sulston, J.E. and S. Brenner. 1974. The DNA of *C. elegans. Genetics 77:* 95.
Waterston, R.H. and S. Brenner. 1978. A suppressor mutation in the nematode acting on specific alleles of many genes. *Nature 275:* 715.
Waterston, R.H., R.M. Fishpool, and S. Brenner. 1977. Mutants affecting paramyosin in *Caenorhabditis elegans. J. Mol. Biol. 117:* 679.
Waterston, R.H., J.N. Thomson, and S. Brenner. 1980. Mutants with altered muscle structure in *Caenorhabditis elegans. Dev. Biol. 77:* 271.
Waterston, R.H., D.G. Moerman, D. Baillie, and T.R. Lane. 1982. Mutations affecting myosin heavy-chain accumulation and function in the nematode *Caenorhabditis elegans.* In *Disorders of the motor unit* (ed. D.L. Shotland), p. 747. J. Wiley, New York.
Wills, N., R.F. Gesteland, J. Karn, L. Barnett, S. Bolten, and R.H. Waterston. 1983. The genes *sup-7* X and *sup-5* III of *C. elegans* suppress amber nonsense mutations via altered transfer RNA. *Cell 33:* 575.
Wilson, E.B. 1925. *The cell in development and heredity.* Macmillan, New York.
Zengel, J.M. and H.F. Epstein. 1980. Mutants altering coordinate synthesis of specific myosins during nematode muscle development. *Proc. Natl. Acad. Sci. 77:* 852.
Zieg, J., M. Silverman, M. Hilmen, and M. Simon. 1977. Recombinational switch for gene expression. *Science 196:* 170.

MECHANISMS THAT RAPIDLY REORGANIZE THE GENOME

(genome reorganizations, transposable gene control systems, restriction enzymes, stabilizing mechanisms)

BARBARA McCLINTOCK

Carnegie Institution of Washington
Cold Spring Harbor Laboratory
Cold Spring Harbor, N.Y. 11724

SUMMARY

Extensive reorganization of components of the genome is initiated in maize by the breakage-fusion-bridge (BFB) cycle. Displacements include chromosomes other than those undergoing the cycle. Restructuring ranges from those rearrangements that are readily viewed with the light microscope, such as reciprocal translocations, inversions, duplications, etc., to apparently short DNA segments that include components of transposable gene-control systems. Much of the restructuring appears to be non-random. Some examples of this are conspicuous, such as centromere to centromere, knob to knob, or centromere to knob attachments. Others are exhibited by attachment of a segment of some chromosome to a newly broken end of a chromosome, or the placement of a piece of one chromosome onto the end of another chromosome that had not undergone the BFB cycle. These modifications suggest participation of "restriction enzyme" systems in their formation.

One system responsible for special types of chromosomal modifications was extensively examined. The chromosome location of its principal genetic element was determined. This genetic element has never given evidence of serving as a component in a gene-control system. Instead, it serves to cut chromosomes. The locations of these cuts do not appear to be random.

The BFB cycles exposed the presence in the maize genome of transposable elements that can serve to control the type and time of gene action. Previous to undergoing the cycle, these elements were quiescent in the genome. One component of a gene control system produces a product that is responsible for transposition of the elements of a system. In this regard, each system operates quite independently of the other. The product can induce modifications in chromosome organizations that are by-products of the transposition mechanism. It also can cause receptive elements located in different regions of the genome to respond by inducing DNA modifications in situ and without altering their subsequent receptivity to the product. Or, the product may cause an element to modify the organization of chromatin to one or the other side of it. Some responses of a receptive element that is located within a gene locus result in removal of sensitivity to the inducing product. Stable, new alleles are so produced. In such instances, the gene product may be altered and the pattern of its expression may also be altered during development of plant and kernel.

It is suspected that the BFB cycle initiates stress within the genome and that the stress calls up reserves to counteract it. In some instances, coping with stress may involve a simple solution, such as gene amplification. Coping with drastic types of stress may initiate seemingly disorderly types of response. It is conceivable that, in some instances, stabilization may follow such disorder. This could provide newly organized genomes with orderly operating gene-control systems while still retaining those components that again can respond to stress.

INTRODUCTION

Recent technical advances that allow determinations of molecular composition, organization, and function of cellular components are initiating a veritable revolution in concepts of organization and action of the eukaryotic genome. Startling new findings responsible for this revolution are accumulating at a rapid rate. These include recognition that DNA components may undergo specific types of modification or reorganization, either during development or in response to stress. It is suspected, also, that some types of stress may initiate rapid displacements of components of the genome. Such stress conditions could be responsible for multiple displacements of some DNA sequences. Earlier, the concept of "stability of the genome" hindered an appreciation of the potential significance of well documented instances of genome manipulation. At the time of discovery of many such instances, DNA was not yet known to be the primary component of the genome. The evidence was registered at the level of the chromosome and with the light microscope. It is now quite clear that in such instances genomes were being manipulated. It is my impression that many geneticists were either unaware of or unimpressed by the varied types of manipulation that were being recorded.

Many years ago it was learned that X-radiation could alter chromosomes by producing translocations, deficiencies, inversions, and ring chromosomes. It was known, also, that ultra-violet light could induce chromosome modifications as well as mutations. It was not considered, however, that organisms might carry within their genomes genetic systems that have the potential to restructure the genome and to do so rapidly and effectively. It will be my purpose today to present evidence of several innate systems in maize that, when triggered, initiate chromosomal rearrangements as well as alterations affecting the structure of a gene locus, the nature of its product, and the mode of control of time and place of its action. Statements about all but one of the inducing mechanisms appear in previous reports. Except for the *Ds-Ac* (Dissociation-Activator) transposable gene-control system (McCLINTOCK, 1951) and for rearrangements instigated by the B-type chromosome (RHOADES and DEMPSEY, 1973, 1975, 1977), supporting evidence based on illustrative examples of chromosome modifications induced by innate systems was not given. I am pleased to have this opportunity to provide a few illustrative examples in support of these statements.

UNORTHODOX TYPES OF CHROMOSOMAL REARRANGEMENTS INDUCED BY THE BREAKAGE-FUSION-BRIDGE CYCLE

In maize both the chromatid and chromosome types of breakage-fusion-bridge cycle (BFB) (McCLINTOCK 1938, 1941, 1942, DOERSCHUG 1973) have exposed innate mechanisms that initiate various grades of genome reorganization. The full extent of their effects was not appreciated until the summer of 1944 and thereafter. For purposes not related to the subject of this report, 677 maize kernels were placed in germinating chambers during the spring of 1944. Each kernel had received a newly broken end of the short arm of chromosome 9 from both the pollen and egg parent. (See Figure 1 for the relative size of the 10 chromosomes of the maize set as well as the potential locations of knobs in each chromosome.) The break was induced by mechanical rupture of a chromatid bridge in the anaphase preceding gamete formation. Following zygote formation, the broken ends fused with each other to form a dicentric chromosome (McCLINTOCK 1942, 1944, 1951). The chromosome type of BFB cycle was then initiated.

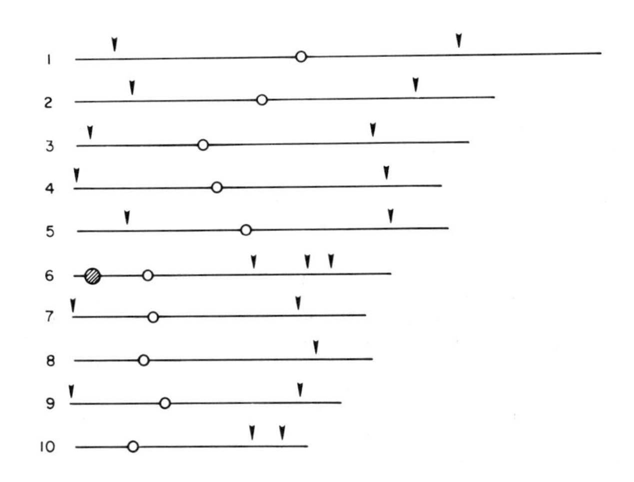

FIGURE 1. Diagram illustrating the relative size of the ten chromosomes of the maize complement. Open circles locate the centromeres. The large hatched circle in chromosome 6 represents the nucleolus organizer. Arrowheads point to locations within each chromosome where knobs may be formed.

Because a break in an anaphase bridge may occur anywhere between the two separating centromeres of a chromatid, a series of duplications, reduplications in direct or reverse order, and of the reciprocal deficiencies of component parts of the short arm of chromosome 9 was anticipated to have occurred during early plant development. Many such modifications were encountered upon examination of the chromosomes at the meiotic stage in plants derived from kernels whose embryos started development with their cells undergoing these cycles.

Among the 677 kernels initially placed in germination chambers, 590 germinated and 87 did not. A high rate of non-germination was anticipated. Of the very young seedlings, 121 were so defective in morphology and growth rate that they did not survive. In addition, 6 kernels produced only roots. The shoot was missing. The seedlings from 7 other kernels did not survive. Two of them were totally devoid of chlorophyll, two were very pale yellow, and three were accidentally injured. This left 456 plants that were transferred to the field. Sporocyte samples were taken from many of these plants to determine what may have happened to their chromosomes 9 during the BFB cycles before "healing" of the broken ends had occurred. The ears of each plant were self-pollinated, and all ears produced by a plant were so treated. The purpose of the test had been to produce minute internal deficiencies of the short arm of chromosome 9 that would give rise to mutant phenotypes when homozygous. Selfed progeny were required for detection of such mutants.

The 590 seedlings that developed from the 677 original kernels could be classified into two main groups. 48 died before a classification of them could be made. In addition, the 13 above-mentioned seedlings that did not survive for the reasons given, were excluded. 245 seedlings appeared to be normal in morphology and chlorophyll content and distribution. It was obvious that the BFB cycles had ceased in them. 284 others were obviously defective. In them the BFB cycles were continuing. These cycles were producing cells whose nuclei were homozygous deficient for parts of the short arm of chromosome 9, and undoubtedly in some cases, for parts of other chromosomes as well. The defective tissues were recognized as sectors having no chlorophyll. Often these defective tissues implied a slowed growth rate. Such sectors exhibited various grades of distortion of leaf shape. Many of these seedlings formed side shoots in addition to the main shoot which often died. From 1 to 14 side shoots were produced by a single seedling. In many instances one side shoot was normal in chlorophyll content and morphology. The BFB cycles had ceased in the cell or cells that initiated this shoot. It was this shoot that finally formed the plant. The defective side shoots finally died. 73 of the seedlings whose cells were still undergoing the BFB cycles did not produce a viable side shoot, and thus these seedlings finally died.

In previous years a number of plants derived from kernels that had undergone BFB cycles in early development were examined at the meiotic stages to determine the nature of change that these cycles had induced in their chromosomes 9. The above detailed description of seedling types produced in one test is included here to place in perspective the origin of plants derived from those that initially underwent the chromosome type of BFB cycle. Observations of chromosomes of some of these plants as well as those of plants grown in earlier years, revealed quite clearly the significance of the varied types of genome modification this cycle could induce.

Besides the expected types of chromosomal modification of the short arm of chromosomes 9, many unexpected chromosome alterations were noted. A careful scanning of each chromosome of the complement was not undertaken because, at the time, such alterations were not anticipated. Other chromosomes of the complement received special attention only if some alterations of one or another of them had been noted. There were such alterations and it is suspected that many more were present than were recorded. Because focus had been centered on chromosome 9, many recorded instances of reorganization of other chromosomes of the complement relate to their association with a chromosome 9.

Gross modifications of chromosomes within the unanticipated class may be assigned to several distinctive categories. One involves "fusion" of centromeres of one chromosome 9 with that of another chromosome of the complement. In each such instance, the short arm of the chromosome 9 involved in the event was lost to the cells as well as loss of either the short or the long arm of the other chromosome of the complement involved in this type of reorganization. Three such instances are diagrammed in Figure 2 and another is shown in a of Figure 3. The homologue of each chromosome is included in the diagrams. A second group resulted from "fusion" of the knob at the end of the short arm of chromosome 9 with a knob in another chromosome of the complement. Examples appear in c of Figure 2 and in b and c of Figure 3. Another category involves fusion of the knob in chromosome 9 short arm with the centromere, producing a ring chromosome similar to that shown in i of Figure 6. Knobs and centromeres were involved in inversion formations. A chromosome 5 was found that had an inversion extending from the centromere to the knob in the long arm. The centromere was now at the former knob position and the knob was at the former centromere position. A similar type of inversion involved a region just distal to the nucleolus organizer in chromosome 6 and the centromere. This placed the centromere at the former location of the nucleolus organizer and the nucleolus organizer at the former location of the centromere. An example of a knob-centromere inversion is diagrammed in the upper homologue in a of Figure 4. In this instance an extension derived from some other chromosome is attached to the centromere.

In another instance the end of the short arm of one chromosome 9 was joined to a region just distal to the nucleolus organizer in chromosome 6. This may have given rise to a dicentric chromosome. The break in an anaphase bridge may be responsible for the reorganizations diagrammed in d of Figure 4.

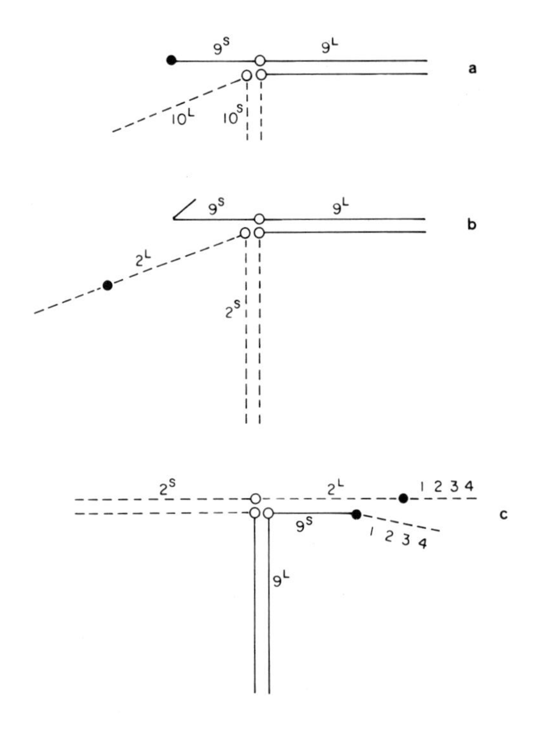

FIGURE 2. Diagrams illustrating two types of chromosomal rearrangements induced in plants whose cells underwent the BFB cycle. Observations were made at the pachytene stage of meiosis. The continuous line refers to chromosome 9 or its parts. The broken line represents another chromosome, the particular one in each instance being designated by number. Superscript S signifies the short arm, superscript L the long arm. Open circles represent centromeres. Closed circles represent knobs. In (a) and (b), only one of the two chromosomes 9 that had undergone the BFB cycle was involved in a centromere-to-centromere association. In (a), the short arm of this chromosome 9 and the long arm of the implicated chromosome 10 were lost to the cell, probably at the time the event occurred. In (b), the short arm of chromosome 9 and the long arm of the implicated chromosome 2 were lost to the cell. In (c), both chromosomes 9 that underwent the BFB cycle were involved in rearrangements with a single chromosome 2. A common centromere unites the short arm of chromosome 2 with the long arm of one chromosome 9, its short arm having been lost. A segment from the knob to the end of the long arm of this same chromosome 2 is attached to the end of the short arm of the other chromosome 9.

Knob to knob "fusions" were not uncommon. One is diagrammed in c of Figure 2. Two others are illustrated in b and c of Figure 3.

A large number of reorganizations of chromosome 9 were noted. The expected duplications, reduplications, and deficiencies of the short arm were common. A telocentric chromosome showing loss of all of the short arm is illustrated by the lower homologue in a of Figure 4. In addition, a number of isochromosomes involving the long arm of chromosome 9 also were noted. There were some pseudo-isochromosomes composed of two chromosomes

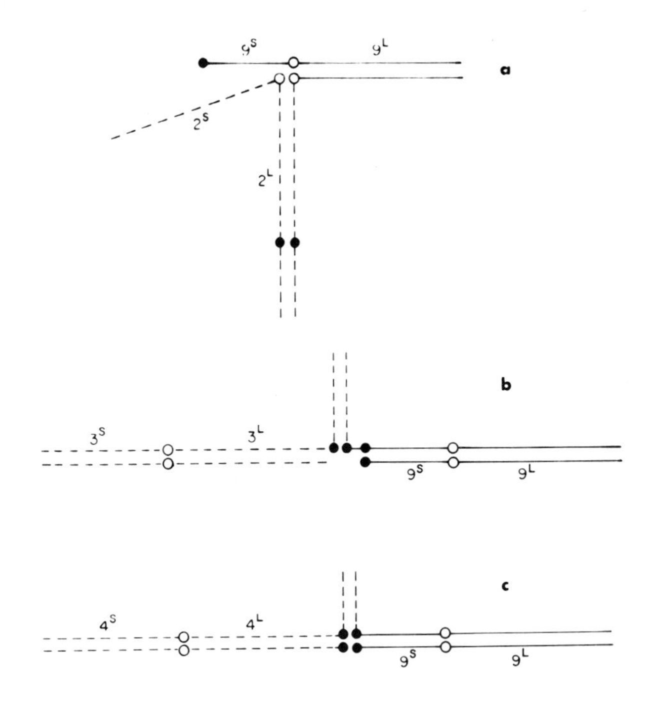

FIGURE 3. Symbolizations are the same as those described for Figure 2. (a) shows a similar configuration as that in (b) of Figure 2 except that in this instance, it is the long arm of one chromosome 2 that shares a centromere with the long arm of one chromosome 9. The short arm of these two chromosomes was lost to the progeny of the cell that produced this association. (b) and (c) illustrate attachments to the end of the short arm of one chromosome 9 of a segment from the knob to the end of the long arm of chromosome 3 in (b), and to the end of the long arm of chromosome 4 in (c). In both instances, the chromosome that contributed the attached segment is readily identifiable. It is deficient for this segment.

9 joined at the ends of their short arms but having only one centromere. The other centromere sometimes appeared within a very tiny ring chromosome. Some pseudo-isochromosomes of the long arm of chromosome 9 had a small extension of the short arm adjacent to the single centromere. Thus, one arm of this pseudo-isochromosome was longer than the other.

There were a number of instances in which a piece of the long arm of chromosome 9 was attached in an inverted order to the knob at the end of the short arm. The lower homologue in b of Figure 4 illustrates one such instance. Figure 4c diagrams another such instance. In this instance, the piece of the long arm attached to the knob in the short arm was derived from the homologue. The remnant of this homologue produced a ring chromosome by fusion of its knob to chromatin at the position of cut in its long arm.

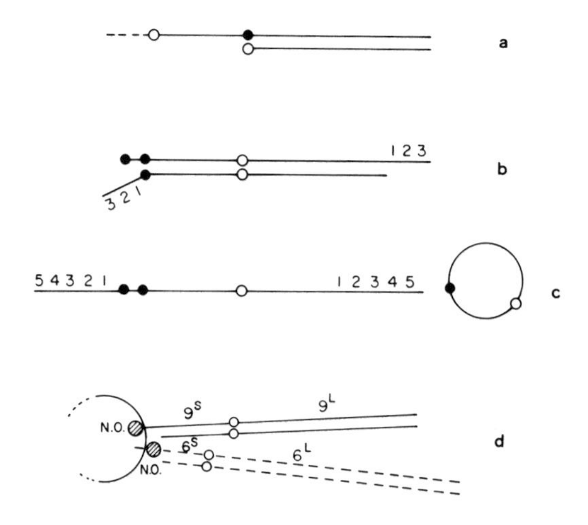

FIGURE 4. (a), (b), and (c) of this Figure illustrate modifications affecting only chromosomes 9. In (a), one chromosome 9 has lost all of its short arm. The homologue has an inversion that reverses the positions of the knob and centromere. A piece of chromatin of unknown origin is attached to the reoriented centromere. In (b), a segment of the long arm of one chromosome 9 was cut and the piece from the cut to the end of the arm was attached at its cut end to the knob terminating the short arm of the same chromosome 9. In (c), both chromosomes 9 were involved in reorganizations. A large terminal segment of the long arm of one chromosome 9 was attached in inverted order to the knob terminating the short arm of its homologue. The donating chromosome then formed a ring chromosome by joining the cut end of its long arm to the knob terminating its short arm. The diagram in (d) is complex but self-explanatory. Its nature is considered in the text.

The described or diagrammed examples represent only a small sample of observed types of unanticipated chromosomal modifications. Many were quite common. One common type was attachment of a piece of the short arm of chromosome 9 to the end of the short arm of chromosome 8. Also, quite often, a piece from another chromosome became attached to the short arm of chromosome 9. The origin of the piece often could be identified. Instances of this are illustrated in c of Figure 2 and in b and c of Figure 3. In other instances, however, identity of the attached piece was not determined. A number of complex rearrangements also were noted. Some resulted in a trisomic constitution of part of one chromosome of the complement. Nevertheless, the chromosome of which it was a part was present in duplicate, and formed a normal bivalent at the meiotic stages. Again, some of the modifications involved three or more chromosomes in complex reorganizations that sometimes were too difficult to resolve.

It must be concluded that the BFB cycle can initiate many distinctive types of genome reorganization. The problem now is to identify, at the molecular level, the components within the genome that this cycle triggers, and the modes of their action.

IDENTIFICATION OF ONE COMPONENT THAT MAY BE RESPONSIBLE FOR SOME OF THE DESCRIBED CHROMOSOMAL REARRANGEMENTS

During the summer of 1953 tests were being conducted with plants having a chromosome 9 that was deficient for all of the short arm distal to the *Sh* locus. The extent of this deficiency is shown in Figure 5. (The location of gene markers in this short arm of chromosome 9 may be judged from the symbols shown in Figure 11). *Yg* refers to green seedlings whereas the recessive, *yg*, when homozygous or hemizygous gives rise to yellow-green seedlings. *C* is required for pigment development in the aleurone layer, which is the outer layer of the endosperm of the kernel. *Sh* action is required for a fully developed endosperm. A kernel lacking *Sh* action is shrunken and the recessive mutant allele is designated *sh*. *Bz* is required for purple or red anthocyanin pigment production. The recessive, *bz*, when homozygous or hemizygous results in production of a bronze color both in the plant tissues and in the aleurone layer of the kernel. For illustrations of the action of *Sh* and *Bz* and the phenotype of their recessive alleles, see Figures 12 and 13. *Wx* is responsible for production of some amylose starch in cells of the gametophytes and the endosperm. The full recessive, called waxy and symbolized *wx*, allows only amylopectin starch to be formed in these tissues. The deficient chromosome 9, illustrated in Figure 5, is transmissible only through the egg parent. In the instance to be described, the pollen parent in the cross that gave rise to plants having a deficient chromosome 9 had two normal chromosomes 9, each carrying the gene markers *Yg*, *C*, *sh*, *bz*, and *wx*. The deficient chromosome carried the markers *Sh*, *Bz*, and *Wx*.

In test crosses of the progeny of the described cross, one plant produced progeny kernels exhibiting an unexpected phenotype. The plant was used as a pollen parent in crosses to plants carrying normal chromosomes 9 and homozygous for the gene markers

FIGURE 5. Camera lucida sketch of the composition of the short arms of chromosome 9 at the pachytene stage of meiosis. The short arm of one completely normal chromosome 9 terminates in a small knob. Its homologue is deficient for a terminal segment of the short arm that starts from a position that is just distal to the *Sh* locus. The fragment chromosome is composed of a terminal segment of the short arm that extends just beyond the locus of the *Bz* gene. It is synapsed with its homologous segment in the normal chromosome 9 almost to the location of the fragment's centromere. The short arm of the fragment is composed of thick, deep-staining chromatin.

c, *sh*, *Bz*, and *wx*. All kernels on the ears so produced showed losses of the *C* gene during endosperm development. Unexpectedly, all kernels were *Sh* and *Wx* in phenotype, and these latter two markers showed no evidence of somatic losses. Cytological examinations and genetic tests of plants derived from kernels produced by the indicated cross showed the presence in them of a fragment chromosome composed of a part of the short arm of chromosome 9. The gene markers *Yg*, *C*, *sh*, and *bz* were carried by this fragment. The modification that had occurred in the parent plant contributing the fragment had inserted a centromere at a location just proximal to the *bz* locus in its *C sh bz wx* carrying chromosome. The remainder of this chromosome was lost. The fragment chromosome had a short arm composed of several wide, deep-staining chromomeres. A camera lucida sketch of the short arms of this chromosome 9 complex is shown in Figure 5. In this Figure, the short arm of one chromosome 9--that contributed by the egg parent--is normal in all respects and terminates in a knob. The deficient chromosome has the *Sh* marker very close to the end of its short arm. The fragment chromosome has a knob terminating its long arm, the abbreviated thick arm forming its short arm. The fragment chromosome carries the gene markers *Yg*, *C*, *sh*, and *bz*, and in this order starting at the knobbed end. In addition it has a new marker, symbolized as X. It is located just proximal to the *bz* locus and adjacent to or at the centromere of the fragment. It is this X component that is responsible for many changes in chromosome organization that were observed over a period of years in plants carrying this X component.

Many X-induced modifications restructured the fragment chromosome itself. A sample of these are diagrammed in Figure 6. The diagram in a of this Figure represents the fragment morphology before changes in it had occurred.

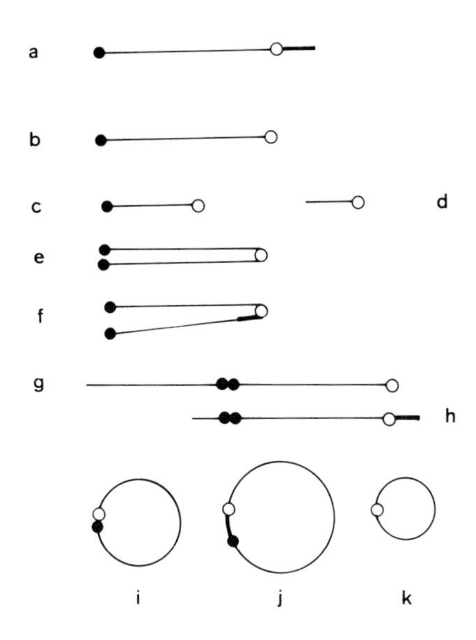

FIGURE 6. The diagrams in (b) to (k) represent types of change in organization and content of the fragment chromosome 9 that its X component induces. The diagram in (a) refers to the configuration of this fragment that was noted originally and is illustrated in Figure 5.

The X component is responsible for many chromosome reorganizations that involve other chromosomes of the complement. One type places the fragment at its centromere location adjacent to the centromere of another chromosome of the complement. Examples of this are diagrammed in a, b, c, and e of Figure 7. Only in e of this Figure was the reciprocal component recovered. It is the telocentric short arm of chromosome 10. Observations of the initiation of such events occurring in a cell just before meiosis have indicated that this part of a chromosome is most often lost to the nucleus. It appears as a pycnotic body in the cytoplasm.

In other instances, the fragment may become attached to the end of an arm of a chromosome. Two instances of this are diagrammed in d and f of Figure 7. In none of the instances shown in Figure 7 was evidence given of action of the X component in plants carrying one or another of these rearrangements. There was one possible exception. Several plant generations after annexation of the fragment chromosome to the centromere of chromosome 10 (e Figure 7) a plant appeared that exhibited somatic losses of the gene markers carried by the chromosome 9 segment of this chromosome. It is suspected that these losses could have been produced by reactivation of the X component or, more likely, by insertion of another component resembling X in its action. An answer could be obtained but study of this case, as well as all others, had to be terminated. Nevertheless, this case deserves analysis.

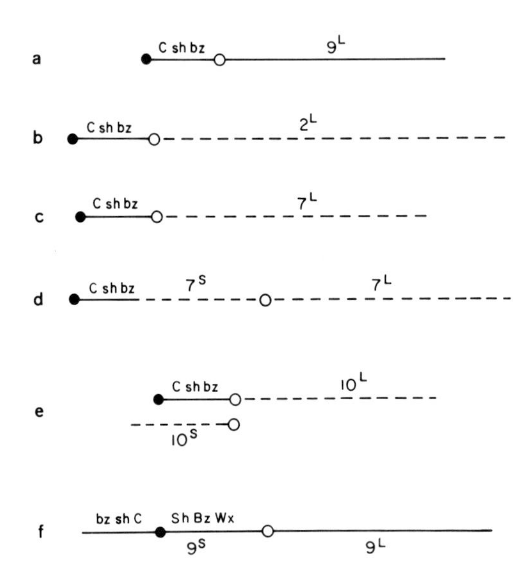

FIGURE 7. Examples of annexation of the fragment chromosome 9 to other chromosomes. In (a), (b), (c), and (e), its attachment is to the centromere of the designated chromosome whose short arm, however, was included in progeny cells only in the case of (e). In (d), the fragment is attached to the end of the short arm of chromosome 7. In all five instances, the position of attachment of the fragment was at or close to its X component although no evidence of X action was noted subsequently. In (f), the fragment is attached by its knob to the end of the short arm of the deficient chromosome 9 shown in Figure 5. The gene markers carried by the fragment after each attachment are indicated in each instance as well as those carried by the deficient chromosome in (f).

In plants carrying the X component, many other types of chromosomal modifications were noted. These were associated with deficiencies, duplications, translocations, and alterations affecting two or more chromosomes. Figures 8a, b, and d illustrate three relatively simple examples of these types, whereas c illustrates one of the more complex examples of reorganization. Two additional modifications are shown in Figure 9. A segment of the end of the long arm of one homologue of each chromosome is depicted as a duplication in inverted order. This Figure, however, is stylized as studies of the modified chromosome 10 showed that it sustained frequent modifications of its long arm, some of which probably were induced by BFB cycles initiated at a meiotic prophase.

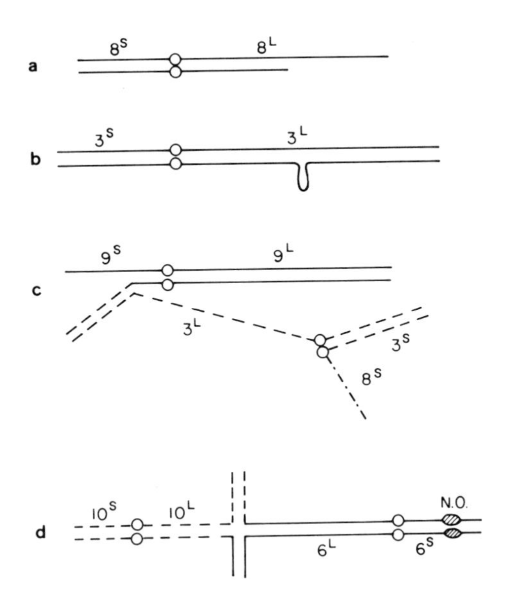

FIGURE 8. Examples of chromosomal rearrangements observed in cells of plants carrying the fragment chromosome 9. It is not known whether these changes were induced by the X component of the fragment or by a similar acting component located elsewhere in the chromosome complement.

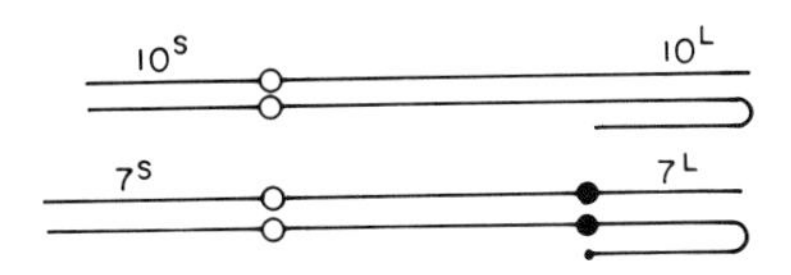

FIGURE 9. The duplications as diagrammed are deceptively stylized as progeny plants showed various other types of modification within the long arm.

One of the modifications that proved to be most effective in illuminating the action of the X component is diagrammed in Figure 10. This Figure illustrates only a sample of the types of change in organization of chromosome 9 that were produced following attachment of the fragment chromosome to the end of the short arm of the deficient chromosome 9. In this instance, the X component was present and active. Initially, this reconstructed chromosome 9 carried the gene markers shown in a of Figure 10 and also *Yg*. The diagrams given in b and f of this Figure are reproduced here to illustrate only some of the changes in organization of this chromosome 9 that were observed among sporocytes of several anthers of a single plant. In each instance the normal homologue is placed adjacent to the modified

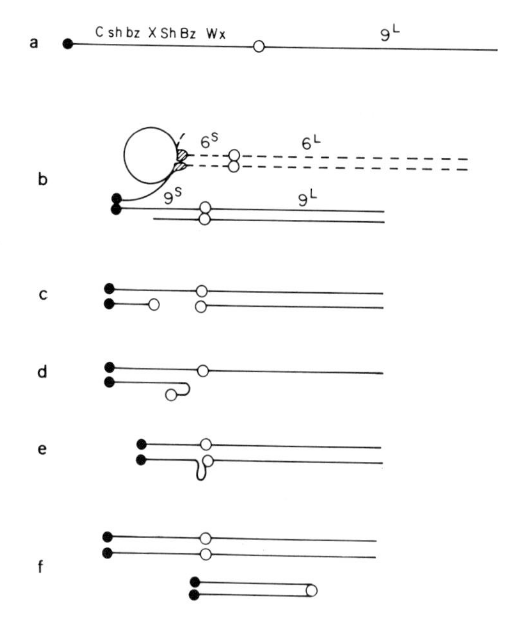

FIGURE 10. Examples of change in organization of a chromosome 9 formed by attachment of the fragment chromosome to the end of the short arm of the deficient chromosome shown in Figure 5. The diagram in (a) illustrates the organization of this chromosome and the markers it carries. The diagrams in (b) to (f) show types of change in this chromosome that were induced by the X component it carries. The normal homologue is included in each diagram. All five examples were observed at the pachytene stage of meiosis among cells of a single plant. Many other types of modification also were observed in this and in other plants. Some are complex and involve chromosomes in addition to chromosome 9. One such example is illustrated in (b).

chromosome 9. Other plants also were examined, and they provided many similar as well as other types of modification of this chromosome 9.

In order to expand and clarify the operation of the X component, allelic markers were introduced into the fragment chromosome by crossing over, and the action of the X component on distributions of these markers was observed in both plant and kernel. In general, its action followed the same rules as those expressed by this component when first identified.

Earlier it was stated that the X component was located close to or at the centromere of the initially isolated fragment chromosome 9. In tests to verify this position, plants were obtained that were homozygous for the deficient chromosome 9, each of which carried the markers *Sh*, *Bz*, and *Wx*, and that had one fragment chromosome 9 with the markers *Yg*, *C*, *sh*, *bz*, and also X. Crossovers that may have occurred between homologous regions of the fragment and the deficient chromosome could be identified on ears produced by crosses to plants that were homozygous for two normal chromosomes 9, each having in its short arm the markers *yg*, *C* or *c*, *sh*, *bz*, and *wx*. The region of overlap where a crossover could be expected to occur is diagrammed in Figure 11.

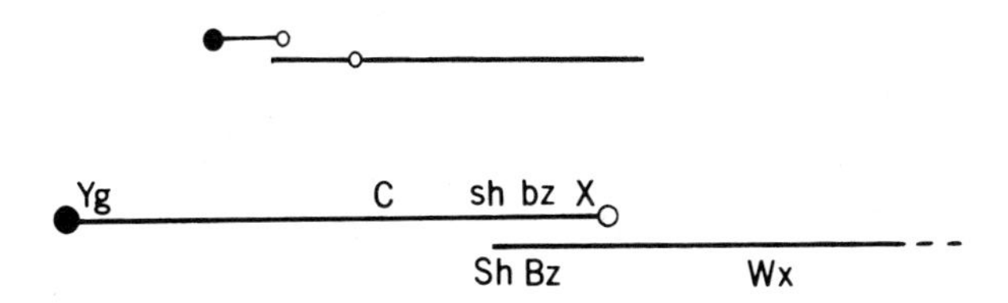

FIGURE 11. Illustration of the synaptic association of a fragment chromosome 9 that had lost its original deep-staining short arm, with the deficient chromosome shown in Figure 5. The upper diagram indicates the relative lengths of these two parts of chromosome 9. The lower diagram is an enlargement of the critical region of synaptic overlap. The genetic markers carried in each segment of chromosome 9 are indicated.

On ears produced by the described crosses, a rare kernel appeared that had the *C sh bz Wx* phenotype. These kernels gave rise to plants that were *Yg*. No *yg* streaks appeared in them, which usually occurs in plants having the original organization of the fragment chromosome. These plants had two normal appearing chromosomes 9. In tests of them, no evidence was given of the presence of the X component in the *Yg C sh bz Wx* carrying chromosome. Its genetic behavior was quite normal. Thus, this chromosome represented a crossover that had occurred between *bz* and the X

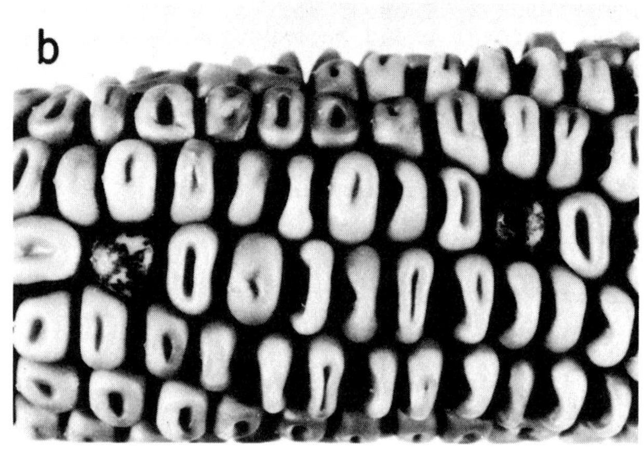

FIGURE 12. Views of two parts of the same ear produced from the cross of a plant that was homozygous for *C*, *sh*, *bz*, and *wx*, when pollen from a plant that carried *C sh bz Wx* in one chromosome 9, *C sh bz wx* in its homologue, and that also had the tiny fragment chromosome carrying *Sh*, *Bz*, and X. The normal chromosome 9 with *C sh bz Wx* and the fragment chromosome with *Sh Bz* and X, represent reciprocal products of a single crossover event. One shrunken (*sh*) kernel that is variegated for deep-colored (*Bz*) and light (*bz*) areas is present in the segment of the ear shown in (a). Loss of *Sh* relates to an event associated with X action, or from a rare crossover. In the ear segment in (b) there are two non-shrunken (*Sh*) kernels that are variegated for *Bz* and *bz* areas. Experience shows that the tiny fragment carrying *Sh*, *Bz*, and X was present at the start of the kernel development. The bronze areas relate to losses of the fragment induced by events initiated by its X component. Most other kernels on this ear are *C sh bz* in phenotype.

component in the fragment, and proximal to *Bz* in the deficient chromosome 9.

In contrast to the above, an occasional kernel appeared on ears produced by the test crosses outlined above, that had a *C Sh Bz Wx* phenotype. These kernels, however, had some bronze (*bz*) areas in them but there was no evidence of variegation for the *C* or the *Wx* phenotype. The expression of both of these genes was quite stable. Variegation for the *Sh* marker would be difficult to detect unless loss of *Sh* had occurred early in kernel development. However, cytological examinations and genetic tests conducted with plants derived from these kernels, and from their progeny, indicated that the kernels had received the reciprocal products of the crossover described above. This crossover produced a normal chromosome 9 carrying *Yg C sh bz Wx* and a tiny fragment chromosome, composed of three small chromomeres, that had within it the markers *Sh*, *Bz*, and the X component as well as the terminal centromere. The X component was responsible for the observed variegation. It could eliminate *Sh* but not *Bz* in a single event, followed later by elimination of *Bz*, or it could eliminate both *Sh* and *Bz* and very often as a consequence of loss of the fragment itself. This latter event was readily determined in plants. The fragment was present in cells of some parts of the plant but absent in those of other parts.

In the plants, sectors appeared exhibiting distinctive and thus readily recognized patterns of loss of *Bz* gene expression. Different patterns often appeared within well-defined sectors of a single plant. Some showed many losses of *Bz* whereas others were characterized by few losses. Patterns falling in between these two extremes also were noted. And, as mentioned, some sectors gave no evidence of the *Bz* phenotype. In some of these, at least, it was known that the fragment had been lost to the cell that gave rise to the bronze sector. Kernels likewise exhibited distinctive patterns of loss of *Bz* expression. Differences in this regard are evident in the *Bz*-expressing kernels on the segments of ears shown in Figures 12 and 13.

Examination of meiotic stages in plants initially receiving the tiny fragment indicated that losses of it sometimes occurred among progenitors of meiotic cells. When the tiny fragment was observed at a meiotic prophase, it often was lost to nuclei subsequently, and quite often during one or the other of the meiotic divisions. Transmission of this fragment, either through the egg cell or sperm cells, was infrequent due both to the premeiotic losses of it and losses occurring during and after the meiotic period. In those instances where it had entered a spore nucleus, its presence could be identified in a kernel derived from functioning of this spore, and the pattern produced by activity of the X component could be compared among *Bz*-expressing kernels on the same ear. The *Bz* carrying kernels on the segments of ears in Figures 12 and 13 are conspicuous because of the very dark pigment that is produced by *Bz* action. All kernels on these ears received from each parent a normal chromosome 9 carrying the markers *C*, *sh*, and *bz*. All but one of the kernels showing *Bz* gene action is also *Sh* in overall appear-

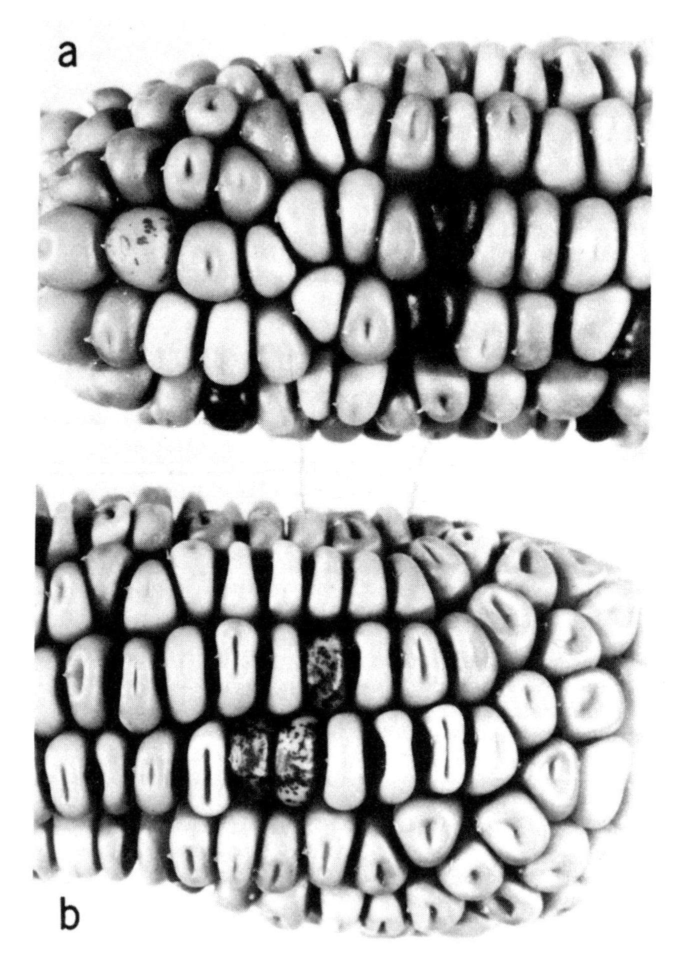

FIGURE 13. The ear, a segment of which appears in (a), arose from the same type of cross that gave rise to the ear in Figure 12 except that a different pollen parent introduced the small fragment carrying *Sh*, *Bz*, and X. Five kernels are nearly solidly *Bz* in phenotype. Each has only a few *bz* areas. One *Sh* kernel, located to the left, has spots of *Bz* in a bronze (*bz*) background. This pattern resembles one that is produced by mutations from *bz* to *Bz*-type expressions. Experience has shown that this pattern results from losses of the tiny fragment or of its *Bz* component. The ear, a segment of which is shown in (b), arose from the same type of cross as that described in Figure 12, and the pollen parent was the same. This segment shows three closely spaced *Sh* kernels that are variegated for *Bz* and *bz* areas. There is little doubt that each of these three kernels received the tiny fragment chromosome carrying *Sh*, *Bz*, and X. Note the similarity in their pattern of variegation.

ance. Two segments of a single ear are shown in Figure 12. A single kernel on this ear showing *Bz* activity is totally *sh* in phenotype. It appears in a of Figure 12. At some stage, early or late but before fertilization, the *Sh* marker had been eliminated. In meiotic prophases of plants carrying this small fragment, reduction in its size was noted occasionally. This suggests that loss of *Sh* expression may sometimes be due to loss of the gene because it had been cut from the fragment by action of the X component.

It should be stated in concluding this section that during the many years of study of the fragment chromosome 9, no instance was found of the X component serving as a gene-control system in the manner of the *Ac*, *Spm*, and *Dt* systems. The frequency of annexations to centromeres, knobs, and ends of chromosomes, as illustrated in the diagrams, suggests some degree of specificity of action of the X component. Thus, the action of the X component and certainly that of *Ac*, *Spm*, and *Dt*, to be discussed in the next section, are difficult to interpret unless the product of each resembles in some essentials the performances of known restriction enzymes. Clearly, the BFB cycles do expose innate systems that are able to restructure the genome and to do so rapidly and precisely.

RESTRUCTURING OF THE GENOME BY TRANSPOSABLE CONTROLLING ELEMENTS

So much has been reported on transposable gene-control systems in maize that little would be served by a recounting of their modes of operation except to emphasize the relation of the so-called regulatory component of a maize control system to transposons in prokaryotic organisms. In this regard NEVERS and SAEDLER (1977) already have outlined basic similarities between them, and have suggested why such systems may be significant in restructuring genomes and at all levels, from modifications of a few base pairs to gross chromosomal rearrangements.

Both transposons in prokaryotic organisms and the regulatory elements in maize that resemble transposons produce several products. One of these is responsible for their transposition from one chromosome location to another, be it within a nucleus of a eukaryote or between organisms sharing the same cell in a prokaryote. In maize, this product can induce DNA modifications that arise by responses to it of an element of a system that had been inserted into a gene locus. As a consequence, the cis element may transpose to a new location in the chromosome complement. Often this leaves behind an altered gene locus. On the other hand, the cis element may remain within the gene locus after its response has induced some locus modification. Such modifications are responsible for the so-called changes in state of a gene locus (McCLINTOCK 1965).

Recently, DOONER and NELSON (1977) supplied critical evidence of changes at a gene locus induced by the *Ds-Ac* system. They showed that cis element-induced alterations of a gene locus can modify the enzymatic product of the gene as well as the time during development when the gene product will be made available.

Because of the many changes in time and type of gene action that these regulatory systems provide, there is now an incentive to determine the molecular structure of such gene loci, and the types of locus reorganizations that are responsible for modified gene products and for altered controls of times of gene action during development.

By means of transpositions, the same regulatory system can operate at a number of gene loci. Because these systems can induce change in DNA organizations within a locus, they provide opportunities to restructure the genome at a fine level. To be effective in evolutionary terms, stabilization of these modifications must occur. It is known that stabilizations do occur and often by single events, such as removal of the cis element from the gene locus, or by loss of its ability to respond to the trans-acting product. Because the BFB cycle is able to expose the presence of these systems that previously were totally concealed, it is more likely that inactivations of the regulatory components result from their placement in quiescent parts of the genome. In this regard, the history of the original mutant of the *A* locus in chromosome 3, designated *a* for its null action, has significance. It was isolated by EMERSON (1918). Its null action remained unchanged over many years, although the mutant was used extensively by a number of maize geneticists. Years later, RHOADES (1936, 1938) discovered in maize imported from Mexico a dominant factor that induced this *a* mutant to produce other mutants exhibiting anthocyanin. RHOADES designated this factor Dotted (*Dt*) because of the dotted pattern of anthocyanin spots in the aleurone layer of the kernel that the *a* mutant produced in response to *Dt*. In his stocks, *Dt* was located in chromosome 9. Nevertheless, in genetic stocks that have not been exposed to *Dt* it is possible to release a *Dt* component from its quiescent state by means of BFB cycles. Sometimes the release occurs within several nuclear generations following initiation of this cycle. When released, it need not appear at any one location in the chromosome complement; and it is subject thereafter to transpositions to new locations (DOERSCHUG 1973).

The *Ds* component of the *Ds-Ac* gene control system can induce gross changes in chromosome organization, two of which are illustrated in McCLINTOCK (1951). Both involve production of duplications that are associated with transposition of *Ds*. Occasionally, a transposition of *Ds* will initiate a reciprocal translocation. *Ds* may also produce modifications of DNA that extend from one or the other side of it, and that can vary greatly in extent. In these instances, *Ds* does not transpose but remains in location (McCLINTOCK 1952, 1953, 1956b).

From the evidence briefly presented in this section, it is possible to conclude that the so-called mutable loci reflect yet another mechanism that is capable of rapidly reorganizing the genome.

CONCLUSIONS

The BFB cycle, upon which emphasis was placed in this report, represents only one mechanism that can trigger innate systems capable of inducing genome reorganizations. The triggering results in various degrees of reorganization, ranging from complex chromosomal translocations to small segments of DNA, possibly composed of a relatively few base pairs. Enzyme systems must be responsible for these modifications. In some instances, a high degree of specificity of action is required. This implies that restriction-type enzymes are included in the maize genome. At present we do not know to what extent or whether these systems normally perform a function in maize, or whether they function only when triggered. It is quite evident, nevertheless, that components of transposable gene-control systems which are quiescent in most genetic stocks may be readily triggered into action by the BFB cycle. This cycle continuously imbalances the genome. It is possible that these imbalances provide the stress required to initiate mechanisms that could reset the genome.

Environmental stress may also trigger a sequence of genomic reorganizations. This is illustrated in animal tissue cultures in which stabilizations often require restructuring of the genome. So far as I am aware, the mechanisms responsible for these restructurings are not known. Species crosses also are known to trigger mechanisms that alter the organization of parental genomes. In some instances it is possible to predict the types of change that will occur, and the chromosomes that will be associated with these alterations (GERSTEL and BURNS 1966, 1967, 1976).

Challenges to overcome needs or responses to special types of stress may trigger gene amplifications. Some of the increased copies may remain within the genome, either at the location of their template(s) or, if freed, at other locations wherein they may enter. There is now a considerable amount of evidence of intraspecific variability in rDNA locations among species in which amplification of this DNA is known to occur. (See NARDI et al. 1977 for recent evidence of this and for a discussion of similar cases.) That redistributions of particular classes of DNA may arise suddenly is suggested by the spectacular differences in distribution of heterochromatic blocks among the chromosomes of three species of *Cyclops* studied by S. BEERMANN (1977). In all three species these blocks are cut-out of the chromosomes during early cleavages in those cells destined to form the soma. Although the distribution of this DNA among the chromosomes of these three species differs greatly, it is suspected that the system responsible for excisions of it may be quite the same in all three species.

I believe there is little reason to question the presence of innate systems that are able to restructure a genome. It is now necessary to learn of these systems and to determine why many of them are quiescent and remain so over very long periods of time only to be triggered into action by forms of stress, the consequences of which vary according to the nature of the challenge to be met.

LITERATURE CITED

BEERMAN, S. 1977 The diminution of heterochromatic chromosome segments in *Cyclops* (Crustacea, Copepoda). Chromosoma (Berl.) *60:* 297-344.

DOERSCHUG, E.B. 1973 Studies of Dotted, a regulatory element in maize. I. Induction of Dotted by chromatid breaks II. Phase variation of Dotted. Theoret. Appl. Genet. *43:* 182-189.

DOONER, H.K. and O.E. NELSON 1977 Controlling element-induced alterations in UDPglucose:flavanoid glucosyltransferase, the enzyme specified by the *bronze* locus in maize. Proc. Nat. Acad. Sci. USA *74:* 5623-5627.

FINCHAM, J.R.S. and G.R.K. SASTRY 1974 Controlling elements in maize. Annual Review of Genetics *8:* 15-50.

GERSTEL, D.U. and J.A. BURNS 1966 Chromosomes of unusual length in hybrids between two species of *Nicotiana*. Chromosomes Today *1:* 41-56.

GERSTEL, D.U. and J.A. BURNS 1967 Phenotypic and chromosomal abnormalities associated with the introduction of heterochromatin from *Nicotiana otophora* into *N. tabacum*. Genetics *56:* 483-502.

GERSTEL, D.U. and J.A. BURNS 1976 Enlarged euchromatic chromosomes ("megachromosomes") in hybrids between *Nicotiana tabacum* and *N. plumbaginifolia*. Genetics *46:* 139-153.

McCLINTOCK, B. 1938 The fusion of broken ends of sister half chromatids following chromatid breakage at meiotic anaphases. Missouri Agr. Exp. Res. Bull. *290:* 1-48.

McCLINTOCK, B. 1941 The stability of broken ends of chromosomes in *Zea mays*. Genetics *26:* 234-282.

McCLINTOCK, B. 1942 The fusion of broken ends of chromosomes following nuclear fusion. Proc. Nat. Acad. Sci. USA *28:* 458-463.

McCLINTOCK, B. 1951 Chromosome organization and genic expression. Cold Spring Harbor Symp. Quant. Biol. *16:* 13-47.

McCLINTOCK, B. 1952 Mutable loci in maize. Carnegie Inst. Wash. Year Book *51:* 212-219.

McCLINTOCK, B. 1953 Mutation in maize. Carnegie Inst. Wash. Year Book *52:* 227-237.

McCLINTOCK, B. 1956a Controlling elements and the gene. Cold Spring Harbor Symp. Quant. Biol. *21:* 197-216.

McCLINTOCK, B. 1956b Mutation in maize. Carnegie Inst. Wash. Year Book *55:* 323-332.

McCLINTOCK, B. 1965 The control of gene action in maize. Brookhaven Symp. Biol. *18:* 162-184.

NARDI, I., G. BARACCHI-PILONE, R. BATISTONI and F. ANDRONICO 1977 Chromosome location of ribosomal RNA genes in *Triturus vulgaris meridionalis* (Amphibia, Urodela). II. Intraspecific variability in number and position of chromosome loci for 18S + 28S ribosomal RNA. Chromosoma (Berl.) *64:* 67-84.

NELSON, O.E. 1968 The *waxy* locus in maize. II. The location of the controlling element alleles. Genetics *60:* 507-524.

NEVERS, P. and H. SAEDLER 1977 Transposable genetic elements as agents of gene instability and chromosomal rearrangements. Nature *268:* 109-115.

RHOADES, M.M. 1936 The effect of varying gene dosage on aleurone colour in maize. Jour. Genet. *33:* 347-354.

RHOADES, M.M. 1938 Effect of the *Dt* gene on the mutability of the a_1 allele in maize. Genetics *23:* 377-397.

RHOADES, M.M. and E. DEMPSEY 1973 Chromatin elimination induced by the B chromosome of maize. Jour. Hered. *64:* 12-18.

RHOADES, M.M. and E. DEMPSEY 1975 Stabilization of freshly broken chromosome ends in the endosperm mitoses. Maize Genet. Coop. News Letter *49:* 53-58.

RHOADES, M.M. and E. DEMPSEY 1977 Chromosome elimination from a structurally modified chromosome 9. Maize Genet. Coop. News Letter *51:* 22-25.

THE NOBEL PRIZE AND A MOLECULAR RETROSPECTIVE

*T*he organization of the chromosomes in the working nuclei of an organism is the standard of reference to which all observed genetic alterations of nuclear origin must be referred. It represents an evolved and necessarily integrated system functioning to control metabolic pathways and developmental patterns.

B. McClintock 1956

The Nobel laureates 1983. *Top from left:* Henry Taube (Chemistry), Barbara McClintock (Medicine), William A. Fowler (Physics), Gerard Debreu (Economics), and Subrahmanyan Chandrasekhar (Physics). *Sitting:* William Golding (Literature)

Display at the DNA Learning Center (CSHL 1989)

THE SIGNIFICANCE OF RESPONSES OF THE GENOME TO CHALLENGE

Nobel lecture, 8 December, 1983

by

BARBARA McCLINTOCK

Carnegie Institution of Washington
Cold Spring Harbor Laboratory
Cold Spring Harbor, New York 11724, U.S.A.

I. *Introduction*

An experiment conducted in the mid-nineteen forties prepared me to expect unusual responses of a genome to challenges for which the genome is unprepared to meet in an orderly, programmed manner. In most known instances of this, the types of response were not predictable in advance of initial observations of them. It was necessary to subject the genome repeatedly to the same challenge in order to observe and appreciate the nature of the changes it induces. Familiar examples of this are the production of mutation by X-rays and by some mutagenic agents. In contrast to such "shocks" for which the genome is unprepared, are those a genome must face repeatedly, and for which it is prepared to respond in a programmed manner. Examples are the "heat shock" responses in eukaryotic organisms, and the "SOS" responses in bacteria. Each of these initiates a highly programmed sequence of events within the cell that serves to cushion the effects of the shock. Some sensing mechanism must be present in these instances to alert the cell to imminent danger, and to set in motion the orderly sequence of events that will mitigate this danger. The responses of genomes to unanticipated challenges are not so precisely programmed. Nevertheless, these are sensed, and the genome responds in a descernible but initially unforeseen manner.

It is the purpose of this discussion to consider some observations from my early studies that revealed programmed responses to threats that are initiated within the genome itself, as well as others similarly initiated, that lead to new and irreversible genomic modifications. These latter responses, now known to occur in many organisms, are significant for appreciating how a genome may reorganize itself when faced with a difficulty for which it is unprepared. Conditions known to provoke such responses are many. A few of these will be considered, along with several examples from nature implying that rapid reorganizations of genomes may underlie some species formations. Our present knowledge would suggest that these reorganizations originated from some "shock" that forced the genome to restructure itself in order to overcome a threat to its survival.

Because I became actively involved in the subject of genetics only twenty-one years after the rediscovery, in 1900, of Mendel's principles of heredity, and at a stage when acceptance of these principles was not general among biologists, I have had the pleasure of witnessing and experiencing the excitement created by revolutionary changes in genetic concepts that have occurred over the past sixty-odd years. I believe we are again experiencing such a revolution. It is altering our concepts of the genome: its component parts, their organizations, mobilities, and their modes of operation. Also, we are now better able to integrate activities of nuclear genomes with those of other components of a cell. Unquestionably, we will emerge from this revolutionary period with modified views of components of cells and how they operate, but only, however, to await the emergence of the next revolutionary phase that again will bring startling changes in concepts.

II. *An experiment with Zea mays conducted in the summer of 1944, and its consequences*

The experiment that alerted me to the mobility of specific components of genomes involved the entrance of a newly ruptured end of a chromosome into a telophase nucleus. This experiment commenced with the growing of approximately 450 plants in the summer of 1944, each of which had started its development with a zygote that had received from each parent a chromosome with a newly ruptured end of one of its arms. The design of the experiment required that each plant be self-pollinated. This was in order to isolate from the self-pollinated progeny new mutants that were expected to appear, and to be confined to locations within the arm of a chromosome whose end had been ruptured. Each mutant was expected to reveal the phenotype produced by a minute homozygous deficiency, and to segregate in a manner resembling that of a recessive in an F_2 progeny. Their modes of origin could be projected from the known behavior of broken ends of chromosomes in successive mitoses. In order to observe those mutants that might express an altered seedling character, forty kernels from each self-pollinated ear were sown in a seedling bench in the greenhouse during the winter of 1944–45.

Some seedling mutants of the type expected did segregate, but they were overshadowed by totally unexpected segregants exhibiting bizarre phenotypes. These segregants were variegated for type and degree of expression of a gene. Those variegated expressions given by genes associated with chlorophyll development were startingly conspicuous. Within any one progeny chlorophyll intensities, and their pattern of distribution in the seedling leaves, were alike. Between progenies, however, both the type and the pattern differed widely. Variegated seedlings from the different progenies were transferred to pots in order to observe the variegated phenomenon in the later developing, larger leaves. It soon became apparent that modified patterns of gene expression were being produced, and that these were confined to sharply defined sectors in a leaf. Thus, the modified expression appeared to relate to an event that had occurred in the ancestor cell that gave rise to the sector. It was this event that was responsible for altering the pattern and/or type of gene expression in descendant cells, often many cell generations removed from the event. It was soon evident that the event was related to some cell component that had been unequally segregated at a mitosis. Twin sectors appeared in which the patterns of gene expression in the two side-by-side sectors were reciprocals of each other. For example, one sector might have a reduced number of uniformly distributed fine green streaks in a white background in comparison to the number and distribution of such streaks initially appearing in the seedling and showing elsewhere on the same leaf. The twin, on the other hand, had a much increased number of such streaks. Because these twin sectors were side-by-side they were assumed to have arisen from daughter cells following a mitosis in which each daughter had been modified in a manner that would differentially regulate the pattern of gene expression in their progeny cells. After observing many such twin sectors, I concluded that regulation of pattern of gene expression in these instances was associated with an event occurring at a mitosis in which one daughter cell had gained something that the other daughter cell had lost. Believing that I was viewing a basic genetic phenomenon, all attention was given, thereafter, to determine just what it was that one cell had gained that the other cell had lost. These proved to be transposable elements that could regulate gene expressions in precise ways. Because of this I called them "controlling elements". Their origins and their actions were a focus of my research for many years thereafter. It is their origin that is important for this discussion, and it is extraordinary. I doubt if this could have been anticipated before the 1944 experiment. It had to be discovered accidently.

III. *Early observations of the effect of X-rays on chromosomes*

The 1944 experiment took place thirteen years after I had begun to examine the behavior of broken ends of chromosomes. It was knowledge gained in these years that led me to conceive this experiment. Initial studies of broken ends of chromosomes began in the summer of 1931. At that time our knowledge of chromosomes and genes was limited. In retrospect we might call it primitive. Genes were "beads" arranged in linear order on the chromosome "string." By 1931, however, means of studying the "string" in some detail was provided by newly developed methods of examining the ten chromosomes of the maize complement in microsporocytes at the pachytene stage of meiosis. At this stage the ten bivalent chromosomes are much elongated in comparison to their metaphase lengths. Each chromosome is identifiable by its relative length, by the location of its centromere, which is readily observed at the pachytene stage, and by the individuality of the chromomeres strung along the length of each chromosome. At that time maize provide the best material for locating known genes along a chromosome arm, and also for precisely determining the break points in chromosomes that had undergone various types of rearrangement, such as translocations, inversions, etc. The usefulness of the salivary gland chromosomes of *Drosophila* for such purposes had not yet been recognized. This came several years later. In the interim, maize chromosomes were revealing, for the first time, some distinctive aspects of chromosome organization and behavior. One of these was the extraordinary effect of X-rays on chromosomes.

The publications of H. J. Muller in 1927 and 1928 (1, 2) and of Hanson in 1928 (3) reporting the use of X-rays for obtaining mutations in *Drosophila*, and similarly that of Stadler in 1928 (4) with the barly plant, produced a profound effect on geneticists. Here was a way of obtaining mutations at will. One did not need to await their spontaneous appearances. Many persons over many years continued to use X-rays for such purposes. But X-rays did not fulfill initial expectations of their usefulness. For other purposes, however, they have been most valuable, particularly for obtaining various types of structural reorganizations of the genome, from minute deficiencies to multiple rearrangements of chromosomes.

It was to observe the effects of X-rays on chromosomes of maize that brought me to the University of Missouri at Columbia in the summer of 1931. Prior to 1931 Dr. Stadler had been using X-rays to obtain mutations in maize. He had developed techniques for isolating those mutations that occur at selected gene loci. One method was to irradiate pollen grains. Pollen grains carry the haploid male gametes. The irradiated male gametes in Stadler's experiments carried wild-type alleles of known recessive mutants. Irradiated pollen was placed on the silks of ears of plants that were homozygous for one or more recessive alleles located in known linkage groups. An X-ray-induced mutation altering the expression of the wild-type allele of one of these recessives should be identifiable in an individual plant derived from such a cross. By the summer of 1931 Stadler had many plants in his field at Columbia, Missouri, from which one could choose those that exhibited one or another of these recessive phenotypes. Stadler had asked me if I would be willing to examine such plants at the meiotic stages to determine what types of events might be responsible for these recessive expressions. I was delighted to do so, as this would be a very new experience. Following my arrival at Columbia in June, 1931, plants were selected whose chromosomes were to be examined. The knowledge gained from these observations was new and impressive. Descriptions and photographs summarizing these observations appeared in a bulletin published by the University of Missouri Agricultural College and Experiment Station (5).

None of the recessive phenotypes in the examined plants arose from "gene mutation". Each reflected loss of a segment of a chromosome that carried the wild-type allele, and X-rays were responsible for inducing these deficiencies. They also were responsible for producing other types of chromosome rearrangements, some of them unexpectedly complex. A conclusion of basic significance could be drawn from these observations: broken ends of chromosomes will fuse, 2-by-2, and any broken end with any other broken end. This principle has been amply proved in a series of experiments conducted over the years. In all such instances the break must sever both strands of the DNA double helix. This is a "double-strand break" in modern terminology. That two such broken ends entering a telophase nucleus will find each other and fuse, regardless of the initial distance that separates them, soon became apparent.

After returning to Cornell University in the fall of 1931, I received a reprint from geneticists located at the University of California, Berkeley. The authors described a pattern of variegation in *Nicotiana* plants that was produced by loss of a fragment chromosome during plant development. The fragment carried the dominant allele of a known recessive present in the normal homologues. Loss of the dominant allele allowed the recessive phenotype to be expressed in the descendants of those cells that had lost this fragment. It occurred to me that the fragment could be a ring-chromosome, and that losses of the fragment were caused by an exchange between sister chromatids following replication of the ring. This would produce a double-size ring with two centromeres. In the following anaphase, passage of the centromeres to opposite poles would produce two chromatid bridges. This, I thought, could prevent the chromosome from being included in either telophase nucleus. I sent my suggestion to the geneticists at Berkeley who then sent me an amused reply. My suggestion, however, was not without logical support. During the summer of 1931 I had seen plants in the maize field that showed variegation patterns resembling the one described for *Nicotiana*. The chromosomes in these plants had not been examined. I then wrote to Dr. Stadler asking if he would be willing to grow more of the same material in the summer of 1932 that had been grown in the summer of 1931. If so, I would like to select the variegated plants to determine the presence of a ring chromosome in each. Thus, in the summer of 1932 with Stadler's generous cooperation, I had the opportunity to examine such plants. Each plant did have a ring chromosome. It was the behavior of this ring that proved to be significant. It revealed several basic phenomena. The following was noted: (*1*) In the majority of mitoses replication of the ring chromosome produced two chromatids that were completely free from each other, and thus could separate without difficulty in the following anaphase. (*2*) Sister strand exchanges do occur between replicated or replicating chromatids, and the frequency of such events increases with increase in the size of the ring. These exchanges produce a double-size ring with two centromeres. (*3*) Mechanical rupture occurs in each of the two chromatid bridges formed at anaphase by passage of the two centromeres on the double-size ring to opposite poles of the mitotic spindle. (*4*) The location of a break can be at any one position along any one bridge. (*5*) The broken ends entering a telophase nucleus then fuse. (*6*) The size and content of each newly constructed ring depend on the position of the rupture that had occurred in each bridge (6, 7, 8).

The conclusion seems inescapable that cells are able to sense the presence in their nuclei of ruptured ends of chromosomes, and then to activate a mechanism that will bring together and then unite these ends, one with another. And this will occur regardless of the initial distance in a telophase nucleus that separated the ruptured ends. The ability of a cell to sense these broken ends, to direct them toward each other, and then to unite them so that the union of the two DNA strands is correctly oriented, is a particularly revealing example of the sensitivity of cells to all that is going on within them. They make wise decisions and act upon them.

Evidence from X-rays, ring chromosomes, and that obtained in later experiments (9, 10, 11, 12), gives unequivocal support for the conclusion that broken ends will find each other and fuse. The challenge is met by a programmed response. This may be necessary, as both accidental breaks and programmed breaks may be frequent. If not repaired, such breaks could lead to genomic deficiencies having serious consequences.

IV. *The entrance into a telophase nucleus of a single broken end of a chromosome*

In the mid-nineteen-thirties another event inducing chromosome rupture was discovered. It revealed why crossing-over should be suppressed between the centromere and the nucleolus organizer in organisms in which chiasmata terminalize, from the initial location of a crossover to the end of the arm of the chromosome. In maize terminalization occurs at the diplotene stage of meiosis. This is before the nucleolus breaks up, which it does at a later stage in the first meiotic prophase. It is known that the force responsible for terminalization is strong. It is enough to induce chromosome breakage should the terminalization process be blocked before the terminalizing chiasma reaches the end of the arm of a chromosome. In maize the centromere and the nucleolus organizer on the nucleolus chromosome are relatively close together. No crossovers have been noted to occur between them. However, if a plant is homozygous for a translocation that places the centromere on the nucleolus chromosome some distance from its nucleolus organizer, crossing over does occur in the interval between them (10). A chiasma so located starts its terminalization process to reach the end of the arm. It is stopped, however, at the nucleolus border. The terminalizing chromatid strands cannot pass through the nucleolus. Instead, the two strands are ruptured at this border. Fusions then occur between the ruptured ends establishing, thereby, a dicentric chromosome deficient for all of the chromatin that runs through the nucleolus and continues beyond to the end of the arm. At the meiotic anaphase, passage of the two centromeres of the dicentric chromosome to opposite poles of the spindle produces a bridge. This bridge is ruptured, and again, the rupture can occur at any one location along the bridge. Now a *single* ruptured end of a chromosome enters the telophase nucleus. How, then, does the cell deal with this novel situation?

In order to determine how a cell responds to the presence of a single ruptured end of a chromosome in its nucleus, tests were conducted with plants that were heterozygous for a relatively long inversion in the long arm of chromosome 4 of maize. It had been known for some time that a crossover within the inverted segment in plants that are heterozygous for an inversion in one arm of a chromosome would result in a dicentric chromosome, and also an acentric fragment composed of all the chromatin from the distal breakpoint of the inversion to the end of the arm. A chromatin bridge would form at the meiotic anaphase by passage of the two centromeres on the dicentric chromosome to opposite poles of the spindle. Mechanical rupture of this bridge as the spindle elongated would introduce a single broken end into the telophase nucleus, as illustrated in a to d, Figure 1. The intent of this experiment was to observe this chromosome in the following mitotic division in order to determine the fate of its ruptured end. This could be accomplished readily by observing the first mitotic division in the microspore. Meiosis on the male side gives rise to four haploid spores, termed microspores. Each spore enlarges. Its nucleus and nucleolus also enlarge. Approximately seven days after completion of meiosis this very enlarged cell prepares for the first post-meiotic mitosis. This mitosis produces two cells, a very large cell with a large, active nucleus and nucleolus, and a small cell with compact chromatin in a small nucleus, surrounded by a

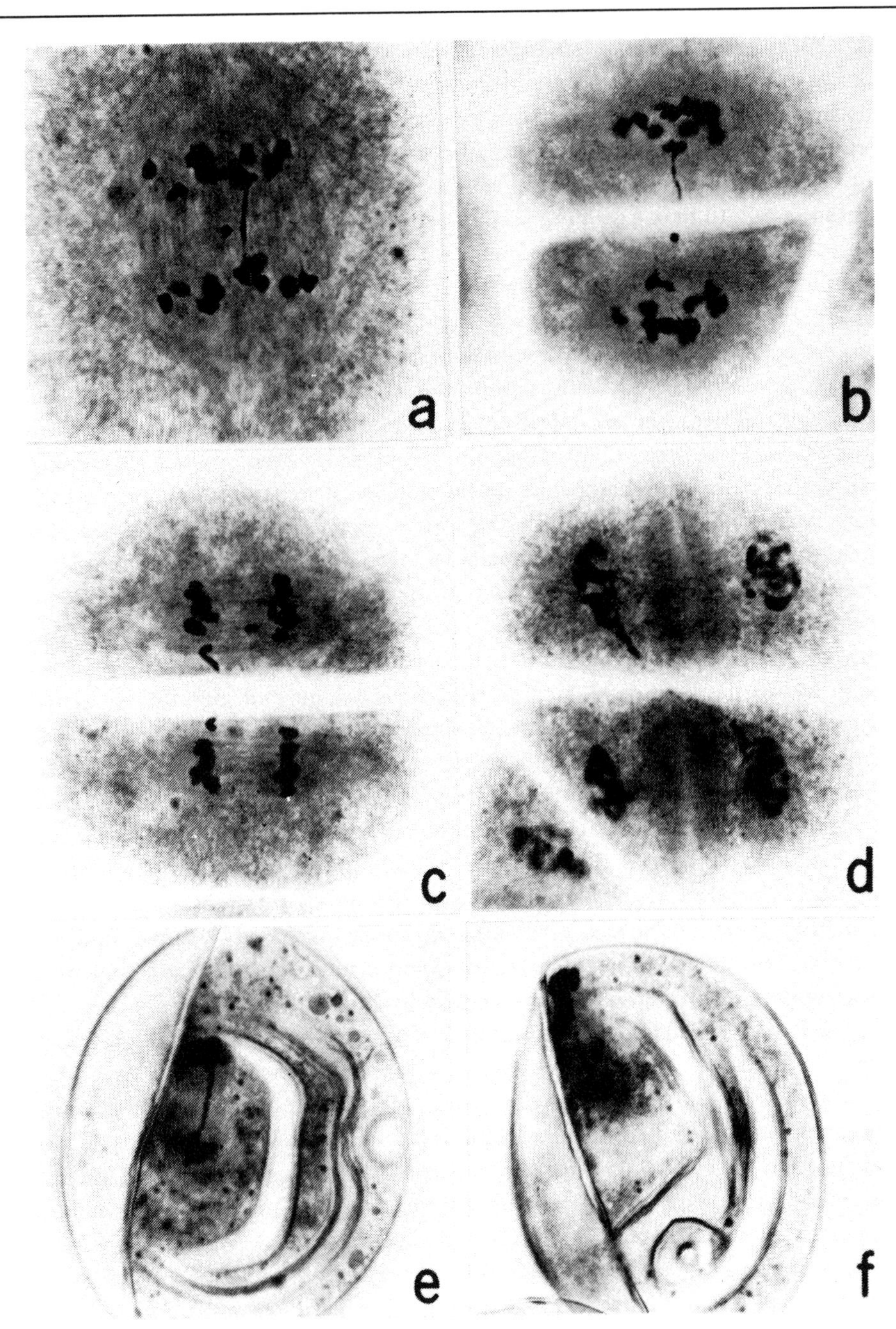

Figure 1. Photographs illustrating the behavior of a newly ruptured end of a chromosome at the meiotic mitoses in microsporocytes and in the post-meiotic mitosis in the microspore. *a.* Chromatin bridge at the first meiotic anaphase, with accompanying acentric fragment. Note thin region in the bridge where rupture probably would have occurred at a slightly later stage. *b.* Two sister cells at a very late prophase of the second meiotic mitosis. The rupture of the chromatid bridge that occurred at the previous anaphase severed the bridge at a non-median position. The larger segment so produced appears in the upper cell and opposite to the shorter segment in the lower cell. Their locations away from the other divalent chromatids relate to late entrances into the previous telophase nuclei, caused by tension on the bridge before its rupture. Their placements show the

thin layer of cytoplasm. This is the generative cell. Sometime later it undergoes a mitosis that will produce two condensed sperm cells. With completion of this division the pollen grain is nearly ready to function. The first division in the microspore may be observed readily merely by using a squash technique. The division of the generative cell, on the other hand, is obscured by the densely packed starch grains that have accumulated during the interval between the two mitoses.

Examination of the first mitotic division in the microspore revealed a strange behavior of the single broken end that had entered a meiotic telophase nucleus. The replicated chromosome again was dicentric. The two chromatids produced by the replication process appeared to be fused at the location of the break that had occurred at the previous meiotic anaphase. In the spore, passage to opposite poles of the two centromeres of this newly created dicentric chromosome again produced a chromatid bridge that again was ruptured (Fig. 1, e, f). Thus, a newly ruptured end of the chromosome again entered each telophase nucleus. How would this newly broken end behave in subsequent mitoses? To determine this requires that the pollen grain with nuclei having such a ruptured end of a chromosome be functional. This could not be in the described instance because pollen grains whose nuclei had such a ruptured end would be deficient for a large terminal segment of the long arm of chromosome 4. Pollen grains whose nuclei have such a deficiency are unable to function.

The problem was resolved by obtaining plants having one chromosome of the maize complement with a duplication of all of its short arm in reverse orientation; its homologue had either a normal organization of its short arm, or better in the test to be performed, a short terminal deficiency of this arm that will not allow pollen grains receiving this chromosome to function. A crossover at the meiotic prophase, as shown in Figure 2, produces a dicentric chromosome that simulates two normal chromosomes attached together at the ends of their short arms, and a fragment chromosome with telomeres at both ends. The dicentric chromosome, produced by the crossover, initiates the chromatid type

positions they occupied in these telophase nuclei. The acentric fragment is in the lower cell, close to the cell membrane that was formed at the end of the first meiotic mitosis. *c*. Anaphase of the second meiotic mitosis. The chromatid with a ruptured end in each cell is placed closer to the newly formed cell wall than are other chromatids, and for reasons given in *b*. Note dissimilar lengths of the arms with ruptured ends. The acentric fragment is near the middle of the spindle in the upper cell. *d*. Telophase of the second mitotic mitosis with extensions in two of the four nuclei pointing toward each other, one in the upper left nucleus and one in the lower right nucleus. The shapes of these nuclei reflect the off-positioning of chromatids having a newly ruptured end. Such off-positioning starts with the first meiotic telophase, continues throughout the interphase and into the prophase, metaphase, and anaphase of the second meiotic mitoses and then into the telophases, as shown here. The acentric fragment is adjacent to but not within the nucleus to the upper right. Note formation of the cell plate in each cell that anticipates the four spores that are products of meiosis. *e*. Mitotic anaphase in the microspore showing a chromatid bridge produced by "fusion" of the replicated broken end. *f*. Same as *e* but a slightly later stage showing rupture of the bridge. (Photographs adapted from Missouri Agricultural Experiment Station Research Bulletin 290, 1938.)

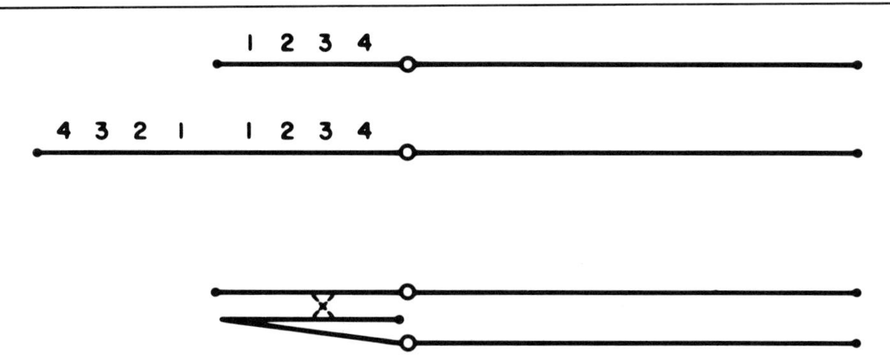

Figure 2. Stylized representation of a crossover between a chromosome 9 with a normal short arm, upper line, and one with a duplication of this arm in reverse orientation, line below. In the lower two lines an exchange between two homologously associated arms is indicated by the cross. Such an exchange would give rise to a dicentric chromosome that simulates two normal chromosomes 9 attached together at the ends of their short arms, plus a small acentric fragment composed of the short arm of chromosome 9. The open circles represent centromeres. Telomeres are depicted as small knobs at the ends of chromosomes.

of breakage-fusion-bridge cycle. This cycle, initially detected at the first mitosis in a microspore, could now be followed in subsequent mitoses. This is because the location of breaks in some of the anaphase bridges gave rise to chromosomes with at least a full complement of genes necessary for pollen functioning. Such functional pollen grains falling on the silks of ears will deliver their two sperm cells to the embryo sac inside a kernel-to-be. One sperm will contribute to the development of the embryo and the other will contribute to the development of the endosperm.

On the female side only a single cell in the kernel-to-be undergoes meiosis, and the embryo sac arises from only one of the four spores produced by the two meiotic mitoses. The other three spores degenerate. This one haploid cell, the megaspore, then undergoes three successive mitoses to form the embryo sac, or female gametophyte. Of the cells in the embryo sac, only the egg cell and the much enlarged central cell need be considered here. The very large central cell has two haploid nuclei positioned close to each other and near the egg cell. Following delivery of the two sperms to the embryo sac, one sperm nucleus fuses with the egg cell nucleus to form the diploid zygote. The other sperm nucleus and the two nuclei in the central cell fuse to form the primary endosperm nucleus, which is triploid. (The term "double fertilization" is commonly employed in referring to these events.) Thus, the embryo and endosperm are formed separately, although both share the same genes, one set from each parent for the embryo, and two sets from the female parent and one set from the male parent for the endosperm. Although developing separately, the two structures are placed side-by-side in the mature kernel, as illustrated in Figure 3.

It was soon learned that the chromatid type of breakage-fusion-bridge cycle, initiated at a meiotic anaphase, will continue during development of the pollen grain and the embryo sac. Whenever a sperm nucleus contributes a chromo-

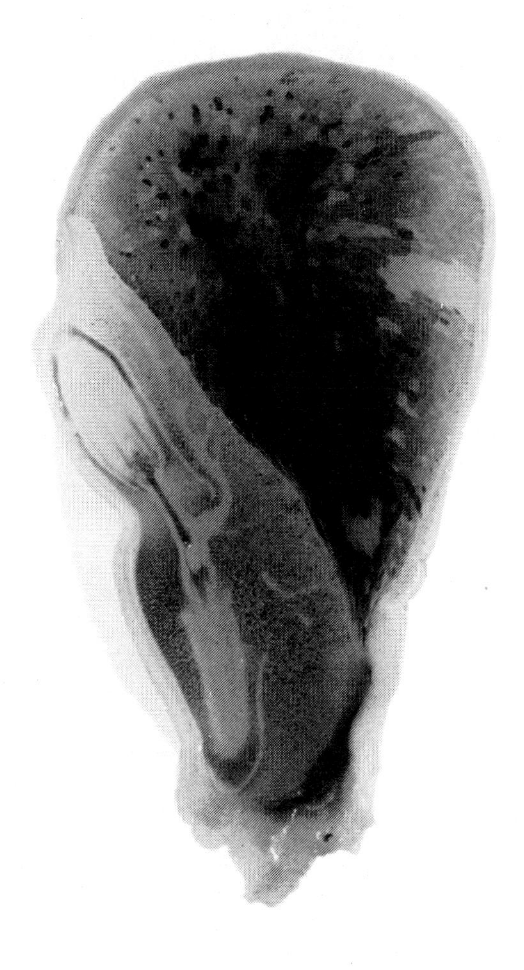

Figure 3. Longitudinal section through a mature maize kernel to show its parts. The cut surface was treated with an iodine-potassium iodide solution to stain amylose in the starch granules of individual cells. The narrow outer layer of the kernel is the pericarp, a maternal tissue. The embryo and endosperm are side-by-side but clearly delimited from each other. The endosperm is above and to the right of the embryo. In this photograph, four parts of the embryo may be noted. To the left and adjacent to the endosperm is the scutellum with its canals. The shoot, to the upper left, and the primary root below it, are connected to each other by the scutellar node. The different staining intensities in the endosperm cells reflect different amounts of amylose in them. These differences relate to the presence and action of a transposable *Ac* element at the *Wx* locus (23). The *Wx* gene is responsible for conversion of amylopectin to amylose, but only in the endosperm, not in the embryo.

some with a newly broken end to the primary endosperm nucleus, this cycle will continue throughout mitoses in the developing endosperm. Similarly, if the two nuclei in the central cell each have such a ruptured end of a chromosome, either the chromosome or chromatid type of breakage-fusion-bridge cycle will occur throughout endosperm development. When, however, a single ruptured end of a chromosome is delivered to the zygote nucleus by either the egg or the sperm nucleus, the ruptured end will "heal" subsequently; the cycle ceases in the developing embryo. Although not yet proven at the molecular level, it is altogether likely that the healing process represents the formation of a new telomere at the ruptured end. This assures that the special requirement for DNA replication at free ends of chromosomes will be satisfied. This new telomere functions normally thereafter. It is as stable in this regard as any other telomere of the maize complement, and tests of this cover many cell and plant generations.

A cell capable of repairing a ruptured end of a chromosome must sense the presence of this end in its nucleus. This sensing activates a mechanism that is required for replacing the ruptured end with a functional telomere. That such a mechanism must exist was revealed by a mutant that arose in my stocks. When homozygous, this mutant would not allow the repair mechanism to operate in the cells of the plant. Entrance of a newly ruptured end of a chromosome into the zygote is followed by the chromatid type of breakage-fusion-bridge cycle throughout mitoses in the developing plant. This suggests that the repair mechanism in the maize strains I have been using is repressed in cells producing the male and female gametophytes and also in the endosperm, but is activated in the embryo. Although all of this was known before the 1944 experiment was conducted, the extent of trauma perceived by cells whose nuclei receive a *single* newly ruptured end of a chromosome that the cell cannot repair, and the speed with which this trauma is registered, was not appreciated until the winter of 1944–45.

V. *Proof that entrance of a newly ruptured end of a chromosome into a telophase nucleus can initiate activations of previously silent genomic elements.* By 1947 it was learned that the bizarre variegated phenotypes that segregated in many of the self-pollinated progenies grown on the seedling bench in the fall and winter of 1944–45, were due to the action of transposable elements. It seemed clear that these elements must have been present in the genome, and in a silent state previous to an event that activated one or another of them. To my knowledge, no progenies derived from self-pollination of plants of the same strain, or related strains, had ever been reported to have produced so many distinctly different variegated expressions of different genes as had appeared in the progenies of these closely related plants, grown in the summer of 1944. It was concluded that some traumatic event was responsible for these activations. The unique event in the history of these plants relates to their origin. Both parents of the plants grown in the summer of 1944 had contributed a chromosome with a newly ruptured end to the zygote that gave rise to each of these plants. The rupture occurred, in the first instance, at a meiotic anaphase in each parent, and the ruptured end then underwent the succession of mitotic anaphase breaks associated with the chromatid type of breakage-fusion-bridge cycle during development of the male and female gametophytes—the pollen grain and the embryo sac. I suspected that an activating event could occur at some time during this phase of the life history of the parent plants. I decided, then, to test if this might be so.

The newly activated elements, isolated from the initial experiment, were observed to regulate gene expressions following insertion of an element, or one of its derivatives, at a gene locus (13, 14, 15). In some instances the general mode of regulation resembled that produced by the *Dotted* "gene" on the standard recessive allele, *a,* of the *A* gene. This *a* allele represents the second recognized gene among a number of others whose action is required for production of anthocyanin pigment, either red or purple, in plant tissues and also in several tissues of the kernel (16). In the mid-nineteen-thirties Marcus Rhoades discovered this *Dotted* (*Dt*) element in a strain of Black Mexican sweet corn (17, 18). It behaved as a dominant gene that caused the otherwise very stable but non-functional *a* allele to mutate to new alleles that allowed anthocyanin pigment to be formed in both plant and kernel. The name *Dotted,* given to it, refers to the pattern of mutations that is expressed in plants and kernels homozygous for the *a* allele on chromosome 3 and having a *Dt* element located elsewhere in the chromosome complement. Small streaks of red or purple pigment appear in plants; the kernels have dots of this pigment distributed over the aleurone layer. (The aleurone layer is the outermost layer of the endosperm.)

Suspecting that this *Dt* had originated from activation of a previously silent element in the maize genome, and also suspecting that such silent elements must be present in all maize genomes, it was decided to test whether the breakage-fusion-bridge cycle would activate one such silent *Dt* element. My stocks that were homozygous for the *a* allele had never given any indication of *Dt* action. Therefore, these stocks were used to test if a presumed silent *Dt* element could be activated by the chromatid type of breakage-fusion-bridge cycle. Plants homozygous for *a,* and having a chromosome 9 constitution similar to that described for Figure 2, were used as pollen parents in crosses to plants that also were homozygous for *a.* These pollen parents had the duplication of the short arm as shown in Figure 2, but its homologue was deficient for a terminal segment of this arm that would not allow pollen grains having it to function. It was determined that 70 to 95 percent of the *functional* pollen grains produced by these plants carried sperms having a chromosome 9 with a newly ruptured end of its short arm, the initial rupture having occurred at the previous meiotic anaphase. Thus, most of the embryos and endosperms in the kernels on ears produced by the described cross started development with a newly ruptured end of the short arm of chromosome 9 in both embryo and endosperm. These kernels were searched for dots of pigment in their aleurone layer. A number of kernels had such dots. Many of these dots were confined to a restricted area of the aleurone layer, suggesting that this area represented a sector derived from a single cell in which a silent *Dt* element had been activated. One kernel had dots distributed over all of the aleurone layer, suggesting that the sperm nucleus contributing to the primary endosperm nucleus already had an activated *Dt* element in it. Tests of the plant arising from this kernel indicated that the sister sperm nucleus that had fused with the egg nucleus did not have such an activated element. Apparently, the activating event had occurred in the nucleus of only one of the two sperms. Significantly, this is only two mitoses removed from initiation of the chromatid type of breakage-fusion-bridge cycle. As mentioned earlier, this cycle continues during successive mitoses in the development of the endosperm. This continuing cycle could explain the presence in some kernels of sectors with pigmented dots, and this, in turn, would imply that activations of silent elements could occur at any time that this cycle remains in operation (19, 20, 21).

A similar test was conducted some years later by Doerschug (22), using the same constitution of the pollen parent as that just described. He obtained similar results. In his tests, however, two kernels with spots of pigment distributed over the entire aleurone layer, proved to have an activated *Dt* element in the plant grown from each of these kernels. The behavior of these two newly activated *Dt* elements was extensively studied by Doerschug. The two elements differed from each other not only in their location in the chromosome complement, but also in their mode of control of the time and place of change in *a* gene action. We now know that such differences in performance of these elements are expected.

Doerschug's two *Dt* isolates are most significant for appreciating the speed of response of a genome to entrance of a newly ruptured end of a chromosome into a telophase nucleus. Each *Dt* element must have been activated in the microspore nucleus or not later than the generative nucleus produced by division of the microspore nucleus. The unexpected event probably is sensed and acted upon from the initial entrance of a single ruptured end of a chromosome into a telophase nucleus, and in each subsequent nucleus that receives such a newly ruptured end. It is recognized that *Dt* is only one among a number of silent but potentially transposable elements that are present in maize genomes. Most probably some of these other silent elements were activated during the described test, but they were not able to be recognized as were activations of *Dt* elements. A similar approach could be used to detect such activations if a proper indicator stock were chosen for the test. Detection of silent elements is now made possible with the aid of DNA cloning method. Silent *Ac* (Activator) elements, as well as modified derivatives of it, have already been detected in several strains of maize (23). When other transposable elements are cloned it will be possible to compare their structural and numerical differences among various strains of maize. Present evidence suggests that wide differences may be found in this regard, as they have been found for middle and highly repetitious DNA sequences (24). In any one strain of maize the number of silent but potentially transposable elements, as well as other repetitious DNAs, may be observed to change, and most probably in response to challenges not yet recognized.

There are clear distinctions in comportment of ends of chromosomes on entering telophase nuclei. These relate to: (*1*) all chromosomes having normal ends, (*2*) two chromosomes, each with a single ruptured end, or one chromosome with both ends ruptured, and (*3*) one chromosome with a single broken end. Both ends of normal, unbroken chromosomes have a normal telomere. No difficulties are experienced. Two ruptured ends, neither with a telomere, will find each other and fuse. In these instances there is no immediate telomere problem. A single broken end has no telomere, and no other broken end with which to fuse. If the cell cannot make a new telomere, which is the case in the maize gametophytes and the endosperm, trauma must be experienced as the evidence indicates. Telomeres are especially adapted to replicate free ends of chromosomes. When no telomere is present, attempts to replicate this uncapped end may be responsible for the apparent "fusions" of the replicated chromatids at the position of the previous break, and be responsible for perpetuating the chromatid type of breakage-fusion-bridge cycle in successive mitoses. Activation of potentially transposable elements, as well as other structural modifications of the chromosomes not considered here, are recognizable consequences of the cell's response to the continuing trauma.

VI. *Further examples of response of genomes to stress*

Cells must be prepared to respond to many sources of stress. Mishaps that affect the operation of a cell must be occurring continuously. Sensing these and instigating repair systems are essential. We are aware of some of the mishaps affecting DNA, and also of their repair mechanisms, but many others could be difficult to recognize. Homeostatic adjustments to various accidents would be required if these accidents occur frequently. Many such mishaps and their adjustments would not be detected unless some event or observation directed attention to them. Some, however, are so conspicuous that they cannot fail to be noted. For example, in *Drosophila,* some sensing device recognizes when the amount of rDNA is above or below the standard amount, and then sets in motion the system that will make the proper adjustment. Similarly, amitotic divisions of macronuclei in ciliates may result in unequal distributions of DNA to daughter nuclei. These deviations are sensed in each daughter cell. To make adjustments, one cell may respond by increasing its DNA content to reach the standard amount. The other cell may discard the excess DNA. There must be numerous homeostatic adjustments required of cells. The sensing devices and the signals that initiate these adjustments are beyond our present ability to fathom. A goal for the future would be to determine the extent of knowledge the cell has of itself, and how it utilizes this knowledge in a "thoughtful" manner when challenged.

One class of programmed responses to stress has received very little attention by biologists. The stress signal induces the cells of a plant to make a wholly new plant structure, and this to house and feed a developing insect, from egg to the emerging adult. A single Vitus plant, for example, may have on its leaves three or more distinctly different galls, each housing a different insect species. The stimulus associated with placement of the insect egg into the leaf will initiate reprogramming of the plant's genome, forcing it to make a unique structure adapted to the needs of the developing insect. The precise structural organization of a gall that gives it individuality must start with an initial stimulus, and each species provides its own specific stimulus. For each insect species the same distinctive reprogramming of the plant genome is seen to occur year-after-year. Some of the most interesting and elaborate plant galls house developing wasps. Each wasp species selects its own responding oak species, and the gall structure that is produced is special for each wasp to oak combination. All of these galls are precisely structured, externally and internally, as a rapid examination of them will show.

The galls on roots of legumes that are associated with nitrogen fixing bacteria are readily available for examination. They illustrate in their own way an example of reprogramming of the plant genome by a stimulus received from a foreign organism. Induction of such reprogrammings by insects, bacteria, fungi, and other organisms, which are not a required response of the plant genome at some stage in its life history, is quite astounding. But it is no more astounding, it would seem, than the sharing of a single genome by two brilliantly designed organisms, the caterpillar and the moth. It is becoming increasingly apparent that we know little of the potentials of a genome. Nevertheless, much evidence tells us that it must be vast.

Many known and explored responses of genomes to stress are not so precisely programmed. Activation of potentially transposable elements in maize is one of these. We do not know when any particular element will be activated. Some responses to stress are especially significant for illustrating how a genome may modify itself when confronted with unfamiliar conditions. Changes induced in genomes when cells are removed from their normal locations and placed in tissue culture surroundings are outstanding examples of this.

The establishment of a successful tissue culture from animal cells, such as those of rat or mouse, is accompanied by readily observed genomic restructuring. None of these animal tissue cultures has given rise to a new animal. Thus, the significance of these changes for the organism as a whole is not yet directly testable. The ability to determine this is a distinct advantage of plant tissue cultures. Many plant tissue cultures have developed new plants and, in some instances, many plants from a single initial cell or tissue isolate. A reason for this difference in behavior of animal and plant tissue cultures is not difficult to find. In many animals the germline is set aside early in cleavage, allowing the soma—a dead-end structure—to develop by any means, including genome restructuring and nonreversible programming. In higher plants, each fertile flower has the equivalent of a "germline" in it. The flower makes the gametes and initiates embryo formation. In this regard, consider the many flowers that may be produced by a bush or a tree. Some system must operate to reprogram the genome in those cells of the flower that will produce the gametes and establish the zygote. This implies that the specific programming sequences, earlier initiated and required for flower production, must be "erased" in order to return the genome to its very early state. If this occurs in so many places in a bush or a tree, then it is not surprising that it may occur in a plant cell or a cluster of cells not within a flower. Also in many plants such resettings are a common means of initiating new individuals from somatic cells. In these instances, however, the process of resetting is regulated, and the genome is not permanently restructured. This is not true for plants arising from many tissue cultures. The treatment, from isolation of the cell or cells of a plant, to callus formation, and then to production of new plants from the cells of these calluses, must inflict on the cells a succession of traumatic experiences. Resetting of the genome, in these instances, may not follow the same orderly sequence that occurs under natural conditions. Instead, the genome is abnormally reprogrammed, or decidedly restructured. These restructurings can give rise to a wide range of altered phenotypic expressions. Some of the altered phenotypes are readily observed in the newly produced plants themselves. Others appear in their progeny. Some initially displayed altered phenotypes do not reappear in the progeny. Their association with genomic change remains problematic. Other altered phenotypes clearly reflect genomic restructuring, and various levels of this have been observed. It may be safe to state that no two of the callus derived plants are exactly alike, and none is just like the plant that donated the cell or cells for the tissue culture. The many levels of genomic modification that already are known and expressed as changed genotypes and phenotypes could be potent sources for selection by the plant breeder, and incidentally, for theoretical ponderings by the biologist.

Modifications in gene expression may be induced in plants when they are infected with an RNA virus. Instances of this may be detected merely by viewing infected plants in the field. For example, patterns of anthocyanin pigment distribution, normally highly regulated and prominently displayed in the flowers of a plant, may appear grossly distorted in those parts of a plant that clearly reveal the virus infection. Recently, it was learned that infection of maize plants with barley stripe mosaic virus, an RNA virus, may traumatize cells to respond by activating potentially transposable elements. These, in turn, may then enter a gene locus and modify its expression (25). Such changes in expression of known genes may be exhibited in the self-pollinated progeny of infected plants. More often they are detected in later generations. Yet, no virus genome has been detected in the immediate progeny of infected plants or in those plants shown to have a transposable element newly inserted at a known gene locus.

Species crosses are another potent source of genomic modification. Plants have provided many excellent examples of this. The advantage of plants is the ease of making crosses to obtain hybrids, the simplicity of growing them, the ready availability of their chromosomes, and the ability to obtain progeny in quantities, if necessary. The alterations produced when the genomes of two species are combined reflect their basic incompatibilities. Evidence of this is the appearance of the same types of genome change whenever the same two species are crossed. Expressions of incompatibilities do differ, but their nature is always in accordance with the particular two species whose genomes are combined. The genus *Nicotiana* has a large number of species that differ from each other in chromosome number, chromosome organization, and phenotypic expressions. Genome incompatibilities have been observed in a large number of

2-by-2 combinations of species. An illustration is the behavior of chromosomes in the hybrid plant produce by the cross of *N. tabacum* by *N. plumbaginifolia*. The chromosomes of *plumbaginifolia* are lost during development of this hybrid plant. Although whole chromosome losses appear to be common, other irregularities in chromosome behavior also occur. These are chromosome fragments, chromosome bridges in somatic anaphases, and the appearance in an occasional metaphase plate of a single, very much elongated chromosome, termed a "megachromosome." The presence of one or two such hugely elongated chromosomes in some somatic metaphase plates characterizes the hybrid derived from the cross of *N. tabacum* × *N. otophora*. In this instance it is known that a heterochromatic segment in each of two chromosomes of the *otophora* set contributes to these linear amplifications (26, 27). Hybrids produced by crosses of distantly related *Nicotiana* species are known to give rise to tumors, some of which resemble teratomas. In one instance it was shown that tumor production relates to a single heritable modification which was initiated in the hybrid.

Major restructuring of chromosome components may arise in a hybrid plant and continue to arise in its progeny, sometimes over successive plant generations. The restructuring may range from apparently simple to obviously complex. These are associated with translocations, inversions, deficiencies, duplications, etc., that are simple in some instances or variously intercalated in others. New stable or relatively stable "species" or "genera" have been derived from such initial hybrids. The commercially useful plant, *Triticale*, is an example. Wheat (*Triticum*) and rye (*Secale*) were crossed and the combined set of chromosomes doubled to provide reproductive stability. Nevertheless, this genome was not altogether stable. Selections continued in later generations for better performances with considerable success, even though instabilities were not eliminated altogether. Some species of *Triticum* undoubtedly arose by a comparable mechanism as that outlined for *Triticale*, and different related genera made their contribution to some of these *Triticum* species. Evidence for this is exceptionally clear (30).

Undoubtedly, new species can arise quite suddenly as the aftermath of accidental hybridizations between two species belonging to different genera. All evidence suggests that genomic modifications of some type would accompany formation of such new species. Some modifications may be slight and involve little more than reassortments of repetitious DNAs, about which we know so little. (The adjective "slight" refers to the apparent simplicity of the restructuring mechanism rather than the significance of its consequences). Major genome restructuring most certainly accompanied formation of some species. Studies of genomes of many different species and genera indicate this. Appreciation of the various degrees of reassortment of components of a genome, that appear during and following various types of genome shock, allows degrees of freedom in considering such origins. It is difficult to resist concluding that some specific "shock" was responsible for origins of new species in the two instances to be described below.

The organization of chromosomes in many closely related species may resemble one another at the light microscope level. Only genetic and molecular analyses would detect those differences in their genomes that could distinguish them as species. In some instances of this type, distinctions relate to the assortment of repetitious DNAs over the genome, as if a response to shock had initiated mobilities of these elements. In other instances, distinctions between related species are readily observed at the light microscope level, such as polyploidizations that are common in plants, or amplifications of DNA that do not alter chromosome number or basic metaphase morphologies. Others relate to chromosome fusions or fragmentations, or readily observed differences in the placement of specific DNA segments. The literature is full of descriptions of differences in chromosome organization among the species of a genus. Two instances of these latter differences warrant special consideration here, because the observed differences in chromosome organization suggest origins from a response to a single event. One response gave rise to extensive fusions of chromosomes. The other placed heterochromatic segments at new locations within the chromosomes of the set.

That such multiple chromosome changes may relate to some initial event occurring in a cell of the germline is proposed and defended in a review article by King (31). An example that would fit his proposal is the organization of chromosomes of the Indian muntjac deer *(Muntiacus muntjak)* (32) when compared with its closely related species, *M. reevesi,* the Chinese muntjac. The latter species has 46 chromosomes as the diploid number, whereas the Indian muntjac has 6 chromosomes in the female and seven chromosomes in the male, and these chromosomes are huge in comparison with those of the Chinese muntjac. Observations of the chromosomes in the hybrid between these two species strongly supports chromosome fusion as the mechanism of origin of the reduced number and huge size of the Indian muntjac chromosomes (33). In general, evidence of fusion of chromosomes is plentiful. When two or three chromosomes of a set appear to have arisen by fusion, the question of simultaneous or sequential events responsible for these fusions cannot be determined with certainty. In the case of the Indian muntjac it is difficult to avoid the conclusion that the fusions of so many chromosomes resulted from some initial shocking event that activated a fusion mechanism already known to exist from the fusions of individual chromosomes in many other organisms. Whatever the cause, the changed chromosome organization is stunning.

Another stunning example of differences in chromosome organization between species is reported by S. Beermann in an extraordinarily thorough and fascinating account (34). This report describes the chromosome organization in three species of the copepod genus *Cyclops*. The main differences among them to be considered here relate to distributions of conspicuous heterochromatic blocks in the chromosomes of each species. In one species, these blocks are confined to the ends of chromosomes. In another species, blocks of heterochromatin are at the ends of chromosomes, but also positioned to each side of the centromere. In the third species, blocks of heterochromatin are distributed all along the chromosomes. An additional feature of this heterochromatin is its unchanged presence in cells of the germline, in contrast to its elimination at cleavages specific for each species and in cells destined to produce the soma.

The elimination process is associated with formation of rings of DNA cut out from the heterochromatin (35). Again it is difficult to avoid concluding that these distinctive distributions of heterochromatin relate to unusual and disturbing events, and that these events activate mechanisms that can redistribute heterochromatin to specific sites.

VII. *Concluding statement*

The purpose of this discussion has been to outline several simple experiments conducted in my laboratory that revealed how a genome may react to conditions for which it is unprepared, but to which it responds in a totally unexpected manner. Among these is the extraordinary response of the maize genome to entrance of a single ruptured end of a chromosome into a telophase nucleus. It was this event that, basically, was responsible for activations of potentially transposable elements that are carried in a silent state in the maize genome. The mobility of these activated elements allows them to enter different gene loci and to take over control of action of the gene wherever one may enter. Because the broken end of a chromosome entering a telophase nucleus can initiate activations of a number of different potentially transposable elements, the modifications these elements induce in the genome may be explored readily. In addition to modifying gene action, these elements can restructure the genome at various levels, from small changes involving a few nucleotides, to gross modifications involving large segments of chromosomes, such as duplications, deficiencies, inversions, and other more complex reorganizations.

It was these various effects of an initial traumatic event that alerted me to anticipate unusual responses of a genome to various shocks it might receive, either produced by accidents occurring within the cell itself, or imposed from without, such as virus infections, species crosses, poisons of various sorts, or even altered surroundings such as those imposed by tissue culture. Time does not allow even a modest listing of known responses of genomes to stress that could or should be included in a discussion aimed at the significance of responses of genomes to challenge. The examples chosen illustrate the importance of stress in instigating genome modification by mobilizing available cell mechanisms that can restructure genomes, and in quite different ways. A few illustrations from nature are included because they support the conclusion that stress, and the genome's reactions to it, may underlie many species formations.

In the future attention undoubtedly will be centered on the genome, and with greater appreciation of its significance as a highly sensitive organ of the cell, monitoring genomic activities and correcting common errors, sensing the unusual and unexpected events, and responding to them, often by restructuring the genome. We know about the components of genomes that could be made available for such restructuring. We know nothing, however, about how the cell senses danger and instigates responses to it that often are truly remarkable.

REFERENCES

1. Muller, H.J., Science *66*, 84–87 (1927).
2. Muller, H.J., Proc. Nat. Acad. Sci. (U.S.A.) *14*, 714–726 (1928).
3. Hanson, F. B., Science *67*, 562–563 (1928).
4. Stadler, L.J., Science *68*, 186–187 (1928).
5. McClintock, B., Missouri Agr. Exp. Sta. Res. Bull. *163*, 1–30 (1931).
6. McClintock, B., Proc. Nat. Acad. Sci (U.S.A.) *18*, 677–681 (1932).
7. McClintock, B., Genetics *23*, 315–376 (1938).
8. McClintock, B., Genetics *26*, 542–571 (1941).
9. McClintock, B., Genetics *26*, 234–282 (1941).
10. McClintock, B., Cold Spring Harbor Symp. Quant. Biol. *9*, 72–80 (1941).
11. McClintock, B., Proc. Nat. Acad. Sci (U.S.A.) *28*, 458–463 (1942).
12. McClintock, B., Stadler Genetics Symp. *10*, 25–48 (1978).
13. McClintock, B., Carnegie Inst. Wash. Year Book No. *47*, 155–169 (1948).
14. McClintock, B., Carnegie Inst. Wash. Year Book No. *48*, 142–154 (1949).
15. McClintock, B., Proc. Nat. Acad. Sci (U.S.A.) *36*, 344–355 (1950).
16. Emerson, R. A., Cornell University Agr. Exp. Sta. Memoir *16*, 231–289 (1918).
17. Rhoades, M. M., Jour. Genetics *33*, 347–354 (1936).
18. Rhoades, M. M., Genetics *23*, 377–397 (1938).
19. McClintock, B., Carnegie Inst. Wash. Year Book No. *49*, 157–167 (1950).
20. McClintock, B., Carnegie Inst. Wash. Year Book No. *50*, 174–181 (1951).
21. McClintock, B., Brookhaven Symp. in Biol. *18*, 162–184 (1965).
22. Doerschug, E. B., Theor. and Appl. Genetics *43*, 182–189 (1973).
23. Fedoroff, N., Wessler S. and Shure, M. Cell *35*, 235–242 (1983).
24. Hake, S. and V. Walbot, Chromosoma (Berl.) *79*, 251–270 (1980).
25. Mottinger, J. P., Dellaporta, S. L. and Keller, P. In press.
26. Gerstel, D. U. and Burns, J. A. Chromosomes Today *1*, 41–56 (1966).
27. Gerstel, D. U. and Burns, J. A. Genetica *46*, 139–153 (1976).
28. Smith, H. H., Brookhaven Symp. in Biol. *6*, 55–78 (1954).
29. Smith, H. H., Brookhaven Lecture Series *52*, 1–8 (1965).
30. Feldman, M., In "Evolution of Crop Plants" (N. W. Simmonds, ed.) pp. 120–128 Longman, London and New York, (1976).
31. King, M., Genetica *59*, 53–60 (1982).
32. Wurster, D. H. and Benirschke, K. Science *168*, 1364–1366 (1970).
33. Shi Liming, Yingying Ye and Xingsheng, Duan. Cytogen. Cell Genet. *26*, 22–27 (1980).
34. Beermann, S., Chromosoma (Berl.) *60*, 297–344 (1977).
35. Beermann, S., and Meyer G. F. Chromosoma (Berl.) *77*, 277–283 (1980).

Broken Chromosomes and Telomeres

ELIZABETH H. BLACKBURN
University of California, San Francisco, California 94143

Barbara McClintock made original and seminal contributions to our understanding of the nature of telomeres, through her studies of the behavior of broken chromosome ends in vivo. Telomeres are the natural ends of intact eukaryotic chromosomes. We now know a great deal about the DNA sequences that are essential for telomere function, the manner in which these sequences are synthesized, the special polymerase that synthesizes them, and the proteins with which they associate. I highlight here the contributions of Barbara McClintock's work, done in the 1930s and 1940s, to our understanding of telomeres and chromosome end behavior. The concepts and tools dominating her papers were genetic and cytological, whereas ours now tend to be molecular and biochemical. Her work impresses me because one can find insights into questions that are still of current interest to researchers. In particular, her work anticipated and laid the conceptual groundwork for our more recent, molecular, view of telomeres. At a personal level, certainly I can say that my own research on telomeres was affected by her scientific insights.

Reading Barbara McClintock's papers written in the 1930s and 1940s and hunting for clues about telomeres is like reading the best kind of detective story in reverse. The fact that I know, at least at some level of understanding, "who done it" does not detract from the rewards of such reading. My favorite detective novels not only reveal the mechanics of how the crime was committed, but also delve into the characters of the protagonists and the underlying human forces involved, which although they may be clearly recognized, remain of course unresolved. Likewise, not only do Barbara McClintock's papers provide explanations for a number of factors about the behavior of chromosomes, but they also delve into the nature of the processes being studied. Although, more recently, some of the riddles of the behavior of telomeres have been solved at the molecular level, our understanding of telomere function is still quite incomplete. Furthermore, Barbara McClintock raised some very profound questions about chromosome ends, broken or natural, that remain unanswered.

First Hints about the Nature of Broken Chromosome Ends

Clues are introduced early in good mystery novels, and indeed the earliest of the series of Barbara McClintock's papers on broken chromosome ends contain important clues about the nature of telomeres as well as broken chromosome ends. In a report that appeared in 1931 (McClintock 1931), whose purpose was to analyze, from a cytological point of view, the nature of X-ray-induced chromosome alterations related to genetic changes, Barbara McClintock found that translocations, deficiencies, and ring chromosomes were frequent outcomes of X-irradiation. All of these involved fusions of broken ends with each other. Significantly, we find the statement: "No case was found of the attachment of a piece of one chromosome to the end of another [intact chromosome]" (McClintock 1931). However, terminal deficiencies were also common, which, for a deficiency that was truly terminal (as opposed to a deletion extending out to near the chromosome end), implies formation of a new chromosome end. With respect to this latter point, in hindsight, we can anticipate future work by Barbara McClintock on chromosome end healing by noting that in some of these experiments, *young embryos* were irradiated. I will deal with the significance of this research below. The main point here is that I believe Barbara McClintock recognized very early that a natural end of an intact, whole chromosome has properties quite distinct from those of broken ends. Although it was to be several years later that such a natural end was given a name ("telomere"), it is clear that she recognized its distinct functional nature.

The Breakage-Fusion-Bridge Cycle

A notable feature of a paper published by Barbara McClintock (1932) is her early recognition of the striking efficiency with which broken ends in the same nucleus fuse with each other. In this paper about ring-shaped chromosomes in maize, she showed that "the size and genetic content of ring-shaped chromosomes are not the same in all the nuclei of the plant" (McClintock 1932). This observation was important not only in upsetting the established notion of genetic constancy in all cells of an organism, but also for setting the stage for understanding the behavior of broken ends. She reported in this paper that "nothing but rings have been observed to come from rings." This conclusion was substantiated and expanded in a full report published in *Genetics* (McClintock 1938a). This remarkable paper constituted the first intensive cytological study of the mitotic behavior of ring chromosomes. An important extension of the generality of the tendency of broken ends to fuse with each other came in a report in the 1938 *Research Bulletin of the Agricultural Experiment Research Station*, University of Missouri (McClintock 1938b), in which Barbara McClintock showed that following chromatid breakage in meiosis, broken ends of sister chromatids also fused with each other. In her words: "The combined evidence points to the conclusion that when a break occurs in a chromatid at Anaphase I or II, the two split halves of this chromatid become fused resulting in a bridge configuration in the anaphase of the first spore mitosis" (McClintock 1938b). These and subsequent papers published in 1939 and 1941 (McClintock 1939a, 1941a,b, 1942) contained careful and thorough cytological observations of anaphase bridges resulting from the presence of ring chromosomes. In a *tour de force* of superb deductive reasoning, she synthesized the first full set of conclusions about the way in which fusion of broken ends generated by ruptures of dicentric chromosomes in anaphase could account for the genetic and cytological behavior of these chromosomes. Taken together, her results led her to formulate the breakage–

fusion–bridge cycle model. This model provided a mechanism to explain how ring chromosomes are perpetuated by repeated breakage–fusion–bridge cycles through successive nuclear divisions. Her conclusions for ring chromosomes in diploid plant tissues can be described briefly as follows. Assuming that in a ring chromosome a crossover occurred after "splitting" (we would now say "replication") of the chromosome, a double-sized ring with two spindle fiber attachment regions (centromeres) would result. In the following nuclear division, as the two spindle fiber attachment regions moved to opposite poles of the mitotic spindle, the continuous chromosome arms between them would become stretched (being visibly manifested as "bridges" connecting the two groups of otherwise separated daughter chromosomes on the spindle in anaphase). The stretched bridges would eventually break, at a variable position, in either late anaphase or telophase. Each daughter nucleus would thus receive two broken ends, which would fuse with each other to regenerate a ring chromosome. The genetic content of each ring would depend on the position of the breaks in each chromosome arm. The next time a crossover occurred between the daughter chromatids of a ring chromosome following chromosome replication, the cycle would repeat itself. The probability of a crossover occurring was found to be proportional to the length of the ring chromosome.

Barbara McClintock was always careful to distinguish between fusions of broken *chromosome* ends, as seen with dicentric ring chromosomes, and fusions between the ends of the two sister *chromatids* produced after a dicentric chromosome had "split" in meiosis. But for each type of situation, she emphasized the efficiency with which broken ends find each other, stating first, with respect to the chromosome type of fusion, that "fusions of broken ends must occur after such breakage, since only ring chromosomes have been found to arise from ring chromosomes" (McClintock 1938a, p. 323), and second, with respect to the chromatid type of fusion, "The evidence suggests that...union [of the two halves of a split chromatid] always results when a chromatid is broken during meiosis" (McClintock 1938b).

Making New Telomeres

What of the natural ends of chromosomes, the telomeres? So far, Barbara McClintock's work had dealt with the behavior of unnatural, broken chromosome ends which were experimentally induced, and which she clearly showed led to deleterious genetic consequences, as a result of the chromosome aberrations generated by their tendency to fuse. However, in 1939, she published a paper (McClintock 1939b) of seminal importance for the concept of telomere function, which ultimately led to the recognition that telomeres are formed by an active process. She reported that under certain conditions, a broken end lost its tendency to fuse with other broken ends. Thus, she concluded that such a broken end had "healed." Importantly, she noted that it thereby became as permanently stable as any normal telomere through all subsequent nuclear divisions. Furthermore, she made the original observation that the tissue type and developmental stage were critical in determining the fate of a broken end: Such healing specifically occurred when the broken end was introduced into a young plant embryo nucleus, at or shortly after the first zygotic division. (Although broken chromosome ends did not usually heal in gametophyte or endosperm tissues, or in the diploid plant tissues later in development, exceptions to the latter under certain experimental situations were discovered in her subsequent work that I will describe below.)

There are two points about these findings that I think are especially important for our understanding of telomeres and for the recognition that telomeres are formed by an active process. First, it has been critical for our subsequent thinking about the molecular nature of the telomere that she states that [in the embryo tissue] "broken ends apparently heal *and remain permanently healed* (my italics). This implies that a permanent molecular modification of the broken end of the chromosome occurs upon healing. Second, the marked tissue and developmental stage specificity of the capability of a broken end to heal was the first real hint of a specialized process, under developmental control, that acts to heal broken ends.

Subsequent observations by Barbara McClintock provide further insights into the nature and control of the healing of broken ends. In the Annual Report of the Carnegie Institution for 1942–1943 (McClintock 1942–1943), she described a genetic cross that she set up in order to introduce a broken chromosome from each parent gamete into the progeny plants. The configuration of these chromosomes was predicted to lead to continued breakage–fusion–bridge cycles in the tissues of these progeny. Evidence of the expected fusions resulting from these broken ends was found in about two thirds of the offspring. However, the remaining one third of the progeny plants did not show evidence of continued fusions. This was in apparent contrast to previous results, in which she had found no evidence of healing during the later stages of plant development. But her previous studies of the fate of broken chromosome ends in later stages of development had involved situations where two broken ends of a dicentric ring chromosome both entered the same daughter nucleus and could thus fuse with each other. To explain this lack of fusions in the new experiments, Barbara McClintock proposed that a freshly broken single end in a nucleus, without another end to fuse to in the same nucleus, can undergo healing even in the later stages of plant (sporophyte) development. The explanation for how this situation might arise came directly from her previous detailed cytological observations of anaphase bridges. She had observed that of the two contiguous anaphase bridges formed by two dicentric chromatids, often only one side would snag, thereby allowing it to be "drawn into" the nucleus, while the other side of the bridge remained intact until the next division cycle. She proposed that in this situation, the first broken end, lacking another end with which to fuse (and therefore, in her terminology, being "unsaturated"), is able to heal. In a subsequent mitosis, the other bridge eventually ruptures and the second broken end, after entering the nucleus, then similarly heals. The importance of this observation is that it shows that plant tissues in late as well as embryonic stages of development have the capability to heal a broken end. This had not been apparent in her previous studies, which had involved experimental situations in which two broken ends were simultaneously introduced into a single nucleus, in which the fusion process usually pre-empted healing. This work represents the first indication that the two outcomes possible for a broken end within a nucleus, fusion or healing, represent competing processes acting on a broken end.

In a second remarkable insight in this article (McClintock 1942–1943), Barbara McClintock proposed that the single broken end, lacking another broken end with which to fuse, becomes healed "during the reproductive cycle of the chromosome." From this, she suggested that "if, as this evidence suggests, [the healing process] is related to the chromosome division cycle, experiments should be focused on this period." This was a prophetic insight, as we now know that permanent healing of broken ends involves the action of telomerase, the enzyme that synthesizes and hence allows complete replication of telomeres.

Barbara McClintock's pioneering work strongly influenced my thinking about telomeres. I was impressed by its implication that a fully functional telomere was generated from a broken chromosome end by an active, developmentally controlled process and was not just the result of some chance event (giving the chromosome a lucky break, so to speak). In a paper published in *Genetics* in 1944 (McClintock 1944), she had briefly mentioned the occasional failure of a single broken end to heal in the young embryo. In these cases, the breakage–fusion–bridge cycle persisted through later developmental stages of the plant. Such instability contrasted with the normally stable behavior of an end that had become healed early in the development of the plant. But I was galvanized by what she said in a letter to me in 1983. In it she told me she had long ago identified a maize mutant that specifically was unable to heal a broken end in the embryo. Since wild-type maize can heal broken ends in embryonic tissue, the phenotype of this mutant carried the strong implication that there is a definite function associated with the ability to heal. In other words, it suggested that healing is a process determined by a gene that can be mutagenized. (Sadly, I understand from her that this mutant stock is now lost.) This information was in my mind when I made the decision to look for an enzymatic activity that adds telomeric DNA to DNA ends, which we ultimately discovered. It has become clear from subsequent work that this enzyme, which we now call telomerase, not only can add telomeric DNA to a pre-existing telomere, but also can heal a broken nontelomeric chromosome end by synthesizing telomeric DNA directly onto it. Barbara McClintock's work contributed to my conviction that such an activity would exist.

Pointing the Way to Future Research

Earlier in this essay, I mentioned that it is clear that Barbara McClintock recognized and, more importantly, formulated some important questions that remain unsolved and still of great interest. Some specifically concern what I regard as a central "plot" of her early papers: the behavior of broken ends and telomeres. The following are three of these questions that strike me forcibly.

First, she recognized that the way in which a broken end was generated might influence its fate. In discussing the results obtained with broken ends resulting from X-irradiation or mechanical rupture, she contrasts these findings with those obtained by others in maize with broken ends resulting from UV-irradiation. The latter treatment of maize results in chromosomes with terminal deficiencies, but unlike those generated by X-irradiation or mechanical rupture, they heal (McClintock 1939a, 1941a). The basis for this intriguing dependence of the fate of a broken end on its method of generation remains unsolved.

Second, in following the cytogenetic fate of an acentric chromosomal fragment generated by a break in meiosis (McClintock 1938b), she noted that such a fragment remained terminally associated with its homolog specifically in anaphase I but was released from this terminal association in anaphase II. We still have no clear idea about what mediates terminal (presumably telomere-telomere) associations between chromosomes, let alone how their specificity for different stages in the meiotic cycle, or any other times, is controlled.

Third, in a conversation we had in 1981, I particularly remember that Barbara McClintock stressed, in talking about broken ends, the efficiency with which they find each other in the large nucleus. She suggested that they might "sense" each other. Now, we do not at present have a molecular paradigm or model for how broken ends might sense each other in a nucleus. But then neither in 1938 was there a molecular explanation for how a broken end might heal to produce a stable telomere.

Like a good detective novel, Barbara McClintock's series of papers on broken chromosome ends are also rich in "subplots." I am struck by how well some of these presage later developments in research on other aspects of chromosome behavior. I will finish my comments on the scientific content of these papers by mentioning two of these subplots. First, she deduced that "splitting" of the chromosome (we would now say DNA replication) occurs before anaphase I of meiosis. Although the question of when the chromosome is replicated in meiosis was actively debated and remained controversial among biologists for many years, in fact, compelling evidence was already provided in two of Barbara McClintock's papers published in 1938: first in a *Genetics* paper (McClintock 1938a) and then in a report in the *Research Bulletin of the Agricultural Experiment Research Station,* University of Missouri (McClintock 1938b), where the reasoning is developed more fully. This deductive reasoning, based on the behavior of broken ends in meiosis, is incisive and a pleasure to read. Characteristically, she carefully discusses the available evidence both for and against the idea that the chromosome has "split" (i.e., replicated its DNA) before anaphase I, although at the end of the paper, she clearly concludes from her own work that it has.

Second, Barbara McClintock is perhaps most celebrated for her pioneering work on transposons, although I hope this essay makes clear how much celebration is also due her for her contributions to our understanding of broken chromosome ends. The papers reporting her first recognition of the chromosomal events that led to the discovery of transposition arose from her studies of mechanically broken chromosomes. One is struck by the fact that right from the outset (McClintock 1945–1946), she spotted that there was something quite distinctive about the break (which originally occurred at a particular site) caused by what she eventually discovered to be the *Ds-Ac* class of transposons. It was her long experience with the phenotypes and chromosomal behavior of plants carrying mechanically ruptured chromosomes that enabled her to recognize that this locus-specific break was very unusual. This sort of insight, and the important discoveries it led to, could probably only have come from someone with her intimate knowledge of this experimental system.

Some Reflections on Barbara McClintock's Papers and Her Science

It is probably evident from the foregoing how scientifically rewarding I have found reading Barbara McClintock's papers dealing with broken chromosomes. Some of her later papers dealing with transposition are popularly considered hard to read, and it might be tempting just to read distillations of some of the conclusions of these earlier papers in reviews or textbooks. Indeed, in a wonderful letter from her, with which she responded to a request from me in 1982 for reprints for a review I was writing, she herself suggested that her papers relating to telomeres and broken ends were too long for current tastes, and for this reason, I may not wish to read them. It says much about her that she would respond to such an ordinary request with first, a carefully annotated list of appropriate references to her papers and then, a beautifully written, lucid account of her cytogenetic findings on telomeres and broken ends, to save me, she explained, the trouble of reading these "long" papers. It was my good fortune that I disregarded her considerate cautioning. In reading this series of papers, I let myself in for one of the great treats in the biological literature.

First of all, let me say that these papers are not hard to read. It is most fun to read them in chronological order, not only to allow the plot unfold, but also to help follow the chromosome constructs, because the papers form a series, with each paper building directly on the previous one. The papers are long, but that is where the rich rewards come in, as they go beyond simply reporting the primary observations. They are filled with the insights that come from meticulously observing and getting to know the experimental system. The thorough observations of the whole plants' phenotypes and the microscopic behavior of their chromosomes are synthesized into masterfully reasoned conclusions. What Barbara McClintock was able to deduce from her observations makes one appreciate the distinction between "seeing" and just "looking." Some of the length of these papers comes from the careful explanations of the logic behind her experimental designs and conclusions. Nothing is overlooked or glossed over. Thus, some of the length also comes from side points which, in themselves, are valuable: For example, her deduction of the timing of chromosome replication in meiosis, to which I alluded briefly above.

The other striking feature of these papers is that, 50–60 years after they were published, they have a rare timelessness. They report the facts unencumbered by that year's fashion in interpretation, so they never sound old-fashioned. For example (astonishingly for a molecular biologist), one can read her papers and be unaware that at the time of their writing, DNA was not known to be the genetic material. The only giveaways are a few hints like the use of the term chromosome splitting; instead, we would say DNA replication, but "chromosome splitting" it in fact remains. In short, reading these papers amply repays the effort and is, simply, a joy.

I share with all those interested in chromosome behavior an indebtedness to Barbara McClintock for her scientific findings. But she also gave me something else, arising from a conversation I had with her in 1977. What has particularly stayed with me from this conversation is that she advised me to trust my intuition in my scientific research. I knew she did not mean at the expense of rigorous experimentation and observation, but as a basis for one's ideas. This simple-sounding piece of advice, which she gave me with considerable emphasis, was, to me, startling at the time and ultimately liberating. I think it touched on an aspect of our intellectual processes which I had not previously admitted as being valid in a biology researcher, and I suspect it influenced me considerably. For this, too, I am very grateful to Barbara McClintock.

References

McClintock, B. 1931. Cytological observations of deficiencies involving known genes, translocations and an inversion in *Zea mays*. *Mo. Agric. Exp. Res. Stn. Res. Bull. 163:* 4.

———. 1932. A correlation of ring-shaped chromosomes with variegation in *Zea mays*. *Proc. Natl. Acad. Sci. 18:* 677.

———. 1938a. The production of homozygous deficient tissues with mutant characteristics by means of the aberrant mitotic behavior of ring-shaped chromosomes. *Genetics 23:* 315.

———. 1938b. The fusion of broken ends of sister half-chromatids following chromatid breakage at meiotic anaphases. *Mo. Agric. Exp. Res. Stn. Res. Bull. 290:* 1.

———. 1939a. The behavior in successive nuclear divisions of a chromosome broken at meiosis. *Genetics 25:* 405.

———. 1939b. The behavior in successive nuclear divisions of a chromosome broken at meiosis. *Proc. Natl. Acad. Sci. 25:* 405.

———. 1941a. The stability of broken ends of chromosomes in *Zea mays*. *Genetics 26:* 234.

———. 1941b. Spontaneous alterations in chromosome size and form in *Zea mays*. *Cold Spring Harbor Symp. Quant. Biol. 9:* 72.

———. 1942. The fusion of broken ends of chromosomes following nuclear fusion. *Genetics 28:* 458.

———. 1942–1943. Maize genetics. *Carnegie Inst. Washington Year Book 42:* 148.

———. 1944. The relation of homozygous deficiencies to mutations and alleic series in maize. *Genetics 29:* 478.

———. 1945–1946. Maize genetics. *Carnegie Inst. Washington Year Book 45:* 176.

Maize Transposable Elements: A Story in Four Parts

NINA V. FEDOROFF
Carnegie Institution of Washington, Baltimore, Maryland 21210

I have never cared much for looking back. What I already understand interests me a great deal less than what I do not yet understand. But it *is* a time to celebrate, to recapture and recapitulate the story of maize transposable elements. My purpose is to sketch in the backdrop of the stage on which Barbara McClintock discovered transposition almost half a century ago. Even more, it is to show how the rich genetic tapestry that she elaborated over more than three decades illuminated the molecular excavations that followed.

Part I: How It All Started

My first encounter with maize transposable elements, like those of many of my contemporaries, was in a seminar course in graduate school. The course was organized by my mentor, Norton Zinder, in the extremely relaxed style characteristic of the young Rockefeller University, and we read only one McClintock paper. I do not remember which paper, but I do remember being confused and not enlightened (experiences to which many still admit). My second encounter with maize transposable elements took place many years later and was so powerfully persuasive that I decided to concentrate my experimental work on their molecular biology. The second time I began at the beginning and read all that McClintock had written about transposable elements: a crucial difference.

I was prompted to revisit transposable elements out of curiosity. At the time, I was finishing a postdoctoral stint in Don Brown's laboratory at the Carnegie Institution of Washington's Department of Embryology in Baltimore, and I had been invited to present my work at the Cold Spring Harbor Laboratory. Late in the day, as the mandatory round of laboratory visits was drawing to a close, I ran into Barbara McClintock in a hall of the Demerec Laboratory. She surprised me with an apology for missing my lecture (why should should she apologize?) and invited me to her laboratory for a talk (I have it firmly in my memory that we drank Cinzano and ate peanuts—Barbara says we didn't). I was very much taken with the lucidity of McClintock's casual discourse—science, scientists, politics, philosophy—the usual stuff of late afternoon laboratory visits. It did not fit her reputation for impenetrability and I was curious—curious enough to find the old Carnegie yearbooks in which she had published and to start reading upon my return to Baltimore. I had to learn McClintock's language (no surprise here, she had given words, a long time ago, to things that no one was ready to see). I had, laboriously, to figure out the genetic crosses, and I had experiments to do, too. But I found myself impatient to read on, to get back each evening to the next year's installment of work in progress and discover how she had figured it out. The stuff was complicated, but it was clear and logical and beautiful. When I tried to explain it to others, I failed altogether and would for a rather long time until I stopped trying to communicate it *all* each time—there was and still is too much to swallow in one gulp.

It seemed to me that working out the molecular biology of this system would be grand fun, but not a practical undertaking for the beginning of a career. Although by that time transposition was no longer heretical, since transposable elements had long since come to light in *Escherichia coli*, very little molecular work had been done in plants. Indeed, not a single maize gene had been cloned, and some of my colleagues were even saying that plant DNA could not be cloned. The notion that I could learn maize genetics, develop molecular cloning techniques for plants, and clone and characterize transposable elements (on the tiny grants that were being awarded for plant research) in addition to teaching and in time to get tenure at a university seemed absurd. But I didn't give up the idea, just tucked it away. About that time, I was offered the staff position held by Igor Dawid, who was leaving Carnegie's Embryology Department to join the National Cancer Institute. It was an unexpected gift and dislodged my tucked-away fantasy. I made an impulsive commitment to the molecular biology of maize transposable elements.

Although nothing much had yet been done to clone and characterize maize genes, the genetic picture could not have been clearer. Transposable element insertion mutations had been identified and studied at many maize loci, primarily by McClintock, but also by maize geneticists both before and after her. Such mutations were not called transposable element insertion mutations until much later, however. The earliest genetic studies were investigations of the inheritance of leaf and flower color variegation in plants. Hugo de Vries appears to have been the first to investigate the inheritance of variegation, developing the concept of "ever-sporting" varieties and concluding that variegation did not show strict Mendelian heritability because variegated plants frequently gave rise to nonvariegated progeny and such nonvariegated plants could, in turn, give variegating progeny.

Some years later, the eminent geneticist R.A. Emerson took up the study of variegation in maize. Writing in the *American Naturalist* in 1914, he commences: "Variegation is distinguished from other color patterns by its incorrigible irregularity." He then begins to reveal the regularities, summarizing both his own work on the maize *P* locus and the earlier work of de Vries, Correns, Hartley, and East and Hayes. He concludes that the inheritance of variegation can, in fact, be understood in Mendelian terms if one assumes that some sort of temporary inhibitor has become associated with a dominant genetic factor necessary for pigmentation. If the inhibitor is lost, as it often is in somatic cells, pigment can be produced. What is perhaps most important about Emerson's study is that he appreciated the relationship between what happened in somatic cells and what was happening in the germ line. The more extensive the area of pigmented tissue, the higher the probability that the wild-type allele of the pigmentation gene would be inherited. Emerson therefore adduced the first evidence that variegation is due to mutation.

Variegation could thus be understood as a phenotype associated with a certain form of a gene. Such genes came to be known as unstable or mutable genes. Mutable genes were studied quite extensively both in plants and later in *Drosophila* over the next few decades, and there was some debate over whether such mutable genes were representative of genes and the mutational process in general or were in some way unique. There were geneticists who argued that mutable genes were exceptional "sick" genes, genes whose properties were fundamentally different from those of normal genes. In 1938, Marcus Rhoades reported a seminal observation. He had identified a genetic locus in whose presence a stable gene became unstable. He found that maize plants homozygous for a stable recessive (colorless) *a* allele of the *A* locus,[1] which encodes an enzyme in the anthocyanin pigment biosynthetic pathway, exhibited sectors of pigmented tissue in both plant and kernels when the dominant *Dotted* (*Dt*) gene was present. He further found that Emerson's *a* tester allele, in use as a stable recessive allele for years, responded to the *Dt* gene in exactly the same way as his own *a* allele. He concluded that whether a gene was stable or unstable could depend on the genetic environment.

Rhoades also gathered more evidence that variegation was due to mutation, in his case of the *a* allele to the *A* allele. He found a number of cases in which a purple anther on an *a* homozygote carrying a *Dt* gene contained equal numbers of *A* and *a* pollen grains, implying that a mutation of *a* to *A* had occurred in a cell that gave rise to both the anther wall and the pollen grains within the anther. Rhoades did not have much use for the notion that mutable genes were "sick" or different from normal genes. His genetic data showed clearly that the presence or absence of a second unlinked gene could make the difference between a gene's stability and instability and furthermore that a mutable gene could mutate to a stable, normal form. In retrospect, it is apparent that Rhoades was also gathering the first clues, not then decipherable, that one mobile element could promote transposition of a second, unlinked element. He focused attention on the specificity of the destabilizing *Dt* gene. It did not cause a general increase in mutation frequency, but destabilized only the *a* allele of the *A* gene. Indeed, he showed that *Dt* did not affect other unstable genes, including Emerson's unstable *P* gene. These were the first hints that mutable genes fell into different groups and that elements within a group could interact but that members of different groups could not.

[1] To avoid confusion in direct quotations, I have generally adhered to the early conventions in maize genetic nomenclature, rather than the contemporary ones. Dominant, usually wild-type, alleles are represented by an italicized, capitalized abbreviation (and serve as the locus designation), whereas recessive alleles are given in lowercase letters. The loci referred to in the text are the *C* (dominant inhibitor *I*, recessive *c*), *Shrunken* (*Sh*), *Bronze* (*Bz*), *Waxy* (*Wx*), *A*, and *A2* loci. *Wd* and *Yg* are the regions covering the chromosome 9 *wd* and *yg* terminal deficiencies. McClintock originally identified transposable elements as loci and referred to the recessive alleles in the same manner (e.g., *Ac* and *ac*), although the recessive allele is the unoccupied site on one homolog corresponding to the element's insertion site on the other. She designated insertion mutations mutable alleles and named them in order of their identification. Thus, *c-m1* is the first mutable allele of the *C* locus. There are two *C*, *A*, *Bz*, and *Sh* loci, originally designated *C1* and *C2*, *A1* and *A2*, etc. As is done in contemporary maize genetic nomenclature, I have omitted the numeral from the designation of the first in the series because it simplifies, if ever so slightly, the task of keeping track of the various mutable alleles.

Part II: How Transposition Was Discovered

McClintock's studies on mutable genes began as an interesting tangent. Making use of what she had learned in her extensive earlier studies on the behavior of broken chromosomes, McClintock devised a method for producing mutations on the short arm of chromosome 9. She initiated cycles of breakage and fusion of the chromosome by introducing a broken chromosome 9 from each parent plant. She then self-pollinated the progeny plants to reveal the presence of new mutations. In 1944 and 1945, she reported detecting a variety of different types of mutations in the selfed progeny of such plants, including the expected terminal deficiencies, some internal deficiencies of various sizes, and some "provocative" mutants that showed variegation from the recessive to the dominant phenotype. From a relatively small number of progenies, she had already recovered 14 "distinctly new expressions of instability of genic action" and observed several more, including variegated white, light green, and luteus seedlings.

She had further identified several cultures in which "an interesting type of chromosomal behavior" had appeared, involving the repeated loss of one of the broken chromosomes from cells during development. By the following year, she had done enough work on these cultures to know that the marker loss she observed resulted from "breakage in chromosome 9 that takes place at a particular locus in the short arm of this chromosome," with the subsequent loss of the acentric fragment at a somatic anaphase. What was unexpected and curious about this phenomenon was its regularity. Judging both from the concomitant loss of several markers and from the absence of a long segment of the same chromosome in cytological preparations of pachytene chromosomes, the breakage was not random but at a specific site on the short arm of chromosome 9. Even at this early stage in her analysis, McClintock noted a similarity in "the factors that control the frequency and timing" between the two different types of variegations that had surfaced in her cultures, one associated with mutable genes and the other, more novel type attributable to chromosome breakage. McClintock concludes her report in the Carnegie Year Book for 1946: "Possibly the resemblance is more than coincidental, in that the underlying phenomena are basically similar."

The picture began to fill in rapidly. McClintock named the mutable locus at which the chromosome broke *Ds* because "the most readily recognizable consequence of its actions is this dissociation." It was already apparent that the *Ds* locus would "undergo dissociation mutations only when a particular dominant factor is present." She named the second factor *Ac* because it activated *Ds*. Chromosome breakage at the *Ds* locus occurred only when the *Ac* locus was present. By 1948, McClintock had come to some startling conclusions about these loci. First, there were alleles that required the presence of *Ac* for instability. Thus, *Ac* not only activated chromosome breakage, but also appeared to be an instability-inducing gene of the type identified by Rhoades. Second, and unprecedented, the chromosome-breaking *Ds* locus could "...change its position in the chromosome..." Moreover, based on altered linkage relationships, McClintock suspected that the *Ac* locus could also move. She hypothesized that *Ac* "...may have been removed from its former position and inserted into a new position in chromosome 9 in a manner similar to that observed for the transposition of the *Ds* locus..."

These conclusions emerged from detailed analyses of the progeny of plants grown from several exceptional kernels appearing on ears of plants containing both a *Ds* locus near the *Wx* locus on chromosome 9 and an *Ac* locus. In the winter and spring of the following year, McClintock typed out her results in tabular form and wrote a full discussion of her observations, inferences, and conclusions. These texts have never been published. The originals remain as typewritten reports in McClintock's files: dozens of tables, diagrams of alternative chromosome structures, extensive discussion. They are lucid and painstaking, and the conclusions, inexorable. Under the subtitle "The origin of transposed *Ds*, Case I," the text commences:

> The first recognized case of transposition of *Ds* arose in the cross of a plant (4108C-1) having the constitution *wd I Sh Bz Wx Ds* in one normal chromosome 9 and *Wd C sh bz wx ds* in a normal homologous chromosome 9. This plant was heterozygous for *Ac* (*Ac ac*). The types of kernels resulting from the cross of this plant to a female plant carrying *C sh bz wx ds ac* are given in table 1.

There follow almost 50 pages of text and 43 tables, many with appendices and diagrams. Step by step, cross by cross, the structure of the rearrangement that accompanied the transposition of *Ds* is revealed. *Ds* moved from its original location (subsequently referred to by McClintock as the "standard" location near the *Wx* locus) to the distal end of chromosome 9 just proximal to the *C* locus, and the entire segment of the chromosome extending from the original location of the *Ds* element to its new location was duplicated in inverted order. By the time the telling is done, there is no doubt. McClintock concludes with the following inferences about the origin of the rearrangement:

> It is possible, now, to reconstruct the events that gave rise to this Duplication chromosome 9 with a transposed *Ds* locus. Three assumptions regarding these events are required. (1) A *Ds* mutation occurred at its usual time—late in the development of the sporophytic tissues—in a cell of plant 4108C-l. The chromosome in which this *Ds* mutation occurred was normal in morphology and carried *I Sh Bz Wx* and *Ds* in its standard location. The *Ds* mutation resulted in breakage of the two sister chromatids at the position of the *Ds* locus in each chromatid. Evidence that a *Ds* mutation brings about breaks in sister chromatids at the locus of *Ds* is well established. This assumption is therefore legitimate. (2) The *Ds* mutation not only caused breaks to occur at the position of the *Ds* locus but resulted in the release of a submicroscopic chromatin segment that carries a *Ds* locus. This released segment carrying *Ds* has unsaturated broken ends. It could be lost from the chromosome complement if fusion with some other broken ends did not occur. Loss of the *Ds* locus as a consequence of *Ds* mutations has been considered in detail elsewhere (see report on *c-m1* mutations, January, 1949). The manner of this loss may be suggested from cases such as the one being described. (3) At the same time that the events described in (1) and (2) occurred, a spontaneous *chromosome* break occurred just to the right of the *I* locus in this chromosome. Both sister chromatids were broken at the same locus. Evidence for frequent spontaneous breaks in maize is good (McClintock, unpublished). This assumption, therefore, is legitimately taken. These three events would give a series of broken ends as shown in A, figure 3 [reproduced in Figure 1]. Fusion of broken ends could readily occur to give rise to the configuration shown in B, figure 3. The resulting chromatids are diagrammed in C, figure 3. A Duplication chromosome 9, with an inverted order of genes in the proximal duplicated segment and having a transposed *Ds* locus just to the right of the *I* locus is now formed.

Figure 3

A.

Position of chromosome breaks

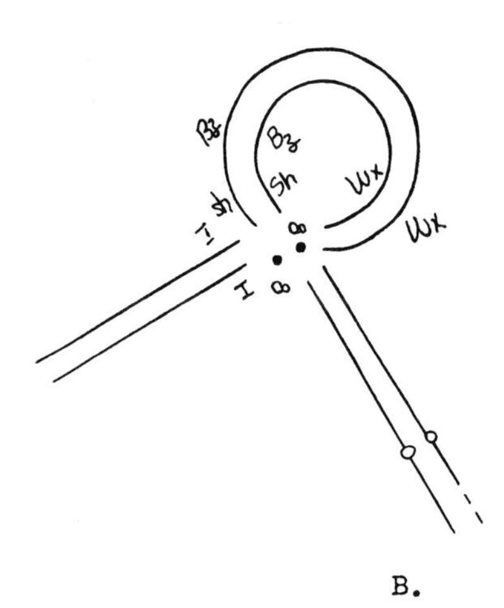

B.

Fusion of broken ends

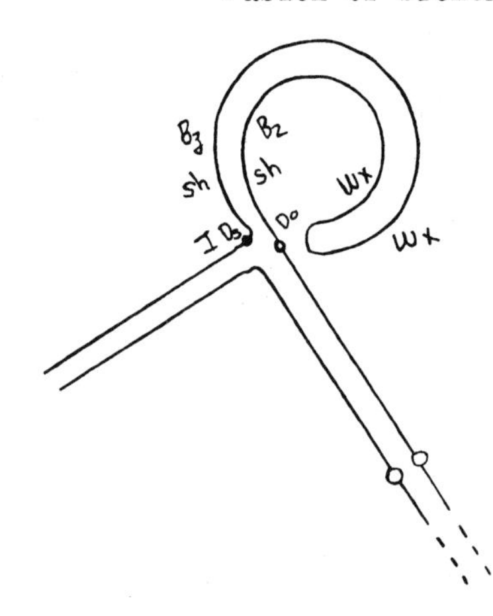

Figure 3 C.

Resulting chromatids

I Ds Sh Bz Wx Wx Bz Sh Ds

I

Figure 1 Enhancement of the original drawing of Figure 3 from Barbara McClintock's notebook. (*A*) Position of chromosome breaks; (*B*) fusion of broken ends; (*C*) resulting chromatids.

It is perhaps not inappropriate to remind the reader that this remarkable passage was composed substantially before the structure and mode of replication of the genic material were understood.

McClintock recovered both the chromosome-breaking *Ds* locus and many new mutable loci from the same cultures. In addition, as noted above, certain characteristics of the chromosome breaking system and the mutable loci made her suspect from the very outset that there was a relationship between them. But the exact nature of the relationship emerged only when she began a careful analysis of the mutable alleles. All of this occurred at about the same time that it began to enter her awareness that the *Ds* locus could move. In 1946, she noticed a single aberrant kernel on an ear of a *yg c sh Bz wx ds* plant crossed by a *Yg C Sh Bz wx Ds* male carrying an *Ac* locus. All of the kernels on this ear, with one exception, were either fully pigmented (no *Ac*) or showed colorless sectors as a consequence of the loss of the dominant C locus after breakage of the chromosome at *Ds* (*Ac* present). One kernel exhibited small pigmented sectors on a colorless background, a phenotype characteristic of an unstable recessive colorless *c* allele mutating to the dominant, fully pigmenting wild-type C allele. McClintock grew a plant from the aberrant kernel in the summer of 1947 and did a number of crosses with it. In an unpublished report composed in January of 1949, McClintock recapitulates the evolution of her thinking about this mutant, which she designated *c-m1*, the first inception of instability at the C locus in her cultures. She says of her initial conclusions about its structure:

> The constitution of this plant was then believed to be *yg c sh wx ds/Yg c-m1 Sh wx Ds; Ac ac*. The data and analysis given below show that the location of *Ds* is not as given. Instead, it is probably at, within or close to the C locus.

She goes on to review the results of the initial crosses and comments that in retrospect, she had failed to take note of an anomalous kernel class. She says parenthetically:

> This latter observation should have made me suspicious about the location of *Ds* but I failed to register a reaction. At the time, I did not know that *Ds* could change its location. Realization of this did not enter my consciousness until late this spring (1948), following the harvest of the winter 1947–1948 greenhouse crop.

It was thus not until plants were grown from some of the kernels on ears obtained from the initial crosses done in the summer of 1947 that the real anomalies surfaced. For example, McClintock selected and grew plants from kernels with *c* to *C* variegation that she believed to have the chromosome 9 constitution *pyd* (or *yg*) *c Sh Bz Wx ds/Yg c-m1 Sh wx Ds* (or *ds*), *Ac ac*. Her first surprise was that all of the plants showed variegation for *pyd* or *yg*, unexpected because some of the selected kernels should have had a recombinant chromosome bearing the *c-m1* locus, but no *Ds*, since the C locus is more than 30 map units from the *Ds* locus. Such plants should not have, but did exhibit chromosome breakage at *Ds* resulting in the loss of the *Yg* allele. She says that this "...did not bother me too much as the sample was small." She made further crosses to plants that were homozygous for the chromosome 9 constitution *c sh wx ds* and lacked *Ac* so as "...to be sure to eliminate such a *Ds* from a *c-m1* carrying chromosome..." The results of this cross, which she refers to in the report as cross (1), were "...strikingly different from expectation." She continues[2]:

> A tabulation of the classes of kernels resulting from cross (1) may interest you for they tell the full story much better than words (table 1). Because the male parent was heterozygous for *Ac*, only half of the kernels having a *c-m1* locus should show *c* to *C* variegation. Also, the kernels that received the stable *c* locus (from the parental *c Sh Wx* chromosome) will not show *c* to *C* variegation with or without *Ac*. You will note that only 14 of the 990 *c Wx* kernels (no visible *c* to *C* mutations) showed *Wx* to *wx* variegation whereas all of the 140 *c* to *C* variegated kernels that were *Wx* likewise showed *Wx* to *wx* variegation. The presence of detectable *sh* sectors in some of these heavily *Wx*→*wx* variegated kernels made it clear that loss of a segment of the short arm of chromosome 9 was responsible, or often associated with, the appearance of the *wx* sectors. Some kind of *Ds*-like action was occurring in these kernels with a *c-m1 Wx* chromosome. This *Ds* action was not associated with a standard *Ds* locus in the *c Sh Wx* chromosome of the male parent. If a *Ds* at the standard location had been present in this chromosome, many of the *c Wx* kernels should have been *Wx*→*wx* variegated but they were not. Secondly, all the *Wx* kernels showing *c* to *C* variegation likewise showed *Wx* to *wx* variegation. These represent the kernels receiving a crossover chromatid that brought *c-m1* and *Wx* onto the same chromosome. Since all of these chromosomes show this *Ds*-like action, a *Ds*-like element must have been present in the *c-m1 Sh wx* chromosome of the male parent. Its location must be in, close to or to the left of the *c-m1* locus. Thirdly, of the 16 *C* kernels arising from germinal mutations of *c-m1*, only one showed any *C* to *c* variegation and this was a peculiar type. (It will be pointed out later that just such new variegations are to be expected. They represent changed location or changed organizations of the inserted element responsible for mutable phenomena.) Unless all such non-variegated *C* kernels have an *ac ac ac* constitution, a *Ds* type variegation should be present in some of them, on the assumption that a *Ds* is in some position other than the locus of *c-m1*. Fourthly, examination of the *C* areas in both the *c* to *C*, *Wx*→*wx* kernels and the *c* to *C*, *wx* kernels was likewise instructive. In these kernels, with some expected exceptions, the *C* areas were not variegated for *C* to *c*. In the *c* to *C*, *Wx* to *wx* kernels, the *C* areas were *Wx*, not *Wx* to *wx* variegated as would be expected if *Ds* were present to the left of the *C* locus. Obviously, then, when a mutation to *C* occurred, the *Ds*-like action in this chromosome ceased...

The conclusions follow:

> The combined observations made me suspect that the *Ds* action was associated with the presence of *Ds* in the *c-m1* locus and that the *Ds* element was usually lost from the chromosome when a mutation to *C* occurred. The following conclusions were formulated: that (1) the *c-m1* locus arose from a *C Sh wx Ds* chromosome following removal of *Ds* from its standard location and its insertion into the normal *C* locus; (2) when a mutation of *c-m1* to *C* occurs, *Ds* disappears; (3) the insertion of *Ds* into the *C* locus brought about an inhibition in the expression of the genes at the *C* locus; consequently the modified *C* locus resembles *c* in phenotypic expression. (4) When this *Ds* element is removed from the chromosome, the normal genic action of the *C* locus can reappear. (5) Two contrasting types of *Ds* action may be recognized when *Ds* is in this new position: (a) one giving the typical dicentric chromatid through breakage and fusion at this new location and (b) one giving the *c* to *C* mutations that usually result in no <u>visible</u> alteration of the chromosome.

[2]McClintock represents variegation by either stating or using an arrow to show the direction of phenotypic change from that dictated by the genotype. Thus, *c* to *C* (or *c*→*C*) variegation denotes somatic reversion from the recessive to the dominant, and *Wx* to *wx* (or *Wx*→*wx*) variegation denotes the sectorial somatic loss of the *Wx* allele in a *Wx*/*wx* heterozygote.

In these brief passages extracted from a report composed in the winter of 1949, McClintock unifies her observations and, in one stroke, clarifies all that had gone before in the study of variegation. The *c-m1* locus is a mutable gene, like Rhoades' *a* allele. It is a stable recessive allele except when the *Ac* locus is present. *Ac*, like Rhoades' *Dt* locus, induces genetic instability, causing the recessive allele to mutate to the dominant, wild-type allele. McClintock explains this by transposition. The *C* locus became mutable when the *Ds* element transposed into it and inhibited its expression, giving the recessive, colorless *c* allele. Mutations to the dominant allele, whether somatic or germinal, remove the inserted transposable element from the locus. Thus, the temporary inhibitor that Emerson had postulated at the *P* locus is a transposable element. (We now know that Emerson's element was *Ac*, but Rhoades' *Dt* belongs to a different element family.) In addition, although *Ds* is the causative agent of both mutability and chromosome breakage, it can do neither on its own, requiring something that the *Ac* element provides, what we today would call the transposase. It is evident from the quoted passages that the ability of McClintock's *Ds* element to both break chromosomes and transpose was extremely important in explicating the relationship between transposition and genetic instability.

By the time that she prepared her paper for the Cold Spring Harbor Symposium of 1951, McClintock had isolated a number of unstable alleles of several genes, including the *A*, *C*, *Bz*, and *Wx* loci. Some she knew to be controlled by *Ac* in the same fashion as the *c-m1* allele (designated *nonautonomous* mutable loci), whereas others were *autonomous*, requiring no second genetic factor for instability. (To maintain continuity with these original designations, I later transferred them to the elements themselves.) She knew that *Ac* itself could transpose and that there were different mutability "systems," later renamed element "families." McClintock recounts that the reactions to her Symposium presentation ranged from perplexed to hostile. Later, she published several papers in refereed journals and, from the paucity of reprint requests, inferred an equally cool reaction on the part of the larger biological community to the astonishing news that genes could move. But McClintock was, by this time, deeply absorbed in the analysis of a new element family she designated *Suppressor-mutator* (*Spm*). It was the subtle complexity of this element family that I found irresistible when I began to understand it more than 25 years later and that I still find so, another dozen years and many molecular and genetic experiments later.

Part III: How Maize Transposable Elements Were Cloned

Extraordinary—and mystifying—as McClintock's assertions about transposition and transposable elements must have seemed in 1951, by the time that I began my work in 1978, the information that she had amassed from her meticulous analysis of genetic crosses made it clear that mobile DNA sequences were the cause and permitted me to make some good guesses about where and how to look for one. I felt quite confident that to isolate a transposable element physically, I needed (simply!) to clone the same gene from a wild-type plant and from one known to have an autonomous transposable element inserted into the gene. Parenthetically, it was by no means the case that all of my colleagues regarded McClintock vindicated by the discovery of transposable elements in bacteria—or even believed her yet in 1978. I remember Frank Stahl, next to me in the Blackford Hall lunch line during one of my early visits to Barbara, leaning down to me from his great height to inquire whether I really believed all of that stuff of hers. (Years later, he was much embarrassed to be reminded of this small incident, long forgotten.)

By the time I began my work, there were unstable alleles of many maize genes. But there were a few problems, too. First, no maize genes had yet been cloned and few biochemical activities had been assigned to gene products. Second, in analyzing the genetic properties of transposable elements, McClintock and other geneticists had made extensive use of mutations in genes coding for enzymes and regulatory factors required for anthocyanin pigment biosynthesis. This pathway was a marvelous genetic choice because not only is it dispensable, but it is also expressed in virtually every plant part, including the kernel aleurone layer, whose genetic constitution is identical (except for ploidy) with that of the embryo contained within the kernel. But a little bit of pigment colors a great deal of plant tissue and the enzymes are not abundant, nor are their transcripts. Among loci that were likely to encode sufficiently abundant gene products, such as starch biosynthetic enzymes, to make cloning their cDNAs feasible with the available technology, few had insertions that behaved like autonomous elements. Indeed, the *Sh* and *Wx* loci appeared to be the only significant candidates and each of these had drawbacks. Sucrose synthase, the enzyme encoded by the *Sh* locus, had been identified and was fairly abundant, but there were no alleles of the locus that were known to behave as if they had either an *Ac* or an *Spm* insertion. There were unstable alleles of the *Sh* locus, but their genetic behavior was quite complex (as their molecular structure turned out to be) and the responsible element was *Ds*. That left the *Wx* locus, which was believed to encode a starch glucosyl transferase, although the enzyme had not yet been purified or even identified with certainty. McClintock had recovered an *Ac* insertion at *Wx*, designated *wx-m9*, as well as a derivative of this allele that behaved genetically as if the *Ac* element had been replaced by or become a *Ds* element; indeed, my guess was that the *Ac* element had mutated, inactivating its transposase.

In due time and with the participation of two postdoctoral fellows, Susan Wessler and Mavis Shure, the protein encoded by the *Wx* locus was purified, its cDNA was cloned, and the gene was cloned from maize plants carrying different alleles of the *Wx* locus: a wild-type allele, the *Ac*-containing allele, a *Ds* derivative of the *Ac* insertion, an independently derived *Ds* insertion, and a revertant wild-type *Wx* allele derived from the *Ac*-containing allele. The results of restriction endonuclease analyses and electron microscopy were immediately informative. When either *Ac* or *Ds* was associated with the gene, it was longer than its wild-type counterpart. When mutant and revertant versions were heteroduplexed, the expected single-stranded loop could easily be seen in the electron microscope, looking to be 4 or 5 kb long (later sequencing revealed the length of *Ac* to be 4563 bp). This meant that *Ac* was a short stretch of foreign DNA inserted into the gene. A revertant *Wx* allele had no insertion, and both *Ds* derivatives proved to be closely related to the *Ac* element. The direct derivative of the *Ac* insertion was located at precisely the same site and differed from *Ac* by a very short internal deletion, later determined to be 194 bp. Thus, a *Ds* element could indeed be a mutational derivative of an *Ac* element. More than that, since the deleted element could no longer transpose without the assistance of an *Ac*, the deletion necessarily disrupted a gene encoding a protein that could act in *trans* and was required for transposition, a transposase. The other *Ds* was inserted at a different site in the gene. It was about 2 kb long and comprised approximately a kilobase of sequence from each end of the element.

Before the isolation of this first *Ac* element from the *Wx* locus, *Ds* elements had been cloned in the laboratories of Jim Peacock in Australia and Peter Starlinger in Germany. The first of these, designated *Ds1*, had a length of about 0.4 kb and does not appear to have been derived from an *Ac* element, resembling it only at the very ends, having virtually identical 11-bp terminal inverted repetitions, as well as some similar short sequences near the termini. The *Ds* element isolated from the *Sh* locus proved to have an extremely interesting structure, quite closely related to *Ac*. This element, which has come to be called the "double" *Ds*, is probably identical or virtually identical with the original *Ds* element that McClintock identified by virtue of its propensity to cause chromosome breakage. It comprises two copies of a short, 2-kb element identical to the one that I had isolated from the *Wx* locus, one inserted in inverted orientation almost exactly into the center of the second copy. Indeed, the 2-kb element recovered from the *Wx* locus may have arisen by transposition into the *Wx* locus of one *Ds* out of a double *Ds* located elsewhere on chromosome 9.

Once an *Ac* element had been cloned, it was used to recover other *Ac* and *Ds* elements, including the element responsible for the instability of the *P* locus studied by Emerson, as well as Brink and his students many years later. Since it had already been demonstrated that the *P* locus element, designated *Mp*, had the genetic properties of an *Ac* element, it was not surprising to discover their molecular identity. What was surprising, given the heterogeneity in the structure of functional bacterial transposons, was that all of the isolated *Ac* elements have so far been identical in length and sequence. In contrast, *Ds* elements are unexpectedly heterogeneous, ranging from one that differs from *Ac* by an extremely small deletion to *Ds1* elements of the type isolated in Peacock's laboratory and which comprise an internally homogeneous group of sequences that resemble each other, but bear little resemblance to *Ac*. In between are *Ds* elements within which short stretches of *Ac* sequences alternate with short unrelated sequences. Thus, the *Ac-Ds* element family comprises the autonomous *Ac* element and a structurally heterogeneous group of sequences that contain, at a minimum, the element's terminal inverted repeats and enough internal sequence to provide the requisite recognition sites demanded by the element's transposase for mobilization of the intervening sequence.

Part IV: How Maize Transposable Elements Look Today

The Ac-Ds element family

One aspect of the early reports on *Ac* and *Ds* that readers find quite difficult is the evidence for frequent changes at mitotic divisions, giving rise to twin sectors of tissue, within each of which the element or the mutable locus exhibits altered genetic properties. Indeed, it was McClintock's appreciation of genetic events that not only changed the subsequent pattern of gene expression, but also differentiated the descendants of the two daughter cells from each other that provided the initial stimulus for her hypothesis that transposable elements were components of nuclear systems that regulate gene expression and prompted her to name them "controlling elements." Although not all of these changes are understood even now, molecular genetic studies have shed considerable light on others. The most easily comprehended and the first to find an explanation (long before the molecular era) is McClintock's observation of twin sectors on variegated plants and kernels in which one twin exhibits a revertant wild-type phenotype and the other exhibits an altered variegation pattern, commonly showing a reduced excision frequency. McClintock discerned early that *Ac* exhibits a negative dosage effect. As the number of copies of the *Ac* element increases, the frequency of excision decreases and excision events are developmentally delayed, giving a lightly variegated phenotype. Much larger twin sectors occur in some genetic backgrounds, and their genetic basis was explored over many years by Brink and his students, especially Greenblatt, using an *Mp* (=*Ac*) element at the *P* locus. It has become clear that twinning of this type reflects the predominant mode of *Ac* transposition. The element transposes out of one of the two sister chromatids after replication of the donor site, restoring *P* gene expression. The element most frequently reinserts nearby on the same chromosome, commonly at a site that has not yet been replicated. When the recipient site is replicated and the sister chromatids segregate, one of the two daughter cells receives a chromosome with two copies of the element and the other daughter cell receives a chromosome with an empty donor site and a transposed element. Each of these daughter cells then gives rise to a sector of tissue with an altered phenotype, a light variegated sector containing the chromosome with two elements and a wild-type sector containing the empty donor site and transposed, but generally phenotypically undetectable, element.

Why a doubling of the *Ac* dosage delays transposition remains to be explained. It does not appear to result from inhibition of element transcription, since transcript concentrations are roughly proportional to *Ac* dosage. It does require an element-encoded gene product, because *Ds* elements do not contribute to the dosage effect, regardless of structure. So far, only a single transcript has been identified for the *Ac* element, and its processed mRNA encodes a DNA-binding protein required for element transposition. It is not yet known whether the same gene product is responsible for the dosage effect.

McClintock reported other, more subtle pattern changes. For example, she noted that the chromosome-breaking *Ds* element commonly changes, generally irreversibly, to a condition that she designated as giving "few, late breaks." She made use of this property of the chromosome-breaking *Ds* in adducing evidence that the original *Ds* was the same as the *Ds* that transposed into the C gene to give rise to the *c-m1* allele. Indeed, she noted that at *c-m1*, a reciprocal relationship exists between chromosome breaks and mutability. Kernels show either a high frequency of somatic reversion from *c* to *C* or a high frequency of chromosome breakage, but not both. A likely explanation for these observations emerged from molecular analyses on *Ds* mutations at the *Sh* locus that I carried out in collaboration with the laboratory of Peter Starlinger.

McClintock had isolated several mutable alleles of the *Sh* locus that had the odd property that the *sh* mutation could revert to *Sh* expression without loss of the *Ds* element from the locus, as judged from the persistence of chromosome breakage at the original position. Our molecular analyses showed the *sh-m5933* allele to be astonishingly complex. The locus contained a large insertion of at least 30 kb, flanked at the 3′ end by a double *Ds* element and at the 5′ end by an incomplete double *Ds* element lacking half of one of the elements. In addition, there was a duplication that included the 5′ half of the insertion, its *Ds*, the 5′ end of the gene, and an unknown length of flanking sequence. I had laboriously selected a number of *Sh* revertants of this allele (boring each kernel that looked like an *Sh* revertant very carefully with a pin to make sure it did not have a hollow center. I then had to rescue them, often as broken tassels, from the tornado that blew over Ed Coe's field, where I was growing my plants that summer). Southern blots on DNA from these held the answer to the genetic puzzle. The large insertion was gone from the gene, but the duplication and its associated *Ds* remained, explaining why chromosome breakage persisted in revertants. One of the revertants showed a typical pattern of "few, late" chromosome breaks. Instead of the *Ds*-element-and-a-half that was in the original duplication, it had only half an element. Thus, concomitant with excision of the large insertion from the gene, a complete 2-kb *Ds* also transposed out of the duplicated segment. The remaining piece of *Ds* was still capable of providing a site of chromosome breakage, but breakage was much less frequent and occurred late in development.

The foregoing observations suggest an explanation for the kinds of changes in *Ds* activity that McClintock noted. First of all, chromosome breakage is very much more frequent at the double *Ds* than at a single element of the same sequence, such as the one recovered from the *Wx* locus. In contrast to the general mode of *Ac* transposition, which involves cleavage of only one of the two daughter chromatids, sister chromatid breakage and the resulting dicentric formation necessitate cleavage of both daughter chromatids at the site of *Ds* insertion. Thus, the normal transposition mechanism distinguishes between the two daughter chromatids of identical sequence and the double *Ds* element disrupts the chromatid selectivity (with markedly deleterious consequences for chromosome structure, a rather important point to consider in thinking about how transposable element damage is minimized). I have suggested that a sensible genetic basis for this is methylation, as has been shown for the bacterial Tn*10* transposon. Perhaps the enzyme that cleaves at element ends can distinguish between the two otherwise identical hemimethylated copies produced after replication of the DNA. The double *Ds* element has a copy of each end in each orientation (reconstruction experiments have shown that subterminal sequences are required for transposition, in addition to the inverted terminal repeats, and that the two ends of the element are not equivalent). After replication, both daughter chromatids will contain cleavable termini, accounting for the aberrant cleavages that result in dicentric chromosome formation. But more than that, a double *Ds*, with its multiple cleavage sites, might be especially unstable. From the kinds of fractional elements that have been recovered, it is evident that the double *Ds* can be cleaved at any of the *Ds* ends during transposition. McClintock noted that the most common kind of change decreased the breakage frequency and that this was generally a stable change. Such changes probably result from deletion or excision of part of the double *Ds*. If the remaining element contains two ends, the normal transposition mechanism might be restored, accounting for McClintock's observation that there is a reciprocal relationship between the frequency of excision and chromosome breakage at *c-m1*.

McClintock noted even finer changes in the frequency and timing of both excision and chromosome breakage events. Some mapped to the *Ds* element and others mapped to *Ac*. These have not been studied systematically in the *Ac-Ds* element family, but many have been worked on or worked out for *Spm* elements and are considered below.

The Spm element family

The *Spm* element family is wonderfully complicated and even stymied McClintock briefly. Given the number of different phenotypes associated with *Spm* insertions, I still marvel that McClintock was able to understand it without so much as a restriction enzyme from our modern molecular tool box. She did it all with her eyes and her intelligence, aided by a microscope and a few simple reagents, and, of course, she did *just* the right crosses. She says:

> In order to study any one element of a system, each of the other variables, as it is recognized, must be removed by crossing and selection, so as to work with the smallest possible number of associated and interacting elements.

McClintock's growing understanding of the remarkable variety of ways that *Spm* elements can affect gene expression strengthened her conviction that transposable elements were "controlling elements." Her view was that these were extragenic devices that regulate gene expression, their mode of action revealed by transposition into genes not previously under their regulatory control. She selected and analyzed—and preserved—a large number of strains, each containing examples of the kinds of genetically transmissible alterations that *Spm* elements can themselves either sustain or impose on a gene. These she provided to anyone who was interested in transposition (even molecular biologists who had never grown a maize plant or done a cross), a legacy of incalculable value.

I was first introduced to *Spm* in the summer of 1979, when Barbara generously selected materials for my first corn crop, which Ben Burr kindly agreed to host in his fields at the Brookhaven National Laboratory. Barbara and I discussed this crop in advance and at great length, but we weren't communicating—we were not even speaking the same language. I was a molecular biologist trying to figure out what materials would be optimal for cloning. Barbara had a very different perspective. I especially remember that she viewed my efforts to obtain plants that were homozygous for a given element with considerable consternation, perhaps even a bit of disdain. It was only much later that I appreciated her point of view. Some of the alleles were so unstable that genetic changes could occur in each of the two insertion alleles of a homozygous plant. How would I ever know what was going on? But we got through the summer somehow and I learned an immense amount. I remember being so exhausted by the end of 2 months of 5 a.m. to 2 a.m. days that I seriously considered abandoning the project. Yet once I began analyzing the results of my crosses, I realized that Barbara had slipped me a treasure trove, bits of virtually everything she had ever worked on, and I was glad I had done every cross I could push my weary body to do, even if all of them were not just the *right* crosses.

In McClintock's cultures, the *Spm* element had never hopped in its entirety into a locus that was convenient for the molecular biologist. Cloning it required a detour through the *Wx* locus and *Ac*, described earlier. I obtained *Spm* using a short, transposition-defective *Spm* (*dSpm*) element extracted from the *wx-m8* allele of the *Wx* locus, a technique whose contemporary name is "transposon tagging." I used the *dSpm* to identify *Spm*-containing clones from a recombinant phage library constructed from DNA isolated from a plant that carried McClintock's *a-m2* allele, known to have a complete *Spm* element insertion. We had first used this approach to clone the *Bz* locus, using the *Ac* element. In the course of those experiments, we had discovered that although there were many sequences in the maize genome that had some homology with the *Ac* element, particularly its ends, there were very few copies of complete *Ac* elements (although more than the single genetically active copy). So cloning a gene that had a genetically characterized *Ac* insertion proved remarkably easy once we knew the structure of the element. The same proved true in the case of the *Spm* element. In the initial group of more than 100 phage clones isolated, only 3 were even likely candidates by virtue of the size and structure of the insertion. At about this time, the *A* locus was cloned in Heinz Saedler's laboratory by a similar method, and members of his laboratory graciously provided a short *A* gene sequence, which we used to quickly confirm our conclusion that the 3 clones we had already identified contained sequences of both the *A* gene and the *Spm* element. The *Spm* element turned out to be twice as big as *Ac* and virtually identical in sequence with the *Enhancer* (*En*) element that had been identified and studied genetically by Peter Peterson and which had already been cloned in Saedler's laboratory.

The fascination of the *Spm* element family is that it can have such diverse effects on expression of a gene. From the outset, my suspicion was that these were all clues, could I but read each correctly, that would allow me to understand how the element worked and was regulated. This view was a clear departure from Barbara's and it caused us a brief period of considerable strain in our relationship, particularly when I first began to articulate it in print. Although I deeply respect the insightful observations underlying McClintock's perspective, the pieces of the puzzle look different to me, working in a different time and a different way. Necessarily, I have renamed some of them and assembled them into a different whole, one that makes the most sense to me, and that is what I offer here.

To begin with, the most common kind of *dSpm* insertion has two markedly different phenotypes, depending on whether a fully functional *Spm* element is or is not present in the genome. In the absence of a *trans*-acting element, many *dSpm* insertion mutations exhibit a phenotype that is intermediate between the null and the wild type. But when an *Spm* element is present, it *suppresses* expression of the gene with the *dSpm* insertion, giving the recessive, null phenotype. It also promotes *mutation* (reversion) of the affected gene to the dominant wild-type allele in some cells, giving sectors of deeply pigmented cells. It was for these reasons that McClintock gave the name *Suppressor-mutator* to the fully functional *trans*-acting element of the family. Because there are also *dSpm* insertion mutations that respond to *Spm* in a markedly different way, I named this original type of *dSpm* insertion mutation *Spm-suppressible.* The first two *Spm-suppressible* alleles that McClintock isolated and characterized were the *a-m1* and *a2-m1*, both unstable alleles of genes in the anthocyanin biosynthetic pathway.

It was Hwa-Yeong Kim in Oliver Nelson's laboratory who first provided a molecular explanation for the ability of a gene with a *dSpm* insertion to function. The *dSpm* can be transcribed and then processed out of the transcript like an intron! Whether or not this restores gene expression depends on the insertion site, the orientation of the *dSpm* with respect to the gene, and the coding properties of the processed transcript. A similar phenomenon was subsequently discovered for the *Ac-Ds* family, and it has been suggested that this is a mechanism for minimizing the genetic damage done by insertion mutations. I suspect that there might be more to it. The *Spm* element's termini bear a strong resemblance at the nucleotide sequence level to its internal splice acceptor sites. Might an element-encoded protein that participates in transposition also serve a regulatory role in alternative splicing? Or might a protein that normally participates in RNA processing have been pressed into service as part of the element's transposase, much as certain RNA bacteriophages incorporate into their replicase proteins that are normally used by the bacterium in translation?

Early in the study of the *Spm-suppressible a-m1* mutant, McClintock accumulated a series of derivatives, which she called altered "states of a locus." Each derivative had a new and different variegation pattern, either by virtue of the number, phenotypic expression, or the developmental timing of the revertant sectors signaling excision of the resident *dSpm* element. Elements have been cloned from both the original *a-m1* allele and many such derivatives. All are simple deletion derivatives of the *Spm* element, most comprising less than 20% of the complete element, but retaining both ends. Almost without exception, the secondary derivatives have further deletions within the sequence. The deletions commonly decrease the ability of the element to excise, and their molecular analysis has assisted in defining the sequences required for transposition. These include the terminal inverted repeats, subterminal repetitive sequences of several hundred base pairs between the inverted repeats and the transcription unit at each element end, and, surprisingly, the element's first exon.

The ability of an *Spm* element to *suppress* expression of a gene with a *dSpm* insertion implies that *Spm* encodes a *trans*-acting gene product that interacts with the insertion in such a way that expression of the mutant gene is disrupted. The simplest explanation is that the element encodes a DNA-binding protein, and indeed, it does. Heinz Saedler and his colleagues found that the most abundant transcript of the very closely related *En* element, designated *tnpA*, encodes a protein that binds to short sequences, called subterminal repeats, reiterated near each element end. In reconstruction experiments, abundant expression of this TnpA protein inhibits expression of a gene under the control of a completely foreign promoter if the promoter has the protein's short binding sequence inserted into it.

One might suspect, then, that the *Spm* element codes for two gene products, TnpA protein being the *suppressor* and a second gene product for the *mutator* function. The element's *mutator* component was defined genetically by its ability to act in *trans* to promote transposition of a *dSpm* insertion out of a gene in some cells during development, restoring wild-type gene function and creating a variegated phenotype. But neither genetic studies nor the early molecular data gave evidence of a second gene product. McClintock reported that the element's two functions always came and went together, whether by a mutation that converted an *Spm* into a *dSpm* or by a kind of reversible genetic event that she called a "change in phase of activity." The element appeared to encode a single transcript, *tnpA*, that spanned almost the entire length of the 8.3-kb element. Its first intron was surprisingly long, half the element's length, but all of its other nine introns were less than 200 bp and of a typical length for plant introns. Moreover, McClintock had identified mutant *Spm* elements that turned out to have deletions in the *tnpA* transcript's first intron. And although McClintock named such elements *weak Spm* (*Spm-w*) elements because they promoted transposition at a lower frequency and later in development than the standard *Spm* element, they still appeared to have both of the element's genetically defined functions. This suggested that all of the element's genetic functions might be attributable to the protein encoded by the *tnpA* transcript.

The sequence of the *tnpA* transcript's first intron was odd, however. It contained two open reading frames (ORF1 and ORF2), the first of which was quite long. Although the *Spm-w* deletions affected one or both ORFs, they left a sizeable part of the coding sequence intact (Fig. 2). To construct a rigorous test for whether the intron-1 ORFs encoded proteins required for transposition, we needed to be able to mutate the element in vitro and reintroduce it into plant cells in such a way that we could detect its activity. This project was undertaken by a postdoctoral fellow in the laboratory, Patrick Masson. Much earlier and not long after I had isolated the first *Ac* element, I had initiated a collaborative effort with Jeff Schell's laboratory to determine whether the element, isolated from a monocot, could function if it were introduced into the cells of a dicot on an *Agrobacterium* Ti plasmid, then and still the easiest means of transforming plant cells. This project, carried out primarily by postdoctoral fellow Barbara Baker in Schell's laboratory and later joined by George Coupland from Peter Starlinger's laboratory, was quite successful. The *Ac* element proved to be very active in tobacco cells. This surprised us somewhat because it was already known that there were differences in transcript processing between monocots and dicots. Masson's initial efforts to design an assay for *Spm* activity in tobacco cells were modeled on the kinds of excision assay plasmids that had been constructed for the *Ac* element, in which the element was inserted between a strong viral promoter and a gene encoding a bacterial drug resistance marker. This did not work with the *Spm* element (we learned much later that this was primarily because tobacco cells often processed the element's transcript differently than did maize cells). Masson was only able to detect the transpositional activity of *Spm* when he replaced the drug resistance marker with a β-glucuronidase (GUS) gene and separated the *trans*-acting *Spm* element from the *dSpm* element disrupting the gene. Making use of a chromogenic substrate for the enzyme, he was then able to detect tiny blue GUS-positive tissue sectors, and these appeared in large numbers only in plant tissues transformed with both the *dSpm*-disrupted GUS gene and the *Spm* element. Thus, the assay could be used to figure out what mutations inactivate *Spm*.

Masson then made frameshift mutations in the ORFs of intron 1, as well as in one of the shorter introns of the *tnpA* transcript. The results were clear-cut.

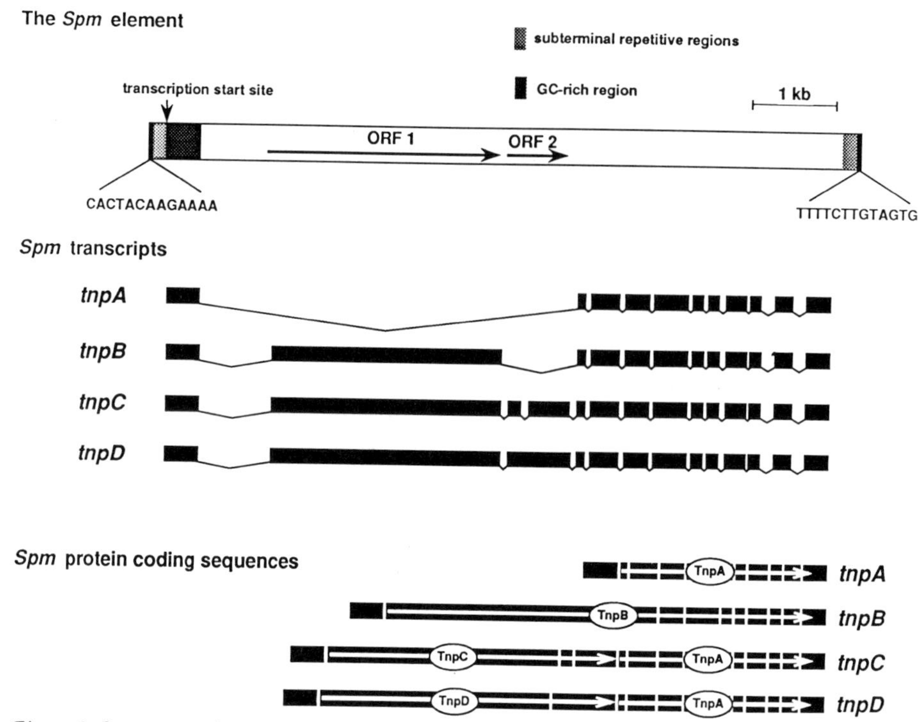

Figure 2 Structure of the *Spm* element and its transcripts.

Frameshift mutations in either intronic ORF disrupted the ability of the *Spm* to promote excision of the *dSpm* element from the GUS gene, whereas mutations in one of the gene's short introns did not. This implied that the ORFs encode an essential protein or proteins, a fact at odds with the absence of the corresponding coding sequences from *tnpA*, then the only characterized element-encoded transcript. In the months that followed, the entire laboratory worked feverishly to resolve this paradox. Overexposed Northern blots of overloaded RNA gels revealed that there were indeed several additional, large transcripts. Amplifying, cloning, and sequencing transcript fragments using the polymerase chain reaction, we began to accumulate evidence for the existence of three large alternatively spliced transcripts, each of which had all 11 of the *tnpA* transcript's exons, but which were much larger because they had one or more additional exons derived almost entirely from the two ORFs of the first intron of *tnpA*. One of the transcripts was monocistronic, containing a continuous coding sequence that extended from ORF1 to the end of *tnpA* transcript's coding sequence, whereas the other two had coding sequences for two proteins, one of which was the *tnpA* protein and the second one of which was encoded by a spliced sequence that connected the two ORFs within intron 1 of the *tnpA* transcript. We named these transcripts *tnpB*, *tnpC*, and *tnpD*, in order of increasing size (see Fig. 2).

The next step was to determine which of this sudden plethora of potential coding sequences were required for transposition. To do this, Masson dissected the processed transcripts into individual coding sequences, attached plant promoters and polyadenylation sequences, and introduced the cDNAs into tobacco cells individually, pairwise, and together with mutated *Spm* elements to test their ability to support excision of the *dSpm* element from the GUS gene. Again, the answer was clear-cut. Element excision occurred only when the *tnpD* cDNA and *tnpA* cDNA were both present. So both TnpA and TnpD participate in transposition (we still do not know what, if anything, TnpB and TnpC do). We could now understand why the different element-encoded genetic functions characterized by McClintock were inseparable: The element had only a single promoter, and its different coding sequences were extracted from the same primary transcript by alternative splicing. And more than that, TnpA appears to be doing double duty, participating in transposition and serving as the "suppressor" gene product.

It struck me as not altogether obvious why the *Spm* element should have a suppressor gene, although one can argue (and I have) that the element's ability to suppress expression of a gene into which a *dSpm* element has inserted is not relevant to regulation of the element and that it simply reflects the DNA-binding properties of an element-encoded protein that normally serves in a different capacity. Recently, however, we have observed that *Spm*'s promoter is also inhibited by high levels of the *tnpA* gene product, even as transposition is stimulated, suggesting that transcription might be inhibited as the concentrations of the transposase proteins reach levels sufficient for transposition. I suspect that this is still an oversimplification of the element's regulatory circuitry. Clues are, once again, to be found in McClintock's genetic treasure box. The most obvious clues are of two kinds. First, the properties of a certain kind of *Spm* insertion mutation (*a-m2*) isolated by McClintock suggest that in some cases, *Spm* can activate expression of a gene with a *dSpm* insertion. Second, she showed that *Spm* elements could be reversibly silenced and suggested that a silent element could be reactivated by an active one, again implying a positive regulatory connection between elements.

It was my fascination with the properties of the *a-m2* alleles of the *A* locus that prompted me to begin my *Spm* studies by analyzing the structure of this extraordinarily interesting series of derivatives. McClintock had identified the original mutation in the early 1950s; indeed, its designation implies that it was the second mutable allele of the *A* locus to have appeared in her cultures. The first isolate had a complete *Spm* element inserted at the locus, and it had an unusual phenotype. The kernels were palely pigmented throughout, as if the *A* gene were active at a reduced level, and the deeply pigmented sectors signaling excision of the element appeared against this pale background. McClintock soon had a series of derivatives, each first identified as a single kernel with an aberrant phenotype. Many of the mutations rendered the inserted element transposition-defective and gave a completely new kind of mutant phenotype for the *Spm* element. In such *dSpm* derivatives, the *A* gene was silent in the absence of a *trans*-acting *Spm* element. In its presence, the phenotype was that of the original *a-m2* mutation: deeply pigmented revertant sectors appearing against the paler background signaling a low level of *A* gene expression. Thus, in contrast to the *Spm-suppressible a-m1* alleles, expression of the *A* gene in these *a-m2* alleles is *dependent* on the presence of a *trans*-acting element! Almost without exception, these transposition-defective derivatives (which I named *Spm-dependent*) proved to have large internal deletions that encompassed part or all of the element's protein-coding sequences.

The simplest hypothesis to explain the *Spm* dependence of *A* gene expression in these alleles is that the gene is expressed from the element's promoter, which requires an element-encoded transcription factor. But molecular analyses showed the insertion site to be a bit upstream of the gene's transcription start site and its transcriptional orientation opposite to that of the *A* gene. More seriously, this simplistic interpretation was belied by a truly unique property of several of the internally deleted *a-m2* alleles, a property McClintock named "presetting." *dSpm* mutants of this type exhibit the expected colorless null kernel phenotype as long as they are continuously propagated in the absence of an *Spm* element, and they give the typical *a-m2* variegation pattern in the presence of an *Spm* element. But a novel phenotype appears regularly among kernels on ears of plants containing both such a *dSpm a-m2* allele and an independently segregating *Spm*. The novel phenotype occurs only in kernels that have the *dSpm*-containing *a-m2* allele and no *Spm*. Many such kernels show irregular areas of *A* gene expression. It is as if the kernel commences development with an active *A* gene, but the gene is inactivated as the kernel develops. McClintock showed clearly that such kernels appear only if the *dSpm* insertion mutant and the *Spm* have been together through at least part of the plant's developmental cycle. She interpreted this to mean that the *Spm* element could preset the *A* gene to be expressed after the element had been removed, either by meiotic segregation or even earlier in development, by a transposition event. The "preset" pattern of *A* gene expression is not heritable, but it can be reproducibly recreated simply by bringing together an allele that can be preset and an *Spm* and then separating them genetically.

The most obvious implication of presetting is that the *A* gene *can* be expressed in the absence of whatever gene product is supplied by *Spm* (there are quite compelling genetic arguments against the explanation that *Spm*-encoded gene products linger in cells after the element is gone). This, in turn, means that the inability of the *A* gene of the *Spm-dependent a-m2* derivatives to function in the absence of a *trans*-acting element must have a different explanation than simple dependence on an element-encoded transcriptional activator. It suggests, instead, that the element sequence inserted at the locus is modified in some reversible way that interferes with expression of the adjacent gene. What the element-encoded gene product might do is to interfere with the modification as long as it is present. After the element is removed by transposition or meiosis, however, the modification that prevents *A* gene expression is restored but not always immediately, permitting expression of the gene through a number of subsequent cell generations.

The foregoing arcane hypothesis so offended a reviewer of the manuscript in which I first articulated it that I was given the choice of including the reviewer's simpler explanation or not publishing. But I thought that it made considerable sense in the light of McClintock's analyses of *Spm* elements undergoing "changes in phase" of activity. She had discovered very early in her analysis of the *Spm-suppressible a2-m1* allele that the *Spm* element in those particular cultures was being alternately silenced and returning to an active state, sometimes within the developmental cycle of a single plant and sometimes at much longer intervals. One of the most remarkable inferences that she was able to draw from studies on plants containing *Spm* elements undergoing such changes in phase of activity was that a silent element could be transiently reactivated by an active one. She teased this conclusion out of the observation that *Spm* elements that cycled between active and inactive phases during plant development conspired to give a dosage effect. Using the *a2-m1* allele, which can be suppressed but gives no excision, she observed that the "unsuppressed" pigmented area decreased with the number of copies of a cycling element in the kernel, in all likelihood because all of the elements need to be inactive to give the pigmented phenotype, whereas only one need be active to give the "suppressed" null phenotype. McClintock noticed that an element that was inactive throughout most or all of a plant's developmental cycle could contribute to this dosage effect in the same manner as a frequently cycling element, implying that both were active when they were together. This, in turn, suggested that the active one could reactivate the inactive one, as long as they were in the same plant.

I suspected that this was what "presetting" was all about. An element-encoded gene product was probably interacting with—perhaps simply by binding to—an inactive element to prevent whatever heritable but reversible genetic modifications were responsible for inactivating the element. I thought that the significance of the *Spm-dependent a-m2* alleles was that the *A* gene was serving much as we used the GUS gene—that it was a reporter gene for the element's regulatory system. My view was that the internal deletions that rendered the resident element transposition-defective had, in essence, separated the target (binding) sequences from the sequences encoding the regulatory molecules, provided by the *trans*-acting *Spm*, thus revealing the operation of what was normally an element-encoded autoregulatory system.

To test these ideas, I needed to identify the genetic modifications responsible for inactivation and discover whether—and how—an active element interacted with an inactive one. These experiments had a major genetic component and took me, at last, beyond the wide limits that McClintock's intellect had ranged (Gerry Neuffer once said to me that every time he thought of a clever experiment to do with transposable elements, he would go back and reread McClintock first, only to discover that she had already done it—I knew what he meant). I began by identifying inactive derivatives of two different *Spm* insertions at the *A* locus, since the locus and an *Spm* element residing there had already been cloned. I then commenced concomitant genetic and molecular analyses of the newly inactivated elements.

I was soon joined in this project by postdoctoral fellow Jody Banks. She had taken on an odd observation I had made earlier, the investigation of which eventually converged on the *Spm* activity phase-change project. In the course of cloning the *a-m2* alleles, I had found that there was an internal deletion in virtually every *dSpm*. There was one exception: the *a-m2-8167B* allele. It behaved like a *dSpm*, but I could find no deletion. I already knew from very early experiments with *Ac* that I could distinguish between the genetically active copy of the element in the genome and others by using restriction enzymes that would not recognize and cut their target sequences if they were methylated. I tried some methylation-sensitive enzymes on genomic DNA containing various *a-m2* alleles. By good fortune, one enzyme I used cut the element within its first exon, most of which is exceedingly GC-rich, and I found that the *a-m2-8167B* allele was resistant to cleavage and that alleles with active *Spm* elements could be cut. Banks extended this observation and began to incorportate genomic DNA from plants containing some of the newly arisen inactive *Spm* elements in her investigation.

Our summers were, of course, devoted to genetic experiments. I had first set myself the seemingly simple task of determining the heritability of the element's inactive phase. Neither the task nor the results were simple. The only consistent answers turned out to be: the heritability varies and (worse yet) it changes. But even as the complexity increased, unifying and simplifying generalizations emerged. To begin with, an inactive element was not transmitted to all progeny in an inactive state, so this was not a simple Mendelian trait. Indeed, the fraction differed not only from plant to plant, but even within a single plant! Fortunately, the differences were sufficiently reproducible so that distinct patterns emerged as the numbers of plants analyzed grew. Whatever the absolute numbers, the probability that an element would be transmitted in an inactive phase to progeny kernels was higher for gametes produced on the main stalk than for tiller gametes. It was also higher for male than for female gametes, whether the gametes were produced by the main stalk or by the tillers. In addition, if one selected kernels from the plants in a given group of siblings that exhibited the highest heritability of the inactive state, its heritability was higher in the next generation. Indeed, after four generations of selection for inactivity, almost all kernels with an inactive element grew into plants that transmitted an inactive element to almost all of the progeny. The heritability of the inactive phase had *increased.* We guessed that the mysterious *a-m2-8167B* allele, which always gave progeny with an inactive element, might contain an extremely inactive element, and we eventually designated this the *cryptic* state. I later did an experiment with the objective of determining the spontaneous reactivation frequency of one such *cryptic Spm* and gave up after examining more than 200,000 kernels without finding anything but an occasional small sector of *Spm* activity.

We thus understood that even after an element had become inactive, the heritability of the inactive state was far from locked in, and it remained capable of further change. The changes seemed to be incremental, and a distribution of heritabilities could be detected in each generation. Only the endpoints in this emerging continuum exhibited high heritability. Once fully active, an element rarely underwent inactivation in the germ line unless it transposed. Indeed, it had taken me several summers to identify my first inactive derivative of an *a-m2 Spm* allele. Conversely, the stability of the inactive state was so great in the case of a *cryptic Spm* that it could be mistaken for a conventional mutant, as we had done in the case of the *a-m2-8167B* allele. (Once we understood the inactivation and reactivation process, we were able to reactivate this element genetically, recovering the original *a-m2* allele, which McClintock had been unable to provide us because it had been lost.) Between these extremes was a variety of heritable states that McClintock and Peterson had first observed and to which we added: elements that were active in one plant part but not another, elements that showed different patterns of inactivation and reactivation during development, and elements that showed differential transmission of the inactive state through different plant parts.

We called elements in this part of the spectrum "programmable" because any given developmental expression pattern was heritable, yet was heritably altered in a small fraction of progeny. McClintock had selected kernels in which the element had undergone a change in activity phase and found the element "reprogrammed" to give a different heritable pattern of expression.

We discovered that "reprogramming" could happen even without a concomitant change of phase. The most conspicuous oddities of the inactivation phenomenon appeared to be these. The element was either on or off, but the heritability of the two states seemed to be capable of changing in small increments. In each generation, there appeared to be a distribution of "settings" around a mean. Although the mean could be displaced to the extremes by selection, this required multiple plant generations.

Yet, because even a cryptic element could be reactivated, the underlying genetic change had to be a reversible one. DNA methylation seemed to be a reasonable guess. Banks first found that active elements could be distinguished from inactive ones by the absence or presence, respectively, of methylated C residues in the short sequence between the end of the element and its transcription start site. It was in the GC-rich sequence encoding the first exon that she found differences between elements that differed in the heritability of the inactive state. The higher the heritability, the more extensive the methylation of sites within this GC-rich region. It was as if methylation commenced upstream of the transcription start site and gradually spread into the GC-rich region, with its many methylatable sites, as the heritability of the inactive state increased. Outside of these two short regions around the transcription start site and representing less than a tenth of its length, the element appeared to be methylated regardless of its activity phase.

The observations that led to McClintock's conviction that an active element could reactivate an inactive one were made with *Spm* elements in the unstably inactive form we had designated programmable. We found, not unexpectedly, that the ability of an active element to reactivate an inactive element decreased as the heritability of the inactive phase increased. Cryptic elements respond least, rarely giving more than sectors of the reactivated phenotype. Genetically silent elements are not transcribed. Genetic reactivation is mirrored in transcriptional reactivation. Unstably inactive elements are transcribed when an active element is present, but cryptic ones are not. Banks further found that both unstably inactive and cryptic elements were less extensively methylated when a *trans*-activating *Spm* was present. But differences in the methylation level persisted, so that cryptic elements remained more extensively methylated than the less stably inactive elements.

It gradually emerged that there was both an immediate and a heritable component in the interaction between an active and an inactive element. As McClintock noted, inactive elements tend to revert to the inactive form when they are separated from the *trans*-activating element by genetic segregation. We found, however, that the interaction leaves a trace in the next generation. Exposure to the *trans*-activating element increases the probability that a plant with an unstably inactive element will transmit it to its progeny in an active phase. Even cryptic elements are not entirely deaf to the introduction of a *trans*-activating element. Although they do not show uniform reactivation, kernels receiving both elements often exhibit small sectors of the phenotype characteristic of the reactivated element.

I noticed that, just occasionally, a kernel with both a cryptic and a *trans*-activating element showed the reactivated phenotype throughout the kernel, and I began to wonder whether the *trans*-activating element could, eventually, reactivate even a cryptic element. The answer is that it could, but just as gradually as elements became cryptic in the first place. The reactivation process required several generations of exposure to the *trans*-activating element (with repeated selection of kernels exhibiting the reactivated phenotype) before the cryptic element became fully and heritably active. In an intermediate step in the reactivation process, the cryptic element became capable of *trans*-activation, just like other unstably inactive elements. Only in the final stage of the reactivation process did the formerly cryptic element remain active both immediately after segregation from the *trans*-activating *Spm* and in subsequent generations.

It was in the progressing reactivation process that I finally found what I was looking for: A *trans*-activating element could "preset" an inactive element! On ears in which a *trans*-activating element was segregating from a reactivating one, the reactivating *cryptic Spm* often appeared to remain active in parts of or throughout kernels that had not received the *trans*-activating elements. It was as if the element no longer required *trans*-activation. I grew a number of such kernels and almost invariably found little or no evidence of an active element: It had either relapsed almost completely into its former cryptic state or it was active only if a *trans*-activating element were reintroduced. Yet exposure to the *trans*-activating element in the previous generation had allowed the element to be expressed during development of the kernel. Thus, the active element had "preset" the inactive one.

By this time, I was convinced that presetting did indeed reflect an interaction between an element-encoded gene product and the element, interfering with its inactivation. It is known that the TnpA protein encoded by *Spm* binds to short sequences repeated many times near the ends of the element. Perhaps the molecular mechanics are these: TnpA binding facilitates transcription either simply by preventing element methylation or because the protein also serves as a transcription factor. But of course it is not really that simple. McClintock spoke not only of presetting, but of "setting" and "erasure" as well. Banks observed that element methylation changes markedly in development. An inactive element is least extensively methylated in embryos, and methylation increases later in development to different extents in different plant parts and tissues. In addition, the extent of methylation is always lower in the presence of a second, active element. In sum, this is a dynamic developmental process. Recall, too, that TnpA can inhibit expression from the element's promoter. Because there are many binding sites for the protein at each element end, perhaps the answer is that TnpA facilitates element expression when its concentration is low enough so that only a few molecules are bound to the element. Occupation of a sufficient number of the sites at both ends might then bring the element ends together, preventing further transcription and promoting transposition (recall, too, that TnpA is required for mobility, as is TnpD). But the cessation of transcription by such autoinhibition might have an alternative outcome as well if the complex dissociates without transposing, opening the door to methylation and locking the element into an inactive state, until the developmental cycle begins again.

These ideas, which are represented diagrammatically in Figure 3, comprise my best effort to understand how and why transposable elements survive in the plant and how the plant survives its transposable elements. I am not fond of the "junk DNA" and "selfish gene" notions, although they undoubtedly contain elements of truth, particularly in the sense that a sequence that has "learned" to outreplicate genomic DNA, as transposable elements have, will

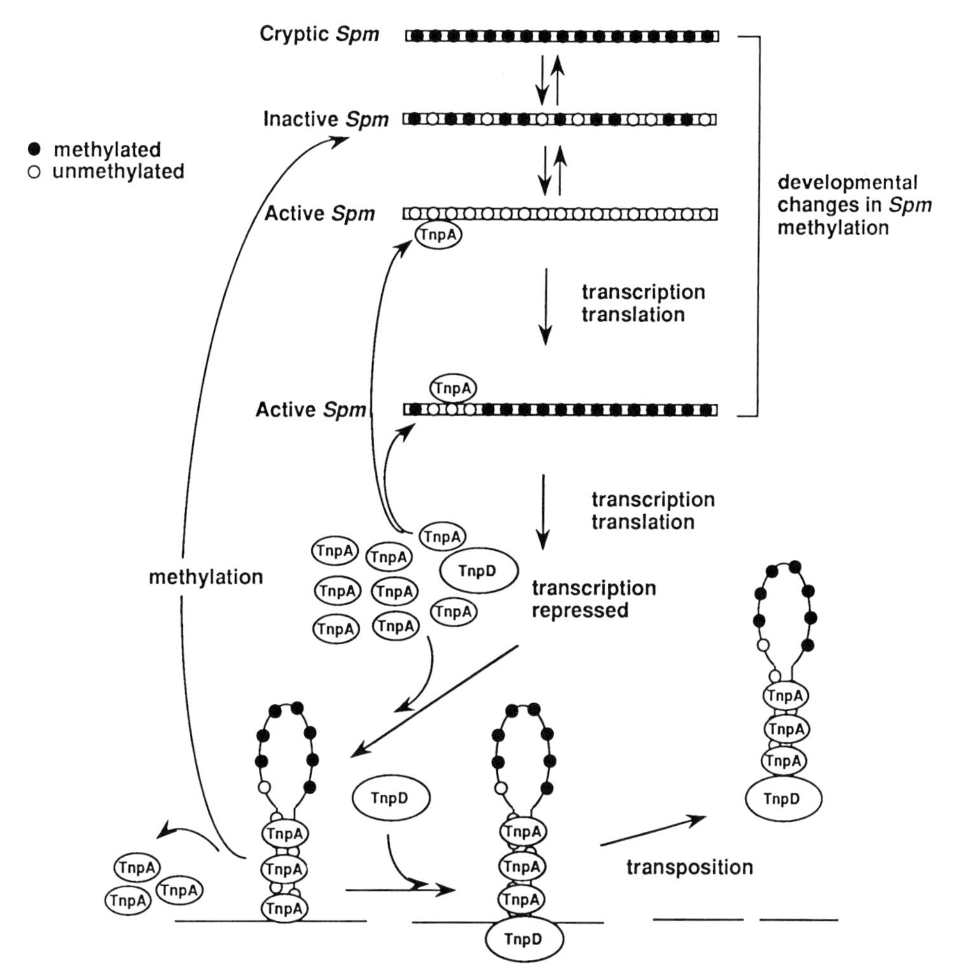

Figure 3 Regulatory scheme for the *Spm* element.

tend to accumulate in the genome. Continuing in this anthropomorphic vein, I suspect that this takes a bit more intelligence than meets the eye. When transposable elements come out to play—and it emerges as a curious and perhaps significant generalization that those are times of stress, when chromosomes are breaking for one reason or another—they turn out to be both destructive and self-destructive. They disrupt gene expression when they insert into genes and often leave behind a mutation because they excise imperfectly, leaving some fraction of the direct duplication generated upon insertion. Transposable elements, whether in maize or in other organisms, promote a variety of chromosomal rearrangements, not a small fraction of which are simply lethal. It is increasingly obvious that transposition is not fail-safe for the element, either. As is perhaps clearest in maize and *Drosophila*, transposable elements frequently sustain internal deletions, a propensity that appears to be inherent in the transposition process itself. Modern molecular archaeology has shown the genome to be littered with the bones of dead transposons. Finally, it has begun to impress itself on plant molecular biologists fiddling with transgenes that plants can count their genes.

Introduction of an extra copy of a resident gene, be it of plant or even bacterial origin, frequently leads to the complete (albeit reversible) inactivation of both gene copies. Given that transposition frequently results in the appearance of additional copies of elements, transposable elements must have become adept at evading the accounting system.

I cannot say for sure that transposable elements are useful to the corn plant, although I can imagine that rapid genomic change (at which transposable elements excel) might have played an important part in corn's history, accounting for their contemporary presence. I do know from experience, however, that corn lines with too many active transposable elements are in trouble: Some of their offspring look more like cabbages than corn plants. If I try to think like a corn plant (although sometimes I'm convinced that I think more like a cabbage), I conclude that my best bet is to keep my options open by hanging on to some of these principles of radical change, but shackling them as securely as possible. Among the sturdiest in a plant's armamentarium must be the molecular devices it uses in developmental determination—the extremely stable, but ultimately reversible, genetic mechanisms that underlie (assure, ensure) the orderly morphological and molecular progressions that are the plant. Indeed, as I have suggested elsewhere in the literature, there is a perhaps more than coincidental resemblance between the *Spm* inactivation mechanism and such fundamental developmental processes in plants as the transition from the juvenile to the adult reproductive phase.

A transposable element must also enter into the dialog and speak the regulatory language of the country in which it finds itself. It is not enough for it to know how to cut and paste and trim; indeed, it is likely that elements need help even to do that much (some may not even encode the enzyme that cleaves at element ends). So it would not be too surprising if transposable elements could themselves regulate various aspects of their own activity. Nor would it be surprising if they could pick up on, incorporate, make use of signals that the plant might have elaborated for completely different purposes. Yet, they have their own purpose, not in a teleological sense, but in the sense that their modus operandi must be understood as part of a developmental fabric, possessed of its own unique history, within which they are embedded and within whose constraints they have survived into the twentieth century present of *Zea mays*. Not the least in importance might be their propensity to cause insertion mutations in genes; there is indeed a bit of evidence that they prefer genes to silent DNA. Could it be that among the most significant genes in their history are those encoding proteins that synthesize or regulate pigments for the simple, almost trivial, reason that people liked the variegated pigmentation patterns of insertion mutations in these genes and propagated the progeny of variegated plants?

I hope I will be forgiven such undisciplined indulgence in anthropomorphic reasoning. I endeavor by this device to draw attention to some aspects of how transposable elements work that seem to me awfully clever, while avoiding pompous pronouncements on evolution (about which I know nothing, except that it happened, and even harbor the deep suspicion that the same is true of everyone else). So look, *Spm knows* how to subvert the plant's efforts to silence it and it even probably *knows* that being silenced is its best chance of long-term survival. It *knows* when it should be transcribed, how much of each transcript and protein it needs, and when to jump (and when not to). If we keep at our molecular dig, we will figure out just how it accomplishes its unique regulatory tasks. We can already guess, however, that the components will not be altogether unique and will therefore necessarily be representative of plant "gene control systems." This isn't quite the way you saw it, Barbara, but it's not far off.

References

I have endeavored to identify the individuals contributing key insights or observations, although I have undoubtedly overlooked some. However, to keep the text as informal and readable as possible in the face of the subject's delicious (ferocious?) complexity, I did not cite individual publications. The interested reader will find all of the McClintock quotations, save those that are identified as unpublished, in Volume 17 of the *Great Books in Experimental Biology* series, titled "The discovery and characterization of transposable elements. The collected papers of Barbara McClintock" (ed. J. Moore, Garland, New York, 1987). Primary (at worst, secondary) sources for the early genetic and all but the most recent molecular observations that I have discussed here can be found in two reviews, one that I wrote just before the molecular work began to bear fruit and the other recent enough to be almost exclusively molecular.

Fedoroff, N.V. 1983. Controlling elements in maize. In *Mobile genetic elements* (ed. J. Shapiro), p. 1. Academic Press, New York.

———. 1989. Maize transposable elements. In *Mobile DNA* (ed. M. Howe and D. Berg), p. 375. American Society for Microbiology, Washington, D.C.

Name Index

Subject Index

12/94

Dear Ann,

I knew that you were special the moment I met you on the curriculum committee. You have a great gift of only speaking when you have something important to say.

Thank you so much for all your support over the past several years. You are such an important role model for women scientists. I know you will continue your encouragement of women in your new position.

We will work to keep the McClintock Society alive and growing - we owe that & more to you. Best of luck in the new year and in your new adventure. I know you will be successful. Thank you, again

Pat Emmanuel